Lecture Notes in Mathematics

Volume 2399

This series reports on new developments in all areas of mathematics and their applications - quickly, informally and at a high level. Mathematical texts analysing new developments in modelling and numerical simulation are welcome. The type of material considered for publication includes:

1. Research monographs
2. Lectures on a new field or presentations of a new angle in a classical field
3. Summer schools and intensive courses on topics of current research.

Texts which are out of print but still in demand may also be considered if they fall within these categories. The timeliness of a manuscript is sometimes more important than its form, which may be preliminary or tentative. Please visit the LNM Editorial Policy (https://drive.google.com/file/d/1MOg4TbwOSokRnFJ3ZR3ciEeKs9hOnNX_/view?usp=sharing)

Titles from this series are indexed by Scopus, Web of Science, Mathematical Reviews, and zbMATH.

Chuchu Chen • Tonghe Dang • Jialin Hong •
Guoting Song

Numerical Analysis of Stochastic Functional Differential Equations

Longtime Asymptotics and Probabilistic Characteristics

Chuchu Chen
SKLMS, Academy of Mathematics and Systems Science
Chinese Academy of Sciences
Beijing, China

School of Mathematical Sciences
University of Chinese Academy of Sciences
Beijing, China

Jialin Hong
SKLMS, Academy of Mathematics and Systems Science
Chinese Academy of Sciences
Beijing, China

School of Mathematical Sciences
University of Chinese Academy of Sciences
Beijing, China

Tonghe Dang
Department of Applied Mathematics
Hong Kong Polytechnic University
Hong Kong, China

Guoting Song
School of Mathematics and Statistics
Changchun University
Changchun, China

ISSN 0075-8434 ISSN 1617-9692 (electronic)
Lecture Notes in Mathematics
ISBN 978-981-92-1591-1 ISBN 978-981-92-1592-8 (eBook)
https://doi.org/10.1007/978-981-92-1592-8

This Springer imprint is published by the registered company Springer Nature Singapore Pte Ltd.
The registered company address is: 152 Beach Road, #21-01/04 Gateway East, Singapore 189721, Singapore

Preface

Stochastic functional differential equations (SFDEs) provide powerful mathematical tools for modeling and analyzing complex memory-dependent dynamical systems that arise in diverse fields such as biology, mechanics, neural networks, and finance, incorporating both randomness and delays. The research on SFDEs dates back to the 1960s, and since then, substantial efforts have been dedicated to their study. It is observed that the dynamical behaviors of SFDEs are often distinct from those of stochastic ordinary differential equations and stochastic partial differential equations. To qualitatively and quantitatively study stochastic phenomena and time-lag effects, numerical methods have emerged as a powerful tool. While many excellent treatises exist on the theoretical analysis of SFDEs, a monograph devoted specifically to their numerical analysis remains scarce. This monograph aims to fill that gap by presenting recent developments in the numerical analysis of SFDEs.

This monograph focuses on the numerical analysis of SFDEs, with a particular emphasis on the longtime asymptotics and probabilistic characteristics of numerical methods used to solve such equations. The longtime behaviors under investigation include the time-independent convergence analysis in both the strong and weak senses, the numerical invariant measure, and the ergodicity of numerical methods. Additionally, the probabilistic characteristics of numerical solutions explored in this monograph encompass the density function, limit theorems, and the Freidlin–Wentzell type large deviation principle. Corresponding numerical simulations are provided to validate these analytical results in this work. The topics presented here lie at the intersection of several fascinating areas: numerical analysis, stochastic analysis, ergodicity theory, Malliavin calculus, large deviation theory, and probability theory, providing a rich framework to deepen our understanding of SFDEs from both theoretical and numerical perspectives.

This monograph consists of six chapters. Chapter 1 starts with a brief review of the SFDE and demonstrates several concrete examples that involve the time-lag effect and randomness. We then present the existence and uniqueness of the solution for the SFDE under the globally Lipschitz continuous condition and the non-globally Lipschitz continuous condition, respectively. It is known that the solution of the SFDE defined in $\mathbb{R}^d$ is no longer Markovian, whereas the functional solution defined in $C([-\tau, 0]; \mathbb{R}^d)$ is Markovian, where τ is the delay. By establishing the moment boundedness and the exponential attractiveness of the solution and the functional solution, we show that the functional solution admits a unique invariant measure. Then we turn to studying the probabilistic characteristics of the SFDE, including limit theorems, density function, and large deviation principle. At the end of this chapter, we provide the construction and solvability of several classes of numerical methods for the SFDE.

In Chap. 2, we focus on a general class of numerical methods, the θ-Euler–Maruyama (θ-EM) method, of the SFDE with superlinearly growing drift coefficient, and study the longtime mean-square convergence of this class of numerical methods. We first establish the time-independent boundedness of high-order moments of the θ-EM functional solution. Then we derive the inheritance of the exponential attractiveness and the exponential stability of the exact functional solution by the θ-EM method. Armed with these preparations, we prove that the mean-square convergence rate on the infinite time horizon is $\frac{1}{2}$.

Chapter 3 studies the numerical invariant measure and weak convergence analysis of the θ-EM method. The existence and uniqueness of the numerical invariant measure, which yields the ergodicity, are proven by means of the time-independent moment boundedness and the exponential attractiveness of the θ-EM functional solution. A common approach to obtaining the convergence rate of the numerical invariant measure is based on the longtime weak convergence of the numerical method, which can be established by means of the Malliavin calculus. The prerequisite is to derive the time-independent estimates for Malliavin derivatives and Gâteaux derivatives of both the exact and numerical functional solutions. We show that the moments of these derivatives decay exponentially in time. Based on these properties, we obtain the weak convergence rate 1 on the infinite time horizon, which is twice the mean-square convergence rate. As a result, the convergence rate for the numerical invariant measure of the θ-EM functional solution is derived, which is the same as the weak convergence rate.

Chapter 4 is devoted to investigating probabilistic limit theorems, in particular, the strong law of large numbers and the central limit theorem of numerical methods for the SFDE. Taking into consideration the fact that the functional solution of the autonomous SFDE is a time-homogeneous Markov process, we extend the study to the general time-homogeneous Markov processes. We propose a unified and transparent approach to investigating probabilistic limit theorems of numerical discretizations of the underlying Markov process. Once the properties of uniformly mixing and convergence for numerical discretizations are satisfied, it is shown that the numerical strong law of large numbers and the numerical central limit theorem hold, namely, the time-averages of numerical discretizations converge to the ergodic limit almost surely and that the normalized time-averages converge in distribution to a normal distribution. The limits coincide with the ones for the underlying Markov process. Based on these results, we obtain the corresponding probabilistic limit theorems of the θ-EM method for the SFDE.

We investigate in Chap. 5 the existence and convergence of the density function of the θ-EM solution. We first present the nondegeneracy of the numerical solution and prove the existence of the corresponding numerical density function by means of the Malliavin calculus. For the SFDE driven by the multiplicative noise, combining the mean-square convergence result of the θ-EM method, the convergence of the numerical density function is proved. While for the case of the additive noise, the convergence rate of the numerical density function is derived, based on the test-functional-independent weak convergence analysis for the θ-EM method. The major obstacle in the proof lies in the estimates of the negative moments of the determinant of the corresponding Malliavin covariance matrix of the numerical solution, which is overcome by presenting a discrete comparison principle for the SFDE with additive noise. By making full use of the Malliavin integration by parts formula, the test-functional-independent weak convergence rate and further the convergence rate of the numerical density function are both proved to be 1.

The Freidlin–Wentzell type large deviation principle of the θ-EM method is presented in Chap. 6, for the SFDE with small noise on the infinite time horizon. A well-known approach to establishing the large deviation principle is through its equivalence to the Laplace principle, which is also known as the weak convergence method. The key to the analysis lies in establishing the compactness of solutions for both the skeleton equation and the stochastic controlled equation in Banach space $C_\xi([-\tau, \infty); \mathbb{R}^d)$. In this regard, by analyzing the time-independent moment estimates of the solution of the stochastic controlled equation, we prove that the θ-EM solution satisfies the large deviation principle on the Banach space with the rate function given by the corresponding skeleton equation. Moreover, combining the technique of the Malliavin calculus, we present the logarithmic estimate of the numerical density function and establish the relation between the logarithmic limit and the rate function of the large deviation principle. Four appendices have also been provided, which are meant to supply supporting details without interfering with the main flow of the monograph. They contain some results on commonly used inequalities, Markov semigroup, invariant measure, ergodicity, Malliavin calculus, and large deviation principle.

We hope that this monograph will be of value to a wide range of readers, including researchers and graduate students with a prior understanding of SFDEs and Markov processes. The topics covered in this monograph will provide useful tools for the readers to have a deeper understanding of numerical analysis, particularly in the context of the longtime asymptotics and probabilistic characteristics of SFDEs.

Beijing, China
February, 2025

Chuchu Chen
Tonghe Dang
Jialin Hong
Guoting Song

Acknowledgments Financial support from the National Key R&D Program of China under Grant (No. 2024YFA1015900 and No. 2020YFA0713701), the National Natural Science Foundation of China (No. 12031020, No. 12461160278, and No. 12471386), the Youth Innovation Promotion Association CAS, and Jilin Province Human Resources and Social Security Department (No. 25JBQ036L014) is gratefully acknowledged.

Contents

Acronyms

SFDE	Stochastic functional differential equation
SLLN	Strong law of large numbers
CLT	Central limit theorem
EM	Euler–Maruyama
BEM	Backward Euler–Maruyama
θ-EM	θ-Euler–Maruyama
LDP	Large deviation principle
$\mathbb{N}$	Set of all non-negative integers
$\mathbb{N}_+$	Set of all positive integers
$\mathbb{R}^d$	d-dimensional Euclidean space
$\mathbb{R}_+$	Set of all non-negative numbers
$\|x\|$	Euclidean norm of $x \in \mathbb{R}^d$ or $x \in \mathbb{R}^{d\times m}$
$\langle x, y\rangle$	Inner product for $x, y \in \mathbb{R}^d$
$\otimes$	Kronecker product
$\|\alpha\|$	$\sum_{i=1}^d \alpha_i$ for a multi-index $\alpha = (\alpha_1, \ldots, \alpha_d)$
$i!$	Factorial of $i \in \mathbb{N}$
$i!!$	Double factorial of $i \in \mathbb{N}_+$
C_p^i	Combinatorial number, i.e., $C_p^i = \frac{p!}{i!(p-i)!}$ for $p, i \in \mathbb{N}$ with $i \le p$
$a \vee b$ (resp. $a \wedge b$)	Maximum (resp. minimum) of $a, b \in \mathbb{R}$
$\mathbf{1}_A$	Indicator function of a set A, defined by $\mathbf{1}_A(x) = 1$ for $x \in A$ and $\mathbf{1}_A(x) = 0$ for $x \in A^c$
$\mathcal{C}^d := (C([-\tau, 0]; \mathbb{R}^d), \\|\cdot\\|)$	Space of continuous functions $\phi : [-\tau, 0] \to \mathbb{R}^d$ equipped with the norm $\\|\phi\\| = \sup_{s\in[-\tau,0]} \|\phi(s)\|$
$(C([-\tau, T]; \mathbb{R}^d), \|\\|\cdot\\|\|)$	Space of continuous functions $\phi : [-\tau, T] \to \mathbb{R}^d$ equipped with the norm $\|\\|\phi\\|\| = \sup_{s\in[-\tau,T]} \|\phi(s)\|$

$\mathcal{B}(E)$	Borel σ-algebra of some given Banach space E
$\mathcal{P}$	Space of probability measures on $(C^d, \mathcal{B}(C^d))$
$C_b^k(\mathbb{R}^d; \mathbb{R})$	Space of continuous functions $f : \mathbb{R}^d \to \mathbb{R}$ with bounded and continuous derivatives of orders up to k
$C_{pol}^\infty(\mathbb{R}^d; \mathbb{R})$	Space of smooth functions $f : \mathbb{R}^d \to \mathbb{R}$ whose partial derivatives have at most polynomial growth
$C_0([0, \infty); \mathbb{R}^m)$	Space of continuous functions $w \in C([0, \infty); \mathbb{R}^m)$ with $w(0) = 0$
$(C_\xi([-\tau, \infty); \mathbb{R}^d), \lVert \cdot \rVert_\mathfrak{a})$	Space of continuous functions $u : [-\tau, \infty) \to \mathbb{R}^d$ with $u(r) = \xi(r)$ for $r \in [-\tau, 0]$, endowed with the norm $$\lVert u \rVert_\mathfrak{a} := \sum_{k=1}^{\infty} e^{-\mathfrak{a} t_k} (\sup_{t \in [-\tau, t_k]} \lvert u(t) \rvert) \Delta$$ for some $\mathfrak{a} > 0$, which is also written as C_ξ for short
$C(\mathbb{S}; \mathbb{R}_+)$	Space of continuous functions $U : \mathbb{S} \to \mathbb{R}_+$, where $\mathbb{S}$ can be taken as $\mathbb{R}^d$, $\mathbb{R}^d \times \mathbb{R}^d$, $\mathbb{R}^d \times [-\tau, \infty)$, or $\mathbb{R}^d \times \mathbb{R}^d \times [-\tau, \infty)$
$C^{2,1}(\mathbb{R}^d \times [-\tau, \infty); \mathbb{R}_+)$	Space of continuous non-negative functions $V(x, t)$ defined on $\mathbb{R}^d \times [-\tau, \infty)$ such that they are continuously twice differentiable in x and once in t
$d_{q,\gamma}$	Quasi-metric on C^d defined by $d_{q,\gamma}(u_1, u_2) := (1 \wedge \lVert u_1 - u_2 \rVert^\gamma)(1 + \lVert u_1 \rVert^q + \lVert u_2 \rVert^q)^{\frac{1}{2}}$, where $q \geq 1$ and $\gamma \in (0, 1]$
$C_{q,\gamma} := C_{q,\gamma}(C^d; \mathbb{R})$	Space of continuous functionals on C^d endowed with the norm $\lVert f \rVert_{q,\gamma} := \sup_{u \in C^d} \frac{\lvert f(u) \rvert}{1 + \lVert u \rVert^{\frac{q}{2}}} + \sup_{u_1, u_2 \in C^d, u_1 \neq u_2} \frac{\lvert f(u_1) - f(u_2) \rvert}{d_{q,\gamma}(u_1, u_2)}$
$(\Omega, \mathcal{F}, \{\mathcal{F}_t\}_{t \geq 0}, \mathbb{P})$	Filtered probability space
$\mathbb{E}[u]$	Expectation of a random variable u
$\mathbb{E}[u \mid \mathcal{G}]$	Conditional expectation of u with respect to a σ-algebra $\mathcal{G} \subset \mathcal{F}$
$\mathcal{N}(0, v^2)$	Normal distribution with mean 0 and variance v^2
$\xrightarrow{d}$	Convergence in distribution with respect to $\mathbb{P}$
$\xrightarrow{\mathbb{P}}$	Convergence in probability with respect to $\mathbb{P}$
$\xrightarrow{\mathbb{P}\text{-a.s.}}$	Convergence almost surely with respect to $\mathbb{P}$

Δ	Stepsize of the θ-EM method, usually defined as $\Delta := \frac{\tau}{N} \in (0, 1]$ for a sufficiently large integer N
K	A generic positive constant independent of Δ whose value may vary at different occurrences
$\lfloor t \rfloor_\Delta$	Largest grid point that no larger than t, i.e., $\max\{t_k \le t : t_k = k\Delta,\ k = -N, -N+1, \ldots\}$
$\lceil t \rceil_\Delta$	Smallest grid point greater than t, i.e., $\min\{t_k > t : t_k = k\Delta, k = 0, 1, \ldots\}$
$\lfloor t \rfloor$	Largest integer that no larger than t
$\lceil t \rceil$	Smallest integer that no less than t
μ	Invariant measure of the SFDE
μ^Δ	Invariant measure of the θ-EM method
ξ	Deterministic initial datum in C^d
$x^\xi(\cdot)$ (resp. $x^\xi_\cdot$)	Solution (resp. functional solution) of the SFDE
$y^{\xi,\Delta}(\cdot)$ (resp. $y^{\xi,\Delta}_\cdot$)	Solution (resp. functional solution) of the θ-EM method
$\varphi(t; t_i, \eta)$	Solution of the SFDE at time t with initial datum η at t_i
$\varphi^{Int}(\cdot; t_i, \eta)$	Linear interpolation of $\varphi(\cdot; t_i, \eta)$ by grids $\{(t_k, \varphi(t_k; t_i, \eta)), k = i, i+1, \ldots\}$
$\Phi(t_k; t_i, \eta^{Int})$	Solution of the θ-EM method at time t_k with initial datum η^{Int} at t_i
μ^ξ_t	Probability measure induced by x^ξ_t
P^ξ_t	Markov semigroup induced by μ^ξ_t
$\mu^{\xi,\Delta}_t$	Probability measure induced by $y^{\xi,\Delta}_t$
$P^{\xi,\Delta}_t$	Markov semigroup induced by $\mu^{\xi,\Delta}_t$
$\mathbb{W}_q$	Bounded-Wasserstein distance for $q \ge 1$, i.e., $$\mathbb{W}_q(\mathfrak{u}_1, \mathfrak{u}_2) := \Big(\inf_{\pi \in \Pi(\mathfrak{u}_1, \mathfrak{u}_2)} \int_{C^d \times C^d} 1 \wedge \|\phi_1 - \phi_2\|^q \pi(\mathrm{d}\phi_1, \mathrm{d}\phi_2) \Big)^{\frac{1}{q}},$$ where $\Pi(\mathfrak{u}_1, \mathfrak{u}_2)$ denotes the collection of all probability measures on $C^d \times C^d$ with marginal measures $\mathfrak{u}_1$ and $\mathfrak{u}_2$
$\mathfrak{p}$	Density function of the solution of the SFDE
$\mathfrak{p}^\Delta$	Density function of the solution of the θ-EM method

$L^p(\Omega, E)$	Space of all E-valued random variables which are p-integrable with respect to $\mathbb{P}$
$L^2_\Delta := (L^2_\Delta([0,\infty);\mathbb{R}^m), \|\cdot\|_{L^2_\Delta})$	Space of functions $v : [0,\infty) \to \mathbb{R}^m$ satisfying that $v(\lfloor\cdot\rfloor)$ is square integrable equipped with some norm denoted by $\|\cdot\|_{L^2_\Delta}$
$\mathcal{S}_M$	Subspace of $L^2_\Delta([0,\infty);\mathbb{R}^m)$ for functions v satisfying $\|v\|^2_{L^2_\Delta} \leq M$
$\mathcal{A}_M$	Space of $\{\mathcal{F}_t\}_{t\geq 0}$-adapted functions $v : \Omega \times [0,\infty) \to \mathbb{R}^m$ satisfying that $v \in \mathcal{S}_M$ $\mathbb{P}$-a.s.
$(\mathcal{L}(E_1;E_2), \|\cdot\|_{\mathcal{L}(E_1;E_2)})$	Space of bounded linear operators from E_1 to E_2 endowed with usual operator norm denoted by $\|\cdot\|_{\mathcal{L}(E_1;E_2)}$
$(H, \langle\cdot,\cdot\rangle_H, \|\cdot\|_H)$	Hilbert space $H := L^2(\mathbb{T};\mathbb{R}^m)$ endowed with the inner product $\langle g, h\rangle_H := \int_{\mathbb{T}} g(t)^\top h(t)\mathrm{d}t$ for $g, h \in H$ and the norm $\|\cdot\|_H := \langle\cdot,\cdot\rangle_H^{\frac{1}{2}}$, where $\mathbb{T} = [0, T]$ or $[0, +\infty)$
$\mathcal{D}$	Gâteaux derivative operator
$\mathcal{D}^\alpha$	Gâteaux derivative operator of order $\alpha \in \mathbb{N}_+$
D	Malliavin derivative operator
D^α	Malliavin derivative operator of order $\alpha \in \mathbb{N}_+$
$(\mathbb{D}^{\alpha,p}(\mathbb{R}^d), \|\cdot\|_{\alpha,p})$	Malliavin Sobolev space consisting of all random variables $f : \Omega \to \mathbb{R}^d$ whose Malliavin derivatives up to order α belong to $L^p(\Omega)$, endowed with the norm $\|f\|_{\alpha,p} := \left(\mathbb{E}\left[\|f\|^p + \sum_{j=1}^{\alpha} \|D^j f\|^p_{H^{\otimes j}}\right]\right)^{\frac{1}{p}}$
$\mathrm{trace}[A]$	Trace of a matrix A

Chapter 1
Stochastic Functional Differential Equation

Since K. Itô and M. Nisio [48] initiated the study of the stochastic functional differential equation (SFDE) in 1964, there have been plenty of related works on its theoretical analysis; see e.g. [2, 6–10, 20, 26, 40, 42, 52, 63, 65, 73, 79, 88, 89]. In this chapter, we present a brief introduction to the SFDE and its numerical methods. Precisely, in Sect. 1.1, we present the general form of the equation and give three types of examples involving memory effects, which are widely used in population dynamics, automatic control theory, and finance. In Sect. 1.2, we establish the existence and uniqueness of the exact solution for the SFDE. In Sect. 1.3, we explore the properties of the exact solution or functional solution of the SFDE, including the moment boundedness, exponential attractiveness, invariant measure, probabilistic limit theorems, density function, and large deviation principle. Finally, in Sect. 1.4, we discuss several widely used numerical methods for solving the SFDE.

1.1 An Overview to SFDE

The SFDE is commonly used to model systems that depend not only on the present state but also on the past ones, which has wide applications in a diverse range of fields such as biology, epidemiology, mechanics, neural networks, economics, and finance; see e.g. [1, 21, 25, 53, 67, 71, 84, 94]. The general form of the SFDE driven by Brownian motion reads as follows:

$$\begin{cases} \mathrm{d}x^{\xi}(t) = b(x_t^{\xi})\mathrm{d}t + \sigma(x_t^{\xi})\mathrm{d}W(t), \quad t > 0, \\ x^{\xi}(t) = \xi(t), \quad t \in [-\tau, 0], \end{cases} \tag{1.1}$$

where the delay $\tau > 0$, the initial datum $\xi \in \mathcal{C}^d := (\mathcal{C}([-\tau, 0]; \mathbb{R}^d), \|\cdot\|)$, $\{W(t), t \geq 0\}$ is an m-dimensional standard Brownian motion on a filtered complete probability space $(\Omega, \mathcal{F}, \{\mathcal{F}_t\}_{t\geq 0}, \mathbb{P})$, and the drift coefficient $b : \mathcal{C}^d \to \mathbb{R}^d$ and

C. Chen et al., *Numerical Analysis of Stochastic Functional Differential Equations*, Lecture Notes in Mathematics 2399, https://doi.org/10.1007/978-981-92-1592-8_1

the diffusion coefficient $\sigma : C^d \to \mathbb{R}^{d\times m}$ are measurable functions. Here for $t \geq 0$, x_t^ξ is defined by the relation $x_t^\xi(r) = x^\xi(t+r)$ for $r \in [-\tau, 0]$, and $(C([-\tau, 0]; \mathbb{R}^d), \|\cdot\|)$ denotes the space of continuous functions $\phi : [-\tau, 0] \to \mathbb{R}^d$ equipped with the norm $\|\phi\| = \sup_{s\in[-\tau,0]} |\phi(s)|$. We refer to the $\mathbb{R}^d$-valued process $\{x^\xi(t)\}_{t\geq-\tau}$ as the *solution*, and to the C^d-valued process $\{x_t^\xi\}_{t\geq 0}$ as the *functional solution* (or *segment process*).

Due to the dependence on the past, it is known that the solution of the SFDE valued in $\mathbb{R}^d$ is no longer Markovian, whereas the functional solution valued in C^d is Markovian; see e.g. [70]. Hence, the SFDE is highly degenerate (infinite-dimensional Markov processes with finite-dimensional noises), which has quite peculiar dynamical behaviors that are different from those of the stochastic ordinary differential equation and the stochastic partial differential equation.

The SFDE offers a rather broad and general description of the dependence on the past, which appears in a wide variety of specific forms in the literature. For example, memory terms of (1.1) may be of the form $x^\xi(t-\tau)$, $x^\xi(\lfloor t/\tau \rfloor \tau)$, $\int_{-\tau}^0 x_t^\xi(r)\mathrm{d}\nu(r)$ with ν being a probability measure on $[-\tau, 0]$, in which cases, the corresponding equation is called the stochastic delay differential equation, the stochastic differential equation with piecewise continuous argument, and the SFDE with distributed delay. Here, $\lfloor t/\tau \rfloor$ represents the largest integer no larger than t/τ. Below, we present concrete formulations and examples of these classes of SFDEs.

Stochastic Delay Differential Equation For $n \in \mathbb{N}_+$, if coefficients b and σ are functions of arguments $x^\xi(t), x^\xi(t-\bar{\tau}_1(t)), \ldots, x^\xi(t-\bar{\tau}_n(t))$, then Eq. (1.1) is also called a stochastic delay differential equation. Precisely,

$$\begin{cases} \mathrm{d}x^\xi(t) = H_1(x^\xi(t), x^\xi(t-\bar{\tau}_1(t)), \ldots, x^\xi(t-\bar{\tau}_n(t)))\mathrm{d}t \\ \qquad\qquad + H_2(x^\xi(t), x^\xi(t-\bar{\tau}_1(t)), \ldots, x^\xi(t-\bar{\tau}_n(t)))\mathrm{d}W(t), \quad t > 0, \\ x^\xi(t) = \xi(t), \quad t \in [-\tau, 0]. \end{cases}$$

Here $H_1 : (\mathbb{R}^d)^{\otimes(n+1)} \to \mathbb{R}^d$, $H_2 : (\mathbb{R}^d)^{\otimes(n+1)} \to \mathbb{R}^{d\times m}$ are measurable functions, functions $\bar{\tau}_i : \mathbb{R}_+ \to (0, \tau]$ for $i \in \{1, \ldots, n\}$, and $W(\cdot)$ is an m-dimensional standard Brownian motion.

Example 1.1 In population ecology, the delay Lotka–Volterra model provides a fundamental framework to describe the dynamics of certain interacting species subject to time-lag effects (e.g., gestation periods or resource regeneration). Its deterministic form is expressed as

$$\frac{\mathrm{d}x(t)}{\mathrm{d}t} = \mathrm{diag}\big(x_1(t), \ldots, x_d(t)\big)\Big(b + A\,x(t) + B\,x(t-\tau)\Big), \quad t > 0$$

with initial data $\xi = (\xi_1, \ldots, \xi_d)^\top \in C^d$, where $\xi_i(r) > 0$ for $r \in [-\tau, 0]$. Here, $x(t) = (x_1(t), \ldots, x_d(t))^\top \in \mathbb{R}^d$, $b = (b_1, \ldots, b_d)^\top \in \mathbb{R}^d$ with $b_i \geq 0$ denotes the intrinsic growth vector, $A, B \in \mathbb{R}^{d\times d}$ are interaction matrices, and $\tau > 0$ is the delay. To account for environmental fluctuations in factors such as resource availability and species response rates, authors in [68] establish the following stochastic delay Lotka–Volterra model,

$$\mathrm{d}x(t) = \mathrm{diag}\left(x_1(t), \ldots, x_d(t)\right)\Big[\left(A(x(t) - \bar{x}) + B(x(t-\tau) - \bar{x})\right)\mathrm{d}t \\ + D\left(x(t) - \bar{x}\right)\mathrm{d}W(t)\Big].$$

Here, $\bar{x}$ is the positive equilibrium of the deterministic version (i.e., $b + A\bar{x} + B\bar{x} = 0$), $W(\cdot)$ is a one-dimensional standard Brownian motion, and $D = (D_{ij})_{d\times d} \in \mathbb{R}^{d\times d}$ represents the noise intensity matrix.

Example 1.2 Consider a large population subject to migration, with the aggregate migration rate modeled as white noise. Assume that each individual undergoes a fixed development period before contributing to population dynamics. Authors in [71] propose the following stochastic delay differential equation to model the evolution of the large population $x(t)$ at time t:

$$\mathrm{d}x(t) = (-\alpha x(t) + \beta x(t-\tau))\mathrm{d}t + \sigma \mathrm{d}W(t),$$

where $\alpha > 0$ denotes the death rate, $\beta > 0$ is the birth rate, $\tau > 0$ represents the development period of each individual (e.g., $\tau = 9$ months), $\sigma > 0$, and $W(\cdot)$ denotes a one-dimensional standard Brownian motion.

Stochastic Differential Equation with Piecewise Continuous Argument If coefficients b and σ are functions of arguments $x(t), x(\lfloor t/\tau \rfloor \tau)$, then Eq. (1.1) is also called a stochastic differential equation with piecewise continuous argument. Precisely,

$$\begin{cases} \mathrm{d}x(t) = H_1(x(t), x(\lfloor t/\tau \rfloor \tau))\mathrm{d}t + H_2(x(t), x(\lfloor t/\tau \rfloor \tau))\mathrm{d}W(t), & t > 0, \\ x(0) \;= x_0 \in \mathbb{R}^d, \end{cases}$$

where $H_1 : \mathbb{R}^d \times \mathbb{R}^d \to \mathbb{R}^d$, $H_2 : \mathbb{R}^d \times \mathbb{R}^d \to \mathbb{R}^{d\times m}$ are measurable functions, the delay $\tau > 0$, and $W(\cdot)$ is an m-dimensional standard Brownian motion.

Example 1.3 In automatic control theory, to stabilize a d-dimensional unstable stochastic system

$$\mathrm{d}x(t) = b(x(t), t)\mathrm{d}t + \sigma(x(t), t)\mathrm{d}W(t),$$

the author in [64] proposes a feedback control strategy based on discrete-time observations. Specifically, the control term is designed as $u(x(t), x(\lfloor t/\tau \rfloor \tau), t)$ and the corresponding controlled stochastic system is of the form

$$\mathrm{d}x(t) = \left(b(x(t), t) + u(x(t), x(\lfloor t/\tau \rfloor \tau), t)\right)\mathrm{d}t + \sigma(x(t), t)\mathrm{d}W(t).$$

Here, coefficients $b : \mathbb{R}^d \times \mathbb{R}_+ \to \mathbb{R}^d$, $u : (\mathbb{R}^d)^{\otimes 2} \times \mathbb{R}_+ \to \mathbb{R}^d$, and $\sigma : \mathbb{R}^d \times \mathbb{R}_+ \to \mathbb{R}^{d\times m}$ are measurable functions, the delay $\tau > 0$, and $W(\cdot)$ is an m-dimensional standard Brownian motion.

Example 1.4 The authors in [27] introduce the following stochastic differential equation with piecewise continuous argument:

$$\begin{cases} \mathrm{d}x(t) = (-x^3(t) - 10x(t) + 2x(\lfloor t \rfloor) + 1)\mathrm{d}t + (x(t) + x(\lfloor t \rfloor))\mathrm{d}W(t), \quad t > 0, \\ x(0) = x_0 \in \mathbb{R}, \end{cases}$$

where $x(t) \in \mathbb{R}$ denotes the state variable, $x(\lfloor t \rfloor)$ is the piecewise constant term which represents the state evaluated at the most recent integer time, and $W(\cdot)$ is a one-dimensional standard Brownian motion. This equation captures nonlinear stochastic dynamics that evolve continuously in time and embeds discrete-time memory through a piecewise constant argument.

SFDE with Distributed Delay If coefficients b and σ are functions of arguments $x^\xi(t)$, $\int_{-\tau}^0 x_t^\xi(r)\mathrm{d}\nu(r)$, then Eq. (1.1) is also called an SFDE with distributed delay. Precisely,

$$\begin{cases} \mathrm{d}x^\xi(t) = H_1\Big(x^\xi(t), \int_{-\tau}^0 x_t^\xi(r)\mathrm{d}\nu(r)\Big)\mathrm{d}t \\ \qquad\qquad + H_2\Big(x^\xi(t), \int_{-\tau}^0 x_t^\xi(r)\mathrm{d}\nu(r)\Big)\mathrm{d}W(t), \ t > 0, \\ x^\xi(t) \quad = \xi(t), \ t \in [-\tau, 0]. \end{cases}$$

Here, $H_1 : \mathbb{R}^d \times \mathbb{R}^d \to \mathbb{R}^d$, $H_2 : \mathbb{R}^d \times \mathbb{R}^d \to \mathbb{R}^{d\times m}$ are measurable functions, the delay $\tau > 0$, ν is a probability measure on $[-\tau, 0]$, and $W(\cdot)$ is an m-dimensional standard Brownian motion.

Example 1.5 The Black–Scholes model is one of the most famous option pricing models in finance, proposed by F. Black, M. Scholes, and R. Merton in 1973. This model provides a mathematical framework for calculating the theoretical price of European options. In recent years, various generalizations of the model have been made to better model real-life options. In particular, the literature [24] is one of the first works that took hereditary structure into consideration in studying the pricing problem of European option, in which the stock process $\{S(t)\}_{t\geq 0}$ satisfies

$$\mathrm{d}S(t) = \int_{-\tau}^0 S(t+r)\mathrm{d}\tilde{\nu}_1(r)\mathrm{d}t + \int_{-\tau}^0 S(t+r)\mathrm{d}\tilde{\nu}_2(r)\mathrm{d}W(t), \quad t \geq 0.$$

Here, $S(t) \in \mathbb{R}_+$ is the stock price at time $t \in \mathbb{R}_+$, $\tilde{\nu}_j$ are probability measures on $[-\tau, 0]$ for $j = 1, 2$, and $W(\cdot)$ is a one-dimensional standard Brownian motion.

Example 1.6 Precisely, in [88], the drift coefficient takes the form

$$b(\phi) = \frac{1}{\tau}\int_{-\tau}^{0} \sin(\phi(s))\mathrm{d}s - \mathrm{sign}(\phi(0))\sqrt{|\phi(0)|} - 2\phi(0),\ \phi \in C^1.$$

And in [57], the authors consider the drift coefficient

$$b(\phi) = \frac{1}{\tau}\int_{-\tau}^{0} \phi(s)\mathrm{d}s - |\phi(0)|^2\phi(0) - 2\phi(0),\ \phi \in C^d.$$

These drift coefficients can be collectively expressed in the form

$$b(\phi) = b_1(\phi) + b_2(\phi), \quad \phi \in C^d.$$

Here,

$$b_1(\phi) = \psi_1(\phi(0)) \text{ and } b_2(\phi) = \int_{-\tau}^{0} \psi_2(\phi(s))k(s)\mathrm{d}\nu(s),$$

where $\psi_1(\cdot)$ and $\psi_2(\cdot)$ are continuous functions from $\mathbb{R}^d$ to $\mathbb{R}^d$, $k(\cdot)$ is a continuous function from $[-\tau, 0]$ to $\mathbb{R}$, and ν is a probability measure on $[-\tau, 0]$.

1.2 Existence and Uniqueness of Exact Solution

In this section, we investigate the existence and uniqueness of the solution for Eq. (1.1) under the globally Lipschitz condition and the non-globally Lipschitz condition, respectively.

1.2.1 Globally Lipschitz Case

A well-known condition imposed for the existence and uniqueness of the solution is the globally Lipschitz condition; see e.g. [63, 70]. We state this result in the following theorem.

Theorem 1.1 ([63, Theorem 5.2.2]) *Let coefficients b and σ satisfy that for any $\phi_1, \phi_2 \in C^d$,*

$$|b(\phi_1) - b(\phi_2)| + |\sigma(\phi_1) - \sigma(\phi_2)| \le \bar{K}\|\phi_1 - \phi_2\|, \tag{1.2}$$

where $\bar{K} > 0$. Then for Eq. (1.1) with the initial datum $\xi \in C^d$, there exists a unique solution $\{x^\xi(t)\}_{t\ge-\tau}$.

Proof It suffices to show that for any $T > 0$, Eq. (1.1) admits a unique solution on $[-\tau, T]$. The proof is divided into three steps.

Step 1. Show that for any $T > 0$,

$$\mathbb{E}\Big[\sup_{-\tau\le t\le T}|x^\xi(t)|^2\Big] \le KT(T+1)(1+\|\xi\|^2)e^{KT(T+1)}. \tag{1.3}$$

For any $M \in \mathbb{N}_+$ with $M > \|\xi\|$, define a stopping time t_M by $\mathrm{t}_M := T \wedge \inf\{t \ge 0 : |x^\xi(t)| \ge M\}$. It follows from (1.1) that for $t \in [0, T]$,

$$x^\xi(t\wedge \mathrm{t}_M) = \xi(0) + \int_0^t b(x_s^\xi)\mathbf{1}_{[0,\mathrm{t}_M]}(s)\mathrm{d}s + \int_0^t \sigma(x_s^\xi)\mathbf{1}_{[0,\mathrm{t}_M]}(s)\mathrm{d}W(s).$$

Using $(c_1+c_2+c_3)^2 \le 3(c_1^2+c_2^2+c_3^2)$ for $c_1, c_2, c_3 \in \mathbb{R}$, and applying the Hölder inequality (see Proposition A.2) and the Burkholder–Davis–Gundy inequality (see Proposition A.5), we derive

$$\begin{aligned}\mathbb{E}\Big[\sup_{0\le t\le T}|x^\xi(t\wedge \mathrm{t}_M)|^2\Big] &\le 3\mathbb{E}\Big[\|\xi\|^2 + T\int_0^T |b(x_s^\xi)\mathbf{1}_{[0,\mathrm{t}_M]}(s)|^2\mathrm{d}s\\ &\quad + 4\int_0^T |\sigma(x_s^\xi)|^2\mathbf{1}_{[0,\mathrm{t}_M]}(s)\mathrm{d}s\Big].\end{aligned} \tag{1.4}$$

By (1.2), we have that for $\phi \in C^d$,

$$|b(\phi)| \vee |\sigma(\phi)| \le |b(\mathbf{0})| \vee |\sigma(\mathbf{0})| + \bar{K}\|\phi\| \le K(1+\|\phi\|),$$

where $\mathbf{0}$ denotes the zero element of C^d. This yields

$$\begin{aligned}&\mathbb{E}\Big[\sup_{0\le t\le T}|x^\xi(t\wedge \mathrm{t}_M)|^2\Big]\\ &\le K\|\xi\|^2 + K(T+1)\mathbb{E}\Big[\int_0^T (1+\|x_s^\xi\|^2)\mathbf{1}_{[0,\mathrm{t}_M]}(s)\mathrm{d}s\Big]\\ &\le K\|\xi\|^2 + K(T+1)\int_0^T (1+\mathbb{E}[\|x_{s\wedge \mathrm{t}_M}^\xi\|^2])\mathrm{d}s\\ &\le KT(T+1)(1+\|\xi\|^2) + K(T+1)\int_0^T \mathbb{E}\Big[\sup_{0\le u\le s}|x^\xi(u\wedge \mathrm{t}_M)|^2\Big]\mathrm{d}s.\end{aligned}$$

Applying the Gronwall inequality (see Proposition A.3), we obtain

$$\mathbb{E}\Big[\sup_{0\le t\le T}|x^\xi(t\wedge \mathrm{t}_M)|^2\Big] \le KT(T+1)(1+\|\xi\|^2)e^{KT(T+1)},$$

which finishes the proof of *Step 1* by letting $M \to \infty$.

Step 2. Uniqueness. Assume that there exist two solutions $x^\xi(\cdot)$ and $\tilde{x}^\xi(\cdot)$. It follows from (1.1) that

$$x^\xi(t) - \tilde{x}^\xi(t) = \int_0^t (b(x_s^\xi) - b(\tilde{x}_s^\xi))\mathrm{d}s + \int_0^t (\sigma(x_s^\xi) - \sigma(\tilde{x}_s^\xi))\mathrm{d}W(s).$$

It follows from *Step 1* that $\mathbb{E}\Big[\sup_{0\le t\le T}|x^\xi(t) - \tilde{x}^\xi(t)|^2\Big] < \infty$. Moreover, using the globally Lipschitz condition (1.2), and applying the Hölder inequality and the Burkholder–Davis–Gundy inequality, we arrive at that for any $T > 0$,

$$\begin{aligned}
&\mathbb{E}\Big[\sup_{0\le t\le T}|x^\xi(t) - \tilde{x}^\xi(t)|^2\Big] \\
&\le K_T\mathbb{E}\Big[\int_0^T \Big(|b(x_s^\xi) - b(\tilde{x}_s^\xi)|^2 + |\sigma(x_s^\xi) - \sigma(\tilde{x}_s^\xi)|^2\Big)\mathrm{d}s\Big] \\
&\le K_T\mathbb{E}\Big[\int_0^T \|x_s^\xi - \tilde{x}_s^\xi\|^2\mathrm{d}s\Big] \le K_T\int_0^T \mathbb{E}\Big[\sup_{0\le u\le s}|x^\xi(u) - \tilde{x}^\xi(u)|^2\Big]\mathrm{d}s.
\end{aligned}$$

Applying the Gronwall inequality, we conclude $\mathbb{E}\Big[\sup_{0\le t\le T}|x^\xi(t) - \tilde{x}^\xi(t)|^2\Big] = 0$, which implies the uniqueness of the solution for (1.1).

Step 3. Existence. The proof of the existence of the solution for (1.1) is based on the Picard iterations. For any $T > 0$, let $x^{\xi,0}(t) = \xi(0)$ for $t \in [0, T]$ and $x^{\xi,0}(t) = \xi(t)$ for $t \in [-\tau, 0]$. For each $n \in \mathbb{N}_+$, let $x^{\xi,n}(t) = \xi(t)$ for $t \in [-\tau, 0]$, and define

$$x^{\xi,n}(t) = \xi(0) + \int_0^t b(x_s^{\xi,n-1})\mathrm{d}s + \int_0^t \sigma(x_s^{\xi,n-1})\mathrm{d}W(s)$$

for $t \in [0, T]$. We aim to show that $x^{\xi,n}(\cdot)$ admits a unique limit $x^\xi(\cdot)$ $\mathbb{P}$-a.s. as $n \to \infty$, and prove that $x^\xi(\cdot)$ is the solution of (1.1). To this end, we estimate the errors between $x^{\xi,n+1}(\cdot)$ and $x^{\xi,n}(\cdot)$.

Claim: for each $n \in \mathbb{N}$ and $t \in [0, T]$,

$$\mathbb{E}\Big[\sup_{0\le u\le t}|x^{\xi,n+1}(u) - x^{\xi,n}(u)|^2\Big] \le \frac{\tilde{K}((2T+8)\bar{K}^2 t)^n}{n!}, \tag{1.5}$$

where $\tilde{K} := (4T^2 + 16T)(|b(\mathbf{0})|^2 + |\sigma(\mathbf{0})|^2 + \bar{K}^2\|\xi\|^2)$.

Similar to the proof of *Step 1*, by the Hölder inequality and the Burkholder–Davis–Gundy inequality, we derive

$$\begin{aligned}
&\mathbb{E}\Big[\sup_{0\le u\le t}|x^{\xi,1}(u)-x^{\xi,0}(u)|^2\Big]\\
&\quad=\mathbb{E}\Big[\sup_{0\le u\le t}\Big|\int_0^u b(x_s^{\xi,0})\mathrm{d}s+\int_0^u\sigma(x_s^{\xi,0})\mathrm{d}W(s)\Big|^2\Big]\\
&\quad\le 2\mathbb{E}\Big[t\int_0^t|b(x_s^{\xi,0})|^2\mathrm{d}s+4\int_0^t|\sigma(x_s^{\xi,0})|^2\mathrm{d}s\Big]\\
&\quad\le 2\mathbb{E}\Big[\int_0^t\Big(2T|b(\mathbf{0})|^2+8|\sigma(\mathbf{0})|^2+(2T+8)\bar{K}^2\|x_s^{\xi,0}\|^2\Big)\mathrm{d}s\Big]\le\tilde{K}.
\end{aligned}$$

Assume that for any $i\in\{1,2,\ldots,n-1\}$, one has $\mathbb{E}\Big[\sup_{0\le u\le t}|x^{\xi,i+1}(u)-x^{\xi,i}(u)|^2\Big]\le\frac{\tilde{K}((2T+8)\bar{K}^2t)^i}{i!}$. Utilizing the Hölder inequality, the Burkholder–Davis–Gundy inequality and the globally Lipschitz condition (1.2) on coefficients, we arrive at

$$\begin{aligned}
&\mathbb{E}\Big[\sup_{0\le u\le t}|x^{\xi,n+1}(u)-x^{\xi,n}(u)|^2\Big]\\
&=\mathbb{E}\Big[\sup_{0\le u\le t}\Big|\int_0^u\Big(b(x_s^{\xi,n})-b(x_s^{\xi,n-1})\Big)\mathrm{d}s+\int_0^u\Big(\sigma(x_s^{\xi,n})-\sigma(x_s^{\xi,n-1})\Big)\mathrm{d}W(s)\Big|^2\Big]\\
&\le(2T+8)\mathbb{E}\Big[\int_0^t|b(x_s^{\xi,n})-b(x_s^{\xi,n-1})|^2\mathrm{d}s+\int_0^t|\sigma(x_s^{\xi,n})-\sigma(x_s^{\xi,n-1})|^2\mathrm{d}s\Big]\\
&\le(2T+8)\bar{K}^2\mathbb{E}\Big[\int_0^t\|x_s^{\xi,n}-x_s^{\xi,n-1}\|^2\mathrm{d}s\Big].
\end{aligned}$$

This, together with the induction assumption yields

$$\begin{aligned}
&\mathbb{E}\Big[\sup_{0\le u\le t}|x^{\xi,n+1}(u)-x^{\xi,n}(u)|^2\Big]\\
&\quad\le(2T+8)\bar{K}^2\int_0^t\mathbb{E}\Big[\sup_{0\le u\le s}|x^{\xi,n}(u)-x^{\xi,n-1}(u)|^2\Big]\mathrm{d}s\\
&\quad\le(2T+8)\bar{K}^2\int_0^t\frac{\tilde{K}((2T+8)\bar{K}^2s)^{n-1}}{(n-1)!}\mathrm{d}s\le\frac{\tilde{K}((2T+8)\bar{K}^2t)^n}{n!},
\end{aligned}$$

which leads to (1.5).

Furthermore, it follows from the Chebyshev inequality and (1.5) that

$$\mathbb{P}\Big\{\sup_{0\le t\le T}|x^{\xi,n+1}(t)-x^{\xi,n}(t)|>\frac{1}{2^n}\Big\}\le\frac{4^n\tilde{K}((2T+8)\bar{K}^2T)^n}{n!}.$$

Applying the Borel–Cantelli lemma leads to that for $\mathbb{P}$-a.s. $\omega\in\Omega$, there exists an $n_0:=n_0(\omega)\in\mathbb{N}_+$ such that for $n\in\mathbb{N}_+$ with $n\ge n_0$,

$$\sup_{0\le t\le T}|x^{\xi,n+1}(t)-x^{\xi,n}(t)|\le\frac{1}{2^n}.$$

Since

$$x^{\xi,n}(t)=\xi(0)+\sum_{i=0}^{n-1}(x^{\xi,i+1}(t)-x^{\xi,i}(t))$$

for $t\in[-\tau,T]$, we obtain that $\lim_{n\to\infty}x^{\xi,n}(t)$ exists $\mathbb{P}$-a.s., which is denoted by $x^{\xi}(t)$. Note that the above convergence of $x^{\xi,n}(\cdot)$ is uniform in $t\in[0,T]$. Combining the continuity of $x^{\xi,n}(\cdot)$, we derive that the limit $x^{\xi}(\cdot)$ is also continuous in $t\in[0,T]$. In addition, it follows from (1.5) that for $t\in[0,T]$, $\{x_t^{\xi,n}\}_{n\in\mathbb{N}}$ is a Cauchy sequence in $L^2(\Omega;C^d)$. According to the completeness of $L^2(\Omega;C^d)$, we derive that $x_t^{\xi,n}\to x_t^{\xi}$ as $n\to\infty$ in $L^2(\Omega;C^d)$. Hence, we have

$$\begin{aligned}&\mathbb{E}\Big[\Big|\int_0^t(b(x_s^{\xi})-b(x_s^{\xi,n}))\mathrm{d}s+\int_0^t(\sigma(x_s^{\xi})-\sigma(x_s^{\xi,n}))\mathrm{d}W(s)\Big|^2\Big]\\&\le(2T+8)\bar{K}^2\int_0^t\mathbb{E}\Big[\sup_{0\le u\le s}|x^{\xi}(u)-x^{\xi,n}(u)|^2\Big]\mathrm{d}s\to0\quad\text{as}\quad n\to\infty.\end{aligned}$$

This implies that $x^{\xi}(\cdot)$ is the solution of (1.1). Therefore, we complete the proof. □

As a direct consequence of Theorem 1.1, the SFDEs in Examples 1.2 and 1.5 are well-posed. Since their drift and diffusion coefficients are globally Lipschitz continuous, Theorem 1.1 ensures that for any initial datum $\xi\in C^d$, each equation in Examples 1.2 and 1.5 admits a unique solution on $[-\tau,\infty)$.

1.2.2 Non-globally Lipschitz Case

In practice, SFDEs may be highly nonlinear, and Theorem 1.1 fails to apply to such equations. As a result, authors in [88] establish the existence and uniqueness of the solution for the nonlinear SFDE.

Theorem 1.2 ([88, Theorem 2.3]) *Assume that the coefficients b and σ satisfy the following conditions.*

- *Weak one-sided locally Lipschitz condition (or monotonicity condition). For any compact subset $\bar{C} \subset C^d$, there exist constants $\bar{K}_{\bar{C}} > 0$ and $\tau_{\bar{C}} \in [0, \tau]$ such that for any $\phi_1, \phi_2 \in \bar{C}$ with $\phi_1(r) = \phi_2(r)$ for $r \in [-\tau, -\tau_{\bar{C}}]$,*

$$2\langle \phi_1(0) - \phi_2(0), b(\phi_1) - b(\phi_2)\rangle + |\sigma(\phi_1) - \sigma(\phi_2)|^2 \leq \bar{K}_{\bar{C}}\|\phi_1 - \phi_2\|^2.$$

- *Weak growth condition. Coefficients b and σ are bounded on bounded subsets of C^d and there exists a non-decreasing function $\Psi : \mathbb{R}_+ \to (0, \infty)$ such that $\int_0^\infty \frac{1}{\Psi(s)}\mathrm{d}s = \infty$ and for any $\phi \in C^d$,*

$$2\langle \phi(0), b(\phi)\rangle + |\sigma(\phi)|^2 \leq \Psi(\|\phi\|^2).$$

Then for Eq. (1.1) *with the initial datum $\xi \in C^d$, there exists a unique solution $\{x^\xi(t)\}_{t\geq -\tau}$.*

Remark 1.1 Authors in [88] prove that the weak one-sided locally Lipschitz condition guarantees the existence and uniqueness of the solution for (1.1) in the local sense. Moreover, they also show that the weak growth condition prevents explosions from occurring, which implies the existence and uniqueness of the global solution for (1.1).

Theorem 1.2 provides a general framework for establishing the well-posedness of SFDEs with non-globally Lipschitz coefficients. We now verify that coefficients in Examples 1.4 and 1.6 fit within this framework.

Remark 1.2 We first demonstrate that the SFDE in Example 1.4 satisfies the hypotheses of Theorem 1.2. Recall that its drift and diffusion coefficients are given by

$$b(\phi) := b(u, v) = -u^3 - 10u + 2v + 1, \quad \sigma(\phi) := \sigma(u, v) = u + v,$$

where $\phi \in C^1$, $u := \phi(0)$ and $v \in \{\phi(r)\}_{r\in[-1,0]}$ is a certain quantity.

Verification of the Weak Growth Condition Observe that

$$\begin{aligned} 2\langle \phi(0), b(\phi)\rangle + |\sigma(\phi)|^2 &\leq -2|u|^4 - (20-2)|u|^2 + 4|u||v| + 2|v|^2 + 2u \\ &\leq 1 + 4\|\phi\|^2. \end{aligned}$$

By choosing $\Psi(s) = 1 + 4s$ for $s \in \mathbb{R}_+$, we conclude that the coefficients b and σ satisfy the weak growth condition.

Verification of the Weak One-Sided Locally Lipschitz Condition For any $\phi_1, \phi_2 \in C^1$,

$$\begin{aligned}
&2\langle \phi_1(0) - \phi_2(0), b(\phi_1) - b(\phi_2)\rangle + |\sigma(\phi_1) - \sigma(\phi_2)|^2 \\
&\quad = 2\langle u_1 - u_2, (-u_1^3 - 10u_1 + 2v_1 + 1) - (-u_2^3 - 10u_2 + 2v_2 + 1)\rangle \\
&\qquad + |(u_1 + v_1) - (u_2 + v_2)|^2 \\
&\quad \le -2|u_1 - u_2|^2(u_1^2 + u_1u_2 + u_2^2) - (20 - 2)|u_1 - u_2|^2 \\
&\qquad + 4|u_1 - u_2||v_1 - v_2| + 2|v_1 - v_2|^2 \\
&\quad \le -|u_1 - u_2|^2(u_1^2 + u_2^2) - (20 - 2)|u_1 - u_2|^2 \\
&\qquad + 4|u_1 - u_2||v_1 - v_2| + 2|v_1 - v_2|^2 \\
&\quad \le 4\|\phi_1 - \phi_2\|^2,
\end{aligned}$$

which confirms that the coefficients also satisfy the weak one-sided locally Lipschitz condition. Therefore, for any initial datum $x_0 \in \mathbb{R}$, the SFDE given by Example 1.4 admits a unique solution on $\mathbb{R}_+$.

Remark 1.3 We now verify that the drift coefficients introduced in Example 1.6, namely,

$$b(\phi) = \frac{1}{\tau}\int_{-\tau}^{0} \sin(\phi(r))\mathrm{d}r - \mathrm{sign}(\phi(0))\sqrt{|\phi(0)|} - 2\phi(0),\ \phi \in C^1 \tag{1.6}$$

and

$$b(\phi) = \frac{1}{\tau}\int_{-\tau}^{0} \phi(r)\mathrm{d}r - |\phi(0)|^2\phi(0) - 2\phi(0),\ \phi \in C^d \tag{1.7}$$

satisfy the conditions stated in Theorem 1.2 when σ is globally Lipschitz continuous. For simplicity, we take $\sigma(\phi) = \phi(0)$ for $\phi \in C^1$.

We begin with the first coefficient b in (1.6). Since the function $s \mapsto -\mathrm{sign}(s)\sqrt{|s|}$ is non-increasing on $s \in \mathbb{R}$, it follows that for any $\phi_1, \phi_2 \in C^1$,

$$\Big\langle \phi_1(0) - \phi_2(0), -\mathrm{sign}(\phi_1(0))\sqrt{|\phi_1(0)|} - \big(-\mathrm{sign}(\phi_2(0))\sqrt{|\phi_2(0)|}\big)\Big\rangle \le 0.$$

Applying the Young inequality (see Proposition A.1), we obtain

$$\begin{aligned}
&2\langle \phi_1(0) - \phi_2(0), b(\phi_1) - b(\phi_2)\rangle + |\sigma(\phi_1) - \sigma(\phi_2)|^2 \\
&\quad \le -3|\phi_1(0) - \phi_2(0)|^2 + 2\Big\langle \phi_1(0) - \phi_2(0), \frac{1}{\tau}\int_{-\tau}^{0} \Big(\sin(\phi_1(r)) - \sin(\phi_2(r))\Big)\mathrm{d}r\Big\rangle
\end{aligned}$$

$$\le -2|\phi_1(0)-\phi_2(0)|^2 + \frac{1}{\tau}\int_{-\tau}^{0} |\sin(\phi_1(r)) - \sin(\phi_2(r))|^2 \mathrm{d}r$$

$$\le -2|\phi_1(0)-\phi_2(0)|^2 + \frac{1}{\tau}\int_{-\tau}^{0} |\phi_1(r) - \phi_2(r)|^2 \mathrm{d}r,$$

which confirms that b satisfies the weak one-sided locally Lipschitz condition. Moreover, for any $\phi \in C^1$,

$$\begin{aligned} 2\langle \phi(0), b(\phi)\rangle + |\sigma(\phi)|^2 &\le \frac{2|\phi(0)|}{\tau}\int_{-\tau}^{0} |\sin(\phi(r))|\mathrm{d}r - 3|\phi(0)|^2 \\ &\le 2\|\phi\|, \end{aligned}$$

where we used $|\sin(s)| \le 1$ and $\langle s, -\mathrm{sign}(s)\sqrt{|s|}\,\rangle \le 0$ for $s \in \mathbb{R}$. By choosing $\Psi(s) = 2s^{\frac{1}{2}}$ for $s \in \mathbb{R}_+$, we conclude that b in (1.6) also satisfies the weak growth condition.

A similar argument can be applied to the drift coefficient b in (1.7), noticing the fact that the mapping $u \mapsto -|u|^2 u$ satisfies

$$\begin{aligned} &\langle u_1 - u_2, -|u_1|^2 u_1 - (-|u_2|^2 u_2)\rangle \\ &= -\frac{1}{2}(|u_1|^2 + |u_2|^2)|u_1 - u_2|^2 - \frac{1}{2}(|u_1|^2 - |u_2|^2)^2 \\ &\le -\frac{1}{2}(|u_1|^2 + |u_2|^2)|u_1 - u_2|^2 \le 0 \quad \text{for } u_1, u_2 \in \mathbb{R}^d. \end{aligned}$$

Another type of sufficient condition to ensure the existence and uniqueness of the solution is presented in [61]. They assume that coefficients b and σ satisfy a locally Lipschitz condition and a Khasminskii-type condition, which are stated below separately.

Denote by $C(\mathbb{S}; \mathbb{R}_+)$ the space of continuous functions $U : \mathbb{S} \to \mathbb{R}_+$, where $\mathbb{S}$ can be taken as $\mathbb{R}^d$, $\mathbb{R}^d \times \mathbb{R}^d$, $\mathbb{R}^d \times [-\tau, \infty)$, and $\mathbb{R}^d \times \mathbb{R}^d \times [-\tau, \infty)$. Denote $C^{2,1}(\mathbb{R}^d \times [-\tau, \infty); \mathbb{R}_+)$ the space of continuous non-negative functions $V(x, t)$ defined on $\mathbb{R}^d \times [-\tau, \infty)$ such that they are continuously twice differentiable in x and once in t.

Assumption 1.1 *For any $M > 0$, there exists a constant $\tilde{K}_M > 0$ such that for any $\phi_1, \phi_2 \in C^d$ with $\|\phi_1\| \vee \|\phi_2\| \le M$,*

$$|b(\phi_1) - b(\phi_2)| + |\sigma(\phi_1) - \sigma(\phi_2)| \le \tilde{K}_M \|\phi_1 - \phi_2\|.$$

Assumption 1.2 *There exist two functions $V \in C^{2,1}(\mathbb{R}^d \times [-\tau, \infty); \mathbb{R}_+)$ and $U \in C(\mathbb{R}^d \times [-\tau, \infty); \mathbb{R}_+)$, and two probability measures $\bar{\nu}_1$ and $\bar{\nu}_2$ on $[-\tau, 0]$ such that*

$$\lim_{|x|\to\infty} \inf_{0\le t<\infty} V(x,t) = \infty,$$

and for any $(\phi, t) \in C^d \times \mathbb{R}_+$,

$$\begin{aligned}\mathcal{L}V(\phi,t) \le K\Big(1 + V(\phi(0),t) + \int_{-\tau}^{0} V(\phi(r), t+r)\mathrm{d}\bar{\nu}_1(r)\Big)\\ - U(\phi(0),t) + \int_{-\tau}^{0} U(\phi(r), t+r)\mathrm{d}\bar{\nu}_2(r),\end{aligned} \tag{1.8}$$

where $K > 0$, and the functional $\mathcal{L}V : C^d \times \mathbb{R}_+ \to \mathbb{R}$ is defined by

$$\mathcal{L}V(\phi,t) = V_t(\phi(0),t) + V_x(\phi(0),t)b(\phi) + \frac{1}{2}trace[\sigma^\top(\phi)V_{xx}(\phi(0),t)\sigma(\phi)]. \tag{1.9}$$

Here, $V_x(x,t) := \big(V_{x_1}(x,t), \ldots, V_{x_d}(x,t)\big)$ and $V_{xx}(x,t) := \big(V_{x_i,x_j}(x,t)\big)_{d\times d}$ for $x \in \mathbb{R}^d, t \in \mathbb{R}_+$.

Assumption 1.1 guarantees the existence and uniqueness of the solution for (1.1) in the local sense, and Assumption 1.2 prevents explosions of the solution from occurring. The function $V(x,t)$ defined in Assumption 1.2, called Lyapunov function, is commonly taken as $V(x,t) := |x|^p$ for some $p \ge 2$. The function $U(x,t)$ in (1.8) is used to control the higher-order terms of $\mathcal{L}V$, commonly taken as $U(x,t) := |x|^{p+l}$ for some $l > 0$.

The existence and uniqueness of the solution for (1.1) based on the locally Lipschitz condition and the Khasminskii-type condition is stated as follows.

Theorem 1.3 ([61, Theorem 2]) *Assume that Assumptions 1.1and 1.2 hold. Then for Eq. (1.1) with the initial datum $\xi \in C^d$, there exists a unique solution $\{x^\xi(t)\}_{t\ge-\tau}$.*

Remark 1.4 The well-posedness of the stochastic delay Lotka–Volterra model presented in Example 1.1 can be established by applying Theorem 1.3, provided the conditions $D_{ii} > 0$ for all $i \in \{1, 2, \ldots, d\}$ and $D_{ij} \ge 0$ for all $i \ne j$ with $i, j \in \{1, 2, \ldots, d\}$ are satisfied. First, it follows from the continuity of derivatives of the coefficients in this model that the coefficients are locally Lipschitz continuous. This implies the existence and uniqueness of the solution in the local sense. To proceed, define the Lyapunov function $V : (0,\infty)^d \to (0,\infty)$ introduced in [68] by

$$V(u) := \sum_{i=1}^{d}\Big(\sqrt{u_i} - 1 - \frac{1}{2}\log(u_i)\Big),$$

where $u = (u_1, u_2, \ldots, u_d)^\top$. For a solution $x(t) = ((x(t))_1, \ldots, (x(t))_d)^\top \in \mathbb{R}^d, t \geq -\tau$ of the model in Example 1.1 that stays in $(0, \infty)^d$, the associated operator $\mathcal{L}V : C^d \to \mathbb{R}$, defined as in (1.9), satisfies

$$\mathcal{L}V(x_t) = \frac{1}{2}g(x(t))\Big(A(x(t) - \bar{x}) + B\big(x(t-\tau) - \bar{x}\big)\Big) + \frac{1}{4}|D\big(x(t) - \bar{x}\big)|^2 - \frac{1}{8}\sum_{i=1}^{d}\sqrt{(x(t))_i}\Big(\sum_{j=1}^{d} D_{ij}\big((x(t))_j - \bar{x}_j\big)\Big)^2,$$

where $g(x(t)) := (\sqrt{(x(t))_1} - 1, \ldots, \sqrt{(x(t))_d} - 1)$. Making use of

$$\begin{aligned}\frac{1}{2}g(x(t))\Big(A(x(t) - \bar{x}) + B(x(t-\tau) - \bar{x})\Big) + \frac{1}{4}|D(x(t) - \bar{x})|^2 \\ \leq \frac{1}{2}\sqrt{d(|x(t)| + 1)}\Big(|A|(|x(t)| + |\bar{x}|) + |B||\bar{x}|\Big) \\ + \frac{1}{4}|B|\Big(d(|x(t)| + 1) + |x(t-\tau)|^2\Big) + \frac{1}{2}|D|^2\Big(|x(t)|^2 + |\bar{x}|^2\Big)\end{aligned}$$

and

$$\begin{aligned}\sum_{i=1}^{d}\sqrt{(x(t))_i}\Big(\sum_{j=1}^{d} D_{ij}\big((x(t))_j - \bar{x}_j\big)\Big)^2 \\ \geq \sum_{i=1}^{d} D_{ii}^2 (x(t))_i^{2.5} + \sum_{i=1}^{d}\sqrt{(x(t))_i}\Big(\sum_{j=1}^{d} D_{ij}\bar{x}_j\Big)\Big(\sum_{j=1}^{d} D_{ij}\bar{x}_j - 2\sum_{j=1}^{d} D_{ij}(x(t))_j\Big),\end{aligned}$$

one has

$$\mathcal{L}V(x_t) \leq G(x(t)) - \frac{1}{4}|B|(|x(t)|^2 - |x(t-\tau)|^2), \tag{1.10}$$

where

$$\begin{aligned}G(u) := \frac{1}{2}\sqrt{d(|u| + 1)}\Big(|A|(|u| + |\bar{x}|) + |B||\bar{x}|\Big) \\ + \frac{1}{4}|B|\Big(d(|u| + 1) + |u|^2\Big) + \frac{1}{2}|D|^2\Big(|u|^2 + |\bar{x}|^2\Big) - \frac{1}{8}\sum_{i=1}^{d} D_{ii}^2 u_i^{2.5} \\ - \frac{1}{8}\sum_{i=1}^{d}\sqrt{u_i}\Big(\sum_{j=1}^{d} D_{ij}\bar{x}_j\Big)\Big(\sum_{j=1}^{d} D_{ij}\bar{x}_j - 2\sum_{j=1}^{d} D_{ij}u_j\Big)\end{aligned}$$

with $u = (u_1, \ldots, u_d)^\top \in \mathbb{R}_+^d$. Utilizing $D_{ii} > 0, i \in \{1, 2, \ldots, d\}$, it is straightforward to see that the function $G(\cdot)$ is bounded above on $\mathbb{R}_+^d$ by a positive constant K. Substituting this bound into (1.10) yields

$$\mathcal{L}V(x_t) \leq K - \frac{1}{4}|B|(|x(t)|^2 - |x(t-\tau)|^2),$$

which coincides with (1.8) with $U(v) = \frac{1}{4}|B||v|^2$ and $\bar{\nu}_2 = \delta_{-\tau}$ when the solution $x = (x_1, \ldots, x_d)^\top$ belongs to $(0, \infty)^d$ in Example 1.1. Here, $\delta_{-\tau}$ represents the Dirac measure at the point $-\tau$. Hence, a standard stopping time argument together with Theorem 1.3 concludes the existence of a unique global positive solution; see [68, Theorem 2.3] for details.

1.3 Properties of Exact Solution

Within this section, we present properties of the exact solution and functional solution. We first demonstrate the moment boundedness and the exponential attractiveness of the exact solution and functional solution. Then, focusing on the functional solution, we prove the existence and uniqueness of the invariant measure and establish probabilistic limit theorems of the functional solution. In the context of the exact solution, we prove the existence of the density function and investigate the large deviation principle.

1.3.1 Moment Boundedness

We begin with estimating $\mathbb{E}[V(x^\xi(t), t)]$ on the finite time horizon, where $V(\cdot)$ is defined in Assumption 1.2.

Lemma 1.4 ([61, Theorem 2]) *Under Assumptions 1.1 and 1.2, the solution of* (1.1) *with the initial datum* $\xi \in \mathcal{C}^d$ *satisfies that for* $T > 0$,

$$\sup_{t\in[-\tau,T]} \mathbb{E}[V(x^\xi(t), t)] \leq K\Big(1 + \sup_{r\in[-\tau,0]} V(\xi(r), r) + \sup_{r\in[-\tau,0]} U(\xi(r), r)\Big)e^{KT}.$$

Proof By the Itô formula and Assumption 1.2, we obtain

$$\begin{aligned}V(x^\xi(t), t) &= V(\xi(0), 0) + \int_0^t \mathcal{L}V(x_s^\xi, s)\mathrm{d}s + \int_0^t V_x(x^\xi(s), s)\sigma(x_s^\xi)\mathrm{d}W(s) \\ &\leq V(\xi(0), 0) + K\int_0^t \Big(1 + V(x^\xi(s), s)\end{aligned}$$

$$
\begin{aligned}
&+\int_{-\tau}^{0} V(x^{\xi}(s+r), s+r)\mathrm{d}\bar{\nu}_1(r) \\
&- U(x^{\xi}(s), s) + \int_{-\tau}^{0} U(x^{\xi}(s+r), s+r)\mathrm{d}\bar{\nu}_2(r)\Big)\mathrm{d}s \\
&+\int_0^t V_x(x^{\xi}(s), s)\sigma(x_s^{\xi})\mathrm{d}W(s).
\end{aligned} \tag{1.11}
$$

Using the Fubini theorem yields

$$
\begin{aligned}
&\int_0^t \int_{-\tau}^{0} V(x^{\xi}(s+r), s+r)\mathrm{d}\bar{\nu}_1(r)\mathrm{d}s \\
&= \int_{-\tau}^{0}\int_0^t V(x^{\xi}(s+r), s+r)\mathrm{d}s\mathrm{d}\bar{\nu}_1(r) \\
&\le \int_{-\tau}^{0}\int_{-\tau}^{t} V(x^{\xi}(u), u)\mathrm{d}u\mathrm{d}\bar{\nu}_1(r) \\
&\le \int_{-\tau}^{t} V(x^{\xi}(s), s)\mathrm{d}s \le \tau \sup_{t\in[-\tau,0]} V(\xi(t), t) + \int_0^t V(x^{\xi}(s), s)\mathrm{d}s,
\end{aligned} \tag{1.12}
$$

and similarly,

$$
\int_0^t \int_{-\tau}^{0} U(x^{\xi}(s+r), s+r)\mathrm{d}\bar{\nu}_2(r)\mathrm{d}s \le \tau \sup_{t\in[-\tau,0]} U(\xi(t), t) + \int_0^t U(x^{\xi}(s), s)\mathrm{d}s.
$$

Inserting the above two inequalities into (1.11), we arrive at

$$
\begin{aligned}
V(x^{\xi}(t), t) &\le K\Big(\sup_{t\in[-\tau,0]} V(\xi(t), t) + \sup_{t\in[-\tau,0]} U(\xi(t), t)\Big) \\
&\quad + K\int_0^t \big(1 + V(x^{\xi}(s), s)\big)\mathrm{d}s + \int_0^t V_x(x^{\xi}(s), s)\sigma(x_s^{\xi})\mathrm{d}W(s).
\end{aligned} \tag{1.13}
$$

Taking expectations on both sides of (1.13), we deduce

$$
\begin{aligned}
\mathbb{E}[V(x^{\xi}(t), t)] &\le K\Big(\sup_{t\in[-\tau,0]} V(\xi(t), t) + \sup_{t\in[-\tau,0]} U(\xi(t), t)\Big) \\
&\quad + K\int_0^t \big(1 + \mathbb{E}[V(x^{\xi}(s), s)]\big)\mathrm{d}s.
\end{aligned}
$$

Applying the Gronwall inequality finishes the proof. □

If the Lyapunov function $V(x,t)$ defined by Assumption 1.2 is taken as $V(x,t) := |x|^p$ for some $p \geq 2$, based on Lemma 1.4, we obtain the pth moment boundedness of the solution $\{x^\xi(t)\}_{t\geq -\tau}$ for (1.1) on the finite time horizon.

Proposition 1.5 *Let Assumption 1.1 hold. Assume that for some $p \geq 2$, there exist a function $U \in C(\mathbb{R}^d; \mathbb{R}_+)$, and two probability measures $\bar{\nu}_1$ and $\bar{\nu}_2$ on $[-\tau, 0]$ such that for any $\phi \in C^d$,*

$$\begin{aligned}&|\phi(0)|^{p-2}\Big(2\langle\phi(0), b(\phi)\rangle + (p-1)|\sigma(\phi)|^2\Big)\\&\leq K\big(1 + |\phi(0)|^p + \int_{-\tau}^0 |\phi(r)|^2 \mathrm{d}\bar{\nu}_1(r)\big) - U(\phi(0)) + \int_{-\tau}^0 U(\phi(r))\mathrm{d}\bar{\nu}_2(r).\end{aligned} \tag{1.14}$$

Then the solution of (1.1) with the initial datum $\xi \in C^d$ satisfies that for $T > 0$,

$$\sup_{t\in[-\tau,T]} \mathbb{E}[|x^\xi(t)|^p] \leq K\Big(1 + \|\xi\|^p + \sup_{r\in[-\tau,0]} U(\xi(r))\Big)e^{KT}.$$

Proof The proof is based on Lemma 1.4 by taking $V(x,t) = |x|^p$. According to the definition (1.9) of $\mathcal{L}V$ and $p \geq 2$, one has

$$\begin{aligned}\mathcal{L}V(\phi,t) &= p|\phi(0)|^{p-4}\Big(|\phi(0)|^2\big(\langle\phi(0), b(\phi)\rangle + \frac{1}{2}|\sigma(\phi)|^2\big)\\&\quad + \frac{(p-2)}{2}|\langle\phi(0), \sigma(\phi)\rangle|^2\Big)\\&\leq \frac{p}{2}|\phi(0)|^{p-2}\Big(2\langle\phi(0), b(\phi)\rangle + (p-1)|\sigma(\phi)|^2\Big).\end{aligned} \tag{1.15}$$

This, along with (1.14), implies that coefficients b and σ satisfy Assumption 1.2 with $V(x,t) = |x|^p$. Then the desired argument follows from Lemma 1.4 immediately. □

The following lemma shows the moment boundedness of the solution $\{x^\xi(t)\}_{t\geq -\tau}$ for (1.1) on the infinite time horizon.

Lemma 1.6 ([61, Theorem 3]) *Let Assumptions 1.1 and 1.2 hold with (1.8) being replaced by*

$$\begin{aligned}\mathcal{L}V(\phi,t) \leq K - \bar{a}_1 V(\phi(0),t) + \bar{a}_2 \int_{-\tau}^0 V(\phi(r), t+r)\mathrm{d}\bar{\nu}_1(r)\\ - U(\phi(0),t) + \bar{a}_3 \int_{-\tau}^0 U(\phi(r), t+r)\mathrm{d}\bar{\nu}_2(r),\end{aligned} \tag{1.16}$$

where $K > 0$, $\bar{a}_1 > \bar{a}_2 \geq 0$, and $\bar{a}_3 \in (0, 1)$. Then the solution of (1.1) *with the initial datum $\xi \in C^d$ satisfies*

$$\sup_{t\geq -\tau} \mathbb{E}[V(x^\xi(t), t)] \leq K\Big(1 + \sup_{t\in[-\tau,0]} V(\xi(t), t) + \sup_{t\in[-\tau,0]} U(\xi(t), t)\Big).$$

Proof For any $\varepsilon > 0$, applying the Itô formula to the function $e^{\varepsilon t} V(x^\xi(t), t)$, we obtain,

$$e^{\varepsilon t} V(x^\xi(t), t) = V(\xi(0), 0) + \int_0^t e^{\varepsilon s}\Big(\varepsilon V(x^\xi(s), s) + \mathcal{L}V(x_s^\xi, s)\Big)\mathrm{d}s$$
$$+ \int_0^t e^{\varepsilon s} V_x(x^\xi(s), s)\sigma(x_s^\xi)\mathrm{d}W(s).$$

Taking the condition (1.16) into consideration, we derive

$$\begin{aligned}
&e^{\varepsilon t} V(x^\xi(t), t)\\
&\leq V(\xi(0), 0) + Ke^{\varepsilon t} - (\bar{a}_1 - \varepsilon)\int_0^t e^{\varepsilon s} V(x^\xi(s), s)\mathrm{d}s\\
&+ \bar{a}_2 \int_0^t \int_{-\tau}^0 e^{\varepsilon s} V(x^\xi(s + r), s + r)\mathrm{d}\bar{\nu}_1(r)\mathrm{d}s\\
&- \int_0^t e^{\varepsilon s}\Big(U(x^\xi(s), s)\mathrm{d}s - \bar{a}_3 \int_{-\tau}^0 U(x^\xi(s + r), s + r)\mathrm{d}\bar{\nu}_2(r)\Big)\mathrm{d}s\\
&+ \int_0^t e^{\varepsilon s} V_x(x^\xi(s), s)\sigma(x_s^\xi)\mathrm{d}W(s).
\end{aligned}\tag{1.17}$$

Similar to the derivation of (1.12), one has

$$\begin{aligned}
&\int_0^t \int_{-\tau}^0 e^{\varepsilon s} V(x^\xi(s + r), s + r)\mathrm{d}\bar{\nu}_1(r)\mathrm{d}s\\
&\leq e^{\varepsilon\tau} \int_{-\tau}^0 \int_0^t e^{\varepsilon(s+r)} V(x^\xi(s + r), s + r)\mathrm{d}s\mathrm{d}\bar{\nu}_1(r)\\
&\leq e^{\varepsilon\tau} \int_{-\tau}^0 e^{\varepsilon s} V(\xi(s), s)\mathrm{d}s + e^{\varepsilon\tau} \int_0^t e^{\varepsilon s} V(x^\xi(s), s)\mathrm{d}s\\
&\leq e^{\varepsilon\tau}\Big(\tau \sup_{t\in[-\tau,0]} V(\xi(t), t) + \int_0^t e^{\varepsilon s} V(x^\xi(s), s)\mathrm{d}s\Big).
\end{aligned}$$

This, along with (1.17), implies that

$$\begin{aligned} e^{\varepsilon t} V(x^{\xi}(t), t) \le K\Big(\sup_{t\in[-\tau,0]} V(\xi(t), t) + \sup_{t\in[-\tau,0]} U(\xi(t), t)\Big) + K e^{\varepsilon t} \\ - \big(\bar{a}_1 - \varepsilon - \bar{a}_2 e^{\varepsilon\tau}\big) \int_0^t e^{\varepsilon s} V(x^{\xi}(s), s)\mathrm{d}s - (1 - \bar{a}_3 e^{\varepsilon\tau}) \\ \times \int_0^t e^{\varepsilon s} U(x^{\xi}(s), s)\mathrm{d}s + \int_0^t e^{\varepsilon s} V_x(x^{\xi}(s), s)\sigma(x^{\xi}_s)\mathrm{d}W(s). \end{aligned}$$

Let $\varepsilon > 0$ be sufficiently small such that $\bar{a}_1 - \varepsilon - \bar{a}_2 e^{\varepsilon\tau} \ge 0$ and $1 - \bar{a}_3 e^{\varepsilon\tau} \ge 0$. Then,

$$\begin{aligned} e^{\varepsilon t} V(x^{\xi}(t), t) \le K\Big(\sup_{t\in[-\tau,0]} V(\xi(t), t) + \sup_{t\in[-\tau,0]} U(\xi(t), t)\Big) + K e^{\varepsilon t} \\ + \int_0^t e^{\varepsilon s} V_x(x^{\xi}(s), s)\sigma(x^{\xi}_s)\mathrm{d}W(s). \end{aligned} \tag{1.18}$$

Taking expectations on both sides of the above inequality, we derive

$$e^{\varepsilon t}\mathbb{E}[V(x^{\xi}(t), t)] \le K\Big(\sup_{t\in[-\tau,0]} V(\xi(t), t) + \sup_{t\in[-\tau,0]} U(\xi(t), t)\Big) + K e^{\varepsilon t},$$

which leads to

$$\sup_{t\ge-\tau} \mathbb{E}[V(x^{\xi}(t), t)] \le K\Big(1 + \sup_{t\in[-\tau,0]} V(\xi(t), t) + \sup_{t\in[-\tau,0]} U(\xi(t), t)\Big). \tag{1.19}$$

The proof is completed. □

Remark 1.5 If the parameter K in (1.16) equals zero, the trivial solution of (1.1) is exponentially stable satisfying

$$\limsup_{t\to\infty} \frac{1}{t} \log\big(\mathbb{E}[U(x^{\xi}(t), t)]\big) \le -\varepsilon,$$

$$\limsup_{t\to\infty} \frac{1}{t} \log\big(U(x^{\xi}(t), t)\big) \le -\varepsilon \quad \mathbb{P}\text{-a.s.},$$

where ε is given in the proof of Lemma 1.6; see e.g. [61, Theorem 3 and Theorem 4] for details. If the parameters $\bar{a}_1, \bar{a}_2$ in (1.16) satisfy $\bar{a}_1, \bar{a}_2 \ge 0$, according to [61, Theorem 5], the solution $\{x^{\xi}(t)\}_{t\ge-\tau}$ has the H_∞ stability, namely,

$$\limsup_{t\to\infty} \frac{1}{t} \int_0^t \mathbb{E}[U(x^{\xi}(s), s)]\mathrm{d}s \le \frac{K}{1 - \bar{a}_3},$$

where $\bar{a}_3 \in (0, 1)$ and K are given in (1.16).

Based on Lemma 1.6, by taking $V(x,t) := |x|^p$, we can obtain the boundedness of the pth moment for the solution $\{x^\xi(t)\}_{t\geq 0}$ on the infinite time horizon. The proof is similar to that of Proposition 1.5 and thus is omitted.

Proposition 1.7 *Let Assumption 1.1 hold. Assume that for some $p \geq 2$, there exist positive constants $\bar{a}_1, \bar{a}_2, \bar{a}_3$ with $\bar{a}_1 > \bar{a}_2$ and $\bar{a}_3 \in (0,1)$, a function $U \in C(\mathbb{R}^d; \mathbb{R}_+)$, and two probability measures $\bar{\nu}_1$ and $\bar{\nu}_2$ on $[-\tau, 0]$ such that for any $\phi \in C^d$,*

$$
\begin{aligned}
&|\phi(0)|^{p-2}\Big(2\langle \phi(0), b(\phi)\rangle + (p-1)|\sigma(\phi)|^2\Big) \\
&\quad \leq K - \bar{a}_1|\phi(0)|^p + \bar{a}_2\int_{-\tau}^0 |\phi(r)|^p \mathrm{d}\bar{\nu}_1(r) - U(\phi(0)) + \bar{a}_3\int_{-\tau}^0 U(\phi(r))\mathrm{d}\bar{\nu}_2(r).
\end{aligned}
\tag{1.20}
$$

Then the solution of (1.1) with the initial datum $\xi \in C^d$ satisfies

$$
\sup_{t\geq 0}\mathbb{E}[|x^\xi(t)|^p] \leq K\Big(1 + \|\xi\|^p + \sup_{r\in[-\tau,0]} U(\xi(r))\Big).
$$

In the following, we investigate the moment boundedness of the functional solution $\{x_t^\xi\}_{t\geq 0}$ for (1.1). Our analysis is based on the identity $\mathbb{E}[\|x_t^\xi\|^p] = \mathbb{E}\Big[\sup_{t-\tau\leq u\leq t}|x^\xi(u)|^p\Big]$, the Burkholder–Davis–Gundy inequality, and the moment boundedness of $\{x^\xi(t)\}_{t\geq -\tau}$. Meanwhile, this property can also be established by using the functional Itô formula of SFDEs; see e.g., [74–76].

Proposition 1.8 *Let the conditions in Proposition 1.5 hold. In addition, assume that there exists a probability measure $\bar{\nu}_3$ on $[-\tau, 0]$ such that for any $\phi \in C^d$,*

$$
|\sigma(\phi)|^2 \leq K\Big(1 + |\phi(0)|^2 + \int_{-\tau}^0 |\phi(r)|^2 \mathrm{d}\bar{\nu}_3(r)\Big). \tag{1.21}
$$

Then the functional solution of (1.1) with the initial datum $\xi \in C^d$ satisfies that for $T > 0$,

$$
\mathbb{E}\Big[\sup_{t\in[0,T]}\|x_t^\xi\|^p\Big] \leq K\Big(1 + \|\xi\|^p + \sup_{t\in[-\tau,0]} U(\xi(t))\Big)e^{KT},
$$

where $U(\cdot)$ is given in (1.14).

Proof Recalling (1.15), according to (1.13) with $V(x,t) = |x|^p$ and the Burkholder–Davis–Gundy inequality, we have

$$
\begin{aligned}
\mathbb{E}\Big[\sup_{0\leq u\leq t}|x^\xi(u)|^p\Big] &\leq K\|\xi\|^p + K\sup_{t\in[-\tau,0]} U(\xi(t)) + K\mathbb{E}\Big[\int_0^t (1 + |x^\xi(s)|^p)\mathrm{d}s\Big] \\
&\quad + K\mathbb{E}\Big[\Big(\int_0^t |x^\xi(s)|^{2(p-2)}|\langle x^\xi(s), \sigma(x_s^\xi)\rangle|^2 \mathrm{d}s\Big)^{\frac{1}{2}}\Big].
\end{aligned}
$$

It follows from Proposition 1.5, the Young inequality, and the Hölder inequality that

$$\begin{aligned}
&\mathbb{E}\Big[\sup_{0\le u\le t}|x^{\xi}(u)|^{p}\Big]\\
&\le K\Big(1+\|\xi\|^{p}+\sup_{t\in[-\tau,0]}U(\xi(t))\Big)e^{KT}\\
&\quad+K\mathbb{E}\Big[\Big(\sup_{0\le u\le t}|x^{\xi}(u)|^{2(p-1)}\int_{0}^{t}|\sigma(x_s^{\xi})|^{2}\mathrm{d}s\Big)^{\frac{1}{2}}\Big]\\
&\le K\Big(1+\|\xi\|^{p}+\sup_{t\in[-\tau,0]}U(\xi(t))\Big)e^{KT}+\frac{1}{2}\mathbb{E}\Big[\sup_{0\le u\le t}|x^{\xi}(u)|^{p}\Big]\\
&\quad+K\mathbb{E}\Big[\int_{0}^{t}|\sigma(x_s^{\xi})|^{p}\mathrm{d}s\Big].
\end{aligned}$$

By means of (1.21) and Proposition 1.5, we obtain

$$\mathbb{E}\Big[\sup_{0\le u\le t}|x^{\xi}(u)|^{p}\Big]\le K\Big(1+\|\xi\|^{p}+\sup_{t\in[-\tau,0]}U(\xi(t))\Big)e^{KT}+\frac{1}{2}\mathbb{E}\Big[\sup_{0\le u\le t}|x^{\xi}(u)|^{p}\Big],$$

which implies the desired argument. □

Furthermore, we give the moment estimate of the functional solution $\{x_t^{\xi}\}_{t\ge0}$ in the infinite time horizon.

Proposition 1.9 *Let* (1.21) *and the conditions in Proposition 1.7 hold. Then the functional solution of* (1.1) *with the initial datum* $\xi\in C^{d}$ *satisfies*

$$\sup_{t\ge0}\mathbb{E}[\|x_t^{\xi}\|^{p}]\le K\Big(1+\|\xi\|^{p}+\sup_{t\in[-\tau,0]}U(\xi(t))\Big).$$

Proof The proof is similar to that of Proposition 1.8, and we only give the outline of the proof. According to (1.18) with $V(x,t)=|x|^{p}$, the Burkholder–Davis–Gundy inequality, the Young inequality, and the Hölder inequality, we obtain

$$\begin{aligned}
&\mathbb{E}\Big[\sup_{(t-\tau)\vee0\le u\le t}e^{\varepsilon u}|x^{\xi}(u)|^{p}\Big]\\
&\le K\|\xi\|^{p}+K\sup_{t\in[-\tau,0]}U(\xi(t))+Ke^{\varepsilon t}\\
&\quad+K\mathbb{E}\Big[\Big(\int_{(t-\tau)\vee0}^{t}e^{2\varepsilon s}|x^{\xi}(s)|^{2(p-2)}|\langle x^{\xi}(s),\sigma(x_s^{\xi})\rangle|^{2}\mathrm{d}s\Big)^{\frac{1}{2}}\Big]
\end{aligned}$$

$$\leq K\big(\|\xi\|^p + \sup_{t\in[-\tau,0]} U(\xi(t))\big) + Ke^{\varepsilon t} + \frac{1}{2}\mathbb{E}\Big[\sup_{(t-\tau)\vee 0\leq u\leq t} e^{\varepsilon u}|x^\xi(u)|^p\Big]$$

$$+ K\mathbb{E}\Big[\int_{(t-\tau)\vee 0}^t e^{\varepsilon s}|\sigma(x_s^\xi)|^p \mathrm{d}s\Big],$$

where ε is given in the proof of Lemma 1.6. Combining (1.21) and Proposition 1.7, we derive

$$\begin{aligned} e^{\varepsilon(t-\tau)}\mathbb{E}[\|x_t^\xi\|^p] &\leq \mathbb{E}\Big[\sup_{(t-\tau)\vee 0\leq u\leq t} e^{\varepsilon u}|x^\xi(u)|^p\Big] \\ &\leq K\|\xi\|^p + K\sup_{t\in[-\tau,0]} U(\xi(t)) + Ke^{\varepsilon t}, \end{aligned}$$

which implies the desired assertion. □

1.3.2 *Attractiveness*

In this subsection, we investigate the attractiveness of both the solution $\{x^\xi(t)\}_{t\geq -\tau}$ and the functional solution $\{x_t^\xi\}_{t\geq 0}$ of (1.1). For a given function $V \in C^{2,1}(\mathbb{R}^d \times [-\tau,\infty);\mathbb{R}_+)$, define the functional $\mathbb{L}V : C^d \times C^d \times \mathbb{R}_+ \to \mathbb{R}$ by

$$\begin{aligned} \mathbb{L}V(\phi_1,\phi_2,t) = {} & V_t(\phi_1(0)-\phi_2(0),t) + V_x(\phi_1(0)-\phi_2(0),t)\big(b(\phi_1)-b(\phi_2)\big) \\ & + \frac{1}{2}\mathrm{trace}[(\sigma(\phi_1)-\sigma(\phi_2))^\top \\ & \times V_{xx}(\phi_1(0)-\phi_2(0),t)(\sigma(\phi_1)-\sigma(\phi_2))], \end{aligned} \tag{1.22}$$

where $V_x(x,t) = \big(V_{x_1}(x,t),\ldots,V_{x_d}(x,t)\big)$ and $V_{xx}(x,t) = \big(V_{x_i,x_j}(x,t)\big)_{d\times d}$ for $x\in\mathbb{R}^d, t\in\mathbb{R}_+$.

Assumption 1.3 *There exist positive constants $\bar{a}_4,\bar{a}_5,\bar{a}_6$ with $\bar{a}_4 > \bar{a}_5$ and $\bar{a}_6 \in (0,1)$, two probability measures $\bar{\nu}_4$ and $\bar{\nu}_5$ on $[-\tau,0]$, and two functions $\bar{V} \in C^{2,1}(\mathbb{R}^d\times[-\tau,\infty);\mathbb{R}_+)$ and $\bar{U}\in C(\mathbb{R}^d\times\mathbb{R}^d\times[-\tau,\infty);\mathbb{R}_+)$ such that $\bar{V}(x,t)$ vanishes only at $x=0$, $\bar{U}(x,y,t)$ vanishes at $x=y$, and for $\phi_1,\phi_2\in C^d$,*

$$\begin{aligned} \mathbb{L}\bar{V}(\phi_1,\phi_2,t) \leq {} & -\bar{a}_4\bar{V}(\phi_1(0)-\phi_2(0),t) + \bar{a}_5\int_{-\tau}^0 \bar{V}(\phi_1(r)-\phi_2(r),t+r)\mathrm{d}\bar{\nu}_4(r) \\ & - \bar{U}(\phi_1(0),\phi_2(0),t) + \bar{a}_6\int_{-\tau}^0 \bar{U}(\phi_1(r),\phi_2(r),t+r)\mathrm{d}\bar{\nu}_5(r). \end{aligned}$$

Lemma 1.10 *Let Assumptions 1.1 and 1.3 hold. Then for the initial data* $\xi, \eta \in C^d$, *the solution of* (1.1) *satisfies*

$$\begin{aligned}&\mathbb{E}[\bar{V}(x^{\xi}(t) - x^{\eta}(t), t)]\\ &\quad \le K e^{-\bar{\lambda} t}\Big(\sup_{r\in[-\tau,0]} \bar{V}(\xi(r) - \eta(r), r) + \sup_{r\in[-\tau,0]} \bar{U}(\xi(r), \eta(r), r)\Big), \quad t \in \mathbb{R}_+,\end{aligned}$$

where the positive constant $\bar{\lambda}$ *satisfies* $\bar{a}_4 - \bar{\lambda} - \bar{a}_5 e^{\tau\bar{\lambda}} \ge 0$ *and* $\bar{a}_6 e^{\tau\bar{\lambda}} \le 1$.

Proof Take a positive constant $\bar{\lambda}$ satisfying $\bar{a}_4 - \bar{\lambda} - \bar{a}_5 e^{\tau\bar{\lambda}} \ge 0$ and $\bar{a}_6 e^{\iota\bar{\lambda}} \le 1$. Applying the Itô formula, we have

$$\begin{aligned}&e^{\bar{\lambda} t}\bar{V}(x^{\xi}(t) - x^{\eta}(t), t)\\ &\quad = \bar{V}(\xi(0) - \eta(0), 0) + \int_0^t e^{\bar{\lambda} s}\Big(\bar{\lambda}\bar{V}(x^{\xi}(s) - x^{\eta}(s), s) + \mathbb{L}\bar{V}(x_s^{\xi}, x_s^{\eta}, s)\Big)\mathrm{d}s\\ &\qquad + \int_0^t e^{\bar{\lambda} s}\bar{V}_x(x_s^{\xi} - x_s^{\eta}, s)\big(\sigma(x_s^{\xi}) - \sigma(x_s^{\eta})\big)\mathrm{d}W(s).\end{aligned}$$

Using Assumption 1.3, we obtain

$$\begin{aligned}&e^{\bar{\lambda} t}\bar{V}(x^{\xi}(t) - x^{\eta}(t), t)\\ &\quad \le \bar{V}(\xi(0) - \eta(0), 0) + e^{\bar{\lambda}\tau}\tau\bar{a}_5 \sup_{r\in[-\tau,0]} \bar{V}(\xi(r) - \eta(r), r)\\ &\qquad + e^{\bar{\lambda}\tau}\tau\bar{a}_6 \sup_{r\in[-\tau,0]} \bar{U}(\xi(r), \eta(r), r)\\ &\qquad - (\bar{a}_4 - \bar{\lambda} - \bar{a}_5 e^{\bar{\lambda}\tau})\int_0^t e^{\bar{\lambda} s}\bar{V}(x^{\xi}(s) - x^{\eta}(s), s)\mathrm{d}s\\ &\qquad - (1 - \bar{a}_6 e^{\bar{\lambda}\tau})\int_0^t e^{\bar{\lambda} s}\bar{U}(x^{\xi}(s), x^{\eta}(s), s)\mathrm{d}s\\ &\qquad + \int_0^t e^{\bar{\lambda} s}\bar{V}_x(x^{\xi}(s) - x^{\eta}(s), s)\big(\sigma(x_s^{\xi}) - \sigma(x_s^{\eta})\big)\mathrm{d}W(s)\\ &\quad \le K\Big(\sup_{t\in[-\tau,0]} \bar{V}(\xi(t) - \eta(t), t) + \sup_{t\in[-\tau,0]} \bar{U}(\xi(t), \eta(t), t)\Big)\\ &\qquad + \int_0^t e^{\bar{\lambda} s}\bar{V}_x(x^{\xi}(s) - x^{\eta}(s), s)\big(\sigma(x_s^{\xi}) - \sigma(x_s^{\eta})\big)\mathrm{d}W(s), \end{aligned} \tag{1.23}$$

where we used

$$\int_0^t \int_{-\tau}^0 e^{\bar{\lambda}s}\bar{V}(x^\xi(s+r)-x^\eta(s+r),s+r)\mathrm{d}\bar{\nu}_4(r)\mathrm{d}s$$

$$\le e^{\tau\bar{\lambda}}\tau \sup_{t\in[-\tau,0]} \bar{V}(x^\xi(t)-x^\eta(t),t)+e^{\tau\bar{\lambda}}\int_0^t e^{\bar{\lambda}s}\bar{V}(x^\xi(s)-x^\eta(s),s)\mathrm{d}s,$$

$$\int_0^t \int_{-\tau}^0 e^{\bar{\lambda}s}\bar{U}(x^\xi(s+r),x^\eta(s+r),s+r)\mathrm{d}\bar{\nu}_5(r)\mathrm{d}s$$

$$\le e^{\tau\bar{\lambda}}\tau \sup_{t\in[-\tau,0]} \bar{U}(x^\xi(t),x^\eta(t),t)+e^{\tau\bar{\lambda}}\int_0^t e^{\bar{\lambda}s}\bar{U}(x^\xi(s),x^\eta(s),s)\mathrm{d}s.$$

We finish the proof by taking expectations on both sides of the above inequality. □

If the Lyapunov function $\bar{V}(x,t)$ in Assumption 1.3 is taken as $\bar{V}(x,t):=|x|^2$, combining Lemma 1.10, we can obtain the exponential attractiveness of the solution $\{x^\xi(t)\}_{t\ge 0}$ for (1.1) in the mean-square sense.

Proposition 1.11 *Let Assumption 1.1 hold. Assume that there exist positive constants $\bar{a}_4,\bar{a}_5,\bar{a}_6$ with $\bar{a}_4>\bar{a}_5$ and $\bar{a}_6\in(0,1)$, two probability measures $\bar{\nu}_4$ and $\bar{\nu}_5$ on $[-\tau,0]$, and a function $\bar{U}\in C(\mathbb{R}^d\times\mathbb{R}^d;\mathbb{R}_+)$ such that $\bar{U}(x,y)$ vanishes at $x=y$, and for $\phi_1,\phi_2\in C^d$,*

$$\begin{aligned}&2\langle\phi_1(0)-\phi_2(0),b(\phi_1)-b(\phi_2)\rangle+|\sigma(\phi_1)-\sigma(\phi_2)|^2\\&\quad\le-\bar{a}_4|\phi_1(0)-\phi_2(0)|^2+\bar{a}_5\int_{-\tau}^0|\phi_1(r)-\phi_2(r)|^2\mathrm{d}\bar{\nu}_4(r)\\&\qquad-\bar{U}(\phi_1(0),\phi_2(0))+\bar{a}_6\int_{-\tau}^0\bar{U}(\phi_1(r),\phi_2(r))\mathrm{d}\bar{\nu}_5(r).\end{aligned}\tag{1.24}$$

Then for the initial data $\xi,\eta\in C^d$, the solution of (1.1) satisfies

$$\mathbb{E}[|x^\xi(t)-x^\eta(t)|^2]\le K\Big(\|\xi-\eta\|^2+\sup_{r\in[-\tau,0]}\bar{U}(\xi(r),\eta(r))\Big)e^{-\bar{\lambda}t},\quad t\in\mathbb{R}_+,$$

where the positive constant $\bar{\lambda}$ satisfies $\bar{a}_4-\bar{\lambda}-\bar{a}_5e^{\tau\bar{\lambda}}\ge 0$ and $\bar{a}_6e^{\tau\bar{\lambda}}\le 1$.

Proof The proof is based on Lemma 1.10 by taking $\bar{V}(x,t):=|x|^2$. According to the definition (1.22) of $\mathbb{L}V$, one has that for $\phi_1,\phi_2\in C^d$,

$$\mathbb{L}V(\phi_1,\phi_2,t)=2\langle\phi_1(0)-\phi_2(0),b(\phi_1)-b(\phi_2)\rangle+|\sigma(\phi_1)-\sigma(\phi_2)|^2.$$

This, along with (1.24), implies that coefficients b and σ satisfy Assumption 1.3 with $\bar{V}(x,t)=|x|^2$. The desired argument follows from Lemma 1.10 immediately. □

Let the coefficient σ satisfy the globally Lipschitz condition. Based on Lemma 1.10, by taking $\bar{V}(x,t):=|x|^2$, we can obtain the exponential attractiveness of the functional solution $\{x_t^{\xi}\}_{t\geq 0}$, whose proof is similar to that of Proposition 1.9 and thus is omitted.

Proposition 1.12 *Let the conditions in Proposition 1.11 hold. In addition, assume that there exists a probability measure $\bar{\nu}_3$ on $[-\tau,0]$ such that for any $\phi_1,\phi_2\in C^d$,*

$$|\sigma(\phi_1)-\sigma(\phi_2)|^2\leq K\Big(|\phi_1(0)-\phi_2(0)|^2+\int_{-\tau}^{0}|\phi_1(r)-\phi_2(r)|^2\mathrm{d}\bar{\nu}_3(r)\Big). \tag{1.25}$$

Then for the initial data $\xi,\eta\in C^d$, the functional solutions of (1.1) *satisfy*

$$\mathbb{E}[\|x_t^{\xi}-x_t^{\eta}\|^2]\leq K\Big(\|\xi-\eta\|^2+\sup_{r\in[-\tau,0]}\bar{U}(\xi(r),\eta(r))\Big)e^{-\bar{\lambda}t},\quad t\in\mathbb{R}_+,$$

where the positive constant $\bar{\lambda}$ is given in Proposition 1.11.

1.3.3 Invariant Measure

The invariant measure is an important characteristic for describing longtime dynamical behaviors of the solution process for stochastic differential equations. In this subsection, we examine the Markov property of the functional solution for SFDEs and establish the existence and uniqueness of its corresponding invariant measure. Relevant foundational definitions are provided in Appendix B.

Due to the dependence on the past, it is known that the $\mathbb{R}^d$-valued solution $\{x^{\xi}(t)\}_{t\geq-\tau}$ of (1.1) is no longer Markovian. In contrast, the functional solution $\{x_t^{\xi}\}_{t\geq 0}$ has the Markov property in the state space C^d, which plays an important role in investigating the existence of the invariant measure of the functional solution.

Proposition 1.13 *For the initial datum $\xi\in C^d$, the functional solution $\{x_t^{\xi}\}_{t\geq 0}$ of* (1.1) *is a C^d-valued time-homogeneous Markov process and satisfies that for any $A\in\mathcal{B}(C^d)$, $t>s\geq 0$, and $\zeta\in C^d$,*

$$\mathbb{P}\{x_t^{\xi}\in A|\mathcal{F}_s\}=\mathbb{P}\{x_t^{\xi}\in A|x_s^{\xi}\},$$
$$\mathbb{P}\{x_t^{\xi}\in A|x_s^{\xi}=\zeta\}=\mathbb{P}\{x_{t-s}^{\zeta}\in A|x_0^{\zeta}=\zeta\}.$$

The proof of this proposition is similar to those of [80, Theorem 6.1] and [92, Theorem 4.2], and is omitted. Next, we study the invariant measure for the functional solution of (1.1). Denote by $\mathcal{P}$ the space of probability measures on $(C^d, \mathcal{B}(C^d))$. For any $\xi \in C^d$ and $t \geq 0$, denote by μ_t^ξ the probability measure induced by x_t^ξ, namely,

$$\mu_t^\xi(A) = \mathbb{P}\{\omega \in \Omega : x_t^\xi \in A\} \quad \forall A \in \mathcal{B}(C^d).$$

Define the bounded-Wasserstein distance $\mathbb{W}_q$ $(q \in [1, \infty))$ as

$$\mathbb{W}_q(\mathfrak{u}_1, \mathfrak{u}_2) := \Big(\inf_{\pi \in \Pi(\mathfrak{u}_1, \mathfrak{u}_2)} \int_{C^d \times C^d} 1 \wedge \|\phi_1 - \phi_2\|^q \pi(\mathrm{d}\phi_1, \mathrm{d}\phi_2)\Big)^{\frac{1}{q}}, \tag{1.26}$$

where $\Pi(\mathfrak{u}_1, \mathfrak{u}_2)$ denotes the collection of all probability measures on $C^d \times C^d$ with marginal measures $\mathfrak{u}_1$ and $\mathfrak{u}_2$; see [87, Chapter 6] for details. In the infinite-dimensional state space C^d, to obtain the tightness of the family of probability measures induced by the functional solutions, we need the following lemma; see [13, Theorem 7.3] and [35, Theorem 2.2] for details.

Lemma 1.14 *The family of probability measures $\{\mu_t^\xi\}_{t \geq 0}$ on C^d is tight if and only if the following two conditions hold:*

(a) For any $\bar{\varepsilon}_1 > 0$, there exists a constant $\bar{M} > 0$ such that

$$\sup_{t \geq 0} \mathbb{P}\Big\{x_t^\xi \in C^d : |x_t^\xi(-\tau)| \geq \bar{M}\Big\} \leq \bar{\varepsilon}_1.$$

(b) For any $\bar{\varepsilon}_2, \bar{\varepsilon}_3 > 0$, there exists a constant $\bar{\delta} \in (0, \tau)$ such that

$$\sup_{t \geq 0} \mathbb{P}\Big\{x_t^\xi \in C^d : \sup_{\substack{|s_1 - s_2| \leq \bar{\delta} \\ s_1, s_2 \in [-\tau, 0]}} |x_t^\xi(s_1) - x_t^\xi(s_2)| \geq \bar{\varepsilon}_2\Big\} \leq \bar{\varepsilon}_3.$$

Based on Lemma 1.14, and using Propositions 1.9 and 1.12, we give the existence and uniqueness of the invariant measure for (1.1).

Theorem 1.15 *Under the conditions in Proposition 1.12, the functional solution of (1.1) admits a unique invariant measure $\mu \in \mathcal{P}$ satisfying*

$$\mathbb{W}_2(\mu, \mu_t^\xi) \leq K(1 + \|\xi\|)e^{-\frac{\bar{\lambda}t}{2}}, \quad t \in \mathbb{R}_+,$$

for any $\xi \in C^d$, where $\bar{\lambda}$ is given in Proposition 1.11.

Proof *Step 1. Prove the uniqueness of the invariant measure.* Assume that there exist two invariant measures μ_1 and μ_2 of the functional solution for Eq. (1.1). According to Proposition 1.12, we have

$$\begin{aligned}\mathbb{W}_2(\mu_1, \mu_2) &= \mathbb{W}_2(\mu_t^{\bar{\xi}}, \mu_t^{\bar{\eta}}) \le \big(\mathbb{E}[\|x_t^{\bar{\xi}} - x_t^{\bar{\eta}}\|^2]\big)^{\frac{1}{2}} \\ &\le K\big(\|\bar{\xi} - \bar{\eta}\|^2 + \sup_{r\in[-\tau,0]} \bar{U}(\bar{\xi}(r), \bar{\eta}(r))\big)^{\frac{1}{2}} e^{\frac{-\bar{\lambda}t}{2}} \to 0\end{aligned}$$

as t tends to ∞, where $\bar{\xi}$ and $\bar{\eta}$ denote C^d-valued random variables with distributions μ_1 and μ_2, respectively. This implies the uniqueness of the invariant measure.

Step 2. We show the existence of the invariant measure. Applying the Krylov–Bogoliubov theorem [32, Section 7.1], it suffices to show that $\{x_t^{\xi}\}_{t\ge0}$ is Feller and $\{\mu_t^{\xi}\}_{t\ge0}$ is tight. It follows from Proposition 1.12 that for any $t > 0$, $x_t^{\xi} \to x_t^{\eta}$ in $L^2(\Omega; C^d)$ as $\xi \to \eta$ with $\xi, \eta \in C^d$. By [81, p. 256, Theorem 2], we derive $x_t^{\xi} \to x_t^{\eta}$ in distribution as $\xi \to \eta$, which implies the Feller property, i.e., for any continuous and bounded functional $\Phi : C^d \to \mathbb{R}$,

$$\lim_{\xi\to\eta} \mathbb{E}[\Phi(x_t^{\xi})] = \mathbb{E}[\Phi(x_t^{\eta})].$$

Now, we apply Lemma 1.14 to prove the tightness of $\{\mu_t^{\xi}\}_{t\ge0}$. Note that conditions in Proposition 1.12 yield those in Proposition 1.9 for $p = 2$. It then follows directly from Proposition 1.9 that Lemma 1.14 (a) holds.

We proceed to prove that Lemma 1.14 (b) holds. For any $t \ge 0$ and $M > \|\xi\|$, define a stopping time by $\zeta_t^M := \inf\{u \ge t : \|x_u^{\xi}\| > M\}$. Let $\bar{\delta} \in (0, \tau]$, using the Burkholder–Davis–Gundy inequality leads to

$$\begin{aligned}&\mathbb{E}\Big[\sup_{s\in[0,\bar{\delta}]} |x^{\xi}((t+s) \wedge \zeta_t^M) - x^{\xi}(t \wedge \zeta_t^M)|^2\Big] \\ &\le K\mathbb{E}\Big[\bar{\delta} \int_{t\wedge\zeta_t^M}^{(t+\bar{\delta})\wedge\zeta_t^M} |b(x_s^{\xi})|^2 \mathrm{d}s + K \int_{t\wedge\zeta_t^M}^{(t+\bar{\delta})\wedge\zeta_t^M} |\sigma(x_s^{\xi})|^2 \mathrm{d}s\Big] \le K(M)\bar{\delta},\end{aligned}$$

where we used the local boundedness of coefficients b and σ, which ensures that $K \sup_{\|\phi\|\le M} \big(|b(\phi)|^2 + |\sigma(\phi)|^2\big)$ is bounded above by some positive constant $K(M)$. Noting $\xi \in C^d$, for any $\bar{\varepsilon}_2 \in (0, 1)$, there exists a constant $\bar{\delta}_1 \in (0, \tau]$ such that $\sup_{|s_1-s_2|\le\bar{\delta}_1} |\xi(s_1) - \xi(s_2)| < \bar{\varepsilon}_2$. This implies that for $\bar{\delta} \in (0, \bar{\delta}_1]$,

$$\begin{aligned}&\sup_{t\ge0} \mathbb{P}\Big\{x_t^{\xi} \in C^d : \sup_{s_1,s_2\in[-\tau,0],|s_1-s_2|\le\bar{\delta}} |x_t^{\xi}(s_1) - x_t^{\xi}(s_2)| \ge \bar{\varepsilon}_2\Big\} \\ &= \sup_{t\ge\tau} \mathbb{P}\Big\{x_t^{\xi} \in C^d : \sup_{s_1,s_2\in[-\tau,0],|s_1-s_2|\le\bar{\delta}} |x_t^{\xi}(s_1) - x_t^{\xi}(s_2)| \ge \bar{\varepsilon}_2\Big\}\end{aligned}$$

$$
\begin{aligned}
&= \sup_{t\geq\tau}\mathbb{P}\Big\{\sup_{s_1,s_2\in[-\tau,0],|s_1-s_2|\leq\bar\delta}|x^\xi(t+s_1)-x^\xi(t-\tau)+x^\xi(t-\tau)\\
&\quad -x^\xi(t+s_2)|\geq\bar\varepsilon_2\Big\}\\
&\leq 2\sup_{t\geq0}\mathbb{P}\Big\{\sup_{0\leq s\leq\bar\delta}|x^\xi(t+s)-x^\xi(t)|\geq\frac{\bar\varepsilon_2}{2}\Big\}\\
&\leq 2\sup_{t\geq0}\mathbb{P}\Big\{\sup_{0\leq s\leq\bar\delta}|x^\xi(t+s)-x^\xi(t)|\geq\frac{\bar\varepsilon_2}{2},\zeta_t^M\geq t+\tau\Big\}\\
&\quad +2\sup_{t\geq0}\mathbb{P}\{\zeta_t^M<t+\tau\}.
\end{aligned}
$$

Combining the Chebyshev inequality yields

$$
\begin{aligned}
&\sup_{t\geq0}\mathbb{P}\Big\{x_t^\xi\in C^d:\sup_{s_1,s_2\in[-\tau,0],|s_1-s_2|\leq\bar\delta}|x_t^\xi(s_1)-x_t^\xi(s_2)|\geq\bar\varepsilon_2\Big\}\\
&\leq\frac{8}{\bar\varepsilon_2^2}\sup_{t\geq0}\mathbb{E}\Big[\sup_{s\in[0,\bar\delta]}|x^\xi((t+s)\wedge\zeta_t^M)-x^\xi(t\wedge\zeta_t^M)|^2\Big]\\
&\quad+2\sup_{t\geq0}\mathbb{P}\{\|x_t^\xi\|+\|x_{t+\tau}^\xi\|>M\}\leq K(M)\frac{\bar\delta}{\bar\varepsilon_2^2}+K\frac{(1+\|\xi\|^2)}{M^2}.
\end{aligned}
\tag{1.27}
$$

Let $M>0$ be sufficiently large such that $K\frac{(1+\|\xi\|^2)}{M^2}<\frac{\bar\varepsilon_3}{2}$, and choose $\bar\delta\in(0,\bar\delta_1)$ such that $K(M)\frac{\bar\delta}{\bar\varepsilon_2^2}<\frac{\bar\varepsilon_3}{2}$, which leads to Lemma 1.14 (b). Therefore, $\{\mu_t^\xi\}_{t\geq0}$ is tight.

Step 3. It follows from Proposition 1.12 that

$$
\mathbb{W}_2(\mu,\mu_t^\xi)=\big(\mathbb{E}[\|x_t^{\bar\eta}-x_t^\xi\|^2]\big)^{\frac12}\leq K(1+\|\xi\|)e^{-\frac{\bar\lambda t}{2}},\quad t\in\mathbb{R}_+.
$$

Combining *Steps 1–3*, we finish the proof. □

Remark 1.6 In this subsection, we study the existence and uniqueness of the invariant measure for (1.1) based on the boundedness and the attractiveness in the mean-square sense for the functional solution on the infinite time horizon. Recently, authors in [91] establish this result under weaker conditions, i.e., the boundedness and the attractiveness in probability sense for the functional solution on the infinite time horizon. Additionally, coupling approaches provide a framework for analyzing the uniqueness of the invariant measure for SFDEs, as demonstrated in works such as [4, 43, 52, 78].

1.3.4 Probabilistic Limit Theorems

Probabilistic limit theorems, including the strong law of large numbers (SLLN) and the central limit theorem (CLT), are fundamental topics in the theory of stochastic processes and are closely related to the invariant measure and ergodic theory of the process (see e.g. [26, 44]). For SFDEs with bounded and invertible diffusion coefficient, the probabilistic limit behavior of the functional solutions has been investigated in [9, 96]. In this subsection, we focus on SFDEs under a weaker condition on the diffusion coefficient, with a particular emphasis on the SLLN and CLT for the functional solution $\{x_t^\xi\}_{t\geq 0}$. Specifically, we investigate the probabilistic limit behaviors of the time-average $\frac{1}{T}\int_0^T f(x_t^\xi)\mathrm{d}t$ as $T \to \infty$ over a suitable class of test functionals f, which is strongly connected with the corresponding invariant measure μ of the functional solution.

Denote by $\xrightarrow{d}$, $\xrightarrow{\mathbb{P}}$, and $\xrightarrow{\mathbb{P}\text{-a.s.}}$ the convergence in distribution, in probability, and in the almost sure sense, respectively. For $\bar{\mu} \in \mathcal{P}$, define $\bar{\mu}(f) := \int_{C^d} f(u)\mathrm{d}\bar{\mu}(u)$ the integral of the measurable functional f with respect to $\bar{\mu}$. For fixed positive constants $q \geq 1$ and $\gamma \in (0, 1]$, define the quasi-metric $d_{q,\gamma}$ on C^d by

$$d_{q,\gamma}(u_1, u_2) := (1 \wedge \|u_1 - u_2\|^\gamma)(1 + \|u_1\|^q + \|u_2\|^q)^{\frac{1}{2}}. \tag{1.28}$$

Then we introduce a test functional space related to this quasi-metric. Define $C_{q,\gamma} := C_{q,\gamma}(C^d; \mathbb{R})$ the set of continuous functionals on C^d endowed with the following norm

$$\|f\|_{q,\gamma} := \sup_{u\in C^d} \frac{|f(u)|}{1 + \|u\|^{\frac{q}{2}}} + \sup_{\substack{u_1,u_2\in C^d\\ u_1\neq u_2}} \frac{|f(u_1) - f(u_2)|}{d_{q,\gamma}(u_1, u_2)}. \tag{1.29}$$

Based on Propositions 1.9 and 1.12, and together with Theorem 1.15, we derive the SLLN and the CLT for the functional solution $\{x_t^\xi\}_{t\geq 0}$. The proofs of these results are analogous to those for the general time-homogeneous Markov processes presented in Chap. 4 and are thus omitted.

Theorem 1.16 *Let the conditions in Propositions 1.9 and 1.12 hold with $p \geq 6$, $U(x) = |x|^{\tilde{p}p}$ for $\tilde{p} \geq 1$, and $\bar{U}(x, y) = |x - y|^2(1 + |x|^{2\beta} + |y|^{2\beta})$ for $\beta \geq 0$. In addition, assume that there exists a constant $q \geq 1$ satisfying $2\tilde{p}(q\tilde{p}+2+3\beta) \leq p$. Then, the functional solution $\{x_t^\xi\}_{t\geq 0}$ fulfills the SLLN and the CLT: for any $f \in C_{q,1}$,*

$$\frac{1}{T}\int_0^T f(x_t^\xi)\mathrm{d}t \xrightarrow{\mathbb{P}\text{-a.s.}} \mu(f) \quad \text{as } T \to \infty,$$

$$\frac{1}{\sqrt{T}}\int_0^T (f(x_t^\xi) - \mu(f))\mathrm{d}t \xrightarrow{d} \mathcal{N}(0, v^2) \quad \text{as } T \to \infty,$$

where μ is the invariant measure of $\{x_t^\xi\}_{t\geq 0}$ *and* $v^2 := 2\mu\big((f-\mu(f))\int_0^\infty \mathbb{E}[f(x_t^\cdot)-\mu(f)]\mathrm{d}t\big)$.

Theorem 1.16 shows that under suitable conditions, the time-average $\frac{1}{T}\int_0^T f(x_t^\xi)\mathrm{d}t$ of the functional solution converges to the ergodic limit $\mu(f)$ in the almost sure sense and the normalized fluctuations around $\mu(f)$ can be described by a centered Gaussian random variable with variance v^2.

The significance of such results lies in their ability to quantify the longtime statistical behavior of infinite-dimensional functional solutions of SFDEs. In particular, they ensure that time-averages over a single trajectory converge to the statistical mean, and that deviations from this limit obey Gaussian statistics. This is crucial for the statistical inference and uncertainty quantification of SFDEs.

For example, in financial modeling, where asset prices, volatility, or interest rates are often described by SFDEs (e.g., Example 1.5), Theorem 1.16 provides a basis for analyzing longtime averages and fluctuations of these quantities. Specifically, in option pricing or risk management under models with memory, the CLT can be used to approximate the distribution of cumulative returns. Moreover, the result justifies the use of ergodic estimators in calibrating delay models from market data, ensuring that temporal averages reliably reflect the underlying invariant distribution.

1.3.5 Density Function

It is known that the density of a solution process at a given time, describing the probability law, is one of the most essential characteristics that reveals behaviors of the solution; see [36, 77] and references therein. The study of density functions for solutions of SFDEs has been an active research topic in recent decades. Particularly, the case of globally Lipschitz drift coefficients has been extensively investigated. For instance, [12, 30] establish the existence and smoothness of the density for exact solutions of SFDEs, and [39] studies the asymptotic behavior of perturbed densities for SFDEs with small noise.

In this subsection, we study the existence of the density function for solutions of SFDEs under the one-sided Lipschitz condition on the drift coefficient, which allows for certain non-Lipschitz examples. For $\alpha \in \mathbb{N}_+$, let $\mathcal{D}$ and D denote the Gâteaux derivative operator and the Malliavin derivative operator, respectively, with $\mathcal{D}^\alpha$ and D^α representing their corresponding operators of order α. Furthermore, $(\mathbb{D}^{\alpha,p}(\mathbb{R}^d), \|\cdot\|_{\alpha,p})$ denotes the Malliavin Sobolev space consisting of all random variables $f:\Omega\to\mathbb{R}^d$ whose Malliavin derivatives up to order α belong to $L^p(\Omega)$, endowed with the norm

$$\|f\|_{\alpha,p} := \Big(\mathbb{E}\Big[|f|^p + \sum_{j=1}^{\alpha}\|D^j f\|^p_{H^{\otimes j}}\Big]\Big)^{\frac{1}{p}}.$$

Here $H := L^2(\mathbb{T}; \mathbb{R}^m)$ is a Hilbert space endowed with the inner product $\langle g, h\rangle_H := \int_{\mathbb{T}} g(t)^\top h(t)\mathrm{d}t$ for $g, h \in H$ and the norm $\|\cdot\|_H := \langle\cdot,\cdot\rangle_H^{\frac{1}{2}}$, where $\mathbb{T} = [0, T]$ or $[0, +\infty)$; see Appendix C for details. We first give the pth moment estimates of both the functional solution of (1.1) and its Gâteaux derivative.

Lemma 1.17 *Let* (1.25) *hold. Assume that coefficients* b, σ *have continuous Gâteaux derivatives. In addition, assume that there exists a positive constant* $\bar{a}_4$ *and a probability measure* $\bar{\nu}_4$ *on* $[-\tau, 0]$ *such that for any* $\phi_1, \phi_2 \in C^d$,

$$\begin{aligned}&\langle b(\phi_1) - b(\phi_2), \phi_1(0) - \phi_2(0)\rangle\\&\quad\le \bar{a}_4\Big(|\phi_1(0) - \phi_2(0)|^2 + \int_{-\tau}^0 |\phi_1(r) - \phi_2(r)|^2 \mathrm{d}\bar{\nu}_4(r)\Big).\end{aligned}\tag{1.30}$$

Then for any $p \ge 2$ *and* $T > 0$,

$$\mathbb{E}\Big[\sup_{t\in[0,T]} \|x_t^\xi\|^p\Big] \le K_T, \qquad \sup_{r\in[0,T]} \mathbb{E}\Big[\sup_{t\in[r,T]} \|D_r x_t^\xi\|^p\Big] \le K_T.$$

Proof Under (1.25) and (1.30), the existence and uniqueness of the global solution of (1.1) can be obtained by using Theorem 1.3. Combining Proposition 1.8, we can obtain the first inequality. For the second inequality, using the chain rule for Malliavin derivatives (see Lemma C.4), when $v \le t$, $D_v x^\xi(t)$ satisfies

$$D_v x^\xi(t) = \int_v^t \mathcal{D}b(x_s^\xi)D_v x_s^\xi \mathrm{d}s + \int_v^t \mathcal{D}\sigma(x_s^\xi)D_v x_s^\xi \mathrm{d}W(s) + \sigma(x_v^\xi)\mathbf{1}_{[0,t]}(v).\tag{1.31}$$

By the Itô formula, we derive

$$\begin{aligned}\mathrm{d}(|D_v x^\xi(t)|^{2p}) \le p|D_v x^\xi(t)|^{2p-2}\Big(&2\langle D_v x^\xi(t), \mathcal{D}b(x_t^\xi)D_v x_t^\xi\rangle \mathrm{d}t\\&+ (2p-1)|\mathcal{D}\sigma(x_t^\xi)D_v x_t^\xi|^2 \mathrm{d}t\\&+ 2\langle D_v x^\xi(t), \mathcal{D}\sigma(x_t^\xi)D_v x_t^\xi \mathrm{d}W(t)\rangle\Big).\end{aligned}\tag{1.32}$$

It follows from (1.25) and (1.30) that for any $\phi, \phi_1 \in C^d$,

$$|\mathcal{D}\sigma(\phi_1)\phi|^2 \le K\Big(|\phi(0)|^2 + \int_{-\tau}^0 |\phi(r)|^2 \mathrm{d}\bar{\nu}_3(r)\Big),$$

$$\langle \mathcal{D}b(\phi_1)\phi, \phi(0)\rangle \le \bar{a}_4\Big(|\phi(0)|^2 + \int_{-\tau}^0 |\phi(r)|^2 \mathrm{d}\bar{\nu}_4(r)\Big).$$

Inserting above inequalities into (1.32), we arrive at

$$
\begin{aligned}
|D_v x^\xi(t)|^{2p} \le |\sigma(x_v^\xi)|^{2p}\mathbf{1}_{[0,t]}(v) + K\int_v^t \Big(|D_v x^\xi(s)|^{2p} + |D_v x^\xi(s)|^{2p-2}&\\
\times \Big(\int_{-\tau}^0 |D_v x_s^\xi(r)|^2 \mathrm{d}\bar{\nu}_3(r) + \int_{-\tau}^0 |D_v x_s^\xi(r)|^2 \mathrm{d}\bar{\nu}_4(r)\Big)\Big)\mathrm{d}s&\\
+ 2p\int_v^t |D_v x^\xi(s)|^{2p-2}\big\langle D_v x^\xi(s), \mathcal{D}\sigma(x_s^\xi)D_v x_s^\xi \mathrm{d}W(s)\big\rangle.&
\end{aligned}
$$

Applying the Burkholder–Davis–Gundy inequality, we deduce

$$
\begin{aligned}
&\mathbb{E}\Big[\sup_{v\le t\le T}|D_v x^\xi(t)|^{2p}\Big]\\
&\le \mathbb{E}[|\sigma(x_v^\xi)|^{2p}] + K\int_v^T \mathbb{E}\Big[|D_v x^\xi(s)|^{2p} + |D_v x^\xi(s)|^{2p-2}\Big(\int_{-\tau}^0 |D_v x_s^\xi(r)|^2 \mathrm{d}\bar{\nu}_3(r)\\
&+ \int_{-\tau}^0 |D_v x_s^\xi(r)|^2 \mathrm{d}\bar{\nu}_4(r)\Big)\Big]\mathrm{d}s\\
&+ K\mathbb{E}\Big[\Big(\int_v^T |D_v x^\xi(s)|^{4p-2}|\mathcal{D}\sigma(x_s^\xi)D_v x_s^\xi|^2 \mathrm{d}s\Big)^{\frac{1}{2}}\Big].
\end{aligned}
$$

According to (1.25) and the Young inequality, we arrive at

$$
\begin{aligned}
&\mathbb{E}\Big[\sup_{v\le t\le T}|D_v x^\xi(t)|^{2p}\Big]\\
&\le K_T + K\int_v^T \mathbb{E}\Big[|D_v x^\xi(s)|^{2p} + \int_{-\tau}^0 |D_v x_s^\xi(r)|^{2p} \mathrm{d}\bar{\nu}_3(r)\\
&\quad + \int_{-\tau}^0 |D_v x_s^\xi(r)|^{2p} \mathrm{d}\bar{\nu}_4(r)\Big]\mathrm{d}s + \frac{1}{2}\mathbb{E}\Big[\sup_{v\le t\le T}|D_v x^\xi(t)|^{2p}\Big]\\
&\quad + K\mathbb{E}\Big[\int_v^T |\mathcal{D}\sigma(x_s^\xi)D_v x_s^\xi|^{2p}\mathrm{d}s\Big]\\
&\le K_T + K\int_v^T \mathbb{E}\Big[|D_v x^\xi(s)|^{2p} + \int_{-\tau}^0 |D_v x_s^\xi(r)|^{2p} \mathrm{d}\bar{\nu}_3(r)\\
&\quad + \int_{-\tau}^0 |D_v x_s^\xi(r)|^{2p} \mathrm{d}\bar{\nu}_4(r)\Big]\mathrm{d}s + \frac{1}{2}\mathbb{E}\Big[\sup_{v\le t\le T}|D_v x^\xi(t)|^{2p}\Big]\\
&\le K_T + K\int_v^T \mathbb{E}\Big[\sup_{v\le r\le s}|D_v x^\xi(r)|^{2p}\Big]\mathrm{d}s + \frac{1}{2}\mathbb{E}\Big[\sup_{v\le t\le T}|D_v x^\xi(t)|^{2p}\Big].
\end{aligned}
$$

Using the Gronwall inequality, one has

$$\sup_{0\le v\le T}\mathbb{E}\Big[\sup_{v\le t\le T}|D_v x^{\xi}(t)|^{2p}\Big]\le K_T.$$

The proof is completed. □

Based on Lemma 1.17 and $D_r x^{\xi}(t)=0$ for $r>t$, we derive that $x^{\xi}(t)\in\mathbb{D}^{1,p}(\mathbb{R}^d)$ for all $p\ge 1$. Hence, according to Theorem C.8, in order to obtain the existence of the density function of the solution of (1.1), it suffices to show that the Malliavin covariance matrix of $x^{\xi}(t)$ for $t\in[0,T]$,

$$\gamma^E(t):=\int_0^t D_r x^{\xi}(t)(D_r x^{\xi}(t))^{\top}\mathrm{d}r$$

is $\mathbb{P}$-a.s. invertible. To this end, we introduce a uniform non-degeneracy condition on the diffusion coefficient (i.e., (1.34)). Under this condition, by virtue of [31, Section 3.3], we prove that for some $q>0$, there exists a small number $\epsilon_0(q)$ such that for any $\epsilon<\epsilon_0(q)$, $\sup_{u\in\mathbb{R}^d,|u|=1}\mathbb{P}(u^{\top}\gamma^E(t)u\le\epsilon)\le K_T\epsilon^q$, $t\in[0,T]$. This is stated as follows by taking $q=1$.

Proposition 1.18 *Let the conditions in Lemma 1.17 hold. In addition, assume that there exist positive constants $K,\bar{\beta},\bar{\sigma}_0$, such that for any $\phi,\phi_1,\phi_2\in C^d$,*

$$|\mathcal{D}b(\phi_1)\phi_2|\le K(1+\|\phi_1\|^{\bar{\beta}})\|\phi_2\|, \tag{1.33}$$

$$\inf_{\phi\in C^d}\min_{u\in\mathbb{R}^d,|u|=1}u^{\top}\sigma(\phi)\sigma(\phi)^{\top}u\ge\bar{\sigma}_0. \tag{1.34}$$

Then for any $\epsilon\in(0,1)$, it holds that

$$\sup_{u\in\mathbb{R}^d,|u|=1}\mathbb{P}(u^{\top}\gamma^E(t)u\le\epsilon)\le K_T\epsilon,\ t\in[0,T].$$

Proof Fix $\epsilon\in(0,1)$ and let $\epsilon_1:=\frac{2\epsilon}{\bar{\sigma}_0}$. For any $r\le t$, it follows from (1.31) that

$$\begin{aligned}u^{\top}\gamma^E(t)u&\ge\int_{t-\epsilon_1}^t u^{\top}D_r x^{\xi}(t)(D_r x^{\xi}(t))^{\top}u\mathrm{d}r\ge\int_{t-\epsilon_1}^t u^{\top}\sigma(x_r^{\xi})(\sigma(x_r^{\xi}))^{\top}u\mathrm{d}r\\&\quad+2\int_{t-\epsilon_1}^t u^{\top}\Big(\int_r^t\mathcal{D}b(x_s^{\xi})D_r x_s^{\xi}\mathrm{d}s+\int_r^t\mathcal{D}\sigma(x_s^{\xi})D_r x_s^{\xi}\mathrm{d}W(s)\Big)\\&\quad\times(\sigma(x_r^{\xi}))^{\top}u\mathrm{d}r.\end{aligned}$$

Using (1.34), we derive

$$\int_{t-\epsilon_1}^{t} u^\top \sigma(x_r^\xi)(\sigma(x_r^\xi))^\top u\mathrm{d}r \geq \epsilon_1\bar{\sigma}_0 = 2\epsilon.$$

This, along with the Chebyshev inequality, implies that for any $u \in \mathbb{R}^d$ with $|u| = 1$,

$$\begin{aligned}
&\mathbb{P}(u^\top \gamma^E(t)u \leq \epsilon)\\
&\leq \mathbb{P}\Big(2\Big|\int_{t-\epsilon_1}^{t} u^\top\Big(\int_r^t \mathcal{D}b(x_s^\xi)D_r x_s^\xi \mathrm{d}s\\
&\quad + \int_r^t \mathcal{D}\sigma(x_s^\xi)D_r x_s^\xi \mathrm{d}W(s)\Big)(\sigma(x_r^\xi))^\top u\mathrm{d}r\Big| \geq \epsilon\Big)\\
&\leq 8\epsilon^{-2}\mathbb{E}\Big[\Big|\int_{t-\epsilon_1}^{t} u^\top \int_r^t \mathcal{D}b(x_s^\xi)D_r x_s^\xi \mathrm{d}s(\sigma(x_r^\xi))^\top u\mathrm{d}r\Big|^2\Big]\\
&\quad + 8\epsilon^{-2}\mathbb{E}\Big[\Big|\int_{t-\epsilon_1}^{t} u^\top \int_r^t \mathcal{D}\sigma(x_s^\xi)D_r x_s^\xi \mathrm{d}W(s)(\sigma(x_r^\xi))^\top u\mathrm{d}r\Big|^2\Big].
\end{aligned}$$

Utilizing the Hölder inequality and the Burkholder–Davis–Gundy inequality yields

$$\begin{aligned}
&\mathbb{P}(u^\top \gamma^E(t)u \leq \epsilon)\\
&\leq 8\epsilon^{-2}\epsilon_1\int_{t-\epsilon_1}^{t} \mathbb{E}\Big[\sup_{r\leq s\leq T}|\mathcal{D}b(x_s^\xi)D_r x_s^\xi|^2|\sigma(x_r^\xi)|^2\Big](t-r)^2\mathrm{d}r\\
&\quad + 8\epsilon^{-2}\epsilon_1\int_{t-\epsilon_1}^{t}\Big(\mathbb{E}\Big[\Big|\int_r^t \mathcal{D}\sigma(x_s^\xi)D_r x_s^\xi \mathrm{d}W(s)\Big|^4\Big]\Big)^{\frac{1}{2}}\Big(\mathbb{E}[|\sigma(x_r^\xi)|^4]\Big)^{\frac{1}{2}}\mathrm{d}r\\
&\leq K\epsilon^2\Big(\sup_{0\leq r\leq T}\mathbb{E}\Big[\sup_{r\leq s\leq T}|\mathcal{D}b(x_s^\xi)D_r x_s^\xi|^4\Big]\Big)^{\frac{1}{2}}\Big(\sup_{0\leq r\leq T}\mathbb{E}[|\sigma(x_r^\xi)|^4]\Big)^{\frac{1}{2}}\\
&\quad + K\epsilon\Big(\sup_{0\leq r\leq T}\mathbb{E}\Big[\sup_{r\leq s\leq T}|\mathcal{D}\sigma(x_s^\xi)D_r x_s^\xi|^4\Big]\Big)^{\frac{1}{2}}\Big(\sup_{0\leq r\leq T}\mathbb{E}[|\sigma(x_r^\xi)|^4]\Big)^{\frac{1}{2}}\\
&\leq \epsilon K_T,
\end{aligned}$$

where in the last inequality we used

$$|\mathcal{D}b(x_s^\xi)D_r x_s^\xi| \leq K(1+\|x_s^\xi\|^{\bar{\beta}})\|D_r x_s^\xi\|, \quad |\mathcal{D}\sigma(x_s^\xi)D_r x_s^\xi| \leq K\|D_r x_s^\xi\|$$

due to (1.33), the condition (1.25), and Lemma 1.17. The proof is finished. □

The existence of the density function of the solution of (1.1) is stated as follows.

Theorem 1.19 *Under the conditions in Proposition 1.18, for any $T > 0$, the law of $x^\xi(T)$ admits a density function.*

Proof Based on Proposition 1.18, we have

$$\sup_{u\in\mathbb{R}^d,|u|=1} \mathbb{P}(u^\top\gamma^E(T)u = 0) \le \sup_{u\in\mathbb{R}^d,|u|=1} \mathbb{P}(u^\top\gamma^E(T)u \le \epsilon) \le K_T\epsilon \;\forall\epsilon \in (0,1),$$

which shows the $\mathbb{P}$-a.s. invertibility of $\gamma^E(T)$ due to the arbitrariness of ϵ. This, together with $x^\xi(T) \in \mathbb{D}^{1,p}(\mathbb{R}^d)$ and Theorem C.8, completes the proof. □

The above result offers practical value across diverse fields, including statistical physics, financial mathematics, biological modeling, and stochastic optimization involving delayed feedback effects. A representative example is the Black–Scholes-type asset price model given in Example 1.5. In this financial setting, the existence and smoothness of the density function for the solution enable the explicit computation of expectations, probabilities, and derivatives of payoff functions, which underlie essential tasks such as option pricing, risk assessment, and sensitivity analysis. Specifically, as shown in [85], the availability of such a density function allows the derivation of closed-form expressions for the Greeks of European and Asian options.

1.3.6 Large Deviation Principle

The theory of large deviations, concerning asymptotics of small probabilities of rare events on an exponential scale, is a fundamental area in probability and statistics (see, for instance, [34, 37, 86]). One main interest in this area is the study of Freidlin–Wentzell type large deviation principle (LDP) for stochastic differential equations driven by small noise. It characterizes the exponential decay rates of probabilities for solution trajectories subject to small random perturbations, describing how sample paths deviate from the deterministic limit. We refer the reader to Appendix D for some standard definitions and results from the LDP. In this subsection, we study the Freidlin–Wentzell LDP for the solution of the SFDE. Let $\epsilon \in (0, 1)$. Consider the SFDE (1.1) driven by small noise

$$\begin{cases} \mathrm{d}x^{\xi,\epsilon}(t) = b(x_t^{\xi,\epsilon})\mathrm{d}t + \sqrt{\epsilon}\sigma(x_t^{\xi,\epsilon})\mathrm{d}W(t), & t > 0, \\ x^{\xi,\epsilon}(t) = \xi(t), \quad t \in [-\tau, 0], \end{cases} \tag{1.35}$$

where $\xi \in C^d$. Denote by $C_0([0,T];\mathbb{R}^m)$ the space of continuous functions $w : [0,T] \to \mathbb{R}^m$ with $w(0) = 0$. Denote by $C_\xi([-\tau,T];\mathbb{R}^d)$ the space of continuous functions $\psi : [-\tau,T] \to \mathbb{R}^d$ with $\psi(r) = \xi(r)$ for $r \in [-\tau,0]$, endowed with the norm $\|\psi\|_{C_\xi} := \sup_{-\tau\le t\le T} |\psi(t)|$. Let $\mathbb{H}_0$ denote the Cameron–Martin space of $C_0([0,T];\mathbb{R}^m)$ endowed with the norm $\|\cdot\|_{\mathbb{H}_0}$. Specifically, $\mathbb{H}_0$ consists of all the absolutely continuous functions $\psi \in C_0([0, T]; \mathbb{R}^m)$ with a square integrable derivative, that is, $\psi(t) = \int_0^t \dot{\psi}(s)ds$ with $\dot{\psi}(\cdot) \in L^2([0, T]; \mathbb{R}^m)$. The norm is defined as $\|\psi\|_{\mathbb{H}_0} := \|\dot{\psi}\|_{L^2([0,T];\mathbb{R}^m)}$ (see [46]). It is known in [34] that the family of Brownian motions $\{\sqrt{\epsilon}W(\cdot)\}_{\epsilon>0}$ satisfies the LDP with the good rate function

$$I_W(f) := \begin{cases} \frac{\|f\|^2_{\mathbb{H}_0}}{2}, & f \in \mathbb{H}_0, \\ \infty, & f \notin \mathbb{H}_0. \end{cases}$$

For $f \in C_0([0, T]; \mathbb{R}^m)$, let $x^{\xi,f}$ be the solution of the skeleton equation

$$dx^{\xi,f}(t) = b(x_t^{\xi,f})dt + \sigma(x_t^{\xi,f})f(t)dt, \quad t > 0,$$

where the initial datum $x^{\xi,f}(t) = \xi(t),\ t \in [-\tau, 0]$. We give the Freidlin–Wentzell LDP for the family of exact solutions $\{x^{\xi,\epsilon}(\cdot)\}_{\epsilon>0}$ under the globally Lipschitz condition on the coefficients.

Theorem 1.20 *Under condition* (1.2), *the family* $\{x^{\xi,\epsilon}(\cdot)\}_{\epsilon>0}$ *satisfies the LDP on* $C_\xi([-\tau, T]; \mathbb{R}^d)$ *with the good rate function*

$$I(g) = \inf\{I_W(f) : f \in \mathbb{H}_0,\ g = x^{\xi,f}(\cdot)\}, \quad g \in C_\xi([-\tau, T]; \mathbb{R}^d).$$

The proof of Theorem 1.20 under the global Lipschitz condition follows directly from the classical results (see e.g. [39, Theorem 2.1] and [72, Theorem 3.1]) and thus is omitted. For the case of superlinearly growing drift coefficient, the Freidlin–Wentzell LDP remains valid, whose proof is based on the weak convergence approach (see [50] for details).

Theorem 1.20 asserts that the probability of the solution path deviating from its typical behavior decays exponentially, with the decay rate characterized by the good rate function I. This result provides a way for quantifying rare events in stochastic applications with time-lag effects. For example, in financial modeling, Theorem 1.20 offers a valuable tool for analyzing rare events in markets with memory or hereditary structure. In this context, rare events correspond to measurable sets in the path space, such as the set of paths where asset prices exit a stable region, drop below a critical threshold, or experience sudden liquidity crises. The LDP can be employed to estimate the probability of these events, providing insight into extreme market behavior. Similar large deviation results have also been obtained for various practical stochastic functional differential models, such as the stochastic Predator–Prey model, the stochastic SIR model, and the stochastic Lotka–Volterra model; see [93] for relevant studies on LDPs of these models.

1.4 Numerical Methods

The exact solution of the SFDE arising in many fields, such as biology, mechanics, neural networks, and finance, seldom admits an explicit expression. Even when the coefficients are linear, a closed-form of the exact solution may still be unavailable. To illustrate this, consider the simple one-dimensional linear stochastic delay differential equation

$$\mathrm{d}x(t) = \sigma x(t-\tau)\,\mathrm{d}W(t), \quad t > 0, \tag{1.36}$$

with the initial datum $x(t) = \xi(t)$, $t \in [-\tau, 0]$. In the non-delay case ($\tau = 0$), (1.36) reduces to a linear stochastic ordinary differential equation with the explicit solution

$$x(t) = x(0)e^{\sigma W(t)-(\sigma^2 t)/2}, \quad t \geq 0.$$

When the delay $\tau > 0$, the structure of the solution of (1.36) becomes more complex. This is because the solution process possesses a finite memory, leading to a piecewise recursive dependence. A clarifying perspective is to examine the solution over the sequence of intervals $[i\tau, (i+1)\tau]$ for $i \in \mathbb{N}$. Specifically, once the solution is determined on the preceding interval $[(i-1)\tau, i\tau]$, Eq. (1.36) reduces to a stochastic ordinary differential equation on the current interval $[i\tau, (i+1)\tau]$. This recursive nature implies that the solution process satisfies the following relation derived from Itô integrals

$$\begin{aligned}
x(t) &= \xi(0) + \sigma \int_0^t \xi(u-\tau)\,\mathrm{d}W(u), && 0 \leq t \leq \tau,\\
x(t) &= x(\tau) + \sigma \int_\tau^t \Big[\xi(0) + \sigma \int_0^{(v-\tau)} \xi(u-\tau)\,\mathrm{d}W(u)\Big]\mathrm{d}W(v), && \tau < t \leq 2\tau,\\
&\cdots && \cdots.
\end{aligned}$$

The above indicates that a closed-form solution on the interval $[0, T]$ with $T = N_0\tau$, $N_0 \in \mathbb{N}_+$ is generally unavailable. Moreover, at each step the solution acquires additional nested Itô integrals, causing its representation to accumulate multiple layers of stochastic integration.

In response, numerical methods for SFDEs have emerged as essential tools, enabling qualitative and quantitative study of prevalent stochastic phenomena and time-lag effect. This capability has motivated substantial research interest in recent years. We refer to [3, 5, 15, 18, 23, 33, 41, 47, 83, 90] for the stochastic delay differential equation, [27, 60, 82, 99] for the stochastic differential equation with piecewise continuous argument, and [57, 80] for SFDEs with distributed delay. In this section, we introduce several widely used Euler–Maruyama-type (EM-type) numerical methods for SFDEs in the globally Lipschitz coefficient case and non-globally Lipschitz drift case, respectively. Without loss of generality, we take a sufficiently large integer N such that the time stepsize $\Delta := \frac{\tau}{N} \in (0, 1]$. Let $t_k = k\Delta$ for $k = -N, -N+1, \ldots$, and $\delta W_k = W(t_{k+1}) - W(t_k)$.

I. Globally Lipschitz Coefficient Case When coefficients b and σ satisfy the globally Lipschitz continuity, the simplest numerical method for solving (1.1) is the following *Euler–Maruyama (EM) method* established in [62],

$$\begin{cases} y^{\xi,\Delta}(t_k) = \xi(t_k), \quad k = -N, -N+1, \ldots, 0, \\ y^{\xi,\Delta}(t_{k+1}) = y^{\xi,\Delta}(t_k) + b(y^{\xi,\Delta}_{t_k})\Delta + \sigma(y^{\xi,\Delta}_{t_k})\delta W_k, \quad k \in \mathbb{N}. \end{cases} \tag{1.37}$$

Here, $y^{\xi,\Delta}_{t_k}$ is a C^d-valued random variable defined by

$$y^{\xi,\Delta}_{t_k}(r) = \frac{t_{j+1} - r}{\Delta} y^{\xi,\Delta}(t_{k+j}) + \frac{r - t_j}{\Delta} y^{\xi,\Delta}(t_{k+j+1})$$

for $r \in [t_j, t_{j+1}]$, $j \in \{-N, \ldots, -1\}$.

The EM method can be viewed as a specific instance (i.e., $\theta = 0$) of a broader class of one-step numerical methods known as the *the θ-EM method*. By introducing a parameter $\theta \in [0, 1]$, this family of methods provides a unified framework that encompasses both explicit and implicit methods, which is defined by

$$\begin{cases} y^{\xi,\Delta}(t_k) = \xi(t_k), \quad k = -N, -N+1, \ldots, 0, \\ y^{\xi,\Delta}(t_{k+1}) = y^{\xi,\Delta}(t_k) + (1-\theta)b(y^{\xi,\Delta}_{t_k})\Delta + \theta b(y^{\xi,\Delta}_{t_{k+1}})\Delta + \sigma(y^{\xi,\Delta}_{t_k})\delta W_k, \quad k \in \mathbb{N}. \end{cases} \tag{1.38}$$

Here, the definition of $y^{\xi,\Delta}_{t_k}$ is formally identical to that in the EM method (1.37), i.e.,

$$y^{\xi,\Delta}_{t_k}(r) = \frac{t_{j+1} - r}{\Delta} y^{\xi,\Delta}(t_{k+j}) + \frac{r - t_j}{\Delta} y^{\xi,\Delta}(t_{k+j+1}) \tag{1.39}$$

for $r \in [t_j, t_{j+1}]$, $j \in \{-N, \ldots, -1\}$. We call $\{y^{\xi,\Delta}(t_k)\}_{k \geq -N}$ the *θ-EM solution* and $\{y^{\xi,\Delta}_{t_k}\}_{k\in\mathbb{N}}$ the *θ-EM functional solution*. In particular, when $\theta = 1$, this method is referred to as the backward EM method, i.e.,

$$\begin{cases} y^{\xi,\Delta}(t_k) = \xi(t_k), \quad k = -N, -N+1, \ldots, 0, \\ y^{\xi,\Delta}(t_{k+1}) = y^{\xi,\Delta}(t_k) + b(y^{\xi,\Delta}_{t_{k+1}})\Delta + \sigma(y^{\xi,\Delta}_{t_k})\delta W_k, \quad k \in \mathbb{N}. \end{cases}$$

In recent decades, numerical analysis for SFDEs with globally Lipschitz coefficients, including the convergence and stability of numerical solutions, has been well established. For instance, strong convergence has been proved for the EM method (1.37) in [62] and for the θ-EM method (1.38) in [16]. Weak convergence of the EM method is examined in [19]. Stability analyses are provided for the EM method in

[95] and for the θ-EM method in [59]. Furthermore, the existence and uniqueness of the numerical invariant measure for the EM method is proved in [11].

In addition, the author in [17] establishes the fundamental convergence theorem of numerical methods for SFDEs with globally Lipschitz coefficients, clarifying the relationship between local and global errors. Following this framework, consider a general class of *one-step methods* of the form

$$\begin{cases} y^{\xi,\Delta}(t_k) = \xi(t_k), \quad k = -N, -N+1, \ldots, 0, \\ y^{\xi,\Delta}(t_{k+1}) = y^{\xi,\Delta}(t_k) + \Phi_b\Big(t_{k+1}, t_k, y^{\xi,\Delta}_{t_{k+1}}, y^{\xi,\Delta}_{t_k}\Big)\Delta \\ \qquad +\Phi_\sigma\Big(t_k, y^{\xi,\Delta}_{t_k}; W(s), s \in [t_k, t_{k+1}]\Big), \quad k \in \mathbb{N}, \end{cases} \tag{1.40}$$

where $y^{\xi,\Delta}_{t_k}$ denotes the approximation of the functional solution $x^{\xi}_{t_k}$, the mapping $\Phi_b : \mathbb{R}_+^{\otimes 2} \times (C^d)^{\otimes 2} \to \mathbb{R}^d$, and the mapping $\Phi_\sigma : \mathbb{R}_+ \times C^d \times \mathbb{R}^m \to \mathbb{R}^d$. For example, for the θ-EM method (1.38), functions Φ_b and Φ_σ are defined as

$$\Phi_b\Big(t_{k+1}, t_k, y^{\xi,\Delta}_{t_{k+1}}, y^{\xi,\Delta}_{t_k}\Big) = (1-\theta)b(y^{\xi,\Delta}_{t_k})\Delta + \theta b(y^{\xi,\Delta}_{t_{k+1}}),$$

$$\Phi_\sigma\Big(t_k, y^{\xi,\Delta}_{t_k}; W(s), s \in [t_k, t_{k+1}]\Big) = \sigma(y^{\xi,\Delta}_{t_k})\delta W_k,$$

where $y^{\xi,\Delta}_{t_k}$ is defined according to (1.39). Based on [17, Theorem 1], the following fundamental convergence theorem is established for the one-step method (1.40).

Theorem 1.21 *Assume that coefficients b and σ of* (1.1) *satisfy the globally Lipschitz continuity. Further suppose that for integers $k, i \in \mathbb{N}$ with $k > i$ and $X_t, Y_t \in C^d$ with $t \geq 0$, the mapping Φ_b in* (1.40) *satisfies*

$$\begin{aligned} &\Big|\Phi_b(t_k, t_i, X_{t_k}, X_{t_i}) - \Phi_b(t_k, t_i, Y_{t_k}, Y_{t_i})\Big| \\ &\quad \leq K\Big(|X_{t_k}(0) - Y_{t_k}(0)| + \sum_{j=-N}^{0} |X_{t_i}(t_j) - Y_{t_i}(t_j)|\Big), \end{aligned}$$

and the mapping Φ_σ in (1.40) *satisfies*

$$\mathbb{E}\Big[\Phi_\sigma\big(t_k, X_{t_k}; W(s), s \in [t_k, t_{k+1}]\big) - \Phi_\sigma\big(t_k, Y_{t_k}; W(s), s \in [t_k, t_{k+1}]\big)\Big|\mathcal{F}_{t_k}\Big] = 0,$$

and

$$\begin{aligned} &\mathbb{E}\Big[\Big|\Phi_\sigma\big(t_k, X_{t_k}; W(s), s \in [t_k, t_{k+1}]\big) - \Phi_\sigma\big(t_k, Y_{t_k}; W(s), s \in [t_k, t_{k+1}]\big)\Big|^2\Big] \\ &\quad \leq K\Delta \sum_{j=-N}^{0} |X_{t_i}(t_j) - Y_{t_i}(t_j)|^2, \end{aligned}$$

where $K > 0$. Then for $T > 0$ and $\Delta \in (0, \Delta^*)$ with some small constant $\Delta^* > 0$,

$$\begin{aligned}&\left(\mathbb{E}\left[\left|x^{\xi}(t_k) - y^{\xi,\Delta}(t_k)\right|^2\right]\right)^{\frac{1}{2}} \\ &\quad \le K_T \sup_{0\le k\Delta\le T}\left(\Delta^{-1}\left(\mathbb{E}\left[\left|\mathbb{E}[e_k|\mathcal{F}_{t_k}]\right|^2\right]\right)^{\frac{1}{2}} + \Delta^{-\frac{1}{2}}\left(\mathbb{E}\left[|e_k|^2\right]\right)^{\frac{1}{2}}\right),\end{aligned}$$

where the local error e_k is defined by

$$\begin{aligned}e_k := x^{\xi}(t_{k+1}) - \Big(&x^{\xi}(t_k) + \Phi_b\big(t_{k+1}, t_k, x^{\xi}_{t_{k+1}}, x^{\xi}_{t_k}\big)\Delta \\ &+ \Phi_\sigma\big(t_k, x^{\xi}_{t_k}; W(s), s \in [t_k, t_{k+1}]\big)\Big).\end{aligned}$$

In addition, if the one-step method (1.40) *is mean-square consistent of order p ($p > 0$), i.e.,*

$$\sup_{0\le k\Delta\le T}\left(\mathbb{E}\left[\left|\mathbb{E}[e_k|\mathcal{F}_{t_k}]\right|^2\right]\right)^{\frac{1}{2}} \le K_T\Delta^{p+1},$$

$$\sup_{0\le k\Delta\le T}\left(\mathbb{E}\left[|e_k|^2\right]\right)^{\frac{1}{2}} \le K_T\Delta^{p+\frac{1}{2}},$$

then

$$\sup_{0\le k\Delta\le T}\left(\mathbb{E}\left[\left|x^{\xi}(t_k) - y^{\xi,\Delta}(t_k)\right|^2\right]\right)^{\frac{1}{2}} \le K_T\Delta^{p}.$$

II. Non-globally Lipschitz Drift Case Many SFDEs of practical interest, however, do not satisfy the globally Lipschitz condition. Consequently, a variety of explicit and implicit EM–type methods have been designed to handle equations with superlinear coefficients. In what follows, we present an introduction to the construction of the numerical methods for the SFDE with the non-globally Lipschitz drift coefficient.

Truncated EM method (see [57]). First, we construct a truncated mapping. Under Assumption 1.1, choose a strictly increasing continuous function $\Phi(\cdot)$ from $[1, \infty)$ to $\mathbb{R}_+$ such that $\lim_{l\to\infty}\Phi(l) = \infty$ and

$$\sup_{\substack{\|\phi_1\|\vee\|\phi_2\|\le l\\ \phi_1\ne\phi_2}}\left(\frac{|b(\phi_1) - b(\phi_2)|}{(\Psi(\phi_1, \phi_2))^{1/2}} \vee \frac{|\sigma(\phi_1) - \sigma(\phi_2)|^2}{\Psi(\phi_1, \phi_2)}\right) \le \Phi(l), \quad l \ge 1,$$

where $\phi_1, \phi_2 \in C^d$, and

$$\Psi(\phi_1, \phi_2) := |\phi_1(0) - \phi_2(0)|^2 + \frac{1}{\tau}\int_{-\tau}^{0} |\phi_1(r) - \phi_2(r)|^2 \mathrm{d}r.$$

Denote by $\Phi^{-1}(\cdot)$ the inverse function of $\Phi(\cdot)$ from $[\Phi(1), \infty)$ to $\mathbb{R}_+$. For the stepsize $\Delta \in (0, 1]$, define a truncated mapping $\mathcal{T}^{\Delta}_{\Phi,\tilde{\gamma}} : \mathbb{R}^d \to \mathbb{R}^d$ by

$$\mathcal{T}^{\Delta}_{\Phi,\tilde{\gamma}}(x) = \Big(|x| \wedge \Phi^{-1}(\mathcal{K}\Delta^{-\tilde{\gamma}})\Big)\frac{x}{|x|},$$

where $\mathcal{K} := \Phi(1) \vee |b(\mathbf{0})| \vee |\sigma(\mathbf{0})|^2$ and $\tilde{\gamma} \in (0, 1/2)$. Then, define the truncated EM method by

$$\begin{cases} \breve{y}^{\xi,\Delta}(t_k) = \xi(t_k), \quad k = -N, \ldots, 0, \\ y^{\xi,\Delta}(t_k) = \mathcal{T}^{\Delta}_{\Phi,\tilde{\gamma}}(\breve{y}^{\xi,\Delta}(t_k)), \quad k = -N, \ldots, 0, 1, \ldots, \\ \breve{y}^{\xi,\Delta}(t_{k+1}) = y^{\xi,\Delta}(t_k) + b(y^{\xi,\Delta}_{t_k})\Delta + \sigma(y^{\xi,\Delta}_{t_k})\delta W_k, \quad k \in \mathbb{N}. \end{cases}$$

The numerical solution $\{y^{\xi,\Delta}(t_k)\}_{k\geq -N}$ approximates the solution $\{x^{\xi}(t)\}_{t\geq -\tau}$. The strong convergence, stability, and existence and uniqueness of the numerical invariant measure for the truncated EM method have been studied; see [56–58] for detailed results.

θ-EM method. In fact, the θ-EM method (1.38) with $\theta \in (\frac{1}{2}, 1]$ is well-suited for SFDEs with locally Lipschitz coefficients. Noting that the θ-EM method (1.38) is an implicit method for $\theta \in (0, 1]$. The following lemma ensures the solvability of (1.38) for superlinear SFDEs.

Lemma 1.22 ([80, Lemma 3.2]) *Assume that there exists a positive constant $\bar{L}$ such that for any $\phi_1, \phi_2 \in C^d$,*

$$\langle \phi_1(0) - \phi_2(0), b(\phi_1) - b(\phi_2)\rangle \leq \bar{L}\|\phi_1 - \phi_2\|^2. \tag{1.41}$$

Then the solution of the θ-EM method (1.38) *exists uniquely for any $\Delta \in (0, \frac{1}{\theta\bar{L}})$ with $\theta \in (0, 1]$.*

Proof Fix $\theta \in (0, 1]$. For any $\phi \in C^d$ and $u \in \mathbb{R}^d$, define continuous functions $\Phi^{\Delta}_{\phi,u} : [-\tau, 0] \to \mathbb{R}^d$ and $F^{\Delta}_{\theta,\phi} : \mathbb{R}^d \to \mathbb{R}^d$, respectively, by

$$\Phi^{\Delta}_{\phi,u}(r) = \begin{cases} \phi(r + \Delta), & r \in [-\tau, -\Delta), \\ \frac{-r}{\Delta}\phi(0) + \frac{\Delta+r}{\Delta}u, & r \in [-\Delta, 0] \end{cases} \tag{1.42}$$

and

$$F^{\Delta}_{\theta,\phi}(u) = u - \theta b(\Phi^{\Delta}_{\phi,u})\Delta.$$

Moreover, by virtue of (1.41), we have that for any $\epsilon > 0$,

$$\begin{aligned}\lim_{|u|\to\infty}\frac{\langle u, F^{\Delta}_{\theta,\phi}(u)\rangle}{|u|^2} &= 1-\theta\Delta\lim_{|u|\to\infty}\frac{\langle \Phi^{\Delta}_{\phi,u}(0), b(\Phi^{\Delta}_{\phi,u})\rangle}{|u|^2}\\ &\geq 1-\theta\Delta\lim_{|u|\to\infty}\frac{\bar{L}\|\Phi^{\Delta}_{\phi,u}\|^2+\langle \Phi^{\Delta}_{\phi,u}(0), b(\mathbf{0})\rangle}{|u|^2}.\end{aligned}$$

It follows from the Young inequality and (1.42) that

$$\begin{aligned}\lim_{|u|\to\infty}\frac{\langle u, F^{\Delta}_{\theta,\phi}(u)\rangle}{|u|^2} &\geq 1-\theta\Delta\lim_{|u|\to\infty}\frac{(\bar{L}+\epsilon)\|\Phi^{\Delta}_{\phi,u}\|^2+K}{|u|^2}\\ &\geq 1-\theta\Delta\lim_{|u|\to\infty}\frac{(\bar{L}+\epsilon)(\|\phi\|^2+|u|^2)+K}{|u|^2}\\ &= 1-\theta\Delta(\bar{L}+\epsilon).\end{aligned}$$

Since $\Delta \in (0, \frac{1}{\theta\bar{L}})$, one has $\lim_{|u|\to\infty}\frac{\langle u, F^{\Delta}_{\theta,\phi}(u)\rangle}{|u|^2} > 0$ by letting ϵ be sufficiently small, which implies that $F^{\Delta}_{\theta,\phi}$ is coercive. In addition, for any $u_1, u_2 \in \mathbb{R}^d$ with $u_1 \neq u_2$, by (1.41) and (1.42), we derive

$$\begin{aligned}&\langle u_1-u_2, F^{\Delta}_{\theta,\phi}(u_1)-F^{\Delta}_{\theta,\phi}(u_2)\rangle\\ &\quad= |u_1-u_2|^2-\theta\Delta\langle \Phi^{\Delta}_{\phi,u_1}(0)-\Phi^{\Delta}_{\phi,u_2}(0), b(\Phi^{\Delta}_{\phi,u_1})-b(\Phi^{\Delta}_{\phi,u_2})\rangle\\ &\quad\geq |u_1-u_2|^2-\theta\Delta\bar{L}\|\Phi^{\Delta}_{\phi,u_1}-\Phi^{\Delta}_{\phi,u_2}\|^2\\ &\quad\geq |u_1-u_2|^2-\theta\Delta\bar{L}\sup_{r\in[-\Delta,0]}\frac{\Delta+r}{\Delta}|u_1-u_2|^2\\ &\quad\geq (1-\theta\Delta\bar{L})|u_1-u_2|^2 > 0,\end{aligned}$$

which shows the monotone property of $F^{\Delta}_{\theta,\phi}$. Hence, according to [66, Lemma 3.1], we arrive at that for any $v \in \mathbb{R}^d$, the equation $F^{\Delta}_{\theta,\phi}(u) = v$ has a unique solution $u \in \mathbb{R}^d$, and the inverse $(F^{\Delta}_{\theta,\phi})^{-1}$ of $F^{\Delta}_{\theta,\phi}$ exists. For the fixed integer $k \geq 0$ and $y^{\xi,\Delta}_{t_k}$, we observe

$$\begin{aligned}F^{\Delta}_{\theta,y^{\xi,\Delta}_{t_k}}(y^{\xi,\Delta}(t_{k+1})) &= y^{\xi,\Delta}(t_{k+1})-\theta b(\Phi^{\Delta}_{y^{\xi,\Delta}_{t_k},y^{\xi,\Delta}(t_{k+1})})\Delta\\ &= y^{\xi,\Delta}(t_{k+1})-\theta b(y^{\xi,\Delta}_{t_{k+1}})\Delta,\end{aligned}$$

where we used (1.39) and (1.42). Then (1.38) can be rewritten as

$$F^{\Delta}_{\theta,y^{\xi,\Delta}_{t_k}}(y^{\xi,\Delta}(t_{k+1}))=y^{\xi,\Delta}(t_k)+(1-\theta)b(y^{\xi,\Delta}_{t_k})\Delta+\sigma(y^{\xi,\Delta}_{t_k})\delta W_k.$$

By virtue of the existence of $(F^{\Delta}_{\theta,y^{\xi,\Delta}_{t_k}})^{-1}$, we derive that $y^{\xi,\Delta}(t_{k+1})$ exists uniquely and satisfies

$$y^{\xi,\Delta}(t_{k+1})=(F^{\Delta}_{\theta,y^{\xi,\Delta}_{t_k}})^{-1}\big(y^{\xi,\Delta}(t_k)+(1-\theta)b(y^{\xi,\Delta}_{t_k})\Delta+\sigma(y^{\xi,\Delta}_{t_k})\delta W_k\big)\quad \mathbb{P}\text{-a.s.}$$

Thus, the proof is finished. □

Split-step θ-EM method. Based on θ-EM method (1.38), we define the split-step θ-EM method as follows:

$$\begin{cases} z^{\xi,\Delta}(t_k)=\ \xi(t_k), \quad k=-N,-N+1,\ldots,-1,\\ z^{\xi,\Delta}(t_k)=\ y^{\xi,\Delta}(t_k)-\theta b(y^{\xi,\Delta}_{t_k})\Delta, \quad k\in\mathbb{N}, \end{cases} \tag{1.43}$$

where $y^{\xi,\Delta}_{t_k}$ is given by (1.39). It follows from (1.43) that

$$z^{\xi,\Delta}(t_k)=\ z^{\xi,\Delta}(t_{k-1})+b(y^{\xi,\Delta}_{t_{k-1}})\Delta+\sigma(y^{\xi,\Delta}_{t_{k-1}})\delta W_{k-1}, \quad k\in\mathbb{N}_+. \tag{1.44}$$

Here, $\{z^{\xi,\Delta}(t_k)\}_{k\geq -N}$, which is called the *split-step θ-EM solution*, serves as an auxiliary process that closely approximates the solution $\{y^{\xi,\Delta}(t_k)\}_{k\geq -N}$ of the θ-EM method. It plays a key role in analyzing the properties of the θ-EM solution and thus is used frequently in the subsequent chapters.

For SFDEs with non-Lipschitz continuous coefficients, the convergence of the θ-EM method in the finite time horizon has been established in [100, 101]. More recently, authors in [80] use the θ-EM method with $\theta = 1$ to approximate the invariant measure of SFDEs. In the present work, we concentrate on the longtime asymptotics and probabilistic characteristics of the θ-EM method with $\theta\in(\frac{1}{2},1]$. By analyzing the relationship between $y^{\xi,\Delta}(t_k)$ and $z^{\xi,\Delta}(t_k)$, we will investigate the mean-square convergence and weak convergence in the infinite time horizon of the θ-EM method, the convergence of the invariant measure, the numerical SLLN and CLT, the existence of the density function of the numerical solution, and the Freidlin–Wentzell LDP of numerical solutions.

To conclude this section, we briefly comment on several numerical methodologies for SFDEs from the perspective of computational performance. From the viewpoint of accuracy, a variety of high-order numerical methods for SFDEs have been proposed, including Milstein-type and Taylor-based methods (see e.g. [23, 47, 69, 99]). In contrast, in the setting of stochastic ordinary and partial differential equations, the adaptive time-stepping strategy and parareal strategy have been extensively studied and shown to significantly enhance computational efficiency (see e.g. [22, 28, 29, 38, 51, 54, 55] for adaptive methods and [14, 45, 49, 97, 98] for parareal methods). For SFDEs, however, analogous efficiency-oriented methodologies, especially adaptive and parareal numerical strategies that account for memory and delay structures, remain largely undeveloped. This highlights an important open problem: the development of such efficient numerical strategies tailored to SFDEs, together with their effective integration with high-order numerical methods, so as to achieve both accuracy and efficiency in the numerical simulation of SFDEs.

References

1. E. Allen, *Modeling with Itô Stochastic Differential Equations*, vol. 22 of Mathematical Modelling: Theory and Applications (Springer, Dordrecht, 2007)
2. S. Bachmann, On the strong Feller property for stochastic delay differential equations with singular drift. Stoch. Process. Appl. **130**, 4563–4592 (2020)
3. C.T.H. Baker, E. Buckwar, Numerical analysis of explicit one-step methods for stochastic delay differential equations. LMS J. Comput. Math. **3**, 315–335 (2000)
4. J. Bao, J. Hao, Uniform-in-time estimates for mean-field type SDEs and applications. J. Differ. Equ. **440**, Paper No. 113445 (2025)
5. J. Bao, C. Yuan, Convergence rate of EM scheme for SDDEs. Proc. Amer. Math. Soc. **141**, 3231–3243 (2013)
6. J. Bao, F.-Y. Wang, C. Yuan, Derivative formula and Harnack inequality for degenerate functionals SDEs. Stoch. Dyn. **13**, 1250013, 22 (2013)
7. J. Bao, F.-Y. Wang, C. Yuan, Hypercontractivity for functional stochastic differential equations. Stoch. Process. Appl. **125**, 3636–3656 (2015)
8. J. Bao, F.-Y. Wang, C. Yuan, Asymptotic log-Harnack inequality and applications for stochastic systems of infinite memory. Stoch. Process. Appl. **129**, 4576–4596 (2019)
9. J. Bao, F.-Y. Wang, C. Yuan, Limit theorems for additive functionals of path-dependent SDEs. Discrete Contin. Dyn. Syst., **40**, 5173–5188 (2020)
10. J. Bao, P. Ren, F. Wang, Bismut formula for Lions derivative of distribution-path dependent SDEs. J. Differ. Equ. **282**, 285–329 (2021)
11. J. Bao, J. Shao, C. Yuan, Invariant probability measures for path-dependent random diffusions. Nonlinear Anal. **228**, Paper No. 113201 (2023)
12. D.R. Bell, S.-E.A. Mohammed, Smooth densities for degenerate stochastic delay equations with hereditary drift. Ann. Probab. **23**, 1875–1894 (1995)
13. P. Billingsley, *Convergence of Probability Measures*, Wiley Series in Probability and Statistics: Probability and Statistics, 2nd edn. (Wiley, New York, 1999)
14. C.-E. Bréhier, X. Wang, On parareal algorithms for semilinear parabolic stochastic PDEs. SIAM J. Numer. Anal. **58**, 254–278 (2020)
15. E. Buckwar, Introduction to the numerical analysis of stochastic delay differential equations. J. Comput. Appl. Math. **125**, 297–307 (2000)
16. E. Buckwar, The Θ–Maruyama scheme for stochastic functional differential equations with distributed memory term. Monte Carlo Methods Appl. **10**, 235–244 (2004)
17. E. Buckwar, One-step approximations for stochastic functional differential equations. Appl. Numer. Math. **56**, 667–681 (2006)
18. E. Buckwar, T. Shardlow, Weak approximation of stochastic differential delay equations. IMA J. Numer. Anal. **25**, 57–86 (2005)
19. E. Buckwar, R. Kuske, S. Mohammed, T. Shardlow, Weak convergence of the Euler scheme for stochastic differential delay equations. LMS J. Comput. Math. **11**, 60–99 (2008)

20. O. Butkovsky, A. Kulik, M. Scheutzow, Generalized couplings and ergodic rates for SPDEs and other Markov models. Ann. Appl. Probab. **30**, 1–39 (2020)
21. Y. Cai, T. Wang, L. Zhao, Relativistic stochastic mechanics II: reduced Fokker-Planck equation in curved spacetime. J. Stat. Phys. **190**, Paper No. 181 (2023)
22. S. Campbell, G. Lord, Adaptive time-stepping for stochastic partial differential equations with non-Lipschitz drift (2019). https://arxiv.org/abs/1812.09036
23. W. Cao, Z. Zhang, G.E. Karniadakis, Numerical methods for stochastic delay differential equations via the Wong–Zakai approximation. SIAM J. Sci. Comput. **37**, A295–A318 (2015)
24. M.-H. Chang, R.K. Youree, The European option with hereditary price structures: basic theory. Appl. Math. Comput. **102**, 279–296 (1999)
25. M.-H. Chang, R.K. Youree, Infinite-dimensional Black-Scholes equation with hereditary structure. Appl. Math. Optim. **56**, 395–424 (2007)
26. X. Chen, Limit theorems for functionals of ergodic Markov chains with general state space. Mem. Amer. Math. Soc. **139**, xiv+203 (1999)
27. C. Chen, J. Hong, Y. Lu, Stochastic differential equation with piecewise continuous arguments: Markov property, invariant measure and numerical approximation. Discrete Contin. Dyn. Syst. Ser. B **28**, 765–807 (2023)
28. C. Chen, T. Dang, J. Hong, An adaptive time-stepping fully discrete scheme for stochastic NLS equation: strong convergence and numerical asymptotics. Stochastic Process. Appl. **173**, Paper No. 104373 (2024)
29. C. Chen, T. Dang, J. Hong, Strong convergence of adaptive time-stepping schemes for the stochastic Allen–Cahn equation. IMA J. Numer. Anal. **45**, 404–450 (2025)
30. R. Chhaibi, I. Ekren, A Hörmander condition for delayed stochastic differential equations. Ann. H. Lebesgue **3**, 1023–1048 (2020)
31. J. Cui, J. Hong, Absolute continuity and numerical approximation of stochastic Cahn–Hilliard equation with unbounded noise diffusion. J. Differ. Equ. **269**, 10143–10180 (2020)
32. G. Da Prato, *An Introduction to Infinite-dimensional Analysis*. Universitext (Springer, Berlin, 2006)
33. K. Dareiotis, C. Kumar, S. Sabanis, On tamed Euler approximations of SDEs driven by Lévy noise with applications to delay equations. SIAM J. Numer. Anal. **54**, 1840–1872 (2016)
34. A. Dembo, O. Zeitouni, *Large Deviations Techniques and Applications*, vol. 38 of Applications of Mathematics, 2nd edn. (Springer, New York, 1998)
35. N.H. Du, N.H. Dang, N.T. Dieu, On stability in distribution of stochastic differential delay equations with Markovian switching. Syst. Control Lett. **65**, 43–49 (2014)
36. A.B. Duncan, T. Lelièvre, G.A. Pavliotis, Variance reduction using nonreversible Langevin samplers. J. Stat. Phys. **163**, 457–491 (2016)
37. R.S. Ellis, Large deviations and statistical mechanics, in *Particle Systems, Random Media and Large Deviations (Brunswick, Maine, 1984)*, vol. 41 of Contemp. Math. (Amer. Math. Soc., Providence, 1985, pp. 101–123
38. W. Fang, M.B. Giles, Adaptive Euler–Maruyama method for SDEs with nonglobally Lipschitz drift. Ann. Appl. Probab. **30**, 526–560 (2020)
39. M. Ferrante, C. Rovira, M. Sanz-Solé, Stochastic delay equations with hereditary drift: estimates of the density. J. Funct. Anal. **177**, 138–177 (2000)
40. Q. Guo, X. Mao, R. Yue, Almost sure exponential stability of stochastic differential delay equations. SIAM J. Control Optim. **54**, 1919–1933 (2016)
41. Q. Guo, X. Mao, R. Yue, The truncated Euler–Maruyama method for stochastic differential delay equations. Numer. Algorithms **78**, 599–624 (2018)
42. M. Hairer, Ergodicity of stochastic differential equations driven by fractional Brownian motion. Ann. Probab. **33**, 703–758 (2005)
43. M. Hairer, J.C. Mattingly, M. Scheutzow, Asymptotic coupling and a general form of Harris' theorem with applications to stochastic delay equations. Probab. Theory Related Fields **149**, 223–259 (2011)

44. J. Hong, X. Wang, *Invariant Measures for Stochastic Nonlinear Schrödinger Equations: Numerical Approximations and Symplectic Structures*, vol. 2251 of Lecture Notes in Mathematics (Springer, Singapore, 2019)
45. J. Hong, X. Wang, L. Zhang, Parareal exponential θ-scheme for longtime simulation of stochastic Schrödinger equations with weak damping. SIAM J. Sci. Comput. **41**, B1155–B1177 (2019)
46. Y. Hu, *Analysis on Gaussian Spaces* (World Scientific Publishing, Hackensack, 2017)
47. Y. Hu, S.-E.A. Mohammed, F. Yan, Discrete-time approximations of stochastic delay equations: the Milstein scheme. Ann. Probab. **32**, 265–314 (2004)
48. K. Itô, M. Nisio, On stationary solutions of a stochastic differential equation. J. Math. Kyoto Univ. **4**, 1–75 (1964)
49. B. Jin, Q. Lin, Z. Zhou, Optimizing coarse propagators in parareal algorithms. SIAM J. Sci. Comput. **47**, A735–A761 (2025)
50. D. Jin, Z. Chen, T. Zhou, Large deviations principle for stochastic delay differential equations with super-linearly growing coefficients. Front. Math. **20**, 699–720 (2025)
51. C. Kelly, G.J. Lord, Adaptive time-stepping strategies for nonlinear stochastic systems. IMA J. Numer. Anal. **38**, 1523–1549 (2018)
52. A. Kulik, M. Scheutzow, Well-posedness, stability and sensitivities for stochastic delay equations: a generalized coupling approach. Ann. Probab. **48**, 3041–3076 (2020)
53. J. Lei, M.C. Mackey, Stochastic differential delay equation, moment stability, and application to hematopoietic stem cell regulation system. SIAM J. Appl. Math. **67**, 387–407 (2006/2007)
54. V. Lemaire, An adaptive scheme for the approximation of dissipative systems. Stochastic Process. Appl. **117**, 1491–1518 (2007)
55. A. Leroy, B. Leimkuhler, J. Latz, D.J. Higham, Adaptive stepsize algorithms for Langevin dynamics. SIAM J. Sci. Comput. **46**, A3574–A3598 (2024)
56. G. Li, X. Li, X. Mao, G. Song, Hybrid stochastic functional differential equations with infinite delay: approximations and numerics. J. Differ. Equ. **374**, 154–190 (2023)
57. X. Li, X. Mao, G. Song, An explicit approximation for super-linear stochastic functional differential equations. Stoch. Process. Appl. **169**, Paper No. 104275 (2024)
58. X. Li, X. Mao, G. Song, Explicit approximation of invariant measure for stochastic delay differential equations with the nonlinear diffusion term. J. Theoret. Probab. **37**, 1850–1881 (2024)
59. M. Liu, W. Cao, Z. Fan, Convergence and stability of the semi-implicit Euler method for a linear stochastic differential delay equation. J. Comput. Appl. Math. **170**, 255–268 (2004)
60. Y. Lu, M. Song, M. Liu, Convergence rate and stability of the split-step theta method for stochastic differential equations with piecewise continuous arguments. Discrete Contin. Dyn. Syst. Ser. B **24**, 695–717 (2019)
61. Q. Luo, X. Mao, Y. Shen, Generalised theory on asymptotic stability and boundedness of stochastic functional differential equations. Autom. J. IFAC **47**, 2075–2081 (2011)
62. X. Mao, Numerical solutions of stochastic functional differential equations. LMS J. Comput. Math. **6**, 141–161 (2003)
63. X. Mao, *Stochastic Differential Equations and Applications*, 2nd edn. (Horwood Publishing, Chichester, 2008)
64. X. Mao, Stabilization of continuous-time hybrid stochastic differential equations by discrete-time feedback control. Autom. J. IFAC **49**, 3677–3681 (2013)
65. X. Mao, M.J. Rassias, Khasminskii-type theorems for stochastic differential delay equations. Stoch. Anal. Appl. **23**, 1045–1069 (2005)
66. X. Mao, L. Szpruch, Strong convergence rates for backward Euler–Maruyama method for non-linear dissipative-type stochastic differential equations with super-linear diffusion coefficients. Stochastics **85**, 144–171 (2013)
67. X. Mao, G. Marion, E. Renshaw, Environmental Brownian noise suppresses explosions in population dynamics. Stoch. Process. Appl. **97**, 95–110 (2002)
68. X. Mao, C. Yuan, J. Zou, Stochastic differential delay equations of population dynamics. J. Math. Anal. Appl. **304**, 296–320 (2005)

69. M. Milošević, M. Jovanović, An application of Taylor series in the approximation of solutions to stochastic differential equations with time-dependent delay. J. Comput. Appl. Math. **235**, 4439–4451 (2011)
70. S.-E.A. Mohammed, *Stochastic Functional Differential Equations*, vol. 99 of Research Notes in Mathematics (Pitman (Advanced Publishing Program), Boston, 1984)
71. S.-E.A. Mohammed, Stochastic differential systems with memory: theory, examples and applications, in *Stochastic Analysis and Related Topics, VI (Geilo, 1996)*, vol. 42 of Progr. Probab. (Birkhäuser Boston, Boston, 1998), pp. 1–77
72. S.-E.A. Mohammed, T. Zhang, Large deviations for stochastic systems with memory. Discrete Contin. Dyn. Syst. Ser. B **6**, 881–893 (2006)
73. P.H.A. Ngoc, Novel criteria for exponential stability in mean square of stochastic functional differential equations. Proc. Amer. Math. Soc. **148**, 3427–3436 (2020)
74. D.H. Nguyen, G. Yin, Stability of stochastic functional differential equations with regime-switching: analysis using Dupire's functional Itô formula. Potential Anal. **53**, 247–265 (2020)
75. D. Nguyen, N. Nguyen, G. Yin, Stochastic functional Kolmogorov equations, I: Persistence. Stoch. Process. Appl. **142**, 319–364 (2021)
76. D.H. Nguyen, N.N. Nguyen, G. Yin, Stochastic functional Kolmogorov equations II: Extinction. J. Differ. Equ. **294**, 1–39 (2021)
77. D. Nualart, *The Malliavin Calculus and Related Topics*, Probability and its Applications, 2nd edn. (Springer, Berlin, 2006)
78. S. Peszat, J. Zabczyk, Strong Feller property and irreducibility for diffusions on Hilbert spaces. Ann. Probab. **23**, 157–172 (1995)
79. L. Shaikhet, *Lyapunov Functionals and Stability of Stochastic Functional Differential Equations* (Springer, Cham, 2013)
80. B. Shi, Y. Wang, X. Mao, F. Wu, Approximation of invariant measures of a class of backward Euler–Maruyama scheme for stochastic functional differential equations. J. Differ. Equ. **389**, 415–456 (2024)
81. A.N. Shiryaev, *Probability*, vol. 95 of Graduate Texts in Mathematics, 2nd edn. (Springer, New York, 1996). Translated from the first (1980) Russian edition by R. P. Boas
82. M. Song, Y. Lu, M. Liu, Convergence of the tamed Euler method for stochastic differential equations with piecewise continuous arguments under non-global Lipschitz continuous coefficients. Numer. Funct. Anal. Optim. **39**, 517–536 (2018)
83. G. Song, J. Hu, S. Gao, X. Li, The strong convergence and stability of explicit approximations for nonlinear stochastic delay differential equations. Numer. Algorithms **89**, 855–883 (2022)
84. Y. Sun, J. Cao, pth moment exponential stability of stochastic recurrent neural networks with time-varying delays. Nonlinear Anal. Real World Appl. **8**, 1171–1185 (2007)
85. A. Takeuchi, Joint distributions for stochastic functional differential equations. Stochastics **88**, 711–736 (2016)
86. S.R.S. Varadhan, *Large Deviations and Applications*, vol. 46 of CBMS-NSF Regional Conference Series in Applied Mathematics (Society for Industrial and Applied Mathematics (SIAM), Philadelphia, 1984)
87. C. Villani, *Optimal Transport: Old and New*, vol. 338 of Grundlehren der mathematischen Wissenschaften [Fundamental Principles of Mathematical Sciences] (Springer, Berlin, 2009)
88. M.-K. von Renesse, M. Scheutzow, Existence and uniqueness of solutions of stochastic functional differential equations. Random Oper. Stoch. Equ. **18**, 267–284 (2010)
89. F.-Y. Wang, *Harnack Inequalities for Stochastic Partial Differential Equations*. Springer Briefs in Mathematics (Springer, New York, 2013)
90. X. Wang, S. Gan, D. Wang, θ-Maruyama methods for nonlinear stochastic differential delay equations. Appl. Numer. Math. **98**, 38–58 (2015)
91. Y. Wang, F. Wu, X. Mao, Stability in distribution of stochastic functional differential equations. Syst. Control Lett. **132**, 104513 (2019)
92. Y. Wang, F. Wu, C. Zhu, Limit theorems for additive functionals of stochastic functional differential equations with infinite delay. J. Differ. Equ. **308**, 421–454 (2022)

93. Y. Wang, F. Wu, C. Zhu, Large deviations for regime-switching diffusions with infinite delay. Stoch. Process. Appl. **176**, Paper No. 104418 (2024)
94. F. Wu, Y. Hu, Existence and uniqueness of global positive solutions to the stochastic functional Kolmogorov-type system. IMA J. Appl. Math. **75**, 317–332 (2010)
95. F. Wu, X. Mao, L. Szpruch, Almost sure exponential stability of numerical solutions for stochastic delay differential equations. Numer. Math. **115**, 681–697 (2010)
96. F. Wu, G. Yin, H. Mei, Stochastic functional differential equations with infinite delay: existence and uniqueness of solutions, solution maps, Markov properties, and ergodicity. J. Differ. Equ. **262**, 1226–1252 (2017)
97. L. Zhang, W. Zhou, L. Ji, Parareal algorithms applied to stochastic differential equations with conserved quantities. J. Comput. Math. **37**, 48–60 (2019)
98. L. Zhang, J. Wang, W. Zhou, L. Liu, L. Zhang, Convergence analysis of parareal algorithm based on Milstein scheme for stochastic differential equations. J. Comput. Math. **38**, 487–501 (2020)
99. Y. Zhang, M. Song, M. Liu, B. Zhao, Convergence and stability of the Milstein scheme for stochastic differential equations with piecewise continuous arguments. Numer. Algorithms **96**, 417–448 (2024)
100. J. Zhao, Y. Yi, Y. Xu, Strong convergence and stability of the split-step theta method for highly nonlinear neutral stochastic delay integro differential equation. Appl. Numer. Math. **157**, 385–404 (2020)
101. X. Zong, F. Wu, C. Huang, Theta schemes for SDDEs with non-globally Lipschitz continuous coefficients. J. Comput. Appl. Math. **278**, 258–277 (2015)

Chapter 2
Mean-Square Convergence Analysis in the Infinite Time Horizon

Mean-square convergence is one of the crucial features in assessing the accuracy of stochastic numerical methods. Over the past few decades, considerable attention has been paid to this topic for SFDEs on the finite time horizon. This chapter extends the discussion to the longtime regime. Specifically, we address the following question: for the superlinearly growing coefficients case, how can one obtain the longtime mean-square convergence of numerical methods for the SFDE?

We focus on the θ-EM method to analyze its longtime mean-square convergence for the SFDE (1.1) with a superlinearly growing drift coefficient. In contrast to the finite time horizon case, the key to obtaining the desired convergence result lies in the time-independent boundedness of high-order moment of the numerical solution, which is addressed in Sect. 2.1. Then, in Sect. 2.2, we derive the exponential attractiveness and the exponential stability of the θ-EM method. Section 2.3 is devoted to the longtime mean-square convergence analysis for the θ-EM method, where we show that the convergence is uniform in time t with a rate of $\frac{1}{2}$. The proof leverages the growth properties of the current terms to effectively control the influence of the delay terms, preventing significant deviations from occurring during the numerical discretization of the delayed components.

Both in this chapter and in the subsequent ones, we will focus on the case $\theta \in (\frac{1}{2}, 1]$ and provide a detailed analysis. For the case $\theta \in [0, \frac{1}{2}]$, analogous results can be obtained by additionally imposing linear growth conditions on the coefficients.

2.1 Time-Independent Moment Boundedness of Numerical Solution

In this section, we investigate the time-independent moment boundedness of the θ-EM functional solution, which is the key to obtaining the longtime mean-square convergence rate of the considered numerical method.

C. Chen et al., *Numerical Analysis of Stochastic Functional Differential Equations*, Lecture Notes in Mathematics 2399, https://doi.org/10.1007/978-981-92-1592-8_2

To begin with, we first propose the globally Lipschitz condition for the coefficient $\sigma : C^d \to \mathbb{R}^{d\times m}$ and the one-sided Lipschitz condition for the coefficient $b : C^d \to \mathbb{R}^d$.

Assumption 2.1 *There exist a constant $L > 0$ and a probability measure ν_1 on $[-\tau, 0]$ such that for any $\phi_1,\ \phi_2 \in C^d$,*

$$|\sigma(\phi_1) - \sigma(\phi_2)|^2 \le L\Big(|\phi_1(0) - \phi_2(0)|^2 + \int_{-\tau}^{0} |\phi_1(r) - \phi_2(r)|^2 \mathrm{d}\nu_1(r)\Big).$$

Assumption 2.2 *The drift coefficient b is continuous and there exist a probability measure ν_2 on $[-\tau, 0]$ and positive constants a_1, a_2 satisfying $a_1 > a_2 + L$ such that for any $\phi_1,\ \phi_2 \in C^d$,*

$$\begin{aligned}
&\langle \phi_1(0) - \phi_2(0), b(\phi_1) - b(\phi_2)\rangle \\
&\quad \le -a_1|\phi_1(0) - \phi_2(0)|^2 + a_2 \int_{-\tau}^{0} |\phi_1(r) - \phi_2(r)|^2 \mathrm{d}\nu_2(r).
\end{aligned}$$

Note that under Assumption 2.2, the drift coefficient b may exhibit superlinear growth.

Example 2.1 We present several typical examples of diffusion and drift coefficients satisfying Assumptions 2.1 and 2.2.

- When σ is a constant matrix, Assumption 2.1 is clearly satisfied with $L = 0$.
- Consider the diffusion coefficient

$$\sigma(\phi) = \frac{1}{\sqrt{2m}}\Big(\int_{-\tau}^{0} \phi(s)\mathrm{d}\bar{\nu}_1(s), \ldots, \int_{-\tau}^{0} \phi(s)\mathrm{d}\bar{\nu}_m(s)\Big) \in \mathbb{R}^{d\times m},$$

where $\phi \in C^d$, and $\bar{\nu}_i$, $i = 1, 2, \ldots, m$ are probability measures on $[-\tau, 0]$. Then,

$$\begin{aligned}
|\sigma(\phi_1) - \sigma(\phi_2)|^2 &= \frac{1}{2m}\sum_{i=1}^{m}\Big|\int_{-\tau}^{0} (\phi_1(s) - \phi_2(s))\mathrm{d}\bar{\nu}_i(s)\Big|^2 \\
&\le \frac{1}{2}\int_{-\tau}^{0} |\phi_1(s) - \phi_2(s)|^2 \mathrm{d}\nu_1(s),
\end{aligned}$$

which verifies Assumption 2.1 with $L = \frac{1}{2}$ and $\nu_1(\cdot) := \frac{1}{m}\sum_{i=1}^{m} \bar{\nu}_i(\cdot)$.

- Let $d = 1$. For the drift coefficient $b(\phi) = -\alpha\phi(0)+\beta\phi(-\tau)$ for $\phi \in C^1$ and the constant diffusion coefficient $\sigma > 0$ given in Example 1.2, the Young inequality yields

$$\begin{aligned}&\langle \phi_1(0) - \phi_2(0), b(\phi_1) - b(\phi_2)\rangle \\ &\quad\le -(\alpha - \frac{\beta}{2})|\phi_1(0) - \phi_2(0)|^2 + \frac{\beta}{2}|\phi_1(-\tau) - \phi_2(-\tau)|^2.\end{aligned}$$

  Thus, Assumption 2.2 is fulfilled whenever $\alpha > \beta$, with $a_1 = \alpha - \frac{\beta}{2}$, $a_2 = \frac{\beta}{2}$, and $L = 0$.
- Consider the drift coefficient $b(\phi) = \frac{1}{\tau}\int_{-\tau}^{0}\phi(s)\mathrm{d}s - |\phi(0)|^2\phi(0) - 2\phi(0)$, $\phi \in C^d$ given in Example 1.6. Using the elementary identity

$$\langle |u_1|^2u_1-|u_2|^2u_2, u_1-u_2\rangle = \frac{1}{2}(|u_1|^2+|u_2|^2)|u_1-u_2|^2+\frac{1}{2}(|u_1|^2-|u_2|^2)^2 \ge 0$$

  for all $u_1, u_2 \in \mathbb{R}^d$, we obtain

$$\begin{aligned}&\langle \phi_1(0) - \phi_2(0), b(\phi_1) - b(\phi_2)\rangle \\ &\quad\le \Big\langle \phi_1(0) - \phi_2(0), \frac{1}{\tau}\int_{-\tau}^{0}(\phi_1(s) - \phi_2(s))\mathrm{d}s\Big\rangle - 2|\phi_1(0) - \phi_2(0)|^2 \\ &\quad\le -\frac{3}{2}|\phi_1(0) - \phi_2(0)|^2 + \frac{1}{2\tau}\int_{-\tau}^{0}|\phi_1(s) - \phi_2(s)|^2\mathrm{d}s,\end{aligned}$$

  where the last step follows from the Young inequality. Hence, if the Lipschitz constant L of σ satisfies $L < 1$, then Assumption 2.2 holds with $a_1 = \frac{3}{2}$, $a_2 = \frac{1}{2}$.
- Recall the function $b(\cdot) = b_1(\cdot) + b_2(\cdot)$ given in Example 1.6 with

$$b_1(\phi) = \psi_1(\phi(0)) \text{ and } b_2(\phi) = \int_{-\tau}^{0}\psi_2(\phi(s))k(s)\mathrm{d}\nu(s),$$

  where ψ_2 is a Lipschitz continuous function with the Lipschitz constant Lip_{ψ_2} and k is a bounded function with bound $\|k\|_\infty$. Assume that for $u_1, u_2 \in \mathbb{R}^d$,

$$\langle \psi_1(u_1) - \psi_1(u_2), u_1 - u_2\rangle \le -K_1|u_1 - u_2|^2$$

  with $K_1 > 0$. It follows from the Young inequality that

$$\begin{aligned}&\langle b_2(\phi_1) - b_2(\phi_2), \phi_1(0) - \phi_2(0)\rangle \\ &\quad\le \frac{1}{2}|\phi_1(0) - \phi_2(0)|^2 + \frac{1}{2}\mathrm{Lip}_{\psi_2}^2\|k\|_\infty^2\int_{-\tau}^{0}|\phi_1(s) - \phi_2(s)|^2\mathrm{d}\nu(s).\end{aligned}$$

Therefore, if $K_1 - \frac{1}{2} > \frac{1}{2}\mathrm{Lip}_{\psi_2}^2 \|k\|_\infty^2 + L$, then b satisfies Assumption 2.2 with $\nu_2 = \nu$.

Under Assumptions 2.1 and 2.2, the SFDE (1.1) with the initial datum $\xi \in C^d$ has a unique solution $x^\xi(\cdot)$ on $[-\tau, \infty)$ according to Theorem 1.2. In addition, by Propositions 1.9 and 1.12, we can obtain the second moment boundedness and the exponential attractiveness for the functional solution of (1.1). Moreover, it follows from Theorem 1.15 that this functional solution of (1.1) admits a unique invariant measure. Precisely, under Assumptions 2.1 and 2.2, the functional solution $x_\cdot^\xi$ of (1.1) satisfies

$$\sup_{t\geq 0} \mathbb{E}[\|x_t^\xi\|^2] \leq K(1 + \|\xi\|^2)$$

and

$$\mathbb{E}[\|x_t^\xi - x_t^\eta\|^2] \leq K\|\xi - \eta\|^2 e^{-\lambda t} \quad \forall\, t \geq 0, \tag{2.1}$$

where $\xi, \eta \in C^d$, $K > 0$, and $\lambda > 0$ from Proposition 1.12 here satisfies

$$c_\lambda := 2a_1 - L - (2a_2 + L)e^{\lambda\tau} - \lambda > 0. \tag{2.2}$$

Moreover, the functional solution is asymptotically stable in distribution and admits a unique invariant measure μ satisfying

$$\mathbb{W}_2(\mu_t^\xi, \mu) \leq K(1 + \|\xi\|)e^{-\frac{\lambda}{2}t}, \tag{2.3}$$

where μ_t^ξ defined in Sect. 1.3.3 is the probability measure induced by x_t^ξ.

2.1.1 Time-Independent Boundedness of Second Moment

This subsection is devoted to proving the time-independent boundedness of the second moment of the θ-EM functional solution. Before that, we first estimate the pth moment of the θ-EM solution on any bounded time interval, which ensures that the stochastic integral related to the numerical solution is a martingale.

For this purpose, define the continuous version $\{z^{\xi,\Delta}(t)\}_{t\geq-\tau}$ of (1.43) as an auxiliary process which is given by

$$z^{\xi,\Delta}(t) = \begin{cases} z^{\xi,\Delta}(t_k) + b(y_{t_k}^{\xi,\Delta})(t - t_k) + \sigma(y_{t_k}^{\xi,\Delta})(W(t) - W(t_k)), \\ \qquad\qquad t \in [t_k, t_{k+1}),\ k \in \mathbb{N}, \\ \frac{t_{j+1}-t}{\Delta} z^{\xi,\Delta}(t_j) + \frac{t-t_j}{\Delta} z^{\xi,\Delta}(t_{j+1}), \quad t \in [t_j, t_{j+1}],\ j \in \{-N, \ldots, -1\}. \end{cases}$$

We see clearly that

$$z^{\xi,\Delta}(t) = z^{\xi,\Delta}(0) + \int_0^t b(y_s^{\xi,\Delta})\mathrm{d}s + \int_0^t \sigma(y_s^{\xi,\Delta})\mathrm{d}W(s), \quad t > 0, \tag{2.4}$$

where

$$y_s^{\xi,\Delta} := \sum_{k=0}^{\infty} y_{t_k}^{\xi,\Delta} \mathbf{1}_{[t_k,t_{k+1})}(s).$$

Denote by $\{z_t^{\xi,\Delta}\}_{t\geq 0}$ the segment process with respect to $\{z^{\xi,\Delta}(t)\}_{t\geq -\tau}$, namely, for any $t \geq 0$ and $r \in [-\tau, 0]$, we have $z_t^{\xi,\Delta}(r) = z^{\xi,\Delta}(t+r)$. For $\theta \in (1/2, 1]$, under Assumptions 2.1 and 2.2, the solvability of (1.38) is guaranteed for $\Delta \in (0, 1]$, whose proof is similar to that of Lemma 1.22.

We first present the moment boundedness of the numerical solution on any finite interval $[0, T]$.

Lemma 2.1 *Let Assumptions 2.1 and 2.2 hold. Then for any $T > 0$ and $p \in \mathbb{N}_+$,*

$$\sup_{\Delta\in(0,1]} \sup_{t_k\in[0,T]} \mathbb{E}\Big[|z^{\xi,\Delta}(t_k)|^{2p} + |y^{\xi,\Delta}(t_k)|^{2p}\Big] < K(\|\xi\|^{2p} + |b(\xi)|^{2p}\Delta^{2p} + T)e^{KT}. \tag{2.5}$$

Proof Let $\Delta \in (0, 1]$ and $k \in \mathbb{N}$. For any $M > \|\xi\|$, define a stopping time $\zeta_M^{\xi,\Delta}$ by

$$\zeta_M^{\xi,\Delta} = \inf\{i \in \mathbb{N} : |y^{\xi,\Delta}(t_i)| \vee |z^{\xi,\Delta}(t_i)| > M\}.$$

Obviously,

$$\{\omega \in \Omega : \zeta_M^{\xi,\Delta}(\omega) \geq i+1\} = \{\omega \in \Omega : \zeta_M^{\xi,\Delta}(\omega) \leq i\}^c \in \mathcal{F}_{t_i} \quad \forall\, i \in \mathbb{N}.$$

To simplify notation, we introduce the following abbreviations

$$\begin{aligned}
&z(t_k) = z^{\xi,\Delta}(t_k), \quad z_{t_k} = z_{t_k}^{\xi,\Delta}, \\
&y(t_k) = y^{\xi,\Delta}(t_k), \quad y_{t_k} = y_{t_k}^{\xi,\Delta}, \\
&b_k = b(y_{t_k}^{\xi,\Delta}), \quad \sigma_k = \sigma(y_{t_k}^{\xi,\Delta}), \quad \zeta_M = \zeta_M^{\xi,\Delta}.
\end{aligned} \tag{2.6}$$

It follows from Assumption 2.2 that for any $\phi \in C^d$,

$$2\langle \phi(0), b(\phi)\rangle \leq K - \frac{3a_1 + a_2 + L}{2}|\phi(0)|^2 + 2a_2\int_{-\tau}^0 |\phi(r)|^2 \mathrm{d}\nu_2(r). \tag{2.7}$$

This, along with (1.43) implies that

$$
\begin{aligned}
|z(t_{k+1})|^2 &= |z(t_k)|^2 + |b_k|^2\Delta^2 + |\sigma_k\delta W_k|^2 + 2\langle y(t_k) - \theta b_k\Delta, b_k\rangle\Delta \\
&\quad + 2\langle z(t_k), \sigma_k\delta W_k\rangle + 2\langle b_k, \sigma_k\delta W_k\rangle\Delta \\
&\le |z(t_k)|^2 + (1-2\theta)|b_k|^2\Delta^2 + |\sigma_k\delta W_k|^2 + K\Delta \\
&\quad - \frac{3a_1 + a_2 + L}{2}\Delta|y(t_k)|^2 + 2a_2\Delta\int_{-\tau}^{0}|y_{t_k}(r)|^2\mathrm{d}\nu_2(r) \\
&\quad + 2\langle z(t_k), \sigma_k\delta W_k\rangle + \frac{2}{\theta}\langle y(t_k) - z(t_k), \sigma_k\delta W_k\rangle.
\end{aligned} \tag{2.8}
$$

By $\theta \in (\frac{1}{2}, 1]$, we have

$$
|z(t_{k+1})|^2 \le |z(t_k)|^2 + |\sigma_k\delta W_k|^2 + K\Delta + 2a_2\Delta\int_{-\tau}^{0}|y_{t_k}(r)|^2\mathrm{d}\nu_2(r) + \mathcal{M}_k, \tag{2.9}
$$

where $\{\mathcal{M}_k\}_{k=0}^{\infty}$ is a local martingale defined by

$$
\mathcal{M}_k = \frac{2(\theta-1)}{\theta}\langle z(t_k), \sigma_k\delta W_k\rangle + \frac{2}{\theta}\langle y(t_k), \sigma_k\delta W_k\rangle. \tag{2.10}
$$

If $\zeta_M \ge k+1$, we have $k \wedge \zeta_M = k$ and

$$
\begin{aligned}
|z(t_{(k+1)\wedge\zeta_M})|^2 &= |z(t_{k+1})|^2 \\
&\le |z(t_{k\wedge\zeta_M})|^2 + |\sigma_k\delta W_k|^2 + K\Delta \\
&\quad + 2a_2\Delta\int_{-\tau}^{0}|y_{t_k}(r)|^2\mathrm{d}\nu_2(r) + \mathcal{M}_k.
\end{aligned}
$$

If $\zeta_M < k+1$, we have $\zeta_M \le k$ and

$$
|z(t_{(k+1)\wedge\zeta_M})|^2 = |z(t_{\zeta_M})|^2 = |z(t_{k\wedge\zeta_M})|^2.
$$

The above two cases imply that

$$
\begin{aligned}
|z(t_{(k+1)\wedge\zeta_M})|^2 &\le |z(t_{k\wedge\zeta_M})|^2 + \Big(|\sigma_k\delta W_k|^2 + K\Delta \\
&\quad + 2a_2\Delta\int_{-\tau}^{0}|y_{t_k}(r)|^2\mathrm{d}\nu_2(r) + \mathcal{M}_k\Big)\mathbf{1}_{\{\zeta_M \ge k+1\}}.
\end{aligned}
$$

Then for any $p \in \mathbb{N}_+$,

$$
\begin{aligned}
\mathbb{E}[|z(t_{(k+1)\wedge\zeta_M})|^{2p}] \le \mathbb{E}[|z(t_{k\wedge\zeta_M})|^{2p}] + \sum_{l=1}^{p} C_p^l \mathbb{E}\Big[|z(t_{k\wedge\zeta_M})|^{2(p-l)}\Big(|\sigma_k \delta W_k|^2 \\
+ K\Delta + 2a_2\Delta \int_{-\tau}^{0} |y_{t_k}(r)|^2 \mathrm{d}\nu_2(r) + \mathcal{M}_k\Big)^l \mathbf{1}_{\{\zeta_M \ge k+1\}}\Big] \\
=: \mathbb{E}[|z(t_{k\wedge\zeta_M})|^{2p}] + \sum_{l=1}^{p} \mathscr{I}_l, \qquad (2.11)
\end{aligned}
$$

where the positive constant C_p^l is the binomial coefficient. Note that

$$
\delta W_k \mathbf{1}_{\{\zeta_M \ge k+1\}} = W(t_{(k+1)\wedge\zeta_M}) - W(t_{k\wedge\zeta_M}).
$$

By virtue of the martingale stopping theorem (see Theorem A.6), we obtain

$$
\begin{aligned}
&\mathbb{E}\Big[A(t_k)\delta W_k \mathbf{1}_{\{\zeta_M \ge k+1\}} \big| \mathcal{F}_{t_{k\wedge\zeta_M}}\Big] = 0, \\
&\mathbb{E}\Big[\big|A(t_k)\delta W_k\big|^2 \mathbf{1}_{\{\zeta_M \ge k+1\}} \big| \mathcal{F}_{t_{k\wedge\zeta_M}}\Big] = |A(t_k)|^2 \mathbf{1}_{\{\zeta_M \ge k+1\}} \Delta,
\end{aligned}
$$

where $A(t_k)$ is an $\mathbb{R}^{d\times m}$-valued $\mathcal{F}_{t_k}$-measurable random variable for $k \in \mathbb{N}$. Then

$$
\begin{aligned}
\mathscr{I}_1 = C_p^1 \mathbb{E}\Big[\mathbb{E}\Big[|z(t_{k\wedge\zeta_M})|^{2(p-1)}\big(|\sigma_k \delta W_k|^2 + K\Delta + 2a_2\Delta \int_{-\tau}^{0} |y_{t_k}(r)|^2 \mathrm{d}\nu_2(r) \\
+ \mathcal{M}_k\big)\mathbf{1}_{\{\zeta_M \ge k+1\}} \big| \mathcal{F}_{t_{k\wedge\zeta_M}}\Big]\Big] \\
= C_p^1 \Delta \mathbb{E}\Big[|z(t_{k\wedge\zeta_M})|^{2(p-1)}\big(|\sigma_k|^2 + K + 2a_2 \int_{-\tau}^{0} |y_{t_k}(r)|^2 \mathrm{d}\nu_2(r)\big)\mathbf{1}_{\{\zeta_M \ge k+1\}}\Big].
\end{aligned}
$$

Making use of Assumption 2.1 and applying the Young inequality, we derive

$$
\begin{aligned}
\mathscr{I}_1 \le C_p^1 \Delta \mathbb{E}\Big[|z(t_{k\wedge\zeta_M})|^{2(p-1)}\Big(2L|y(t_k)|^2 + 2L \int_{-\tau}^{0} |y_{t_k}(r)|^2 \mathrm{d}\nu_1(r) + K \\
+ 2a_2 \int_{-\tau}^{0} |y_{t_k}(r)|^2 \mathrm{d}\nu_2(r)\Big)\mathbf{1}_{\{\zeta_M \ge k+1\}}\Big] \\
\le K\Delta + K\Delta \mathbb{E}[|z(t_{k\wedge\zeta_M})|^{2p}] + K\Delta \mathbb{E}\Big[|y(t_{k\wedge\zeta_M})|^{2p} \mathbf{1}_{\{\zeta_M \ge k+1\}}\Big] \\
+ K\Delta \mathbb{E}\Big[\Big(\int_{-\tau}^{0} |y_{t_k}(r)|^{2p} \mathrm{d}\nu_1(r) + \int_{-\tau}^{0} |y_{t_k}(r)|^{2p} \mathrm{d}\nu_2(r)\Big)\mathbf{1}_{\{\zeta_M \ge k+1\}}\Big].
\end{aligned}
\qquad (2.12)
$$

Moreover, for $l \in \mathbb{N}_+$ with $2 \le l \le p$, we obtain

$$\begin{aligned}&\Big(|\sigma_k \delta W_k|^2 + K\Delta + 2a_2\Delta\int_{-\tau}^{0}|y_{t_k}(r)|^2 \mathrm{d}\nu_2(r) + \mathcal{M}_k\Big)^l\\&\le K\Big(|\sigma_k\delta W_k|^{2l} + \Delta^l + \Delta^l\int_{-\tau}^{0}|y_{t_k}(r)|^{2l}\mathrm{d}\nu_2(r) + |z(t_k)|^l|\sigma_k\delta W_k|^l\\&\quad + |y(t_k)|^l|\sigma_k\delta W_k|^l\Big).\end{aligned}$$

This, along with the following fact

$$\mathbb{E}\Big[\big|A(t_k)\delta W_k\big|^i \mathbf{1}_{\{\zeta_M \ge k+1\}}\big|\mathcal{F}_{t_{k\wedge\zeta_M}}\Big] \le K|A(t_k)|^i\Delta^{\frac{i}{2}}\mathbf{1}_{\{\zeta_M\ge k+1\}} \quad \forall i \in \mathrm{N} \text{ with } i \ge 2$$

implies that

$$\begin{aligned}\sum_{l=2}^{p}\mathscr{I}_l &\le \sum_{l=2}^{p} K\mathbb{E}\Big[\mathbb{E}\Big[|z(t_{k\wedge\zeta_M})|^{2(p-l)}\Big(|\sigma_k\delta W_k|^{2l} + \Delta^l + \Delta^l\int_{-\tau}^{0}|y_{t_k}(r)|^{2l}\mathrm{d}\nu_2(r)\\&\quad + |z(t_k)|^l|\sigma_k\delta W_k|^l + |y(t_k)|^l|\sigma_k\delta W_k|^l\Big)\mathbf{1}_{\{\zeta_M\ge k+1\}}\big|\mathcal{F}_{t_{k\wedge\zeta_M}}\Big]\Big]\\&\le K\Delta\sum_{l=2}^{p}\mathbb{E}\Big[|z(t_{k\wedge\zeta_M})|^{2(p-l)}\Big(|\sigma_k|^{2l} + 1 + \int_{-\tau}^{0}|y_{t_k}(r)|^{2l}\mathrm{d}\nu_2(r)\\&\quad + |z(t_k)|^l|\sigma_k|^l + |y(t_k)|^l|\sigma_k|^l\Big)\mathbf{1}_{\{\zeta_M\ge k+1\}}\Big].\end{aligned}$$

Applying Assumption 2.1 and the Young inequality again, we arrive at

$$\begin{aligned}\sum_{l=2}^{p}\mathscr{I}_l &\le K\Delta + K\Delta\mathbb{E}[|z(t_{k\wedge\zeta_M})|^{2p}] + K\Delta\mathbb{E}\Big[|y(t_{k\wedge\zeta_M})|^{2p}\mathbf{1}_{\{\zeta_M\ge k+1\}}\Big]\\&\quad + K\Delta\mathbb{E}\Big[\Big(\int_{-\tau}^{0}|y_{t_k}(r)|^{2p}\mathrm{d}\nu_1(r) + \int_{-\tau}^{0}|y_{t_k}(r)|^{2p}\mathrm{d}\nu_2(r)\Big)\mathbf{1}_{\{\zeta_M\ge k+1\}}\Big].\end{aligned} \tag{2.13}$$

Inserting (2.12) and (2.13) into (2.11) and by the recursive calculation, we obtain

$$\begin{aligned}&\mathbb{E}[|z(t_{(k+1)\wedge\zeta_M})|^{2p}]\\&\le \mathbb{E}[|z(t_{k\wedge\zeta_M})|^{2p}] + K\Delta + K\Delta\mathbb{E}[|z(t_{k\wedge\zeta_M})|^{2p}]\\&\quad + K\Delta\Big[|y(t_{k\wedge\zeta_M})|^{2p}\mathbf{1}_{\{\zeta_M\ge k+1\}}\Big] + K\Delta\mathbb{E}\Big[\Big(\int_{-\tau}^{0}|y_{t_k}(r)|^{2p}\mathrm{d}\nu_1(r)\end{aligned}$$

$$
\begin{aligned}
&+\int_{-\tau}^{0}|y_{t_k}(r)|^{2p}\mathrm{d}\nu_2(r)\big)\mathbf{1}_{\{\zeta_M\geq k+1\}}\Big]\\
&\leq \mathbb{E}[|z(0)|^{2p}]+K(k+1)\Delta+K\Delta\sum_{i=0}^{k}\mathbb{E}[|z(t_{i\wedge\zeta_M})|^{2p}]\\
&+K\Delta\sum_{i=0}^{k}\mathbb{E}\Big[|y(t_{i\wedge\zeta_M})|^{2p}\mathbf{1}_{\{\zeta_M\geq k+1\}}\Big]+K\Delta\sum_{i=0}^{k}\mathbb{E}\Big[\big(\int_{-\tau}^{0}|y_{t_i}(r)|^{2p}\mathrm{d}\nu_1(r)\\
&+\int_{-\tau}^{0}|y_{t_i}(r)|^{2p}\mathrm{d}\nu_2(r)\big)\mathbf{1}_{\{\zeta_M\geq i+1\}}\Big].
\end{aligned}
\tag{2.14}
$$

It follows from (1.39) and the convex property of $|\cdot|^{2p}$ that

$$
\begin{aligned}
&\sum_{i=0}^{k}\int_{-\tau}^{0}|y_{t_i}(r)|^{2p}\mathrm{d}\nu_1(r)\mathbf{1}_{\{\zeta_M\geq i+1\}}\\
&=\sum_{j=-N}^{-1}\sum_{i=0}^{k}\int_{t_j}^{t_{j+1}}\Big|\frac{t_{j+1}-r}{\Delta}y(t_{i+j})+\frac{r-t_j}{\Delta}y(t_{i+j+1})\Big|^{2p}\mathrm{d}\nu_1(r)\mathbf{1}_{\{\zeta_M\geq i+1\}}\\
&\leq\sum_{j=-N}^{-1}\sum_{i=0}^{k}\int_{t_j}^{t_{j+1}}\frac{t_{j+1}-r}{\Delta}\mathrm{d}\nu_1(r)|y(t_{i+j})|^{2p}\mathbf{1}_{\{\zeta_M\geq i+j+1\}}\\
&+\sum_{j=-N}^{-1}\sum_{i=0}^{k}\int_{t_j}^{t_{j+1}}\frac{r-t_j}{\Delta}\mathrm{d}\nu_1(r)|y(t_{i+j+1})|^{2p}\mathbf{1}_{\{\zeta_M\geq i+j+2\}}.
\end{aligned}
$$

We apply the change of variables for indices $i+j+1$ and $i+j+2$, respectively, to obtain

$$
\begin{aligned}
&\sum_{i=0}^{k}\int_{-\tau}^{0}|y_{t_i}(r)|^{2p}\mathrm{d}\nu_1(r)\mathbf{1}_{\{\zeta_M\geq i+1\}}\\
&\leq\sum_{j=-N}^{-1}\sum_{l=-N}^{k-1}\int_{t_j}^{t_{j+1}}\frac{t_{j+1}-r}{\Delta}\mathrm{d}\nu_1(r)|y(t_l)|^{2p}\mathbf{1}_{\{\zeta_M\geq l+1\}}\\
&+\sum_{j=-N}^{-1}\sum_{l=-N+1}^{k}\int_{t_j}^{t_{j+1}}\frac{r-t_j}{\Delta}\mathrm{d}\nu_1(r)|y(t_l)|^{2p}\mathbf{1}_{\{\zeta_M\geq l+1\}}
\end{aligned}
$$

$$\leq \sum_{j=-N}^{-1} \sum_{l=-N}^{k} \int_{t_j}^{t_{j+1}} \mathrm{d}\nu_1(r)|y(t_l)|^{2p}\mathbf{1}_{\{\zeta_M \geq l+1\}}$$

$$\leq N\|\xi\|^{2p} + \sum_{i=0}^{k} |y(t_i)|^{2p}\mathbf{1}_{\{\zeta_M \geq i+1\}}. \tag{2.15}$$

Similarly,

$$\sum_{i=0}^{k} \int_{-\tau}^{0} |y_{t_i}(r)|^{2p}\mathrm{d}\nu_2(r)\mathbf{1}_{\{\zeta_M \geq i+1\}} \leq N\|\xi\|^{2p} + \sum_{i=0}^{k} |y(t_i)|^{2p}\mathbf{1}_{\{\zeta_M \geq i+1\}}. \tag{2.16}$$

Inserting (2.15) and (2.16) into (2.14), we have

$$\mathbb{E}[|z(t_{(k+1)\wedge\zeta_M})|^{2p}] \leq \mathbb{E}[|z(0)|^{2p}] + K(k+1)\Delta + K\Delta\sum_{i=0}^{k}\mathbb{E}[|z(t_{i\wedge\zeta_M})|^{2p}]$$
$$+ K\tau\|\xi\|^{2p} + K\Delta\sum_{i=0}^{k}\mathbb{E}\big[|y(t_i)|^{2p}\mathbf{1}_{\{\zeta_M \geq i+1\}}\big]. \tag{2.17}$$

It is straightforward to see from $z(t_i) = y(t_i) - \theta b_i \Delta$ and (2.7) that

$$|z(t_i)|^2 = |y(t_i)|^2 + \theta^2\Delta^2 b_i^2 - 2\theta\Delta\langle y(t_i), b_i\rangle$$
$$\geq -K\Delta + (1 + \theta\Delta\frac{3a_1 + a_2 + L}{2})|y(t_i)|^2 - 2a_2\theta\Delta\int_{-\tau}^{0} |y_{t_i}(r)|^2\mathrm{d}\nu_2(r). \tag{2.18}$$

According to [36, Lemma 4.1], we know that for any $u, v \in \mathbb{R}$, $q \geq 1$, and $\varepsilon > 0$, there exists a constant $K_\varepsilon > 1$ such that $|u + v|^q \leq K_\varepsilon|u|^q + (1+\varepsilon)|v|^q$. This, along with (2.18) and $|u|^q + |v|^q \leq (|u| + |v|)^q$ implies that

$$\Big(1 + \big(\theta\Delta\frac{3a_1 + a_2 + L}{2}\big)^p\Big)\sum_{i=0}^{k} |y(t_i)|^{2p}\mathbf{1}_{\{\zeta_M \geq i+1\}}$$
$$\leq \Big(1 + \theta\Delta\frac{3a_1 + a_2 + L}{2}\Big)^p \sum_{i=0}^{k} |y(t_i)|^{2p}\mathbf{1}_{\{\zeta_M \geq i+1\}}$$
$$\leq \sum_{i=0}^{k}\Big(K\Delta + |z(t_i)|^2 + 2a_2\theta\Delta\int_{-\tau}^{0} |y_{t_i}(r)|^2\mathrm{d}\nu_2(r)\Big)^p\mathbf{1}_{\{\zeta_M \geq i+1\}}$$

$$
\begin{aligned}
&\le K_\varepsilon \sum_{i=0}^{k} \Big(K\Delta + |z(t_i)|^2 \mathbf{1}_{\{\zeta_M \ge i+1\}} \Big)^p \\
&\quad + (1+\varepsilon)(2a_2\theta\Delta)^p \sum_{i=0}^{k} \int_{-\tau}^{0} |y_{t_i}(r)|^{2p} \mathrm{d}\nu_2(r) \mathbf{1}_{\{\zeta_M \ge i+1\}}.
\end{aligned}
$$

By (2.16), we derive

$$
\begin{aligned}
&\Big(1 + \big(\theta\Delta \frac{3a_1 + a_2 + L}{2}\big)^p\Big) \sum_{i=0}^{k} |y(t_i)|^{2p} \mathbf{1}_{\{\zeta_M \ge i+1\}} \\
&\le K K_\varepsilon (k+1)\Delta + K K_\varepsilon \sum_{i=0}^{k} |z(t_{i\wedge\zeta_M})|^{2p} + K\Delta^{p-1} \|\xi\|^{2p} \\
&\quad + (1+\varepsilon)(2a_2\theta\Delta)^p \sum_{i=0}^{k} |y(t_i)|^{2p} \mathbf{1}_{\{\zeta_M \ge i+1\}}.
\end{aligned}
$$

Choose a sufficiently small $\varepsilon > 0$ such that

$$
\Big(\frac{3a_1 + a_2 + L}{2}\Big)^p > (1+\varepsilon)(2a_2)^p.
$$

Then,

$$
\begin{aligned}
\sum_{i=0}^{k} |y(t_i)|^{2p} \mathbf{1}_{\{\zeta_M \ge i+1\}} &\le K(k+1)\Delta + K \sum_{i=0}^{k} |z(t_{i\wedge\zeta_M})|^{2p} + K\Delta^{p-1} \|\xi\|^{2p} \\
&\quad - \Big(\big(\frac{3a_1 + a_2 + L}{2}\big)^p - (1+\varepsilon)(2a_2)^p\Big)\theta^p \Delta^p \sum_{i=0}^{k} |y(t_i)|^{2p} \mathbf{1}_{\{\zeta_M \ge i+1\}} \\
&< K(k+1)\Delta + K \sum_{i=0}^{k} |z(t_{i\wedge\zeta_M})|^{2p} + K\|\xi\|^{2p}. \qquad (2.19)
\end{aligned}
$$

Inserting this into (2.17) yields

$$
\begin{aligned}
\mathbb{E}[|z(t_{(k+1)\wedge\zeta_M})|^{2p}] &\le K\Big(|b(\xi)|\Delta + \|\xi\|\Big)^{2p} + K(k+1)\Delta \\
&\quad + K\Delta \sum_{i=0}^{k} \mathbb{E}[|z(t_{i\wedge\zeta_M})|^{2p}].
\end{aligned}
$$

By the discrete Gronwall inequality (see Proposition A.4), we have that for any $T > 0$,

$$\sup_{0\le t_k\le T} \mathbb{E}[|z(t_{k\wedge\zeta_M})|^{2p}] \le K(\|\xi\|^{2p} + |b(\xi)|^{2p}\Delta^{2p} + T)e^{KT}.$$

Letting $M \to \infty$, one has

$$\sup_{0\le t_k\le T} \mathbb{E}[|z(t_k)|^{2p}] \le K(\|\xi\|^{2p} + |b(\xi)|^{2p}\Delta^{2p} + T)e^{KT}. \tag{2.20}$$

Similarly, it follows from (2.18) that

$$\begin{aligned}
&(1 + (\theta\Delta\frac{3a_1 + a_2 + L}{2})^p)|y(t_k)|^{2p} \\
&\le K_\varepsilon\big(K\Delta + |z(t_k)|^2\big)^p + (1+\varepsilon)(2a_2\theta\Delta)^p \int_{-\tau}^0 |y_{t_k}(r)|^{2p}\mathrm{d}\nu_2(r).
\end{aligned}$$

Note that

$$\begin{aligned}
\sup_{0\le t_k\le T} \mathbb{E}\Big[\int_{-\tau}^0 |y_{t_k}(r)|^{2p}\mathrm{d}\nu_2(r)\Big] &\le \sup_{0\le t_k\le T}\int_{-\tau}^0 \sup_{-\tau\le u\le 0} \mathbb{E}[|y_{t_k}(u)|^{2p}]\mathrm{d}\nu_2(r) \\
&\le \sup_{0\le t_k\le T} \mathbb{E}[|y(t_k)|^{2p}] + \|\xi\|^{2p}.
\end{aligned}$$

Hence, we derive

$$\begin{aligned}
&(1 + (\theta\Delta\frac{3a_1 + a_2 + L}{2})^p)\sup_{0\le t_k\le T} \mathbb{E}[y(t_k)|^{2p}] \\
&\le KK_\varepsilon\Big(\Delta + \sup_{0\le t_k\le T} \mathbb{E}[|z(t_k)|^{2p}]\Big) \\
&\quad + (1+\varepsilon)(2a_2\theta\Delta)^p\Big(\sup_{0\le t_k\le T} \mathbb{E}[|y(t_k)|^{2p}] + \|\xi\|^{2p}\Big).
\end{aligned}$$

This yields

$$\sup_{0\le t_k\le T} \mathbb{E}[y(t_k)|^{2p}] \le K(1 + \|\xi\|^{2p}) + \sup_{0\le t_k\le T} \mathbb{E}[|z(t_k)|^{2p}],$$

where we used $(\theta\Delta\frac{3a_1+a_2+L}{2})^p > (1+\varepsilon)(2a_2\theta\Delta)^p$. Combining (2.20), we complete the proof. □

When making no distinction among initial data of numerical solutions, we use the abbreviations in (2.6).

Proposition 2.2 *Let Assumptions 2.1 and 2.2 hold. Then for any $T > 0$ and $p \in \mathbb{N}_+$,*

$$\sup_{\Delta\in(0,1]} \mathbb{E}\Big[\sup_{t_k\in[0,T]} \big(|z^{\xi,\Delta}(t_k)|^{2p} + |y^{\xi,\Delta}(t_k)|^{2p}\big)\Big] \le K_T(1 + \|\xi\|^{2p} + |b(\xi)|^{2p}\Delta^{2p}),$$

where K_T is a positive constant.

Proof Our analysis focuses on integers $p \in \mathbb{N}_+$ with $p \ge 2$, since the case $p = 1$ is an immediate consequence of the Hölder inequality

$$\begin{aligned}
&\mathbb{E}\Big[\sup_{t_k\in[0,T]} \big(|z^{\xi,\Delta}(t_k)|^{2} + |y^{\xi,\Delta}(t_k)|^{2}\big)\Big] \\
&\le K\Big(\mathbb{E}\Big[\sup_{t_k\in[0,T]} \big(|z^{\xi,\Delta}(t_k)|^{2p} + |y^{\xi,\Delta}(t_k)|^{2p}\big)\Big]\Big)^{\frac{1}{p}}.
\end{aligned}$$

Similar to the proof of Lemma 2.1, by (2.9) and the recursive calculation, we obtain that for $p \in \mathbb{N}_+$ with $p \ge 2$,

$$\begin{aligned}
&|z(t_{k+1})|^{2p} \\
&\le |z(0)|^{2p} + \sum_{i=0}^{k} C_p^1 |z(t_i)|^{2(p-1)}\Big(|\sigma_i \delta W_i|^2 + K\Delta + 2a_2\Delta \int_{-\tau}^{0} |y_{t_i}(r)|^2 \mathrm{d}\nu_2(r)\Big) \\
&\quad + \sum_{i=0}^{k}\sum_{l=2}^{p} C_p^l |z(t_i)|^{2(p-l)} \Big||\sigma_i \delta W_i|^2 + K\Delta + 2a_2\Delta \int_{-\tau}^{0} |y_{t_i}(r)|^2 \mathrm{d}\nu_2(r) + \mathcal{M}_i\Big|^l \\
&\quad + \sum_{i=0}^{k} C_p^1 |z(t_i)|^{2(p-1)} \mathcal{M}_i.
\end{aligned}$$

Then

$$\mathbb{E}\Big[\sup_{0\le t_k\le T} |z(t_k)|^{2p}\Big] \le |z(0)|^{2p} + I_1 + I_2 + I_3, \tag{2.21}$$

where

$$\begin{aligned}
I_1 := \sum_{k=0}^{\lfloor T/\Delta\rfloor} &\Big\{C_p^1 \mathbb{E}\Big[|z(t_k)|^{2(p-1)}\Big(|\sigma_k \delta W_k|^2 + K\Delta + 2a_2\Delta \int_{-\tau}^{0} |y_{t_k}(r)|^2 \mathrm{d}\nu_2(r)\Big)\Big] \\
&+ \sum_{l=2}^{p} C_p^l \mathbb{E}\Big[|z(t_k)|^{2(p-l)} \Big||\sigma_k \delta W_k|^2 + K\Delta + 2a_2\Delta \int_{-\tau}^{0} |y_{t_k}(r)|^2 \mathrm{d}\nu_2(r) \\
&+ \mathcal{M}_k\Big|^l\Big]\Big\},
\end{aligned}$$

$$I_2 := \mathbb{E}\Big[\sup_{0\le t_k\le T}\sum_{i=0}^{k} C_p^1\frac{2(\theta-1)}{\theta}|z(t_i)|^{2(p-1)}\langle z(t_i),\sigma_i\delta W_i\rangle\Big],$$

$$I_3 := \mathbb{E}\Big[\sup_{0\le t_k\le T}\sum_{i=0}^{k} C_p^1\frac{2}{\theta}|z(t_i)|^{2(p-1)}\langle y(t_i),\sigma_i\delta W_i\rangle\Big].$$

By similar arguments to those in the proof of Lemma 2.1, we derive

$$I_1 \le K\Delta\sum_{k=0}^{\lfloor T/\Delta\rfloor}\mathbb{E}[|z(t_k)|^{2p}] + K\Delta + K\|\xi\|^{2p} + K\Delta\sum_{k=0}^{\lfloor T/\Delta\rfloor}\mathbb{E}[|y(t_k)|^{2p}]. \tag{2.22}$$

Applying the Burkholder–Davis–Gundy inequality, we arrive at

$$\begin{aligned} I_2 &\le K\mathbb{E}\Big[\Big(\sum_{k=0}^{\lfloor T/\Delta\rfloor}|z(t_k)|^{4p-2}|\sigma_k|^2\Delta\Big)^{\frac{1}{2}}\Big] \\ &\le \mathbb{E}\Big[\sup_{0\le t_i\le T}|z(t_i)|^{2p-1}\big(K\Delta\sum_{k=0}^{\lfloor T/\Delta\rfloor}|\sigma_k|^2\big)^{\frac{1}{2}}\Big]. \end{aligned}$$

Using the Young inequality, the Hölder inequality, and Assumption 2.1, one has

$$\begin{aligned} I_2 &\le \frac{1}{4}\mathbb{E}\Big[\sup_{0\le t_k\le T}|z(t_k)|^{2p}\Big] + K\mathbb{E}\Big[\big(\Delta\sum_{k=0}^{\lfloor T/\Delta\rfloor}|\sigma_k|^2\big)^p\Big] \\ &\le \frac{1}{4}\mathbb{E}\Big[\sup_{0\le t_k\le T}|z(t_k)|^{2p}\Big] + K\Delta(T^{p-1}+1)\sum_{k=0}^{\lfloor T/\Delta\rfloor}\mathbb{E}[|\sigma_k|^{2p}] \\ &\le \frac{1}{4}\mathbb{E}\Big[\sup_{0\le t_k\le T}|z(t_k)|^{2p}\Big] + K(T^{p-1}+1) \\ &\quad + K\Delta(T^{p-1}+1)\sum_{k=0}^{\lfloor T/\Delta\rfloor}\mathbb{E}\Big[|y(t_k)|^{2p} + \int_{-\tau}^{0}|y_{t_k}(r)|^{2p}\mathrm{d}\nu_1(r)\Big], \end{aligned}$$

which together with (2.15) with $M\to\infty$ yields

$$\begin{aligned} I_2 \le{}& \frac{1}{4}\mathbb{E}\Big[\sup_{0\le t_k\le T}|z(t_k)|^{2p}\Big] + K\Delta(T^{p-1}+1)\sum_{k=0}^{\lfloor T/\Delta\rfloor}\mathbb{E}[|y(t_k)|^{2p}] \\ &+ K(T^{p-1}+1)(1+\|\xi\|^{2p}). \end{aligned} \tag{2.23}$$

Similarly, for $p \in \mathbb{N}_+$ with $p \geq 2$, by the Young inequality and the Hölder inequality, we obtain

$$
\begin{aligned}
I_3 &\leq K\mathbb{E}\Big[\Big(\sum_{k=0}^{\lfloor T/\Delta \rfloor} |z(t_k)|^{4p-4}|y(t_k)|^2|\sigma_k|^2\Delta\Big)^{\frac{1}{2}}\Big] \\
&\leq \frac{1}{4}\mathbb{E}\Big[\sup_{0\leq t_k\leq T} |z(t_k)|^{2p}\Big] + K\mathbb{E}\Big[\big(\Delta\sum_{k=0}^{\lfloor T/\Delta \rfloor} |y(t_k)|^2|\sigma_k|^2\big)^{\frac{p}{2}}\Big] \\
&\leq \frac{1}{4}\mathbb{E}\Big[\sup_{0\leq t_k\leq T} |z(t_k)|^{2p}\Big] + K\Delta(T^{\frac{p}{2}-1}+1)\sum_{k=0}^{\lfloor T/\Delta \rfloor} \mathbb{E}[|y(t_k)|^p|\sigma_k|^p].
\end{aligned}
$$

Using the Young inequality again and combining Assumption 2.1 and (2.15) with $M \to \infty$, we derive

$$
\begin{aligned}
I_3 &\leq \frac{1}{4}\mathbb{E}\Big[\sup_{0\leq t_k\leq T} |z(t_k)|^{2p}\Big] + K\Delta(T^{\frac{p}{2}-1}+1)\sum_{k=0}^{\lfloor T/\Delta \rfloor} \mathbb{E}[|y(t_k)|^{2p}] \\
&\quad + K\Delta(T^{\frac{p}{2}-1}+1)\sum_{k=0}^{\lfloor T/\Delta \rfloor} \mathbb{E}[|\sigma_k|^{2p}] \\
&\leq \frac{1}{4}\mathbb{E}\Big[\sup_{0\leq t_k\leq T} |z(t_k)|^{2p}\Big] + K\Delta(T^{\frac{p}{2}-1}+1)\sum_{k=0}^{\lfloor T/\Delta \rfloor} \mathbb{E}[|y(t_k)|^{2p}] \\
&\quad + K(T^{\frac{p}{2}-1}+1)(1+\|\xi\|^{2p}).
\end{aligned} \tag{2.24}
$$

Plugging (2.22)–(2.24) into (2.21), a direct computation derives

$$
\begin{aligned}
&\mathbb{E}\Big[\sup_{0\leq t_k\leq T} |z(t_k)|^{2p}\Big] \\
&\leq \Big(K + K\Delta\sum_{k=0}^{\lfloor T/\Delta \rfloor} \mathbb{E}[|z(t_k)|^{2p}] + K\Delta + K(T^{p-1}+1)(1+\|\xi\|^{2p}) \\
&\quad + K\Delta(T^{p-1}+1)\sum_{k=0}^{\lfloor T/\Delta \rfloor} \mathbb{E}[|y(t_k)|^{2p}]\Big) + \frac{1}{2}\mathbb{E}\Big[\sup_{0\leq t_k\leq T} |z(t_k)|^{2p}\Big].
\end{aligned}
$$

This leads to

$$\begin{aligned}
&\mathbb{E}\Big[\sup_{0\le t_k\le T}|z(t_k)|^{2p}\Big]\\
&\le 2\Big(K+K\Delta\sum_{k=0}^{\lfloor T/\Delta\rfloor}\mathbb{E}[|z(t_k)|^{2p}]+K\Delta+K(T^{p-1}+1)(1+\|\xi\|^{2p})\\
&\quad+K\Delta(T^{p-1}+1)\sum_{k=0}^{\lfloor T/\Delta\rfloor}\mathbb{E}[|y(t_k)|^{2p}]\Big)\\
&\le K(T^{p-1}+1)(1+\|\xi\|^{2p})+K(T+1)\sup_{0\le t_k\le T}\mathbb{E}[|z(t_k)|^{2p}]\\
&\quad+K(T^p+1)\sup_{0\le t_k\le T}\mathbb{E}[|y(t_k)|^{2p}]\\
&\le K_T(1+\|\xi\|^{2p}+|b(\xi)|^{2p}\Delta^{2p}),
\end{aligned}$$

where K_T is a positive constant. From the above inequality and (2.18), one obtains that

$$\mathbb{E}\big[\sup_{0\le t_k\le T}|y(t_k)|^{2p}\big]\le K_T(1+\|\xi\|^{2p}+|b(\xi)|^{2p}\Delta^{2p}).$$

Thus the proof is completed. □

The following lemma establishes the boundedness of the second moment of the numerical solution in the infinite time horizon.

Lemma 2.3 *Let Assumptions 2.1 and 2.2 hold. Then*

$$\sup_{\Delta\in(0,1]}\sup_{k\ge -N}(\mathbb{E}[|z^{\xi,\Delta}(t_k)|^2]+\mathbb{E}[|y^{\xi,\Delta}(t_k)|^2])\le K(1+\|\xi\|^2+|b(\xi)|^2\Delta^2).$$

Proof For any $\Delta\in(0,1]$ and $k\in\mathbb{N}$, according to (2.8) and $z(t_k)=y(t_k)-\theta b_k\Delta$, we have

$$\begin{aligned}
|z(t_{k+1})|^2&\le|z(t_k)|^2+\frac{1-2\theta}{\theta^2}|y(t_k)-z(t_k)|^2+|\sigma_k\delta W_k|^2+K\Delta\\
&\quad-\frac{3a_1+a_2+L}{2}\Delta|y(t_k)|^2+2a_2\Delta\int_{-\tau}^0|y_{t_k}(r)|^2\mathrm{d}\nu_2(r)+\mathcal{M}_k\\
&=\frac{(1-\theta)^2}{\theta^2}|z(t_k)|^2+\frac{1-2\theta}{\theta^2}|y(t_k)|^2\\
&\quad+\frac{2(2\theta-1)}{\theta^2}\langle z(t_k),y(t_k)\rangle+|\sigma_k\delta W_k|^2
\end{aligned}$$

$$
\begin{aligned}
&+ K\Delta - \frac{3a_1 + a_2 + L}{2}\Delta|y(t_k)|^2 \\
&+ 2a_2\Delta\int_{-\tau}^{0}|y_{t_k}(r)|^2 \mathrm{d}\nu_2(r) + \mathcal{M}_k,
\end{aligned}
$$

where $\mathcal{M}_k$ is defined by (2.10). By $a_1 > a_2 + L$ given in Assumption 2.2, one can choose a sufficiently small $\kappa > 0$ such that

$$
\tilde{c}_\kappa := \big(a_1 + a_2 - (\frac{a_1 - a_2 + L}{2} + 2a_2)e^{\kappa\tau}\big) > 0.
$$

For the term $\langle z(t_k), y(t_k)\rangle$, applying the Young inequality yields

$$
2\langle z(t_k), y(t_k)\rangle \le \frac{1}{1+\tilde{c}_\kappa\Delta}|z(t_k)|^2 + (1+\tilde{c}_\kappa\Delta)|y(t_k)|^2.
$$

Recall that $\theta \in (\frac{1}{2}, 1]$. Then we have

$$
\begin{aligned}
|z(t_{k+1})|^2 \le& \Big(\frac{(1-\theta)^2}{\theta^2} + \frac{2\theta - 1}{\theta^2(1+\tilde{c}_\kappa\Delta)}\Big)|z(t_k)|^2 + |\sigma_k\delta W_k|^2 + K\Delta \\
&- \Big(\frac{3a_1 + a_2 + L}{2} - \tilde{c}_\kappa\Big)\Delta|y(t_k)|^2 + 2a_2\Delta\int_{-\tau}^{0}|y_{t_k}(r)|^2 \mathrm{d}\nu_2(r) + \mathcal{M}_k,
\end{aligned}
$$

where we used $0 \le \frac{2\theta-1}{\theta^2} \le 1$. Hence, for any $\varepsilon \in (0, \kappa]$,

$$
\begin{aligned}
e^{\varepsilon t_{k+1}}|z(t_{k+1})|^2 =& |z(0)|^2 + \sum_{i=0}^{k}(e^{\varepsilon t_{i+1}}|z(t_{i+1})|^2 - e^{\varepsilon t_i}|z(t_i)|^2) \\
\le& |z(0)|^2 + \Big(\frac{(1-\theta)^2}{\theta^2} + \frac{2\theta - 1}{\theta^2(1+\tilde{c}_\kappa\Delta)} - e^{-\varepsilon\Delta}\Big)\sum_{i=0}^{k} e^{\varepsilon t_{i+1}}|z(t_i)|^2 \\
&+ \sum_{i=0}^{k} e^{\varepsilon t_{i+1}}|\sigma_i\delta W_i|^2 + K\Delta\sum_{i=0}^{k} e^{\varepsilon t_{i+1}} \\
&- (\frac{3a_1 + a_2 + L}{2} - \tilde{c}_\kappa)\Delta\sum_{i=0}^{k} e^{\varepsilon t_{i+1}}|y(t_i)|^2 \\
&+ 2a_2\Delta\sum_{i=0}^{k} e^{\varepsilon t_{i+1}}\int_{-\tau}^{0}|y_{t_i}(r)|^2 \mathrm{d}\nu_2(r) + \sum_{i=0}^{k} e^{\varepsilon t_{i+1}}\mathcal{M}_i. \qquad (2.25)
\end{aligned}
$$

According to Proposition 2.2, we obtain that $\mathcal{M}_k$ is a martingale and hence $\mathbb{E}\Big[\sum_{i=0}^{k} e^{\varepsilon t_{i+1}}\mathcal{M}_i\Big] = 0$. By Assumption 2.1, we derive that for $\iota \in (0, 1)$,

$$|\sigma(\phi)|^2 \le K_\iota|\sigma(\mathbf{0})|^2 + (1+\iota)L\Big(|\phi(0)|^2 + \int_{-\tau}^{0}|\phi(r)|^2\mathrm{d}\nu_1(r)\Big),$$

where the constant $K_\iota > 0$ depends on ι. Noting $a_1 > a_2 + L$, we choose $\iota = \frac{a_1-a_2+L}{2L} - 1$ and derive

$$|\sigma(\phi)|^2 \le K_\iota|\sigma(\mathbf{0})|^2 + \frac{a_1-a_2+L}{2}\Big(|\phi(0)|^2 + \int_{-\tau}^{0}|\phi(r)|^2\mathrm{d}\nu_1(r)\Big).$$

Taking expectations on both sides of (2.25) and the above inequality, we arrive at

$$\begin{aligned}
&e^{\varepsilon t_{k+1}}\mathbb{E}[|z(t_{k+1})|^2] \\
&\le |z(0)|^2 + \Big(\frac{(1-\theta)^2}{\theta^2} + \frac{2\theta-1}{\theta^2(1+\tilde{c}_\kappa\Delta)} - e^{-\varepsilon\Delta}\Big)\sum_{i=0}^{k} e^{\varepsilon t_{i+1}}\mathbb{E}[|z(t_i)|^2] \\
&\quad + K\Delta\sum_{i=0}^{k} e^{\varepsilon t_{i+1}} \\
&\quad - (a_1+a_2-\tilde{c}_\kappa)\Delta\sum_{i=0}^{k} e^{\varepsilon t_{i+1}}\mathbb{E}[|y(t_i)|^2] + \frac{a_1-a_2+L}{2}\Delta\sum_{i=0}^{k} e^{\varepsilon t_{i+1}} \\
&\quad \times \mathbb{E}\Big[\int_{-\tau}^{0}|y_{t_i}(r)|^2\mathrm{d}\nu_1(r)\Big] + 2a_2\Delta\sum_{i=0}^{k} e^{\varepsilon t_{i+1}}\mathbb{E}\Big[\int_{-\tau}^{0}|y_{t_i}(r)|^2\mathrm{d}\nu_2(r)\Big].
\end{aligned} \tag{2.26}$$

Similar to (2.15), we deduce that for $\iota = 1, 2$,

$$\begin{aligned}
&\sum_{i=0}^{k} e^{\varepsilon t_{i+1}}\int_{-\tau}^{0}|y_{t_i}(r)|^2\mathrm{d}\nu_\iota(r) \\
&\quad \le e^{\varepsilon\tau}\sum_{j=-N}^{-1}\sum_{i=0}^{k} e^{\varepsilon t_{i+j+1}}\int_{t_j}^{t_{j+1}}\frac{t_{j+1}-r}{\Delta}\mathrm{d}\nu_\iota(r)|y(t_{i+j})|^2 \\
&\qquad + e^{\varepsilon\tau}\sum_{j=-N}^{-1}\sum_{i=0}^{k} e^{\varepsilon t_{i+j+2}}\int_{t_j}^{t_{j+1}}\frac{r-t_j}{\Delta}\mathrm{d}\nu_\iota(r)|y(t_{i+j+1})|^2
\end{aligned}$$

$$
\begin{aligned}
&\le e^{\varepsilon\tau}\sum_{j=-N}^{-1}\sum_{l=-N}^{k-1}e^{\varepsilon t_{l+1}}\int_{t_j}^{t_{j+1}}\frac{t_{j+1}-r}{\Delta}\mathrm{d}\nu_l(r)|y(t_l)|^2\\
&\quad+e^{\varepsilon\tau}\sum_{j=-N}^{-1}\sum_{l=-N}^{k}e^{\varepsilon t_{l+1}}\int_{t_j}^{t_{j+1}}\frac{r-t_j}{\Delta}\mathrm{d}\nu_l(r)|y(t_l)|^2\\
&\le e^{\varepsilon\tau}\sum_{j=-N}^{-1}\sum_{l=-N}^{k}e^{\varepsilon t_{l+1}}\int_{t_j}^{t_{j+1}}\mathrm{d}\nu_l(r)|y(t_l)|^2\\
&\le Ne^{\varepsilon\tau}\|\xi\|^2+e^{\varepsilon\tau}\sum_{i=0}^{k}e^{\varepsilon t_{i+1}}|y(t_i)|^2, \qquad (2.27)
\end{aligned}
$$

which yields

$$
\begin{aligned}
&\frac{a_1-a_2+L}{2}\Delta\sum_{i=0}^{k}e^{\varepsilon t_{i+1}}\int_{-\tau}^{0}|y_{t_i}(r)|^2\mathrm{d}\nu_1(r)+2a_2\Delta\sum_{i=0}^{k}e^{\varepsilon t_{i+1}}\int_{-\tau}^{0}|y_{t_i}(r)|^2\mathrm{d}\nu_2(r)\\
&\le\Big(\frac{a_1-a_2+L}{2}+2a_2\Big)e^{\varepsilon\tau}\tau\|\xi\|^2+\Big(\frac{a_1-a_2+L}{2}+2a_2\Big)\\
&\quad\times e^{\varepsilon\tau}\Delta\sum_{i=0}^{k}e^{\varepsilon t_{i+1}}|y(t_i)|^2.
\end{aligned}
$$

Inserting this into (2.26) yields

$$
\begin{aligned}
&e^{\varepsilon t_{k+1}}\mathbb{E}[|z(t_{k+1})|^2]\\
&\quad\le|z(0)|^2+\Big(\frac{(1-\theta)^2}{\theta^2}+\frac{2\theta-1}{\theta^2(1+\tilde{c}_\kappa\Delta)}-e^{-\varepsilon\Delta}\Big)\sum_{i=0}^{k}e^{\varepsilon t_{i+1}}\mathbb{E}[|z(t_i)|^2]+\frac{K}{\varepsilon}e^{\varepsilon t_{k+2}}\\
&\qquad-\Big(a_1+a_2-\tilde{c}_\kappa-\big(\frac{a_1-a_2+L}{2}+2a_2\big)e^{\varepsilon\tau}\Big)\Delta\sum_{i=0}^{k}e^{\varepsilon t_{i+1}}\mathbb{E}[|y(t_i)|^2]\\
&\qquad+\Big(\frac{a_1-a_2+L}{2}+2a_2\Big)e^{\varepsilon\tau}\tau\|\xi\|^2.
\end{aligned}
$$

It follows from the definition of $\tilde{c}_\kappa$ that $a_1 + a_2 - \tilde{c}_\kappa - (\frac{a_1-a_2+L}{2} + 2a_2)e^{\varepsilon\tau} \geq 0$. Moreover, using $e^{-\varepsilon\Delta} > 1 - \varepsilon\Delta$, we conclude that

$$\frac{(1-\theta)^2}{\theta^2} + \frac{2\theta - 1}{\theta^2(1+\tilde{c}_\kappa\Delta)} - e^{-\varepsilon\Delta} < \frac{(1-\theta)^2}{\theta^2} + \frac{2\theta-1}{\theta^2(1+\tilde{c}_\kappa\Delta)} - (1-\varepsilon\Delta)$$
$$= -\frac{(2\theta-1)\tilde{c}_\kappa}{\theta^2(1+\tilde{c}_\kappa\Delta)}\Delta + \varepsilon\Delta < (\varepsilon - \frac{(2\theta-1)\tilde{c}_\kappa}{\theta^2(1+\tilde{c}_\kappa)})\Delta. \tag{2.28}$$

Hence, letting $\varepsilon = \frac{(2\theta-1)\tilde{c}_\kappa}{\theta^2(1+\tilde{c}_\kappa)} \wedge \kappa$ yields

$$e^{\varepsilon t_{k+1}}\mathbb{E}[|z(t_{k+1})|^2] \leq K(\|\xi\|^2 + |b(\xi)|^2\Delta^2) + Ke^{\varepsilon t_{k+2}},$$

which implies

$$\sup_{\Delta\in(0,1]}\sup_{k\geq -N}\mathbb{E}[|z(t_k)|^2] \leq K(1 + \|\xi\|^2 + |b(\xi)|^2\Delta^2).$$

In addition, by (2.18), we obtain

$$(1+\theta\Delta\frac{3a_1+a_2+L}{2})\sup_{\Delta\in(0,1]}\sup_{k\geq -N}\mathbb{E}[|y(t_k)|^2]$$
$$\leq K\Delta + \sup_{\Delta\in(0,1]}\sup_{k\geq -N}\mathbb{E}[|z(t_k)|^2] + 2a_2\theta\Delta\sup_{\Delta\in(0,1]}\sup_{k\geq -N}\mathbb{E}[|y(t_k)|^2]. \tag{2.29}$$

Together with $a_1 > a_2 + L$, we arrive at

$$\sup_{\Delta\in(0,1]}\sup_{k\geq -N}\mathbb{E}[|y(t_k)|^2] \leq K(1 + \|\xi\|^2 + |b(\xi)|^2\Delta^2).$$

The proof is completed. □

Proposition 2.4 *Let Assumptions 2.1 and 2.2 hold. Then*

$$\sup_{\Delta\in(0,1]}\sup_{k\geq 0}\mathbb{E}\big[\|z_{t_k}^{\xi,\Delta}\|^2 + \|y_{t_k}^{\xi,\Delta}\|^2\big] \leq K(1 + \|\xi\|^2 + |b(\xi)|^2\Delta^2).$$

Proof It follows from (2.25), (2.27), (2.28), and $\varepsilon = \frac{(2\theta-1)\tilde{c}_\kappa}{\theta^2(1+\tilde{c}_\kappa)} \wedge \kappa$ that

$$e^{\varepsilon t_{i+1}}|z(t_{i+1})|^2$$
$$\leq |z(0)|^2 + K\|\xi\|^2 + \sum_{i=0}^{k} e^{\varepsilon t_{i+1}}|\sigma_i\delta W_i|^2 + K\Delta\sum_{i=0}^{k} e^{\varepsilon t_{i+1}}$$

$$- \Big(\frac{3a_1 + a_2 + L}{2} - \tilde{c}_\kappa - 2a_2 e^{\varepsilon\tau}\Big)\Delta \sum_{i=0}^{k} e^{\varepsilon t_{i+1}} |y(t_i)|^2 + \sum_{i=0}^{k} e^{\varepsilon t_{i+1}} \mathcal{M}_i$$

$$\leq |z(0)|^2 + K\|\xi\|^2 + \sum_{i=0}^{k} e^{\varepsilon t_{i+1}} |\sigma_i \delta W_i|^2 + K\Delta \sum_{i=0}^{k} e^{\varepsilon t_{i+1}} + \sum_{i=0}^{k} e^{\varepsilon t_{i+1}} \mathcal{M}_i,$$

where we used $\frac{3a_1+a_2+L}{2} - \tilde{c}_\kappa - 2a_2 e^{\varepsilon\tau} > a_1 + a_2 - \tilde{c}_\kappa - 2a_2 e^{\varepsilon\tau} > 0$ in the last inequality. Hence,

$$\mathbb{E}\Big[\sup_{(k-N)\vee 0 \leq i \leq k} e^{\varepsilon t_{i+1}} |z(t_{i+1})|^2\Big]$$

$$\leq |z(0)|^2 + K\|\xi\|^2 + \mathbb{E}\Big[\sup_{(k-N)\vee 0 \leq i \leq k} \sum_{l=0}^{i} e^{\varepsilon t_{l+1}} |\sigma_l \delta W_l|^2\Big] + K\Delta \sum_{i=0}^{k} e^{\varepsilon t_{i+1}}$$

$$+ \mathbb{E}\Big[\sup_{(k-N)\vee 0 \leq i \leq k} \sum_{l=0}^{i} e^{\varepsilon t_{l+1}} \mathcal{M}_l\Big]$$

$$\leq |z(0)|^2 + K\|\xi\|^2 + \Delta\mathbb{E}\Big[\sum_{l=0}^{k} e^{\varepsilon t_{l+1}} |\sigma_l|^2\Big] + \frac{K}{\varepsilon} e^{\varepsilon t_{k+2}}$$

$$+ \mathbb{E}\Big[\sup_{(k-N)\vee 0 \leq i \leq k} \sum_{l=(k-N)\vee 0}^{i} e^{\varepsilon t_{l+1}} \mathcal{M}_l\Big]. \tag{2.30}$$

Using Assumption 2.1 and Lemma 2.3, we obtain

$$\Delta\mathbb{E}\Big[\sum_{l=0}^{k} e^{\varepsilon t_{l+1}} |\sigma_l|^2\Big] \leq K\Big(\Delta \sum_{l=0}^{k} e^{\varepsilon t_{l+1}} \mathbb{E}\Big[|y(t_l)|^2 + \int_{-\tau}^{0} |y_{t_l}(r)|^2 \mathrm{d}\nu_1(r)\Big] + e^{\varepsilon t_{k+2}}\Big)$$

$$\leq K\Big(\Delta \sup_{i \geq -N} \mathbb{E}[|y(t_i)|^2] \sum_{l=0}^{k} e^{\varepsilon t_{l+1}} + e^{\varepsilon t_{k+2}}\Big)$$

$$\leq K(1 + \|\xi\|^2 + |b(\xi)|^2\Delta^2) e^{\varepsilon t_{k+2}}. \tag{2.31}$$

Similar to the estimates of I_2 and I_3 in Proposition 2.2, one has

$$\mathbb{E}\Big[\sup_{(k-N)\vee 0 \leq i \leq k} \sum_{l=(k-N)\vee 0}^{i} e^{\varepsilon t_{l+1}} \mathcal{M}_l\Big]$$

$$\leq \mathbb{E}\Big[\Big(\big(\sup_{(k-N)\vee 0 \leq i \leq k} e^{\varepsilon t_i} |z(t_i)|^2\big)\big(K e^{\varepsilon\Delta} \Delta \sum_{l=(k-N)\vee 0}^{k} e^{\varepsilon t_{l+1}} |\sigma_l|^2\big)\Big)^{\frac{1}{2}}\Big]$$

$$
\begin{aligned}
&+\mathbb{E}\Big[\Big(\big(\sup_{(k-N)\vee 0\le i\le k} e^{\varepsilon t_i}|y(t_i)|^2\big)\big(Ke^{\varepsilon\Delta}\Delta\sum_{l=(k-N)\vee 0}^{k} e^{\varepsilon t_{l+1}}|\sigma_l|^2\big)\Big)^{\frac{1}{2}}\Big]\\
&\le \frac{1}{4}\mathbb{E}\Big[\sup_{(k-N)\vee 0\le i\le k} e^{\varepsilon t_i}|z(t_i)|^2\Big]+\frac{1}{4}\mathbb{E}\Big[\sup_{(k-N)\vee 0\le i\le k} e^{\varepsilon t_i}|y(t_i)|^2\Big]\\
&+K\Delta\mathbb{E}\Big[\sum_{i=(k-N)\vee 0}^{k} e^{\varepsilon t_{i+1}}|\sigma_i|^2\Big]\\
&\le \frac{1}{4}\mathbb{E}\Big[\sup_{(k-N)\vee 0\le i\le k} e^{\varepsilon t_i}|z(t_i)|^2\Big]+\frac{1}{4}\mathbb{E}\Big[\sup_{(k-N)\vee 0\le i\le k} e^{\varepsilon t_i}|y(t_i)|^2\Big]\\
&+K(1+\|\xi\|^2+|b(\xi)|^2\Delta^2)e^{\varepsilon t_{k+2}}.
\end{aligned}
\tag{2.32}
$$

From (2.18), we derive

$$
\begin{aligned}
&(1+\theta\Delta\frac{3a_1+a_2+L}{2})\mathbb{E}\Big[\sup_{(k-N)\vee 0\le i\le k} e^{\varepsilon t_i}|y(t_i)|^2\Big]\\
&< K\Delta e^{\varepsilon t_k}+\mathbb{E}\Big[\sup_{(k-N)\vee 0\le i\le k} e^{\varepsilon t_i}|z(t_i)|^2\Big]\\
&+2a_2\theta\Delta\mathbb{E}\Big[\sup_{(k-N)\vee 0\le i\le k} e^{\varepsilon t_i}\int_{-\tau}^{0}|y_{t_i}(r)|^2 \mathrm{d}\nu_2(r)\Big]\\
&\le K\Delta e^{\varepsilon t_k}+\mathbb{E}\Big[\sup_{(k-N)\vee 0\le i\le k} e^{\varepsilon t_i}|z(t_i)|^2\Big]+2a_2\theta\Delta\mathbb{E}\Big[\sup_{(k-N)\vee 0\le i\le k}\ \sup_{-N\le j\le -1}\\
&\int_{t_j}^{t_j+1}\Big(e^{-\varepsilon t_j}e^{\varepsilon t_{i+j}}\frac{t_{j+1}-r}{\Delta}|y(t_{i+j})|^2\\
&+e^{-\varepsilon t_{j+1}}e^{\varepsilon t_{i+j+1}}\frac{r-t_j}{\Delta}|y(t_{i+j+1})|^2\Big)\mathrm{d}\nu_2(r)\Big]\\
&\le K\Delta e^{\varepsilon t_k}+\mathbb{E}\Big[\sup_{(k-N)\vee 0\le i\le k} e^{\varepsilon t_i}|z(t_i)|^2\Big]\\
&+2a_2\theta\Delta e^{\varepsilon\tau}\sum_{i=(k-2N)\vee(-N)}^{k} e^{\varepsilon t_i}\mathbb{E}[|y(t_i)|^2]\\
&\le K\Delta e^{\varepsilon t_k}+\mathbb{E}\Big[\sup_{(k-N)\vee 0\le i\le k} e^{\varepsilon t_i}|z(t_i)|^2\Big]+4a_2\theta e^{\varepsilon\tau}\tau e^{\varepsilon t_k}\sup_{i\ge -N}\mathbb{E}[|y(t_i)|^2]\\
&\le K\Delta e^{\varepsilon t_k}+\mathbb{E}\Big[\sup_{(k-N)\vee 0\le i\le k} e^{\varepsilon t_i}|z(t_i)|^2\Big]+e^{\varepsilon t_k}K(1+\|\xi\|^2+|b(\xi)|^2\Delta^2),
\end{aligned}
$$

which implies

$$
\begin{aligned}
&\mathbb{E}\Big[\sup_{(k-N)\vee 0\le i\le k} e^{\varepsilon t_i}|y(t_i)|^2\Big] \\
&\quad\le \mathbb{E}\Big[\sup_{(k-N)\vee 0\le i\le k} e^{\varepsilon t_i}|z(t_i)|^2\Big] + e^{\varepsilon t_k}K(1+\|\xi\|^2+|b(\xi)|^2\Delta^2).
\end{aligned}
\tag{2.33}
$$

Inserting (2.33) into (2.32) leads to

$$
\begin{aligned}
&\mathbb{E}\Big[\sup_{(k-N)\vee 0\le i\le k} \sum_{l=(k-N)\vee 0}^{i} e^{\varepsilon t_{l+1}}\mathcal{M}_l\Big] \\
&\le \frac{1}{2}\mathbb{E}\Big[\sup_{(k-N)\vee 0\le i\le k} e^{\varepsilon t_i}|z(t_i)|^2\Big] + K(1+\|\xi\|^2+|b(\xi)|^2\Delta^2)e^{\varepsilon t_{k+2}} \\
&\le \frac{1}{2}\mathbb{E}\Big[\sup_{(k-N)\vee 0\le i\le k} e^{\varepsilon t_{i+1}}|z(t_{i+1})|^2\Big] + K(1+\|\xi\|^2+|b(\xi)|^2\Delta^2)e^{\varepsilon t_{k+2}} \\
&\quad + |z(0)|^2 + \mathbb{E}[e^{\varepsilon t_{k-N}}|z(t_{k-N})|^2] \\
&\le \frac{1}{2}\mathbb{E}\Big[\sup_{(k-N)\vee 0\le i\le k} e^{\varepsilon t_{i+1}}|z(t_{i+1})|^2\Big] + K(1+\|\xi\|^2+|b(\xi)|^2\Delta^2)e^{\varepsilon t_{k+2}}.
\end{aligned}
\tag{2.34}
$$

Substituting (2.31) and (2.34) into (2.30), we deduce

$$
\begin{aligned}
e^{\varepsilon t_{k-N}}\mathbb{E}\Big[\sup_{(k-N)\vee 0\le i\le k} |z(t_{i+1})|^2\Big] &\le \mathbb{E}\Big[\sup_{(k-N)\vee 0\le i\le k} e^{\varepsilon t_{i+1}}|z(t_{i+1})|^2\Big] \\
&\le K(1+\|\xi\|^2+|b(\xi)|^2\Delta^2)e^{\varepsilon t_{k+2}},
\end{aligned}
$$

which implies

$$
\sup_{\Delta\in(0,1]}\sup_{k\ge 0}\mathbb{E}\Big[\sup_{(k-N)\vee 0\le i\le k} |z(t_{i+1})|^2\Big] \le K(1+\|\xi\|^2+|b(\xi)|^2\Delta^2).
$$

Combining (2.33), we finish the proof. □

2.1.2 Time-Independent Boundedness of High-Order Moment

For the SFDE (1.1) with superlinearly growing coefficients, the time-independent boundedness of high-order moment of the numerical solution is required in the longtime mean-square convergence analysis of the numerical method. Note that Assumption 2.2 ensures

$$\langle \phi(0), b(\phi)\rangle \leq K - a_1|\phi(0)|^2 + a_2 \int_{-\tau}^{0} |\phi(r)|^2 \mathrm{d}\nu_2(r),$$

which is used to establish the time-independent mean-square boundedness of the θ-EM functional solution. To estimate the time-independent boundedness of higher-order moment, we introduce the following stronger assumptions on the coefficient b and the initial datum ξ.

Assumption 2.3 *There exist positive constants K, a_3, a_4, ℓ, and a probability measure ν_2 on $[-\tau, 0]$ such that for any $\phi \in C^d$,*

$$\langle \phi(0), b(\phi)\rangle \leq K - a_3|\phi(0)|^{2+\ell} + a_4 \int_{-\tau}^{0} |\phi(r)|^2 \mathrm{d}\nu_2(r).$$

Assumption 2.4 *There exist positive constants K and β such that for any $\phi_1, \phi_2 \in C^d$,*

$$|b(\phi_1) - b(\phi_2)| \leq K\|\phi_1 - \phi_2\|(1 + \|\phi_1\|^{\beta} + \|\phi_2\|^{\beta}).$$

Assumption 2.5 *There exist positive constants K and $\rho \geq 1/2$ such that*

$$|\xi(s_1) - \xi(s_2)| \leq K|s_1 - s_2|^{\rho}, \quad s_1, s_2 \in [-\tau, 0].$$

Example 2.2 We now verify that the drift coefficient b presented in Example 1.6, given by

$$b(\phi) = \frac{1}{\tau}\int_{-\tau}^{0} \phi(r)\mathrm{d}r - |\phi(0)|^2\phi(0) - 2\phi(0), \ \phi \in C^d$$

satisfies Assumptions 2.3 and 2.4. For $\phi, \phi_1, \phi_2 \in C^d$, direct computations yield

$$\begin{aligned}\langle \phi(0), b(\phi)\rangle &\leq \Big\langle \phi(0), \frac{1}{\tau}\int_{-\tau}^{0} \phi(r)\mathrm{d}r\Big\rangle - |\phi(0)|^4 - 2|\phi(0)|^2 \\ &\leq -|\phi(0)|^4 + \frac{1}{2\tau}\int_{-\tau}^{0} |\phi(r)|^2 \mathrm{d}r\end{aligned}$$

and

$$\begin{aligned}&|b(\phi_1) - b(\phi_2)| \\ &\leq \frac{1}{\tau}\int_{-\tau}^{0} |\phi_1(r) - \phi_2(r)|\mathrm{d}r + \big||\phi_1(0)|^2\phi_1(0) - |\phi_2(0)|^2\phi_2(0)\big|\end{aligned}$$

$$
\begin{aligned}
&+ 2|\phi_1(0) - \phi_2(0)| \\
&\leq 3\|\phi_1 - \phi_2\| + |\phi_1(0) - \phi_2(0)||\phi_1(0) \\
&+ \phi_2(0)||\phi_1(0)| + |\phi_2(0)|^2|\phi_1(0) - \phi_2(0)| \\
&\leq \|\phi_1 - \phi_2\|\Big(3 + \frac{3}{2}\|\phi_1\|^2 + \frac{3}{2}\|\phi_2\|^2\Big) \leq 3\|\phi_1 - \phi_2\|\Big(1 + \|\phi_1\|^2 + \|\phi_2\|^2\Big).
\end{aligned}
$$

The preceding inequalities confirm that Assumptions 2.3 and 2.4 are fulfilled with the parameters $a_3 = 1$, $a_4 = \frac{1}{2}$, $\ell = 2$, and $\beta = 2$.

Note that under Assumptions 2.1 and 2.3, the conditions in Proposition 1.7 Theorem 1.15 are fulfilled. As a result, for integer $p \geq 2$,

$$
\sup_{t\geq 0} \mathbb{E}[\|x^{\xi}(t)\|^p] \leq K(1 + \|\xi\|^p). \tag{2.35}
$$

Moreover, the invariant measure μ given in (2.3) satisfies

$$
\mu(\|\cdot\|^p) := \int_{C^d} \|\phi\|^p \mathrm{d}\mu(\phi) \leq K. \tag{2.36}
$$

We begin with showing the time-independent boundedness of the auxiliary process $\{z^{\xi,\Delta}(t_k)\}_{k\in\mathbb{N}}$. The proof is first performed for the stepsize $\Delta \in (0, \Delta^*]$ with some $\Delta^* \in (0, 1)$ (see Lemma 2.5) and then is extended to $\Delta \in (\Delta^*, 1]$ (see Lemma 2.6).

Lemma 2.5 *Let Assumptions 2.1, 2.3, and 2.4 hold. Then there exists $\Delta^* \in (0, 1)$ such that for integer $p \geq 2$,*

$$
\sup_{\Delta\in(0,\Delta^*]} \sup_{k\geq 0} \mathbb{E}[|z^{\xi,\Delta}(t_k)|^p] \leq K(1 + \|\xi\|^{p(\beta+1)}) \quad \forall p \geq 2.
$$

Proof Similar to (2.8), by Assumption 2.3, $z(t_k) = y(t_k) - \theta b_k \Delta$, and

$$
2\langle z(t_k), y(t_k)\rangle \leq \frac{1}{1+\Delta}|z(t_k)|^2 + (1+\Delta)|y(t_k)|^2,
$$

we derive that for any $\Delta \in (0, 1]$ and $k \in \mathbb{N}$,

$$
\begin{aligned}
|z(t_{k+1})|^2 \leq\ & A_{\theta,\Delta}|z(t_k)|^2 + 2K\Delta - 2a_3\Delta|y(t_k)|^{2+\ell} + 2a_4\Delta \int_{-\tau}^{0} |y_{t_k}(r)|^2 \mathrm{d}\nu_2(r) \\
&+ \frac{2\theta - 1}{\theta^2}\Delta|y(t_k)|^2 + |\sigma_k \delta W_k|^2 + \mathcal{M}_k,
\end{aligned} \tag{2.37}
$$

where $A_{\theta,\Delta} := \frac{(1-\theta)^2}{\theta^2} + \frac{2\theta-1}{\theta^2(1+\Delta)}$, and $\mathcal{M}_k$ is defined by (2.10). Taking $R_2 > 2L$, and by the Young inequality, there exists a sufficiently large $R_1 > 0$ such that

$$\begin{aligned} &2K\Delta - 2a_3\Delta|y(t_k)|^{2+\ell} + \frac{2\theta-1}{\theta^2}\Delta|y(t_k)|^2 \\ &\le 2K\Delta - 2a_3\Delta|y(t_k)|^{2+\ell} + \Delta|y(t_k)|^2 \\ &= 2K\Delta - 2a_3\Delta|y(t_k)|^{2+\ell} - R_2\Delta|y(t_k)|^2 + (R_2+1)\Delta|y(t_k)|^2 \\ &\le R_1\Delta - a_3\Delta|y(t_k)|^{2+\ell} - R_2\Delta|y(t_k)|^2, \end{aligned}$$

where we used $(R_2+1)\Delta|y(t_k)|^2 \le (R_2+1)K_{\tilde{\varepsilon}}\Delta + (R_2+1)\tilde{\varepsilon}\Delta|y(t_k)|^{2+\ell}$ with $(R_2+1)\tilde{\varepsilon} < a_3$. Inserting the above inequality into (2.37), we obtain

$$\begin{aligned} |z(t_{k+1})|^2 \le A_{\theta,\Delta}|z(t_k)|^2 + R_1\Delta - a_3\Delta|y(t_k)|^{2+\ell} + 2a_4\Delta\int_{-\tau}^{0}|y_{t_k}(r)|^2\mathrm{d}\nu_2(r) \\ - R_2\Delta|y(t_k)|^2 + |\sigma_k\delta W_k|^2 + \mathcal{M}_k. \end{aligned} \tag{2.38}$$

For any integer $p \ge 2$,

$$\mathbb{E}\Big[\big(|z(t_{k+1})|^2 + a_3\Delta|y(t_k)|^{2+\ell}\big)^p\Big] \le A_{\theta,\Delta}^p\mathbb{E}[|z(t_k)|^{2p}] + \sum_{i=1}^{p} C_p^i\mathbb{E}[I_i], \tag{2.39}$$

where

$$\begin{aligned} I_i := A_{\theta,\Delta}^{p-i}|z(t_k)|^{2(p-i)}\Big(R_1\Delta - R_2\Delta|y(t_k)|^2 + 2a_4\Delta\int_{-\tau}^{0}|y_{t_k}(r)|^2\mathrm{d}\nu_2(r) \\ + |\sigma_k\delta W_k|^2 + \mathcal{M}_k\Big)^i. \end{aligned}$$

For any $\varepsilon_1 \in (0,1)$, it follows from Assumption 2.1 and the Young inequality

$$ab \le \frac{p-1}{p}\varepsilon_1 a^p + \frac{1}{p\varepsilon_1^{p-1}}b^p, a, b > 0$$

that

$$\begin{aligned} \mathbb{E}[I_1] \le \mathbb{E}\Big[A_{\theta,\Delta}^{p-1}|z(t_k)|^{2(p-1)}\Big(R_1\Delta - (R_2-2L)\Delta|y(t_k)|^2 + 2|\sigma(\mathbf{0})|^2\Delta \\ + 2a_4\Delta\int_{-\tau}^{0}|y_{t_k}(r)|^2\mathrm{d}\nu_2(r) + 2L\Delta\int_{-\tau}^{0}|y_{t_k}(r)|^2\mathrm{d}\nu_1(r)\Big)\Big] \\ \le 3\varepsilon_1\Delta A_{\theta,\Delta}^p\mathbb{E}[|z(t_k)|^{2p}] + R(\varepsilon_1)\Delta \end{aligned}$$

$$- (R_2 - 2L)\Delta\mathbb{E}\big[A_{\theta,\Delta}^{p-1}|z(t_k)|^{2(p-1)}|y(t_k)|^2\big]$$
$$+ \frac{(2a_4)^p\Delta}{\varepsilon_1^{p-1}}\mathbb{E}\Big[\int_{-\tau}^0 |y_{t_k}(r)|^{2p}\mathrm{d}\nu_2(r)\Big] + \frac{(2L)^p\Delta}{\varepsilon_1^{p-1}}\mathbb{E}\Big[\int_{-\tau}^0 |y_{t_k}(r)|^{2p}\mathrm{d}\nu_1(r)\Big]. \tag{2.40}$$

By (1.43) and Assumption 2.3, one has

$$\begin{aligned}|y(t_k)|^2 &\leq |z(t_k)|^2 + 2K\theta\Delta - 2a_3\theta\Delta|y(t_k)|^{2+\ell} + 2a_4\theta\Delta\int_{-\tau}^0 |y_{t_k}(r)|^2\mathrm{d}\nu_2(r)\\ &\leq |z(t_k)|^2 + 2K\theta\Delta + 2a_4\theta\Delta\int_{-\tau}^0 |y_{t_k}(r)|^2\mathrm{d}\nu_2(r).\end{aligned} \tag{2.41}$$

It is straightforward to see that

$$\begin{aligned}|y(t_k)|^{2p} &\leq \Big(|z(t_k)|^2 + 2K\theta\Delta + 2a_4\theta\Delta\int_{-\tau}^0 |y_{t_k}(r)|^2\mathrm{d}\nu_2(r)\Big)^{p-1}|y(t_k)|^2\\ &\leq 3^{p-1}|z(t_k)|^{2(p-1)}|y(t_k)|^2 + (6K\theta\Delta)^{p-1}|y(t_k)|^2\\ &\quad + (6a_4\theta\Delta)^{p-1}|y(t_k)|^2\int_{-\tau}^0 |y_{t_k}(r)|^{2(p-1)}\mathrm{d}\nu_2(r).\end{aligned}$$

This implies

$$\begin{aligned}&- |z(t_k)|^{2(p-1)}|y(t_k)|^2\\ &\leq -\frac{1}{3^{p-1}}|y(t_k)|^{2p} + (2K\theta\Delta)^{p-1}|y(t_k)|^2\\ &\quad + (2a_4\theta\Delta)^{p-1}|y(t_k)|^2\int_{-\tau}^0 |y_{t_k}(r)|^{2(p-1)}\mathrm{d}\nu_2(r)\\ &\leq -\frac{1}{3^{p-1}}|y(t_k)|^{2p} + (2K\Delta)^{p-1} + (2\Delta)^{p-1}(K^{p-1} + a_4^{p-1})|y(t_k)|^{2p}\\ &\quad + (2a_4\Delta)^{p-1}\int_{-\tau}^0 |y_{t_k}(r)|^{2p}\mathrm{d}\nu_2(r),\end{aligned} \tag{2.42}$$

where we used the Young inequality in the last step. Inserting (2.42) into (2.40) and using $A_{\theta,\Delta} \in (0, 1)$ yield

$$\begin{aligned}\mathbb{E}[I_1] \leq\ & 3\varepsilon_1\Delta A_{\theta,\Delta}^p\mathbb{E}[|z(t_k)|^{2p}] + R(\varepsilon_1)\Delta - \frac{(R_2 - 2L)A_{\theta,\Delta}^{p-1}}{3^{p-1}}\Delta\mathbb{E}[|y(t_k)|^{2p}]\\ &+ (R_2 - 2L)2^{p-1}(K^{p-1} + a_4^{p-1})\Delta^p\mathbb{E}[|y(t_k)|^{2p}]\end{aligned}$$

$$+ (R_2 - 2L)(2a_4)^{p-1}\Delta^p \mathbb{E}\Big[\int_{-\tau}^{0} |y_{t_k}(r)|^{2p} \mathrm{d}\nu_2(r)\Big]$$

$$+ \frac{(2a_4)^p}{\varepsilon_1^{p-1}}\Delta \mathbb{E}\Big[\int_{-\tau}^{0} |y_{t_k}(r)|^{2p} \mathrm{d}\nu_2(r)\Big] + \frac{(2L)^p\Delta}{\varepsilon_1^{p-1}}\mathbb{E}\Big[\int_{-\tau}^{0} |y_{t_k}(r)|^{2p} \mathrm{d}\nu_1(r)\Big]. \tag{2.43}$$

For the term I_i with $i \in \{2, \dots, p\}$, by $(\sum_{l=1}^{6} c_l)^i \le 6^i(\sum_{l=1}^{6} |c_l|^i)$, $c_l \in \mathbb{R}$ and $\mathbb{E}[|g_k\delta W_k|^i] \le m(i-1)!!|g_k|^i\Delta^{\frac{i}{2}}$ with g_k being $\mathcal{F}_{t_k}$-measurable, we derive

$$\begin{aligned}
\mathbb{E}[I_i] \le\ & 6^i A_{\theta,\Delta}^{p-i}\mathbb{E}\Big[|z(t_k)|^{2(p-i)}\Big(R_1\Delta^i + R_2\Delta^i|y(t_k)|^{2i} + m(2i-1)!!|\sigma_k|^{2i}\Delta^i \\
& + (2a_4)^i\Delta^i\int_{-\tau}^{0}|y_{t_k}(r)|^{2i}\mathrm{d}\nu_2(r) + 2^i m(i-1)!!A_{\theta,\Delta}^{\frac{i}{2}}|z(t_k)|^i|\sigma_k|^i\Delta^{\frac{i}{2}} \\
& + 4^i m(i-1)!!|y(t_k)|^i|\sigma_k|^i|\Delta^{\frac{i}{2}}\Big)\Big] \\
\le\ & 6\varepsilon_1\Delta A_{\theta,\Delta}^{p}\mathbb{E}[|z(t_k)|^{2p}] + R(\varepsilon_1)\Delta + \frac{6^p i R_2^{\frac{p}{i}}\Delta^2}{p\varepsilon_1^{(p-i)/i}}\mathbb{E}[|y(t_k)|^{2p}] \\
& + \frac{i(12a_4)^p\Delta}{p\varepsilon_1^{(p-i)/i}}\mathbb{E}\Big[\int_{-\tau}^{0}|y_{t_k}(r)|^{2p}\mathrm{d}\nu_2(r)\Big] + \frac{6^p(m(2i-1)!!)^{\frac{p}{i}}i\Delta}{p\varepsilon_1^{(p-i)/i}}\mathbb{E}[|\sigma_k|^{2p}] \\
& + \frac{12^{2p}(m(i-1)!!)^{\frac{2p}{i}}i\Delta}{2p\varepsilon_1^{(2p-i)/i}}\mathbb{E}[|\sigma_k|^{2p}] \\
& + \frac{24^p(m(i-1)!!)^{\frac{p}{i}}i\Delta}{p\varepsilon_1^{(p-i)/i}}\mathbb{E}[|y(t_k)|^p|\sigma_k|^p],
\end{aligned}$$

where $i!!$ represents the double factorial of i, and we used the Young inequalities

$$ab \le \frac{p-i}{p}\varepsilon_1 a^{\frac{p}{p-i}} + \frac{i}{p\varepsilon_1^{(p-i)/i}}b^{\frac{p}{i}},$$

$$ab \le \frac{2p-i}{2p}\varepsilon_1 a^{\frac{2p}{2p-i}} + \frac{i}{2p\varepsilon_1^{(2p-i)/i}}b^{\frac{2p}{i}}, \quad a, b > 0.$$

Together with $\varepsilon_1^{(2p-i)/i} \le \varepsilon_1^{(p-i)/i}$, the Young inequality, and Assumption 2.1, we obtain

$$
\begin{aligned}
\mathbb{E}[I_i] &\le 6\varepsilon_1 \Delta A_{\theta,\Delta}^{p} \mathbb{E}[|z(t_k)|^{2p}] + R(\varepsilon_1)\Delta + \frac{6^p R_2^{\frac{p}{i}} \Delta^2}{\varepsilon_1^{(p-i)/i}} \mathbb{E}[|y(t_k)|^{2p}] \\
&\quad + \frac{(12a_4)^p \Delta}{\varepsilon_1^{(p-i)/i}} \mathbb{E}\Big[\int_{-\tau}^{0} |y_{t_k}(r)|^{2p} \mathrm{d}\nu_2(r)\Big] + \frac{12^{2p}(m(2i-1)!!)^{\frac{2p}{i}} \Delta}{\varepsilon_1^{(2p-i)/i}} \mathbb{E}[|\sigma_k|^{2p}] \\
&\quad + \frac{24^p (m(i-1)!!)^{\frac{p}{i}} \Delta}{\varepsilon_1^{(p-i)/i}} \mathbb{E}\big[|y(t_k)|^{2p} + |\sigma_k|^{2p}\big] \\
&\le 6\varepsilon_1 \Delta A_{\theta,\Delta}^{p} \mathbb{E}[|z(t_k)|^{2p}] + R(\varepsilon_1)\Delta + \frac{6^p R_2^{\frac{p}{i}} \Delta^2}{\varepsilon_1^{(p-i)/i}} \mathbb{E}[|y(t_k)|^{2p}] \\
&\quad + \frac{(12a_4)^p \Delta}{\varepsilon_1^{(p-i)/i}} \mathbb{E}\Big[\int_{-\tau}^{0} |y_{t_k}(r)|^{2p} \mathrm{d}\nu_2(r)\Big] \\
&\quad + B_i \Delta \mathbb{E}[|y(t_k)|^{2p}] + B_i \Delta \mathbb{E}\Big[\int_{-\tau}^{0} |y_{t_k}(r)|^{2p} \mathrm{d}\nu_1(r)\Big],
\end{aligned}
\tag{2.44}
$$

where

$$
B_i := \frac{(12)^{2p}(2L)^p 2^{2p-2}(m(2i-1)!!)^{\frac{2p}{i}}}{\varepsilon_1^{(2p-i)/i}} + \frac{24^p(1+(2L)^p 2^{2p-2})(m(i-1)!!)^{\frac{p}{i}}}{\varepsilon_1^{(p-i)/i}},
$$

and we used

$$
\begin{aligned}
|\sigma_k|^{2p} &\le \Big(2|\sigma_k - \sigma(0)|^2 + 2|\sigma(0)|^2\Big)^p \\
&\le \Big(2L\big(|y(t_k)|^2 + \int_{-\tau}^{0} |y_{t_k}(r)|^2 \mathrm{d}\nu_1(r)\big) + 2|\sigma(0)|^2\Big)^p \\
&\le 2^{2p-2}(2L)^p\Big(|y(t_k)|^{2p} + \int_{-\tau}^{0} |y_{t_k}(r)|^{2p} \mathrm{d}\nu_1(r)\Big) + 2^{2p-1}|\sigma(0)|^{2p}.
\end{aligned}
\tag{2.45}
$$

Plugging (2.43) and (2.44) into (2.39) leads to

$$
\begin{aligned}
&\mathbb{E}[|z(t_{k+1})|^{2p}] \\
&\le \Big(1 + 3\varepsilon_1 p\Delta + 6\varepsilon_1\Delta\sum_{i=2}^{p} C_p^i\Big) A_{\theta,\Delta}^p \mathbb{E}[|z(t_k)|^{2p}] \\
&\quad + R(\varepsilon_1)\Delta + J_{k,1}(R_2) + J_{k,2}(R_2),
\end{aligned}
$$

where

$$
\begin{aligned}
J_{k,1}(R_2) = & -\frac{p(R_2 - 2L)A_{\theta,\Delta}^{p-1}}{3^{p-1}}\Delta\mathbb{E}[|y(t_k)|^{2p}] \\
& + \frac{p(2a_4)^p}{\varepsilon_1^{p-1}}\Delta\mathbb{E}\Big[\int_{-\tau}^{0}|y_{t_k}(r)|^{2p}\mathrm{d}v_2(r)\Big] \\
& + \frac{p(2L)^p\Delta}{\varepsilon_1^{p-1}}\mathbb{E}\Big[\int_{-\tau}^{0}|y_{t_k}(r)|^{2p}\mathrm{d}v_1(r)\Big] + \sum_{i=2}^{p} C_p^i\Delta\Big(B_i\mathbb{E}[|y(t_k)|^{2p}] \\
& + \frac{(12a_4)^p}{\varepsilon_1^{(p-i)/i}}\mathbb{E}\Big[\int_{-\tau}^{0}|y_{t_k}(r)|^{2p}\mathrm{d}v_2(r)\Big] + B_i\mathbb{E}\Big[\int_{-\tau}^{0}|y_{t_k}(r)|^{2p}\mathrm{d}v_1(r)\Big]\Big), \\
J_{k,2}(R_2) = & \; p2^{p-1}(R_2 - 2L)(K^{p-1} + a_4^{p-1})\Delta^p\mathbb{E}[|y(t_k)|^{2p}] + p(R_2 - 2L) \\
& \times (2a_4)^{p-1}\Delta^p\mathbb{E}\Big[\int_{-\tau}^{0}|y_{t_k}(r)|^{2p}\mathrm{d}v_2(r)\Big] + \sum_{i=2}^{p} C_p^i\frac{6^p R_2^{\frac{p}{i}}\Delta^2}{\varepsilon_1^{(p-i)/i}}\mathbb{E}[|y(t_k)|^{2p}].
\end{aligned}
$$

It follows from $\sum_{i=1}^{p} C_p^i = 2^p$ that for any $\varepsilon > 0$,

$$
\begin{aligned}
& e^{\varepsilon t_{k+1}}\mathbb{E}[|z(t_{k+1})|^{2p}] - e^{\varepsilon t_k}\mathbb{E}[|z(t_k)|^{2p}] \\
& \le \big((1 + 2^p 6\varepsilon_1\Delta)A_{\theta,\Delta}^p e^{\varepsilon\Delta} - 1\big)e^{\varepsilon t_k}\mathbb{E}[|z(t_k)|^{2p}] + R(\varepsilon_1)e^{\varepsilon t_{k+1}}\Delta \\
& \quad + e^{\varepsilon t_{k+1}} J_{k,1}(R_2) + e^{\varepsilon t_{k+1}} J_{k,2}(R_2),
\end{aligned}
$$

which leads to

$$
\begin{aligned}
e^{\varepsilon t_{k+1}}\mathbb{E}[|z(t_{k+1})|^{2p}] \le & \; |z(0)|^{2p} + \sum_{l=0}^{k}\big((1 + 2^p 6\varepsilon_1\Delta)A_{\theta,\Delta}^p e^{\varepsilon\Delta} - 1\big)e^{\varepsilon t_l}\mathbb{E}[|z(t_l)|^{2p}] \\
& + \sum_{l=0}^{k}\big(R(\varepsilon_1)e^{\varepsilon t_{l+1}}\Delta + e^{\varepsilon t_{l+1}} J_{l,1}(R_2) + e^{\varepsilon t_{l+1}} J_{l,2}(R_2)\big).
\end{aligned}
\tag{2.46}
$$

Note that

$$\begin{aligned}(1+2^p6\varepsilon_1\Delta)A^p_{\theta,\Delta}e^{\varepsilon\Delta} &= (1+2^p6\varepsilon_1\Delta)\Big(\frac{(1-\theta)^2}{\theta^2}+\frac{2\theta-1}{\theta^2(1+\Delta)}\Big)^p e^{\varepsilon\Delta}\\ &= (1+2^p6\varepsilon_1\Delta)(1-\frac{(2\theta-1)\Delta}{\theta^2(1+\Delta)})^p e^{\varepsilon\Delta}\\ &\le (1-\frac{(2\theta-1)\Delta}{2\theta^2})e^{(2^p6\varepsilon_1+\varepsilon)\Delta}\le e^{(-\frac{(2\theta-1)}{2\theta^2}+2^p6\varepsilon_1+\varepsilon)\Delta},\end{aligned}$$

where we used $e^{-x}\ge 1-x$ for $x\in[0,1]$ in the last step. Letting $2^p6\varepsilon_1=\varepsilon=\frac{(2\theta-1)}{4\theta^2}$, we derive

$$(1+2^p6\varepsilon_1\Delta)A^p_{\theta,\Delta}e^{\varepsilon\Delta}\le 1. \tag{2.47}$$

This, along with (2.46) implies that

$$\begin{aligned}&e^{\varepsilon t_{k+1}}\mathbb{E}[|z(t_{k+1})|^{2p}]\\ &\quad\le |z(0)|^{2p}+\sum_{l=0}^{k}\big(R(\varepsilon_1)e^{\varepsilon t_{l+1}}\Delta+e^{\varepsilon t_{l+1}}J_{l,1}(R_2)+e^{\varepsilon t_{l+1}}J_{l,2}(R_2)\big).\end{aligned} \tag{2.48}$$

By the definition of $A_{\theta,\Delta}$ with $\theta\in(1/2,1]$, we have

$$0<A_\theta:=\frac{(1-\theta)^2}{\theta^2}+\frac{2\theta-1}{2\theta^2}<\frac{(1-\theta)^2}{\theta^2}+\frac{2\theta-1}{\theta^2(1+\Delta)}=A_{\theta,\Delta}<1.$$

It follows from (2.27) that

$$\begin{aligned}&\sum_{l=0}^{k}e^{\varepsilon t_{l+1}}J_{l,1}(R_2)\\ &\quad\le -\Big(\frac{p(R_2-2L)A_\theta^{p-1}}{3^{p-1}}-e^{\varepsilon\tau}\tilde{B}-\sum_{l=2}^{p}C_p^iB_i\Big)\Delta\sum_{l=0}^{k}e^{\varepsilon t_{l+1}}\mathbb{E}[|y(t_l)|^{2p}]\\ &\qquad+e^{\varepsilon\tau}\tilde{B}\tau\|\xi\|^{2p},\end{aligned}$$

where

$$\tilde{B}:=\frac{p(2a_4)^p}{\varepsilon_1^{p-1}}+\frac{p(2L)^p}{\varepsilon_1^{p-1}}+\sum_{i=2}^{p}C_p^i\Big(\frac{(12a_4)^p}{\varepsilon_1^{(p-i)/i}}+B_i\Big).$$

Choose a sufficiently large number $R_2^* > 0$ such that

$$\frac{p(R_2^* - 2L)A_\theta^{p-1}}{2 \cdot 3^{p-1}} - e^{\varepsilon\tau}\tilde{B} - \sum_{i=2}^{p} C_p^i B_i > 0.$$

This leads to

$$\sum_{l=0}^{k} e^{\varepsilon t_{l+1}} J_{l,1}(R_2^*) \leq -\frac{p(R_2^* - 2L)A_\theta^{p-1}}{2 \cdot 3^{p-1}} \Delta \sum_{l=0}^{k} e^{\varepsilon t_{l+1}} \mathbb{E}[|y(t_l)|^{2p}] + e^{\varepsilon\tau}\tilde{B}\tau\|\xi\|^{2p}. \tag{2.49}$$

Similarly,

$$\sum_{l=0}^{k} e^{\varepsilon t_{l+1}} J_{l,2}(R_2^*) \leq D_\Delta \Delta \sum_{l=0}^{k} e^{\varepsilon t_{l+1}} \mathbb{E}[|y(t_l)|^{2p}] + e^{\varepsilon\tau}\tau p(R_2^* - 2L)(2a_4)^{p-1}\|\xi\|^{2p},$$

where

$$D_\Delta := p2^{p-1}(R_2^* - 2L)(K^{p-1} + a_4^{p-1})\Delta^{p-1} + e^{\varepsilon\tau} p(R_2^* - 2L)(2a_4)^{p-1}\Delta^{p-1} + \sum_{i=2}^{p} C_p^i \frac{6^p (R_2^*)^{\frac{p}{i}}\Delta}{\varepsilon_1^{(p-i)/i}}.$$

Choose a sufficiently small number $\Delta^* \in (0, 1)$ such that $D_{\Delta^*} \leq \frac{p(R_2^*-2L)A_\theta^{p-1}}{2 \cdot 3^{p-1}}$. Together with (2.49), we obtain that for any $\Delta \in (0, \Delta^*]$,

$$\sum_{l=0}^{k} e^{\varepsilon t_{l+1}} (J_{l,1}(R_2^*) + J_{l,2}(R_2^*)) \leq e^{\varepsilon\tau}\tau\big(\tilde{B} + p(R_2^* - 2L)(2a_4)^{p-1}\big)\|\xi\|^{2p}. \tag{2.50}$$

Inserting (2.50) into (2.48) and using Assumption 2.4 yield

$$\begin{aligned} &e^{\varepsilon t_{k+1}} \mathbb{E}[|z(t_{k+1})|^{2p}] \\ &\quad \leq |\xi - \theta b(\xi)\Delta|^{2p} + \sum_{l=0}^{k} R(\varepsilon_1) e^{\varepsilon t_{l+1}} \Delta \\ &\qquad + e^{\varepsilon\tau}\tau\big(\tilde{B} + p(R_2^* - 2L)(2a_4)^{p-1}\big)\|\xi\|^{2p} \end{aligned}$$

$$
\begin{aligned}
&\leq K(1+\|\xi\|^{2p(\beta+1)}) + \frac{R(\varepsilon_1)e^{\varepsilon t_{k+2}}}{\varepsilon} \\
&\quad + e^{\varepsilon\tau}\tau\big(\tilde{B} + p(R_2^* - 2L)(2a_4)^{p-1}\big)\|\xi\|^{2p},
\end{aligned}
$$

which finishes the proof. □

Lemma 2.6 *Let conditions of Lemma 2.5 hold. Then for integer $p \geq 2$,*

$$
\sup_{\Delta\in(0,1]}\sup_{k\geq -N}\mathbb{E}[|z^{\xi,\Delta}(t_k)|^p] \leq K(1+\|\xi\|^{p(\beta+1)}).
$$

Proof Combining Lemma 2.5, we only need to show that for any $\Delta \in (\Delta^*, 1]$,

$$
\sup_{k\geq -N}\mathbb{E}[|z^{\xi,\Delta}(t_k)|^p] \leq K(1+\|\xi\|^{p(\beta+1)}).
$$

Recalling (2.39), we reestimate terms $I_i, i = 1, \ldots, p$.

In fact, similar to (2.43), we have that for any $\kappa_1 \in (0, 1)$,

$$
\begin{aligned}
\mathbb{E}[I_1] \leq\ & 3\kappa_1 A_{\theta,\Delta}^p\mathbb{E}[|z(t_k)|^{2p}] + R(\kappa_1)\Delta - \frac{(R_2-2L)A_{\theta,\Delta}^{p-1}}{3^{p-1}}\Delta\mathbb{E}[|y(t_k)|^{2p}] \\
&+ (R_2-2L)2^{p-1}(K^{p-1}+a_4^{p-1})\Delta^p\mathbb{E}[|y(t_k)|^{2p}] \\
&+ \Big((R_2-2L)(2a_4)^{p-1} + \frac{(2a_4)^p}{\kappa_1^{p-1}}\Big)\Delta^p\mathbb{E}\Big[\int_{-\tau}^0 |y_{t_k}(r)|^{2p}\mathrm{d}\nu_2(r)\Big] \\
&+ \frac{(2L\Delta)^p}{\kappa_1^{p-1}}\mathbb{E}\Big[\int_{-\tau}^0 |y_{t_k}(r)|^{2p}\mathrm{d}\nu_1(r)\Big].
\end{aligned}
\tag{2.51}
$$

Similar to the proof of (2.44), for $i \in \{2, \ldots, p\}$, we derive

$$
\begin{aligned}
&\mathbb{E}[I_i] \\
&\leq 6\kappa_1 A_{\theta,\Delta}^p\mathbb{E}[|z(t_k)|^{2p}] + R(\kappa_1)\Delta^p + \frac{6^p i R_2^{\frac{p}{i}}\Delta^p}{p\kappa_1^{(p-i)/i}}\mathbb{E}[|y(t_k)|^{2p}] \\
&\quad + \frac{i(12a_4\Delta)^p}{p\kappa_1^{(p-i)/i}}\mathbb{E}\Big[\int_{-\tau}^0 |y_{t_k}(r)|^{2p}\mathrm{d}\nu_2(r)\Big] + \frac{6^p(m(2i-1)!!)^{\frac{p}{i}} i\Delta^p}{p\kappa_1^{(p-i)/i}}\mathbb{E}[|\sigma_k|^{2p}] \\
&\quad + \frac{12^{2p}(m(i-1)!!)^{\frac{2p}{i}} i\Delta^p}{2p\kappa_1^{(2p-i)/i}}\mathbb{E}[|\sigma_k|^{2p}] \\
&\quad + \frac{24^p(m(i-1)!!)^{\frac{p}{i}} i\Delta^{\frac{p}{2}}}{p\kappa_1^{(p-i)/i}}\mathbb{E}[|y(t_k)|^p|\sigma_k|^p]
\end{aligned}
$$

$$
\begin{aligned}
&\leq 6\kappa_1 A_{\theta,\Delta}^p \mathbb{E}[|z(t_k)|^{2p}] + R(\kappa_1)\Delta^p + R(\kappa_1)\Delta^p \mathbb{E}[|y(t_k)|^{2p}] \\
&\quad \frac{(12a_4\Delta)^p}{\kappa_1^{(p-i)/i}} \mathbb{E}\Big[\int_{-\tau}^0 |y_{t_k}(r)|^{2p} \mathrm{d}\nu_2(r)\Big] + \frac{12^{2p}(m(2i-1)!!)^{\frac{2p}{i}}\Delta^p}{\kappa_1^{(2p-i)/i}} \mathbb{E}[|\sigma_k|^{2p}] \\
&\quad + \frac{24^p(m(i-1)!!)^{\frac{p}{i}}\Delta}{\kappa_1^{(p-i)/i}} \mathbb{E}\Big[|y(t_k)|^{2p} + |\sigma_k|^{2p}\Big].
\end{aligned}
$$

It follows from (2.45) with $\iota = 1$ that

$$
\begin{aligned}
&\mathbb{E}[I_i] \\
&\leq 6\kappa_1 A_{\theta,\Delta}^p \mathbb{E}[|z(t_k)|^{2p}] + R(\kappa_1)\Delta + R(\kappa_1)\Delta^p \mathbb{E}[|y(t_k)|^{2p}] \\
&\quad + \frac{(12a_4\Delta)^p}{\kappa_1^{(p-i)/i}} \mathbb{E}\Big[\int_{-\tau}^0 |y_{t_k}(r)|^{2p} \mathrm{d}\nu_2(r)\Big] \\
&\quad + 2^{2p-2}(2L)^p \frac{12^{2p}(m(2i-1)!!)^{\frac{2p}{i}}\Delta^p}{\kappa_1^{(2p-i)/i}} \mathbb{E}\Big[\int_{-\tau}^0 |y_{t_k}(r)|^{2p} \mathrm{d}\nu_1(r)\Big] \\
&\quad + (1 + 2^{2p-2}(2L)^p) \frac{24^p(m(2i-1)!!)^{\frac{p}{i}}\Delta}{\kappa_1^{(p-i)/i}} \mathbb{E}\Big[|y(t_k)|^{2p} + \int_{-\tau}^0 |y_{t_k}(r)|^{2p} \mathrm{d}\nu_1(r)\Big],
\end{aligned} \tag{2.52}
$$

where $A_{\theta,\Delta} = \frac{(1-\theta)^2}{\theta^2} + \frac{2\theta-1}{\theta^2(1+\Delta)}$. Plugging (2.51) and (2.52) into (2.39) leads to

$$
\begin{aligned}
&\mathbb{E}[|z(t_{k+1})|^{2p}] + \frac{(a_3\Delta)^p}{2^p} \mathbb{E}[|y(t_k)|^{(2+\ell)p}] \\
&\quad \leq (1 + 3\kappa_1 p + 6\kappa_1 \sum_{i=2}^p C_p^i) A_{\theta,\Delta}^p \mathbb{E}[|z(t_k)|^{2p}] + R(\kappa_1)\Delta + J_{k,3} + J_{k,4},
\end{aligned}
$$

where

$$
\begin{aligned}
J_{k,3} := &-\Big(\frac{p(R_2 - 2L)A_{\theta,\Delta}^{p-1}}{3^{p-1}} - \sum_{i=2}^p C_p^i \big(1 + 2^{2p-2}(2L)^p\big) \frac{24^p(m(2i-1)!!)^{\frac{p}{i}}}{\kappa_1^{(p-i)/i}}\Big) \\
&\times \Delta \mathbb{E}[|y(t_k)|^{2p}] + \sum_{i=2}^p C_p^i \big(1 + 2^{2p-2}(2L)^p\big) \frac{24^p(m(2i-1)!!)^{\frac{p}{i}}\Delta}{\kappa_1^{(p-i)/i}} \\
&\times \mathbb{E}\Big[\int_{-\tau}^0 |y_{t_k}(r)|^{2p} \mathrm{d}\nu_1(r)\Big],
\end{aligned}
$$

$$
\begin{aligned}
J_{k,4} := {} & R(\kappa_1)\Delta^p\mathbb{E}[|y(t_k)|^{2p}] + \Delta^p\Big(p(R_2-2L)(2a_4)^{p-1} + p\frac{(2a_4)^p}{\kappa_1^{p-1}} \\
& + \sum_{i=2}^{p} C_p^i \frac{(12a_4)^p}{\kappa_1^{(p-i)/i}}\Big)\mathbb{E}\Big[\int_{-\tau}^{0} |y_{t_k}(r)|^{2p}\mathrm{d}v_2(r)\Big] + \Big(\frac{p(2L)^p}{\kappa_1^{p-1}} \\
& + \sum_{i=2}^{p} C_p^i 2^{2p-2}(2L)^p \frac{12^{2p}(m(2i-1)!!)^{\frac{2p}{i}}}{\kappa_1^{(2p-i)/i}}\Big)\Delta^p\mathbb{E}\Big[\int_{-\tau}^{0} |y_{t_k}(r)|^{2p}\mathrm{d}v_1(r)\Big].
\end{aligned}
$$

It is clear that for any $\kappa > 0$,

$$
\begin{aligned}
& e^{\kappa t_{k+1}}\mathbb{E}[|z(t_{k+1})|^{2p}] \\
& \le |z(0)|^{2p} + \sum_{l=0}^{k}\big((1+2^p 6\kappa_1)A_{\theta,\Delta}^p e^{\kappa\Delta} - 1\big)e^{\kappa t_l}\mathbb{E}[|z(t_l)|^{2p}] \\
& \quad + \sum_{l=0}^{k}\Big(R(\kappa_1)e^{\kappa t_{l+1}}\Delta + e^{\kappa t_{l+1}}J_{l,3} + e^{\kappa t_{l+1}}J_{l,4} \\
& \quad - \frac{(a_3\Delta)^p}{2^p}e^{\kappa t_{l+1}}\mathbb{E}[|y(t_l)|^{(2+\ell)p}]\Big).
\end{aligned}
\tag{2.53}
$$

By the definition of $A_{\theta,\Delta}$, we have that for any $\Delta \in [\Delta^*, 1]$,

$$
A_\theta = \frac{(1-\theta)^2}{\theta^2} + \frac{2\theta-1}{2\theta^2} < \frac{(1-\theta)^2}{\theta^2} + \frac{2\theta-1}{\theta^2(1+\Delta)} = A_{\theta,\Delta} \le A_{\theta,\Delta^*} < 1.
$$

Choose a sufficiently small number $\kappa_1 > 0$ such that

$$
(1+2^p 6\kappa_1)A_{\theta,\Delta^*}^p < 1.
$$

For such κ_1, choose another sufficiently small number $\kappa > 0$ such that

$$
(1+2^p 6\kappa_1)A_{\theta,\Delta^*}^p e^{\kappa} \le 1. \tag{2.54}
$$

This, along with (2.53) implies that

$$
\begin{aligned}
e^{\kappa t_{k+1}}\mathbb{E}[|z(t_{k+1})|^{2p}] \le {} & |z(0)|^{2p} + \sum_{l=0}^{k}\Big(R(\kappa_1)e^{\kappa t_{l+1}}\Delta + e^{\kappa t_{l+1}}J_{l,3} \\
& + e^{\kappa t_{l+1}}J_{l,4} - \frac{(a_3\Delta)^p}{2^p}e^{\kappa t_{l+1}}\mathbb{E}[|y(t_l)|^{(2+\ell)p}]\Big).
\end{aligned}
\tag{2.55}
$$

By (2.27), we deduce

$$
\begin{aligned}
&\sum_{l=0}^{k} e^{\kappa t_{l+1}} J_{l,3} \\
&\le -\Big(\frac{p(R_2-2L)A_{\theta,\Delta}^{p-1}}{3^{p-1}} - \sum_{i=2}^{p} C_p^i\big(1+2^{2p-2}(2L)^p\big) \\
&\quad \times \frac{24^p(m(2i-1)!!)^{\frac{p}{i}}}{\kappa_1^{(p-i)/i}}(1+e^{\kappa\tau})\Big) \\
&\quad \times \Delta \sum_{l=0}^{k} e^{\kappa t_{l+1}} \mathbb{E}[|y(t_l)|^{2p}] + \sum_{i=2}^{p} C_p^i\big(1+2^{2p-2}(2L)^p\big) \\
&\quad \times \frac{24^p(m(2i-1)!!)^{\frac{p}{i}}}{\kappa_1^{(p-i)/i}} e^{\kappa\tau}\tau\|\xi\|^{2p}.
\end{aligned}
$$

Letting $R_2 > 0$ be sufficiently large such that

$$
\frac{p(R_2-2L)A_{\theta}^{p-1}}{3^{p-1}} - \sum_{i=2}^{p} C_p^i\big(1+2^{2p-2}(2L)^p\big)\frac{24^p(m(2i-1)!!)^{\frac{p}{i}}}{\kappa_1^{(p-i)/i}}(1+e^{\kappa\tau}) > 0,
$$

we obtain

$$
\sum_{l=0}^{k} e^{\kappa t_{l+1}} J_{l,3} \le \sum_{i=2}^{p} C_p^i\big(1+2^{2p-2}(2L)^p\big)\frac{24^p(m(2i-1)!!)^{\frac{p}{i}}}{\kappa_1^{(p-i)/i}} e^{\kappa\tau}\tau\|\xi\|^{2p}. \tag{2.56}
$$

Similarly,

$$
\begin{aligned}
&\sum_{l=0}^{k}\Big(e^{\kappa t_{l+1}} J_{l,4} - \frac{(a_3\Delta)^p}{2^p} e^{\kappa t_{l+1}} \mathbb{E}[|y(t_l)|^{(2+\ell)p}]\Big) \\
&\le G\Delta^p \sum_{l=0}^{k} e^{\kappa t_{l+1}} \mathbb{E}[|y(t_l)|^{2p}] - \frac{(a_3\Delta)^p}{2^p}\sum_{l=0}^{k} e^{\kappa t_{l+1}} \mathbb{E}[|y(t_l)|^{(2+\ell)p}] \\
&\quad + K\|\xi\|^{2p} \le K\|\xi\|^{2p} + Ke^{\kappa t_{k+1}},
\end{aligned} \tag{2.57}
$$

where

$$G := R(\kappa_1) + e^{\kappa\tau}\Big(p\big((R_2 - 2L)(2a_4)^{p-1} + \frac{(2a_4)^p}{\kappa_1^{p-1}}\big) + \sum_{i=2}^{p} C_p^i \frac{(12a_4)^p}{\kappa_1^{(p-i)/i}}\Big)$$
$$+ e^{\kappa\tau}\Big(\frac{p(2L)^p}{\kappa_1^{p-1}} + \sum_{i=2}^{p} C_p^i 2^{2p-2}(2L)^p \frac{12^{2p}(m(2i-1)!!)^{\frac{2p}{i}}}{\kappa_1^{(2p-i)/i}}\Big).$$

Then the desired result follows from (2.55)–(2.57). □

Lemma 2.7 *Let Assumptions 2.1, 2.3, and 2.4 hold. Then for integer $p \geq 2$, it holds that*

$$\sup_{\Delta\in(0,1]}\sup_{k\geq -N} \mathbb{E}[|y^{\xi,\Delta}(t_k)|^p] \leq K(1 + \|\xi\|^{p(\beta+1)}).$$

Proof It follows from (2.41) that

$$\begin{aligned}
&\sup_{k\geq 0}\mathbb{E}\Big[|y(t_k)|^{2p} + (2a_3\theta\Delta)^p|y(t_k)|^{p(2+\ell)}\Big] \\
&\leq \sup_{k\geq 0}\mathbb{E}\Big[\Big(|y(t_k)|^2 + 2a_3\theta\Delta|y(t_k)|^{2+\ell}\Big)^p\Big] \\
&\leq K\sup_{k\geq 0}\mathbb{E}[|z(t_k)|^{2p}] + K\Delta^p\sup_{k\geq 0}\mathbb{E}\Big[\int_{-\tau}^{0}|y_{t_k}(r)|^{2p}\mathrm{d}\nu_2(r)\Big] + K\Delta^p \\
&\leq K\sup_{k\geq 0}\mathbb{E}[|z(t_k)|^{2p}] + K\Delta^p(1 + \|\xi\|^{2p}) + K\Delta^p\sup_{k\geq 0}\mathbb{E}[|y(t_k)|^{2p}].
\end{aligned}$$

This, along with the Young inequality and Lemma 2.6 implies that

$$\begin{aligned}
&\sup_{k\geq -N}\mathbb{E}[|y^{\xi,\Delta}(t_k)|^p] \\
&\leq K\sup_{k\geq 0}\mathbb{E}[|z(t_k)|^{2p}] + K(1 + \|\xi\|^{2p}) + K\Delta^p\sup_{k\geq 0}\mathbb{E}[|y(t_k)|^{2p}] \\
&\quad - (2a_3\theta\Delta)^p\mathbb{E}[|y(t_k)|^{p(2+\ell)}] \\
&\leq K\sup_{k\geq 0}\mathbb{E}[|z(t_k)|^{2p}] + K(1 + \|\xi\|^{2p}) \\
&\leq K(1 + \|\xi\|^{p(1+\beta)}) \quad \forall\, p \geq 2.
\end{aligned}$$

This proof is therefore completed. □

Proposition 2.8 *Let Assumptions 2.1, 2.3, and 2.4 hold. Then for any integer $p \ge 2$, it holds that*

$$\sup_{\Delta\in(0,1]}\sup_{k\ge 0}\mathbb{E}\big[\|z_{t_k}^{\xi,\Delta}\|^p+\|y_{t_k}^{\xi,\Delta}\|^p\big]\le K(1+\|\xi\|^{p(\beta+1)}). \tag{2.58}$$

Proof We divide the proof into two cases. Recall that $\Delta^* \in (0,1)$ and $\varepsilon = \frac{2\theta-1}{4\theta^2}$ are defined in the proof of Lemma 2.5.

Case 1: $\Delta \in (0, \Delta^*]$. It follows from (2.38) that for $k \in \mathbb{N}$,

$$\begin{aligned}|z(t_{k+1})|^{2p} &\le \Big(A_{\theta,\Delta}|z(t_k)|^2+R_1\Delta+2a_4\Delta\int_{-\tau}^0|y_{t_k}(r)|^2\mathrm{d}\nu_2(r)\\ &\quad+|\sigma_k\delta W_k|^2+\mathcal{M}_k\Big)^p\\ &\le A_{\theta,\Delta}^p|z(t_k)|^{2p}+\sum_{i=1}^p C_p^i\tilde{I}_{k,i},\end{aligned}$$

where $A_{\theta,\Delta}=\frac{(1-\theta)^2}{\theta^2}+\frac{2\theta-1}{\theta^2(1+\Delta)}$, $\mathcal{M}_k$ is defined by (2.10), and

$$\tilde{I}_{k,i}:=A_{\theta,\Delta}^{p-i}|z(t_k)|^{2(p-i)}\Big(R_1\Delta+2a_4\Delta\int_{-\tau}^0|y_{t_k}(r)|^2\mathrm{d}\nu_2(r)+|\sigma_k\delta W_k|^2+\mathcal{M}_k\Big)^i.$$

This yields

$$\begin{aligned}e^{\varepsilon t_{k+1}}|z(t_{k+1})|^{2p} &= \sum_{l=0}^k\Big(e^{\varepsilon t_{l+1}}|z(t_{l+1})|^{2p}-e^{\varepsilon t_l}|z(t_l)|^{2p}\Big)+|z(0)|^{2p}\\ &\le |z(0)|^{2p}+(A_{\theta,\Delta}^pe^{\varepsilon\Delta}-1)\sum_{l=0}^k e^{\varepsilon t_l}|z(t_l)|^{2p}\\ &\quad+\sum_{l=0}^k\sum_{i=1}^p C_p^ie^{\varepsilon t_{l+1}}\tilde{I}_{l,i}.\end{aligned} \tag{2.59}$$

Applying the Young inequality derives

$$\begin{aligned}\tilde{I}_{l,1} &\le 3\varepsilon_1\Delta A_{\theta,\Delta}^p|z(t_l)|^{2p}\\ &\quad+\frac{\Delta}{\varepsilon_1^{p-1}}\Big(R_1^p+(2a_4)^p\int_{-\tau}^0|y_{t_l}(r)|^{2p}\mathrm{d}\nu_2(r)+\frac{|\sigma_l\delta W_l|^{2p}}{\Delta^p}\Big)\\ &\quad+A_{\theta,\Delta}^{p-1}|z(t_l)|^{2(p-1)}\mathcal{M}_l\end{aligned} \tag{2.60}$$

and

$$
\begin{aligned}
\tilde{I}_{l,i} &\le 5^i \Delta A_{\theta,\Delta}^{p-i} |z(t_l)|^{2(p-i)} \Big(R_1^i + (2a_4)^i \int_{-\tau}^{0} |y_{t_l}(r)|^{2i} \mathrm{d}\nu_2(r) + \frac{|\sigma_l \delta W_l|^{2i}}{\Delta} \\
&\quad + \frac{A_{\theta,\Delta}^{\frac{i}{2}} (2|z(t_l)||\sigma_l \delta W_l|)^i}{\Delta} + \frac{(4|y(t_l)||\sigma_l \delta W_l|)^i}{\Delta} \Big) \\
&\le 5\varepsilon_1 \Delta A_{\theta,\Delta}^{p} |z(t_l)|^{2p} + \frac{5^p \Delta}{\varepsilon_1^{(p-i)/i}} \Big(R_1^p + (2a_4)^p \int_{-\tau}^{0} |y_{t_l}(r)|^{2p} \mathrm{d}\nu_2(r) \\
&\quad + \frac{|\sigma_l \delta W_l|^{2p}}{\Delta^{\frac{p}{i}}} + \frac{(4|y(t_l)||\sigma_l \delta W_l|)^p}{\Delta^{\frac{p}{i}}} \Big) + \frac{(10|\sigma_l \delta W_l|)^{2p}}{\varepsilon_1^{(2p-i)/i} \Delta^{\frac{2p}{i}-1}} \quad \forall i \in \{2, \ldots, p\},
\end{aligned} \tag{2.61}
$$

where ε_1 is given in the proof of Lemma 2.5, and we used $0 \le \frac{2(\theta-1)}{\theta} \le 2A_{\theta,\Delta}^{\frac{1}{2}}$, $0 < \frac{2}{\theta} \le 4$. Inserting (2.60) and (2.61) into (2.59), we arrive at

$$
\begin{aligned}
&e^{\varepsilon t_{k+1}} |z(t_{k+1})|^{2p} \\
&\le |z(0)|^{2p} + \Big((1 + 3p\varepsilon_1 \Delta + 5\varepsilon_1 \Delta \sum_{i=2}^{p} C_p^i) A_{\theta,\Delta}^{p} e^{\varepsilon \Delta} - 1 \Big) \sum_{l=0}^{k} e^{\varepsilon t_l} |z(t_l)|^{2p} \\
&\quad + \sum_{l=0}^{k} e^{\varepsilon t_{l+1}} \tilde{J}_{l,1} + p A_{\theta,\Delta}^{p-1} \sum_{l=0}^{k} e^{\varepsilon t_{l+1}} |z(t_l)|^{2(p-1)} \mathcal{M}_l,
\end{aligned}
$$

where

$$
\begin{aligned}
\tilde{J}_{l,1} &:= \Delta \Big(\frac{p}{\varepsilon_1^{p-1}} + \sum_{i=2}^{p} \frac{5^p C_p^i}{\varepsilon_1^{(p-i)/i}} \Big) \Big(R_1^p + (2a_4)^p \int_{-\tau}^{0} |y_{t_l}(r)|^{2p} \mathrm{d}\nu_2(r) \Big) \\
&\quad + \frac{p|\sigma_l \delta W_l|^{2p}}{(\varepsilon_1 \Delta)^{p-1}} + 5^p \Delta \sum_{i=2}^{p} \frac{C_p^i}{\varepsilon_1^{(p-i)/i}} \Big(\frac{|\sigma_l \delta W_l|^{2p}}{\Delta^{\frac{p}{i}}} + \frac{(4|y(t_l)||\sigma_l \delta W_l|)^p}{\Delta^{\frac{p}{i}}} \Big) \\
&\quad + \sum_{i=2}^{p} \frac{C_p^i (10|\sigma_l \delta W_l|)^{2p}}{\varepsilon_1^{(2p-i)/i} \Delta^{\frac{2p}{i}-1}}.
\end{aligned}
$$

It follows from (2.47) that

$$
(1 + 3p\varepsilon_1 \Delta + 5\varepsilon_1 \Delta \sum_{i=2}^{p} C_p^i) A_{\theta,\Delta}^{p} e^{\varepsilon \Delta} - 1 \le (1 + 2^p 6\varepsilon_1 \Delta) A_{\theta,\Delta}^{p} e^{\varepsilon \Delta} - 1 \le 0,
$$

which implies

$$e^{\varepsilon t_{k+1}}|z(t_{k+1})|^{2p} \le |z(0)|^{2p} + \sum_{l=0}^{k} e^{\varepsilon t_{l+1}} \tilde{J}_{l,1} + p \sum_{l=0}^{k} A_{\theta,\Delta}^{p-1} e^{\varepsilon t_{l+1}} |z(t_l)|^{2(p-1)} \mathcal{M}_l .$$

Taking the supremum over the index set $\{k - N, \ldots, k\}$ and then taking the expectations, on both sides of the above inequality, we clearly see that

$$\begin{aligned}
&\mathbb{E}\Big[\sup_{(k-N)\vee 0 \le j \le k} e^{\varepsilon t_{j+1}} |z(t_{j+1})|^{2p}\Big] \\
&\le |z(0)|^{2p} + \mathbb{E}\Big[\sum_{l=0}^{k} e^{\varepsilon t_{l+1}} \tilde{J}_{l,1}\Big] + p\mathbb{E}\Big[\sup_{(k-N)\vee 0 \le j \le k} \sum_{l=0}^{(k-N)\vee 0} e^{\varepsilon t_{l+1}} A_{\theta,\Delta}^{p-1} \times \\
&\quad |z(t_l)|^{2(p-1)} \mathcal{M}_l + \sup_{(k-N)\vee 0 \le j \le k} \sum_{l=(k-N)\vee 0}^{j} e^{\varepsilon t_{l+1}} A_{\theta,\Delta}^{p-1} |z(t_l)|^{2(p-1)} \mathcal{M}_l\Big] \\
&= |z(0)|^{2p} + \mathbb{E}\Big[\sum_{l=0}^{k} e^{\varepsilon t_{l+1}} \tilde{J}_{l,1}\Big] \\
&\quad + p\mathbb{E}\Big[\sup_{(k-N)\vee 0 \le j \le k} \sum_{l=(k-N)\vee 0}^{j} e^{\varepsilon t_{l+1}} A_{\theta,\Delta}^{p-1} |z(t_l)|^{2(p-1)} \mathcal{M}_l\Big].
\end{aligned} \tag{2.62}$$

By the linear growth condition on coefficient σ, one has

$$\begin{aligned}
&\mathbb{E}\Big[\sum_{l=0}^{k} e^{\varepsilon t_{l+1}} \tilde{J}_{l,1}\Big] \\
&\le K_{\varepsilon_1} \Delta \mathbb{E}\Big[\sum_{l=0}^{k} e^{\varepsilon t_{l+1}} \Big(1 + \int_{-\tau}^{0} |y_{t_l}(r)|^{2p} \mathrm{d}\nu_2(r) + |\sigma_l|^{2p} + |y(t_l)|^{2p}\Big)\Big] \\
&\le K_{\varepsilon_1} \Delta \mathbb{E}\Big[\sum_{l=0}^{k} e^{\varepsilon t_{l+1}} \big(1 + \int_{-\tau}^{0} |y_{t_l}(r)|^{2p} \mathrm{d}\nu_2(r) + |y(t_l)|^{2p} \\
&\quad + \int_{-\tau}^{0} |y_{t_l}(r)|^{2p} \mathrm{d}\nu_1(r)\big)\Big].
\end{aligned}$$

Utilizing Lemma 2.7 yields that

$$
\begin{aligned}
&\mathbb{E}\Big[\sum_{l=0}^{k} e^{\varepsilon t_{l+1}} \tilde{J}_{l,1}\Big] \\
&\leq K_{\varepsilon_1}\|\xi\|^{2p} + K_{\varepsilon_1}\Delta \sup_{j\geq -N}\mathbb{E}[|y(t_j)|^{2p}]\sum_{l=0}^{k} e^{\varepsilon t_{l+1}} + K_{\varepsilon_1}\Delta\sum_{l=0}^{k} e^{\varepsilon t_{l+1}} \\
&\leq K_{\varepsilon_1}(1+\|\xi\|^{2p(\beta+1)})\frac{e^{\varepsilon t_{k+2}}}{\varepsilon}.
\end{aligned} \tag{2.63}
$$

Recalling the definition of $\mathcal{M}_k$, we apply the Burkholder–Davis–Gundy inequality to obtain

$$
\begin{aligned}
&p\mathbb{E}\Big[\sup_{(k-N)\vee 0\leq j\leq k}\sum_{l=(k-N)\vee 0}^{j} e^{\varepsilon t_{l+1}} A_{\theta,\Delta}^{p-1}|z(t_l)|^{2(p-1)}\mathcal{M}_l\Big] \\
&\leq \sqrt{32}p\mathbb{E}\Big[\Big(\sum_{l=(k-N)\vee 0}^{k}\big(2e^{\varepsilon t_{l+1}} A_{\theta,\Delta}^{p-\frac{1}{2}}|z(t_l)|^{2p-1}|\sigma_l|\big)^2\Delta\Big)^{\frac{1}{2}}\Big] \\
&\quad + \sqrt{32}p\mathbb{E}\Big[\Big(\sum_{l=(k-N)\vee 0}^{k}\big(4e^{\varepsilon t_{l+1}} A_{\theta,\Delta}^{p-1}|z(t_l)|^{2(p-1)}|y(t_l)||\sigma_l|\big)^2\Delta\Big)^{\frac{1}{2}}\Big].
\end{aligned}
$$

By virtue of $A_{\theta,\Delta}\in(0,1)$ and the Young inequality, we obtain

$$
\begin{aligned}
&p\mathbb{E}\Big[\sup_{(k-N)\vee 0\leq j\leq k}\sum_{l=(k-N)\vee 0}^{j} e^{\varepsilon t_{l+1}} A_{\theta,\Delta}^{p-1}|z(t_l)|^{2(p-1)}\mathcal{M}_l\Big] \\
&\leq 2\sqrt{32}p\mathbb{E}\Big[\Big(\sup_{(k-N)\vee 0\leq j\leq k}(e^{\varepsilon t_{j+1}})^{\frac{2p-1}{2p}}|z(t_j)|^{2p-1}\Big)\Big(\sum_{l=(k-N)\vee 0}^{k}(e^{\varepsilon t_{l+1}})^{\frac{1}{p}}|\sigma_l|^2\Delta\Big)^{\frac{1}{2}}\Big] \\
&\quad + 4\sqrt{32}p\mathbb{E}\Big[\sup_{(k-N)\vee 0\leq j\leq k}(e^{\varepsilon t_{j+1}})^{\frac{(p-1)}{p}}|z(t_j)|^{2(p-1)}\times \\
&\quad \Big(\sum_{l=(k-N)\vee 0}^{k}(e^{\varepsilon t_{l+1}})^{\frac{2}{p}}|y(t_l)|^2|\sigma_l|^2\Delta\Big)^{\frac{1}{2}}\Big] \\
&\leq \frac{1}{2e^{\varepsilon\Delta}}\mathbb{E}\Big[\sup_{(k-N)\vee 0\leq j\leq k} e^{\varepsilon t_{j+1}}|z(t_j)|^{2p}\Big] \\
&\quad + (2\sqrt{32}p)^{2p}(4e^{\varepsilon\Delta})^{2p-1}\mathbb{E}\Big[\Big(\sum_{l=(k-N)\vee 0}^{k} e^{\frac{\varepsilon t_{l+1}}{p}}|\sigma_l|^2\Delta\Big)^{p}\Big]
\end{aligned}
$$

$$+ (4\sqrt{32}p)^p(4e^{\varepsilon\Delta})^{p-1}\mathbb{E}\Big[\Big(\sum_{l=(k-N)\vee 0}^{k}(e^{\varepsilon t_{l+1}})^{\frac{2}{p}}|y(t_l)|^2|\sigma_l|^2\Delta\Big)^{\frac{p}{2}}\Big].$$

Then it follows from the Hölder inequality and the fact $N\Delta = \tau$ that

$$\begin{aligned}
&p\mathbb{E}\Big[\sup_{(k-N)\vee 0\le j\le k}\sum_{l=(k-N)\vee 0}^{j} e^{\varepsilon t_{l+1}}A_{\theta,\Delta}^{p-1}|z(t_l)|^{2(p-1)}\mathcal{M}_l\Big]\\
&\le e^{\varepsilon t_{(k-N)\vee 0}}\mathbb{E}[|z(t_{(k-N)\vee 0})|^{2p}] + \frac{1}{2}\mathbb{E}\Big[\sup_{(k-N)\vee 0\le j\le k} e^{\varepsilon t_{j+1}}|z(t_{j+1})|^{2p}\Big]\\
&\quad + K(N+1)^{p-1}\Delta^p\sum_{l=(k-N)\vee 0}^{k} e^{\varepsilon t_{l+1}}\mathbb{E}[|\sigma_l|^{2p}]\\
&\quad + K(N+1)^{\frac{p}{2}-1}\Delta^{\frac{p}{2}}\sum_{l=(k-N)\vee 0}^{k} e^{\varepsilon t_{l+1}}\mathbb{E}[|y(t_l)|^p|\sigma_l|^p]\\
&\le e^{\varepsilon t_{(k-N)\vee 0}}\mathbb{E}[|z(t_{(k-N)\vee 0})|^{2p}] + \frac{1}{2}\mathbb{E}\Big[\sup_{(k-N)\vee 0\le j\le k} e^{\varepsilon t_{j+1}}|z(t_{j+1})|^{2p}\Big]\\
&\quad + K(\tau+1)^{p-1}\Delta\sum_{l=(k-N)\vee 0}^{k} e^{\varepsilon t_{l+1}}\mathbb{E}[|\sigma_l|^{2p}]\\
&\quad + K(\tau+1)^{\frac{p}{2}-1}\Delta\sum_{l=(k-N)\vee 0}^{k} e^{\varepsilon t_{l+1}}\mathbb{E}[|y(t_l)|^{2p}].
\end{aligned}$$

According to Assumption 2.1 and Lemmas 2.6 and 2.7, we deduce

$$\begin{aligned}
&p\mathbb{E}\Big[\sup_{(k-N)\vee 0\le j\le k}\sum_{l=(k-N)\vee 0}^{j} e^{\varepsilon t_{l+1}}A_{\theta,\Delta}^{p-1}|z(t_l)|^{2(p-1)}\mathcal{M}_l\Big]\\
&\le K(1+\|\xi\|^{2p(\beta+1)})e^{\varepsilon t_{k+1}} + \frac{1}{2}\mathbb{E}\Big[\sup_{(k-N)\vee 0\le j\le k} e^{\varepsilon t_{j+1}}|z(t_{j+1})|^{2p}\Big]\\
&\quad + K\Delta\sum_{l=(k-N)\vee 0}^{k} e^{\varepsilon t_{l+1}}\mathbb{E}\Big[|y(t_l)|^{2p} + \int_{-\tau}^{0}|y_{t_l}(r)|^{2p}\mathrm{d}v_1(r)\Big]\\
&\le K(1+\|\xi\|^{2p(\beta+1)})e^{\varepsilon t_{k+1}} + \frac{1}{2}\mathbb{E}\Big[\sup_{(k-N)\vee 0\le j\le k} e^{\varepsilon t_{j+1}}|z(t_{j+1})|^{2p}\Big]
\end{aligned}$$

$$+ K \sup_{j \geq -N} \mathbb{E}[|y(t_j)|^{2p}] \sum_{l=(k-N)\vee 0}^{k} e^{\varepsilon t_{l+1}} \Delta$$

$$\leq K_{\varepsilon}(1 + \|\xi\|^{2p(\beta+1)}) e^{\varepsilon t_{k+2}} + \frac{1}{2} \mathbb{E}\Big[\sup_{(k-N)\vee 0 \leq j \leq k} e^{\varepsilon t_{j+1}} |z(t_{j+1})|^{2p} \Big]. \tag{2.64}$$

Inserting (2.63) and (2.64) into (2.62) yields

$$\mathbb{E}\Big[\sup_{(k-N)\vee 0 \leq j \leq k} e^{\varepsilon t_{j+1}} |z(t_{j+1})|^{2p} \Big] \leq K_{\varepsilon_1, \varepsilon}(1 + \|\xi\|^{2p(\beta+1)}) e^{\varepsilon t_{k+2}}.$$

This implies that for any $\Delta \in (0, \Delta^*]$,

$$\begin{aligned}
\mathbb{E}\Big[\sup_{(k-N)\vee 0 \leq j \leq k} |z(t_{j+1})|^{2p} \Big] &\leq e^{-\varepsilon t_{(k-N)\vee 0+1}} K_{\varepsilon_1, \varepsilon}(1 + \|\xi\|^{2p(\beta+1)}) e^{\varepsilon t_{k+2}} \\
&\leq K(1 + \|\xi\|^{2p(\beta+1)}).
\end{aligned}$$

Case 2: $\Delta \in [\Delta^*, 1]$. We reestimate $\tilde{I}_{l,i}$ in (2.59). Similar arguments to Case 1, we deduce

$$\begin{aligned}
\tilde{I}_{l,1} \leq\ & 3\kappa_1 A_{\theta,\Delta}^{p} |z(t_l)|^{2p} + \frac{1}{\kappa_1^{p-1}} \Big((R_1 \Delta)^p + (2a_4 \Delta)^p \int_{-\tau}^{0} |y_{t_l}(r)|^{2p} \mathrm{d}\nu_2(r) \\
&+ |\sigma_l \delta W_l|^{2p} \Big) + A_{\theta,\Delta}^{p-1} |z(t_l)|^{2(p-1)} \mathcal{M}_l
\end{aligned}$$

and

$$\begin{aligned}
\tilde{I}_{l,i} \leq\ & 5\kappa_1 A_{\theta,\Delta}^{p} |z(t_l)|^{2p} + \frac{5^p}{\kappa_1^{(p-i)/i}} \Big((R_1 \Delta)^p + (2a_4 \Delta)^p \int_{-\tau}^{0} |y_{t_l}(r)|^{2p} \mathrm{d}\nu_2(r) \\
&+ |\sigma_l \delta W_l|^{2p} + (4|y(t_l)||\sigma_l \delta W_l|)^p \Big) + \frac{(10|\sigma_l \delta W_l|)^{2p}}{\kappa_1^{(2p-i)/i}} \quad \forall i \in \{2, \ldots, p\},
\end{aligned}$$

where κ_1 is given in the proof of Lemma 2.6. Hence, for any $\Delta \in [\Delta^*, 1]$,

$$\begin{aligned}
& e^{\kappa t_{k+1}} |z(t_{k+1})|^{2p} \\
&\leq |z(0)|^{2p} + \Big((1 + 3p\kappa_1 + 5\kappa_1 \sum_{i=2}^{p} C_p^i) A_{\theta,\Delta}^{p} e^{\kappa \Delta} - 1 \Big) \sum_{l=0}^{k} e^{\kappa t_l} |z(t_l)|^{2p} \\
&\quad + \sum_{l=0}^{k} e^{\kappa t_{l+1}} \tilde{J}_{l,2} + p A_{\theta,\Delta}^{p-1} \sum_{l=0}^{k} e^{\kappa t_{l+1}} |z(t_l)|^{2(p-1)} \mathcal{M}_l,
\end{aligned} \tag{2.65}$$

where $\kappa > 0$ is given in the proof of Lemma 2.6, and

$$\tilde{J}_{l,2} := \Delta^p\Big(\frac{p}{\kappa_1^{p-1}} + \sum_{i=2}^{p}\frac{5^p C_p^i}{\kappa_1^{(p-i)/i}}\Big)\Big(R_1^p + (2a_4)^p\int_{-\tau}^{0}|y_{t_l}(r)|^{2p}\mathrm{d}\nu_2(r)\Big)$$
$$+ \frac{p|\sigma_l\delta W_l|^{2p}}{\kappa_1^{p-1}} + \sum_{i=2}^{p}\frac{5^p C_p^i}{\kappa_1^{(p-i)/i}}\big(|\sigma_l\delta W_l|^{2p} + (4|y(t_l)||\sigma_l\delta W_l|)^p\big)$$
$$+ \sum_{i=2}^{p}\frac{C_p^i(10|\sigma_l\delta W_l|)^{2p}}{\kappa_1^{(2p-i)/i}}.$$

Similar to (2.63) and (2.64), we derive

$$\mathbb{E}\Big[\sum_{l=0}^{k} e^{\kappa t_{l+1}}\tilde{J}_{l,2}\Big] \le K_{\kappa,\kappa_1}(1 + \|\xi\|^{2p(\beta+1)})e^{\kappa t_{k+2}},$$

$$p\mathbb{E}\Big[\sup_{(k-N)\vee 0<j<k}\sum_{l=(k-N)\vee 0}^{j} A_{\theta,\Delta}^{p-1}e^{\kappa t_{l+1}}|z(t_l)|^{2(p-1)}\mathcal{M}_l\Big]$$
$$\le K_\kappa(1 + \|\xi\|^{2p(\beta+1)})e^{\kappa t_{k+2}}$$
$$+ \frac{1}{2}\mathbb{E}\Big[\sup_{(k-N)\vee 0\le j\le k} e^{\kappa t_{j+1}}|z(t_{j+1})|^{2p}\Big].$$

Plugging these into (2.65), and combining (2.54), we have

$$\mathbb{E}\Big[\sup_{(k-N)\vee 0\le j\le k} e^{\kappa t_{j+1}}|z(t_{j+1})|^{2p}\Big]$$
$$\le |z(0)|^{2p} + \big(1 + 2^p 6\kappa_1)A_{\theta,\Delta^*}^p e^{\kappa\Delta} - 1\big)\sum_{l=0}^{k} e^{\kappa t_l}\mathbb{E}[|z(t_l)|^{2p}]$$
$$+ K_{\kappa,\kappa_1}(1 + \|\xi\|^{2p(\beta+1)})e^{\kappa t_{k+2}} + \frac{1}{2}\mathbb{E}\Big[\sup_{(k-N)\vee 0\le j\le k} e^{\kappa t_{j+1}}|z(t_{j+1})|^{2p}\Big]$$
$$\le K_{\kappa,\kappa_1}(1 + \|\xi\|^{2p(\beta+1)})e^{\kappa t_{k+2}} + \frac{1}{2}\mathbb{E}\Big[\sup_{(k-N)\vee 0\le j\le k} e^{\kappa t_{j+1}}|z(t_{j+1})|^{2p}\Big].$$

This implies that for any $\Delta \in [\Delta^*, 1]$,

$$\mathbb{E}\Big[\sup_{(k-N)\vee 0\le j\le k}|z(t_{j+1})|^{2p}\Big] \le K(1 + \|\xi\|^{2p(\beta+1)}).$$

Consequently,

$$\sup_{\Delta\in(0,1]}\sup_{k\geq 0}\mathbb{E}\Big[\sup_{k-N\leq j\leq k}|z(t_{j+1})|^{2p}\Big]\leq K(1+\|\xi\|^{2p(\beta+1)}). \tag{2.66}$$

In addition, by virtue of (2.41), Lemma 2.7, and (2.66), we obtain that for any $\Delta\in(0,1]$,

$$\begin{aligned}
&\mathbb{E}\Big[\sup_{(k-N)\vee 0\leq j\leq k}|y(t_j)|^{2p}\Big]\\
&\leq K\mathbb{E}\Big[\sup_{(k-N)\vee 0\leq j\leq k}|z(t_j)|^{2p}\Big]+K\Delta\\
&\quad +K\Delta\mathbb{E}\Big[\sup_{(k-N)\vee 0\leq j\leq k}\int_{-\tau}^{0}|y_{t_j}(r)|^{2p}\mathrm{d}\nu_2(r)\Big]\\
&\leq K(1+\|\xi\|^{2p(\beta+1)})+K\Delta\mathbb{E}\Big[\sup_{(k-2N)\vee(-N)\leq j\leq k}|y(t_j)|^{2p}\Big]\\
&\leq K(1+\|\xi\|^{2p(\beta+1)})+K\Delta\cdot(2N)\sup_{k\geq -N}\mathbb{E}[|y(t_k)|^{2p}]\leq K(1+\|\xi\|^{2p(\beta+1)}),
\end{aligned}$$

where we used

$$\begin{aligned}
\mathbb{E}\Big[\sup_{(k-2N)\vee -N\leq j\leq k}|y(t_j)|^{2p}\Big]&\leq \mathbb{E}\Big[\sum_{j=(k-2N)\vee(-N)}^{k}|y(t_j)|^{2p}\Big]\\
&\leq 2N\sup_{k\geq -N}\mathbb{E}\Big[|y(t_k)|^{2p}\Big].
\end{aligned}$$

The proof is finished. □

The following lemma gives the pth moment estimate of $z_t^{\xi,\Delta}-y_t^{\xi,\Delta}$ on the infinite time horizon.

Lemma 2.9 *Let Assumptions 2.1–2.5 hold. Then for any $\Delta\in(0,1]$ and $p\geq 1$,*

$$\sup_{t\in\mathbb{R}_+}\mathbb{E}[\|z_t^{\xi,\Delta}-y_t^{\xi,\Delta}\|^p]\leq K(1+\|\xi\|^{p(1+\beta)^2})\Delta^{\frac{p}{2}}\quad\forall\xi\in C^d.$$

Proof For any $t\geq 0$ and $r\in[-\tau,0]$, there exist $k\in\mathbb{N}$ and $j\in\{-N,-N+1,\ldots,-1\}$ such that $t\in[t_k,t_{k+1})$ and $r\in[t_j,t_{j+1}]$. It is straightforward to see that

$$t+r\in[t_{k+j},t_{k+j+2}),\quad t_k+r\in[t_{k+j},t_{k+j+1}).$$

Recall $z_t^{\xi,\Delta}(r) = z^{\xi,\Delta}(t+r)$ and $y_t^{\xi,\Delta}(r) = y_{t_k}^{\xi,\Delta}(r) = y^{\xi,\Delta}(t_k + r)$. We first give estimates of $|z_t^{\xi,\Delta}(r) - y_t^{\xi,\Delta}(r)|$, whose proof is divided into four cases.

Case 1: $t + r \in [t_{k+j}, t_{k+j+1}) \in \mathbb{R}_+$. By (1.39) and (2.4), we derive

$$\begin{aligned}
z_t^{\xi,\Delta}(r) - y_t^{\xi,\Delta}(r) &= z^{\xi,\Delta}(t+r) - y_{t_k}^{\xi,\Delta}(r) \\
&= z^{\xi,\Delta}(t_{k+j}) + b(y_{t_{k+j}}^{\xi,\Delta})(t + r - t_{k+j}) \\
&\quad + \sigma(y_{t_{k+j}}^{\xi,\Delta})(W(t+r) - W(t_{k+j})) \\
&\quad - \Big(\frac{t_{j+1} - r}{\Delta} y^{\xi,\Delta}(t_{k+j}) + \frac{r - t_j}{\Delta} y^{\xi,\Delta}(t_{k+j+1})\Big) \\
&= \frac{t_{j+1} - r}{\Delta}\big(z^{\xi,\Delta}(t_{k+j}) - y^{\xi,\Delta}(t_{k+j})\big) \\
&\quad + \frac{r - t_j}{\Delta}\big(z^{\xi,\Delta}(t_{k+j}) - y^{\xi,\Delta}(t_{k+j+1})\big) \\
&\quad + b(y_{t_{k+j}}^{\xi,\Delta})(t + r - t_{k+j}) + \sigma(y_{t_{k+j}}^{\xi,\Delta})(W(t+r) - W(t_{k+j})).
\end{aligned}$$

Combining $z^{\xi,\Delta}(t_l) = y^{\xi,\Delta}(t_l) - \theta b(y_{t_l}^{\xi,\Delta})\Delta$ for all $l \in \mathbb{N}$ yields

$$\begin{aligned}
&z_t^{\xi,\Delta}(r) - y_t^{\xi,\Delta}(r) \\
&= -\frac{t_{j+1} - r}{\Delta}\theta b(y_{t_{k+j}}^{\xi,\Delta})\Delta - \frac{r - t_j}{\Delta}\theta b(y_{t_{k+j+1}}^{\xi,\Delta})\Delta + \frac{r - t_j}{\Delta}\big(z^{\xi,\Delta}(t_{k+j}) \\
&\quad - z^{\xi,\Delta}(t_{k+j+1})\big) + b(y_{t_{k+j}}^{\xi,\Delta})(t + r - t_{k+j}) + \sigma(y_{t_{k+j}}^{\xi,\Delta})(W(t+r) - W(t_{k+j})) \\
&= -\frac{t_{j+1} - r}{\Delta}\theta b(y_{t_{k+j}}^{\xi,\Delta})\Delta - \frac{r - t_j}{\Delta}\theta b(y_{t_{k+j+1}}^{\xi,\Delta})\Delta + b(y_{t_{k+j}}^{\xi,\Delta})(t - t_k) \\
&\quad + \sigma(y_{t_{k+j}}^{\xi,\Delta})\big(W(t+r) - W(t_{k+j}) - \frac{r - t_j}{\Delta}(W(t_{k+j+1}) - W(t_{k+j})\big).
\end{aligned}$$

This implies that for any $p \geq 1$,

$$\begin{aligned}
&|z_t^{\xi,\Delta}(r) - y_t^{\xi,\Delta}(r)|^p \\
&\quad \leq K\Delta^p\big(|b(y_{t_{k+j}}^{\xi,\Delta})|^p + |b(y_{t_{k+j+1}}^{\xi,\Delta})|^p\big) + K\big|\sigma(y_{t_{k+j}}^{\xi,\Delta})\big(W(t+r) - W(t_{k+j})\big)\big|^p \\
&\quad + K\big|\sigma(y_{t_{k+j}}^{\xi,\Delta})\big(W(t_{k+j+1}) - W(t_{k+j})\big)\big|^p.
\end{aligned}$$

Case 2: $t + r \in [t_{k+j}, t_{k+j+1}) \in [-\tau, 0)$. It is obvious that $y_t^{\xi,\Delta}(r) = y_{t_k}^{\xi,\Delta}(r)$, and

$$
\begin{aligned}
|z_t^{\xi,\Delta}(r) - y_t^{\xi,\Delta}(r)|^p &= \Big|\frac{t_{k+j+1} - (t+r)}{\Delta}\xi(t_{k+j}) + \frac{t + r - t_{k+j}}{\Delta}\xi(t_{k+j+1}) \\
&\quad - \frac{t_{j+1} - r}{\Delta}\xi(t_{k+j}) - \frac{r - t_j}{\Delta}\xi(t_{k+j+1})\Big|^p \\
&= (\frac{t - t_k}{\Delta})^p |\xi(t_{k+j+1}) - \xi(t_{k+j})|^p \le K\Delta^{\frac{p}{2}},
\end{aligned}
$$

where we used Assumption 2.5.

Case 3: $t + r \in [t_{k+j+1}, t_{k+j+2}) \in \mathbb{R}_+$. Similar to Case 1, we obtain

$$
\begin{aligned}
&|z_t^{\xi,\Delta}(r) - y_t^{\xi,\Delta}(r)|^p \\
&= \Big|z^{\xi,\Delta}(t_{k+j+1}) + b(y_{t_{k+j+1}}^{\xi,\Delta})(t + r - t_{k+j+1}) \\
&\quad + \sigma(y_{t_{k+j+1}}^{\xi,\Delta})\big(W(t+r) - W(t_{k+j+1})\big) \\
&\quad - \big(\frac{t_{j+1} - r}{\Delta}y^{\xi,\Delta}(t_{k+j}) + \frac{r - t_j}{\Delta}y^{\xi,\Delta}(t_{k+j+1})\big)\Big|^p \\
&\le K\Delta^p |b(y_{t_{k+j+1}}^{\xi,\Delta})|^p + K\Big|\sigma(y_{t_{k+j+1}}^{\xi,\Delta})\big(W(t+r) - W(t_{k+j+1})\big)\Big|^p \\
&\quad + K|z^{\xi,\Delta}(t_{k+j+1}) \\
&\quad - y^{\xi,\Delta}(t_{k+j+1})|^p + K|z^{\xi,\Delta}(t_{k+j+1}) - z^{\xi,\Delta}(t_{k+j})|^p \\
&\quad + K|z^{\xi,\Delta}(t_{k+j}) - y^{\xi,\Delta}(t_{k+j})|^p \\
&\le K\Delta^p\big(|b(y_{t_{k+j}}^{\xi,\Delta})|^p + |b(y_{t_{k+j+1}}^{\xi,\Delta})|^p\big) + K\Big|\sigma(y_{t_{k+j+1}}^{\xi,\Delta})\big(W(t+r) - W(t_{k+j+1})\big)\Big|^p \\
&\quad + K|z^{\xi,\Delta}(t_{k+j+1}) - z^{\xi,\Delta}(t_{k+j})|^p\big(\mathbf{1}_{\{t_{k+j+1}>0\}} + \mathbf{1}_{\{t_{k+j+1}=0\}}\big) \\
&\le K\Delta^p\big(|b(y_{t_{k+j}}^{\xi,\Delta})|^p + |b(y_{t_{k+j+1}}^{\xi,\Delta})|^p\big) + K\Big|\sigma(y_{t_{k+j+1}}^{\xi,\Delta})\big(W(t+r) - W(t_{k+j+1})\big)\Big|^p \\
&\quad + K\Big|\sigma(y_{t_{k+j}}^{\xi,\Delta})\big(W(t_{k+j+1}) - W(t_{k+j})\big)\Big|^p\mathbf{1}_{\{t_{k+j+1}>0\}} + K\Delta^{\frac{p}{2}},
\end{aligned}
$$

where we used Assumption 2.5.

Case 4: $t + r \in [t_{k+j+1}, t_{k+j+2}) \in [-\tau, 0)$. Similar to Case 2, it can be calculated that

$$
|z_t^{\xi,\Delta}(r) - y_t^{\xi,\Delta}(r)|^p \le K\Delta^{\frac{p}{2}}.
$$

Together with Cases 1–4, we have

$$\mathbb{E}\Big[\sup_{r\in[-\tau,0]}|z_t^{\xi,\Delta}(r)-y_t^{\xi,\Delta}(r)|^p\Big]$$
$$\le K(1+\|\xi\|^{p(1+\beta)})\Delta^{\frac{p}{2}}+K(I_1+I_2+I_3), \tag{2.67}$$

where

$$I_1:=\mathbb{E}\Big[\sup_{j\in\{-N,\dots,-1\}}\big(|b(y^{\xi,\Delta}_{t_{k+j}\vee 0})|^p\Delta^p+|b(y^{\xi,\Delta}_{t_{k+j+1}\vee 0})|^p\Delta^p\big)\Big],$$

$$I_2:=\mathbb{E}\Big[\sup_{j\in\{-N,\dots,-1\}}\sup_{r\in[t_j,t_{j+1}]}\Big(\big|\sigma(y^{\xi,\Delta}_{t_{k+j}\vee 0})\big(W((t+r)\vee 0)-W(t_{k+j}\vee 0)\big)\big|^p$$
$$\times\mathbf{1}_{\{t+r\in[t_{k+j},t_{k+j+1})\}}+\big|\sigma(y^{\xi,\Delta}_{t_{k+j+1}\vee 0})\big(W((t+r)\vee 0)-W(t_{k+j+1}\vee 0)\big)\big|^p$$
$$\times\mathbf{1}_{\{t+r\in[t_{k+j+1},t_{k+j+2})\}}\Big)\Big],$$

$$I_3:=\mathbb{E}\Big[\sup_{j\in\{-N,\dots,-1\}}\big|\sigma(y^{\xi,\Delta}_{t_{k+j}\vee 0})\big(W(t_{k+j+1}\vee 0)-W(t_{k+j}\vee 0)\big)\big|^p\Big].$$

It follows from Assumption 2.4 that

$$I_1\le\ 2\Delta^p\mathbb{E}\Big[\sup_{j\in\{-N,\dots,0\}}|b(y^{\xi,\Delta}_{t_{k+j}\vee 0})|^p\Big]$$
$$\le K\Delta^p\mathbb{E}\Big[\sup_{j\in\{-N,\dots,0\}}(1+\|y^{\xi,\Delta}_{t_{k+j}\vee 0}\|^{p(\beta+1)})\Big].$$

Since

$$\sup_{j\in\{-N,\dots,0\}}\|y^{\xi,\Delta}_{t_{k+j}}\|\le\|y^{\xi,\Delta}_{t_{k-N}}\|+\|y^{\xi,\Delta}_{t_k}\|,$$

by (2.58), we have

$$I_1\le K\Delta^p\mathbb{E}[1+\|\xi\|^{p(\beta+1)}+\|y^{\xi,\Delta}_{t_{k-N}}\|^{p(1+\beta)}+\|y^{\xi,\Delta}_{t_k}\|^{p(1+\beta)}]$$
$$\le K(1+\|\xi\|^{p(1+\beta)^2})\Delta^p. \tag{2.68}$$

Applying the Hölder inequality yields

$$I_2\le\Big(\mathbb{E}\Big[\sup_{j\in\{-N,\dots,-1\}}|\sigma(y^{\xi,\Delta}_{t_{k+j}\vee 0})|^{2p}\Big]\Big)^{\frac12}\Big(\mathbb{E}\Big[\sup_{s_1-s_2\in[0,\Delta]}\big|W(s_1)-W(s_2)\big|^{2p}\Big]\Big)^{\frac12}$$
$$+\Big(\mathbb{E}\Big[\sup_{j\in\{-N,\dots,-1\}}|\sigma(y^{\xi,\Delta}_{t_{k+j+1}\vee 0})|^{2p}\Big]\Big)^{\frac12}\Big(\mathbb{E}\Big[\sup_{s_1-s_2\in[0,\Delta]}\big|W(s_1)-W(s_2)\big|^{2p}\Big]\Big)^{\frac12}.$$

Similar to (2.68), by virtue of Assumption 2.1 and (2.58), we arrive at

$$I_2 \le K(1+\|\xi\|^{p(\beta+1)})\Big(\mathbb{E}\Big[\sup_{s_1-s_2\in[0,\Delta]}\big|W(s_1)-W(s_2)\big|^{2p}\Big]\Big)^{\frac{1}{2}}.$$

Applying the Burkholder–Davis–Gundy inequality, we obtain

$$I_2 \le K(1+\|\xi\|^{p(\beta+1)})\Delta^{\frac{p}{2}}. \tag{2.69}$$

Similarly, we have

$$I_3 \le K(1+\|\xi\|^{p(\beta+1)})\Delta^{\frac{p}{2}}. \tag{2.70}$$

The desired argument follows from (2.67)–(2.70). The proof is completed. □

2.2 Attractiveness and Stability of Numerical Solution

In this subsection, we demonstrate that the numerical solution obtained via the θ-EM method can inherit simultaneously the exponential attractiveness and exponential stability of the exact solutions. The inheritance of these longtime asymptotic behaviors ensures that the numerical method can effectively characterize the qualitative dynamics of the underlying stochastic system.

Proposition 2.10 *Let Assumptions 2.1 and 2.2 hold. Then for any* $\Delta \in (0,1]$ *and* $k \in \mathbb{N}$,

$$\mathbb{E}[\|y^{\xi,\Delta}_{t_k} - y^{\eta,\Delta}_{t_k}\|^2] \le K\Big(|b(\xi)-b(\eta)|^2 + \|\xi-\eta\|^2\Big)e^{-\lambda_0 t_k} \quad \forall \xi,\eta \in C^d,$$

where $\lambda_0 \in (0, \frac{(2\theta-1)c_\lambda}{\theta^2(1+c_\lambda)} \wedge \lambda)$, *and positive constants* λ *and* c_λ *are defined by* (2.2).

Proof Let $\xi,\eta \in C^d$, $\Delta \in (0,1]$, and $k \in \mathbb{N}$. Similar to the proof in Lemma 2.3, we deduce that for $\lambda_0 \in (0,\lambda]$,

$$\begin{aligned}
&e^{\lambda_0 t_{k+1}}|z^{\xi,\Delta}(t_{k+1}) - z^{\eta,\Delta}(t_{k+1})|^2\\
&= \sum_{i=0}^{k}\Big(e^{\lambda_0 t_{i+1}}|z^{\xi,\Delta}(t_{i+1}) - z^{\eta,\Delta}(t_{i+1})|^2 - e^{\lambda_0 t_i}|z^{\xi,\Delta}(t_i) - z^{\eta,\Delta}(t_i)|^2\Big)\\
&\quad + |z^{\xi,\Delta}(0) - z^{\eta,\Delta}(0)|^2\\
&\le \Big(\frac{(1-\theta)^2}{\theta^2} + \frac{2\theta-1}{\theta^2(1+c_\lambda\Delta)} - e^{-\lambda_0\Delta}\Big)\sum_{i=0}^{k} e^{\lambda_0 t_{i+1}}|z^{\xi,\Delta}(t_i) - z^{\eta,\Delta}(t_i)|^2
\end{aligned}$$

$$
\begin{aligned}
&+ |z^{\xi,\Delta}(0) - z^{\eta,\Delta}(0)|^2 + \sum_{i=0}^{k} e^{\lambda_0 t_{i+1}} \Big| \Big(\sigma(y_{t_i}^{\xi,\Delta}) - \sigma(y_{t_i}^{\eta,\Delta})\Big)\delta W_i \Big|^2 \\
&- (2a_1 - c_\lambda)\Delta \sum_{i=0}^{k} e^{\lambda_0 t_{i+1}} |y^{\xi,\Delta}(t_i) - y^{\eta,\Delta}(t_i)|^2 \\
&+ 2a_2 \Delta \sum_{i=0}^{k} e^{\lambda_0 t_{i+1}} \int_{-\tau}^{0} |y_{t_i}^{\xi,\Delta}(r) - y_{t_i}^{\eta,\Delta}(r)|^2 \mathrm{d}\nu_2(r) + \sum_{i=0}^{k} e^{\lambda_0 t_{i+1}} \hat{\mathcal{M}}_i \\
\le\ & |z^{\xi,\Delta}(0) - z^{\eta,\Delta}(0)|^2 + 2a_2 e^{\lambda_0 \tau} \tau \|\xi - \eta\|^2 \\
&+ \Big(\frac{(1-\theta)^2}{\theta^2} + \frac{2\theta - 1}{\theta^2(1 + c_\lambda \Delta)} - e^{-\lambda_0 \Delta}\Big) \sum_{i=0}^{k} e^{\lambda_0 t_{i+1}} |z^{\xi,\Delta}(t_i) - z^{\eta,\Delta}(t_i)|^2 \\
&+ \sum_{i=0}^{k} e^{\lambda_0 t_{i+1}} \Big| \Big(\sigma(y_{t_i}^{\xi,\Delta}) - \sigma(y_{t_i}^{\eta,\Delta})\Big)\delta W_i \Big|^2 + \sum_{i=0}^{k} e^{\lambda_0 t_{i+1}} \hat{\mathcal{M}}_i \\
&- (2a_1 - c_\lambda - 2a_2 e^{\lambda_0 \tau})\Delta \sum_{i=0}^{k} e^{\lambda_0 t_{i+1}} |y^{\xi,\Delta}(t_i) - y^{\eta,\Delta}(t_i)|^2,
\end{aligned} \tag{2.71}
$$

where

$$
\begin{aligned}
\hat{\mathcal{M}}_k := & 2\Big\langle z^{\xi,\Delta}(t_k) - z^{\eta,\Delta}(t_k), \Big(\sigma(y_{t_k}^{\xi,\Delta}) - \sigma(y_{t_k}^{\eta,\Delta})\Big)\delta W_k \Big\rangle \\
& + 2\Big\langle b(y_{t_k}^{\xi,\Delta}) - b(y_{t_k}^{\eta,\Delta}), \Big(\sigma(y_{t_k}^{\xi,\Delta}) - \sigma(y_{t_k}^{\eta,\Delta})\Big)\delta W_k \Big\rangle.
\end{aligned}
$$

It follows from Assumptions 2.1 and 2.2 that

$$
\begin{aligned}
& e^{\lambda_0 t_{k+1}} \mathbb{E}[|z^{\xi,\Delta}(t_{k+1}) - z^{\eta,\Delta}(t_{k+1})|^2] \\
&\le |z^{\xi,\Delta}(0) - z^{\eta,\Delta}(0)|^2 + (L + 2a_2)\tau e^{\lambda_0 \tau} \|\xi - \eta\|^2 \\
&\quad + \Big(\frac{(1-\theta)^2}{\theta^2} + \frac{2\theta - 1}{\theta^2(1 + c_\lambda \Delta)} - e^{-\lambda_0 \Delta}\Big) \sum_{i=0}^{k} e^{\lambda_0 t_{i+1}} \mathbb{E}[|z^{\xi,\Delta}(t_i) - z^{\eta,\Delta}(t_i)|^2] \\
&\quad - (2a_1 - L - c_\lambda - (L + 2a_2)e^{\lambda_0 \tau})\Delta \sum_{i=0}^{k} e^{\lambda_0 t_{i+1}} \mathbb{E}[|y^{\xi,\Delta}(t_i) - y^{\eta,\Delta}(t_i)|^2].
\end{aligned}
$$

It follows from (2.2) that

$$
2a_1 - L - c_\lambda - (L + 2a_2)e^{\lambda_0 \tau} > 2a_1 - L - c_\lambda - (L + 2a_2)e^{\lambda \tau} = \lambda > 0.
$$

Similar to (2.28), we obtain

$$\frac{(1-\theta)^2}{\theta^2}+\frac{2\theta-1}{\theta^2(1+c_\lambda\Delta)}-e^{-\lambda_0\Delta}<(\lambda_0-\frac{(2\theta-1)c_\lambda}{\theta^2(1+c_\lambda)})\Delta.$$

Letting $\lambda_0\in(0,\frac{(2\theta-1)c_\lambda}{\theta^2(1+c_\lambda)}\wedge\lambda]$, we have

$$\begin{aligned}&\mathbb{E}[|z^{\xi,\Delta}(t_k)-z^{\eta,\Delta}(t_k)|^2]\\&\le(1\vee((L+2a_2)\tau e^{\lambda_0\tau}))\big(|z^{\xi,\Delta}(0)-z^{\eta,\Delta}(0)|^2+\|\xi-\eta\|^2\big)e^{-\lambda_0 t_k}.\end{aligned}\tag{2.72}$$

Next, we shall show that $\{y^{\cdot,\Delta}(t_k)\}_{k\in\mathbb{N}}$ has a similar property by an induction argument. In fact, it follows from $z^{\xi,\Delta}(t_k)=y^{\xi,\Delta}(t_k)-\theta b(y^{\xi,\Delta}_{t_k})\Delta$ and Assumption 2.2 that

$$\begin{aligned}&(1+2a_1\theta\Delta)|y^{\xi,\Delta}(t_k)-y^{\eta,\Delta}(t_k)|^2\\&\le|z^{\xi,\Delta}(t_k)-z^{\eta,\Delta}(t_k)|^2+2a_2\theta\Delta\int_{-\tau}^0|y^{\xi,\Delta}_{t_k}(r)-y^{\eta,\Delta}_{t_k}(r)|^2\mathrm{d}\nu_2(r).\end{aligned}\tag{2.73}$$

By virtue of (1.39), we arrive at

$$\begin{aligned}&(1+2a_1\theta\Delta)\mathbb{E}[|y^{\xi,\Delta}(t_k)-y^{\eta,\Delta}(t_k)|^2]\\&\le\mathbb{E}[|z^{\xi,\Delta}(t_k)-z^{\eta,\Delta}(t_k)|^2]+2a_2\theta\Delta\int_{-\Delta}^0\frac{r+\Delta}{\Delta}\mathrm{d}\nu_2(r)\\&\quad\times\mathbb{E}[|y^{\xi,\Delta}(t_k)-y^{\eta,\Delta}(t_k)|^2]\\&\quad+2a_2\theta\Delta\Big(\nu_2([-\tau,-\Delta])+\int_{-\Delta}^0\frac{-r}{\Delta}\mathrm{d}\nu_2(r)\Big)\\&\quad\times\sup_{k-N\le l\le k-1}\mathbb{E}[|y^{\xi,\Delta}(t_l)-y^{\eta,\Delta}(t_l)|^2].\end{aligned}\tag{2.74}$$

For $k=1$, according to $a_2e^{\lambda_0\Delta}<a_2e^{\lambda_0\tau}<a_1$, (2.72), and (2.73), we derive

$$\begin{aligned}&\Big(1+2a_1\theta\Delta\big(1-\int_{-\Delta}^0\frac{r+\Delta}{\Delta}\mathrm{d}\nu_2(r)\big)\Big)\mathbb{E}[|y^{\xi,\Delta}(t_1)-y^{\eta,\Delta}(t_1)|^2]\\&\le\mathbb{E}[|z^{\xi,\Delta}(t_1)-z^{\eta,\Delta}(t_1)|^2]+2a_2\theta\Delta\big(\nu_2([-\tau,-\Delta])\\&\quad+\int_{-\Delta}^0\frac{-r}{\Delta}\mathrm{d}\nu_2(r)\big)\|\xi-\eta\|^2\end{aligned}$$

$$\leq \Big(1 + 2a_1\theta\Delta\big(1 - \int_{-\Delta}^{0} \frac{r+\Delta}{\Delta} \mathrm{d}\nu_2(r)\big)\Big)\Big(1 \vee \big((L + 2a_2)\tau e^{\lambda_0 t_1}\big)\Big)$$
$$\times \big(|z^{\xi,\Delta}(0) - z^{\eta,\Delta}(0)|^2 + \|\xi - \eta\|^2\big)e^{-\lambda_0 t_1},$$

where we used

$$\nu_2([-\tau, -\Delta]) + \int_{-\Delta}^{0} \frac{-r}{\Delta} \mathrm{d}\nu_2(r) = 1 - \int_{-\Delta}^{0} \frac{r+\Delta}{\Delta} \mathrm{d}\nu_2(r).$$

It is clear that

$$\mathbb{E}[|y^{\xi,\Delta}(t_1) - y^{\eta,\Delta}(t_1)|^2]$$
$$\leq (1 \vee ((L + 2a_2)\tau e^{\lambda_0 \tau}))\big(|z^{\xi,\Delta}(0) - z^{\eta,\Delta}(0)|^2 + \|\xi - \eta\|^2\big)e^{-\lambda_0 t_1}. \tag{2.75}$$

Assume that for some integer $k > 1$ and $l \in \{1, \ldots, k\}$,

$$\mathbb{E}[|y^{\xi,\Delta}(t_l) - y^{\eta,\Delta}(t_l)|^2]$$
$$\leq (1 \vee ((L + 2a_2)\tau e^{\lambda_0 \tau}))\big(|z^{\xi,\Delta}(0) - z^{\eta,\Delta}(0)|^2 + \|\xi - \eta\|^2\big)e^{-\lambda_0 t_l}. \tag{2.76}$$

Then for $k + 1$, taking into account (2.72), (2.74), and (2.76), we deduce

$$\Big(1 + 2a_1\theta\Delta\big(1 - \int_{-\Delta}^{0} \frac{r+\Delta}{\Delta} \mathrm{d}\nu_2(r)\big)\Big)\mathbb{E}[|y^{\xi,\Delta}(t_{k+1}) - y^{\eta,\Delta}(t_{k+1})|^2]$$
$$\leq \mathbb{E}[|z^{\xi,\Delta}(t_{k+1}) - z^{\eta,\Delta}(t_{k+1})|^2] + 2a_2\theta\Delta\big(1 - \int_{-\Delta}^{0} \frac{r+\Delta}{\Delta} \mathrm{d}\nu_2(r)\big)$$
$$\times \sup_{k-N+1 \leq l \leq k} \mathbb{E}[|y^{\xi,\Delta}(t_l) - y^{\eta,\Delta}(t_l)|^2]$$
$$\leq \big(1 \vee ((L + 2a_2)\tau e^{\lambda_0 \tau})\big)\Big(1 + 2a_2\theta e^{\lambda_0 \tau}\Delta\big(1 - \int_{-\Delta}^{0} \frac{r+\Delta}{\Delta} \mathrm{d}\nu_2(r)\big)\Big)$$
$$\times \big(|z^{\xi,\Delta}(0) - z^{\eta,\Delta}(0)|^2 + \|\xi - \eta\|^2\big)e^{-\lambda_0 t_{k+1}},$$

which implies

$$\mathbb{E}[|y^{\xi,\Delta}(t_{k+1}) - y^{\eta,\Delta}(t_{k+1})|^2]$$
$$\leq (1 \vee ((L + 2a_2)\tau e^{\lambda_0 \tau}))\big(|z^{\xi,\Delta}(0) - z^{\eta,\Delta}(0)|^2 + \|\xi - \eta\|^2\big)e^{-\lambda_0 t_{k+1}}. \tag{2.77}$$

Hence, we conclude from (2.75)–(2.77) that for any $k \in \mathbb{N}$,

$$\begin{aligned}&\mathbb{E}[|y^{\xi,\Delta}(t_k) - y^{\eta,\Delta}(t_k)|^2]\\&\quad\le (1 \vee ((L + 2a_2)\tau e^{\lambda_0\tau}))\big(|z^{\xi,\Delta}(0) - z^{\eta,\Delta}(0)|^2 + \|\xi - \eta\|^2\big)e^{-\lambda_0 t_k}.\end{aligned} \tag{2.78}$$

Furthermore, by similar arguments to those in Proposition 2.4, it follows from (2.71) that

$$\begin{aligned}&\mathbb{E}\Big[\sup_{(k-N)\vee 0\le i\le k} e^{\lambda_0 t_{i+1}}|z^{\xi,\Delta}(t_{i+1}) - z^{\eta,\Delta}(t_{i+1})|^2\Big]\\&\le |z^{\xi,\Delta}(0) - z^{\eta,\Delta}(0)|^2 + 2a_2 e^{\lambda_0\tau}\tau\|\xi - \eta\|^2\\&\quad + \Delta\mathbb{E}\Big[\sum_{i=0}^{k} e^{\lambda_0 t_{i+1}}\Big|\sigma(y_{t_i}^{\xi,\Delta}) - \sigma(y_{t_i}^{\eta,\Delta})\Big|^2\Big]\\&\quad + \mathbb{E}\Big[\sup_{(k-N)\vee 0\le i\le k}\sum_{l=(k-N)\vee 0}^{i} e^{\lambda_0 t_{l+1}}\hat{\mathcal{M}}_l\Big].\end{aligned} \tag{2.79}$$

Making use of Assumption 2.1, we derive

$$\begin{aligned}&\Delta\mathbb{E}\Big[\sum_{i=0}^{k} e^{\lambda_0 t_{i+1}}\Big|\sigma(y_{t_i}^{\xi,\Delta}) - \sigma(y_{t_i}^{\eta,\Delta})\Big|^2\Big]\\&\quad\le \Delta L\mathbb{E}\Big[\sum_{i=0}^{k} e^{\lambda_0 t_{i+1}}\Big(|y^{\xi,\Delta}(t_i) - y^{\eta,\Delta}(t_i)|^2 + \int_{-\tau}^{0}|y_{t_i}^{\xi,\Delta}(r) - y_{t_i}^{\eta,\Delta}(r)|^2\mathrm{d}\nu_1(r)\Big)\Big]\\&\quad\le K\Delta\sum_{i=0}^{k} e^{\lambda_0 t_{i+1}}\mathbb{E}[|y^{\xi,\Delta}(t_i) - y^{\eta,\Delta}(t_i)|^2] + K\|\xi - \eta\|^2.\end{aligned} \tag{2.80}$$

Similar to (2.32), one has

$$\begin{aligned}&\mathbb{E}\Big[\sup_{(k-N)\vee 0\le i\le k}\sum_{l=(k-N)\vee 0}^{i} e^{\lambda_0 t_{l+1}}\hat{\mathcal{M}}_l\Big]\\&\quad\le \frac{1}{4}\mathbb{E}\Big[\sup_{(k-N)\vee 0\le i\le k} e^{\lambda_0 t_i}|z^{\xi,\Delta}(t_i) - z^{\eta,\Delta}(t_i)|^2\Big]\\&\qquad + \frac{1}{4}\mathbb{E}\Big[\sup_{(k-N)\vee 0\le i\le k} e^{\lambda_0 t_i}|y^{\xi,\Delta}(t_i) - y^{\eta,\Delta}(t_i)|^2\Big]\end{aligned}$$

$$
\begin{aligned}
&+ K\Delta\mathbb{E}\Big[\sum_{i=(k-N)\vee 0}^{k} e^{\lambda_0 t_{i+1}}\Big|\sigma(y_{t_i}^{\xi,\Delta}) - \sigma(y_{t_i}^{\eta,\Delta})\Big|^2\Big] \\
&\le \frac{1}{4}\mathbb{E}\Big[\sup_{(k-N)\vee 0\le i\le k} e^{\lambda_0 t_i}|z^{\xi,\Delta}(t_i) - z^{\eta,\Delta}(t_i)|^2\Big] \\
&\quad + \frac{1}{4}\mathbb{E}\Big[\sup_{(k-N)\vee 0\le i\le k} e^{\lambda_0 t_i}|y^{\xi,\Delta}(t_i) - y^{\eta,\Delta}(t_i)|^2\Big] \\
&\quad + K\Delta\sum_{i=0}^{k} e^{\lambda_0 t_{i+1}}\mathbb{E}[|y^{\xi,\Delta}(t_i) - y^{\eta,\Delta}(t_i)|^2] + K\|\xi - \eta\|^2.
\end{aligned}
\tag{2.81}
$$

Inserting (2.80) and (2.81) into (2.79) derives

$$
\begin{aligned}
&\mathbb{E}\Big[\sup_{(k-N)\vee 0\le i\le k} e^{\lambda_0 t_{i+1}}|z^{\xi,\Delta}(t_{i+1}) - z^{\eta,\Delta}(t_{i+1})|^2\Big] \\
&\le K\big(|z^{\xi,\Delta}(0) - z^{\eta,\Delta}(0)|^2 + \|\xi - \eta\|^2\big) \\
&\quad + K\Delta\sum_{i=0}^{k} e^{\lambda_0 t_{i+1}}\mathbb{E}[|y^{\xi,\Delta}(t_i) - y^{\eta,\Delta}(t_i)|^2] \\
&\quad + \frac{1}{4}\mathbb{E}\Big[\sup_{(k-N)\vee 0\le i\le k} e^{\lambda_0 t_i}|z^{\xi,\Delta}(t_i) - z^{\eta,\Delta}(t_i)|^2\Big] \\
&\quad + \frac{1}{4}\mathbb{E}\Big[\sup_{(k-N)\vee 0\le i\le k} e^{\lambda_0 t_i}|y^{\xi,\Delta}(t_i) - y^{\eta,\Delta}(t_i)|^2\Big].
\end{aligned}
\tag{2.82}
$$

Similar to (2.33), we arrive at

$$
\begin{aligned}
&\mathbb{E}\Big[\sup_{(k-N)\vee 0\le i\le k} e^{\lambda_0 t_i}|y^{\xi,\Delta}(t_i) - y^{\eta,\Delta}(t_i)|^2\Big] \\
&\le \mathbb{E}\Big[\sup_{(k-N)\vee 0\le i\le k} e^{\lambda_0 t_i}|z^{\xi,\Delta}(t_i) - z^{\eta,\Delta}(t_i)|^2\Big] \\
&\quad + 2a_2\theta\Delta e^{\lambda_0\tau}\sum_{i=(k-2N)\vee(-N)}^{k} e^{\lambda_0 t_i}\mathbb{E}\Big[|y^{\xi,\Delta}(t_i) - y^{\eta,\Delta}(t_i)|^2\Big].
\end{aligned}
$$

This, along with (2.82) implies that

$$
\begin{aligned}
&\mathbb{E}\Big[\sup_{(k-N)\vee 0\le i\le k} e^{\lambda_0 t_{i+1}}|z^{\xi,\Delta}(t_{i+1}) - z^{\eta,\Delta}(t_{i+1})|^2\Big] \\
&\le K\big(|z^{\xi,\Delta}(0) - z^{\eta,\Delta}(0)|^2 + \|\xi - \eta\|^2\big) + K\Delta\sum_{i=0}^{k} e^{\lambda_0 t_i}\mathbb{E}[|y^{\xi,\Delta}(t_i) - y^{\eta,\Delta}(t_i)|^2]
\end{aligned}
$$

$$+\frac{1}{2}\mathbb{E}\Big[\sup_{(k-N)\vee 0\le i\le k} e^{\lambda_0 t_{i+1}}|z^{\xi,\Delta}(t_{i+1})-z^{\eta,\Delta}(t_{i+1})|^2\Big]$$

$$+\frac{1}{2}\mathbb{E}\Big[e^{\lambda_0 t_{k-N}}|z^{\xi,\Delta}(t_{k-N})-z^{\eta,\Delta}(t_{k-N})|^2\Big].$$

Using (2.72) and (2.78), we obtain

$$\mathbb{E}\Big[\sup_{(k-N)\vee 0\le i\le k} e^{\lambda_0 t_{i+1}}|z^{\xi,\Delta}(t_{i+1})-z^{\eta,\Delta}(t_{i+1})|^2\Big]$$
$$\le K(|b(\xi)-b(\eta)|^2+\|\xi-\eta\|^2)(1+t_k),$$

which yields that for any $\varepsilon\in(0,\lambda_0)$,

$$\mathbb{E}\Big[\sup_{(k-N)\vee 0\le i\le k}|z^{\xi,\Delta}(t_{i+1})-z^{\eta,\Delta}(t_{i+1})|^2\Big]$$
$$\le K(|b(\xi)-b(\eta)|^2+\|\xi-\eta\|^2)e^{-(\lambda_0-\varepsilon)t_k}. \tag{2.83}$$

Without loss of generality, we redefine $\lambda_0\in(0,\frac{(2\theta-1)c_\lambda}{\theta^2(1+c_\lambda)}\wedge\lambda)$. Then, it follows from (2.83) that

$$\mathbb{E}\Big[\sup_{(k-N)\vee 0\le i\le k}|z^{\xi,\Delta}(t_{i+1})-z^{\eta,\Delta}(t_{i+1})|^2\Big]$$
$$\le K(|b(\xi)-b(\eta)|^2+\|\xi-\eta\|^2)e^{-\lambda_0 t_k}. \tag{2.84}$$

Using (2.73) leads to

$$(1+2a_1\theta\Delta)\mathbb{E}\Big[\sup_{(k-N)\vee 0\le i\le k}|y^{\xi,\Delta}(t_i)-y^{\eta,\Delta}(t_i)|^2\Big]$$
$$\le\mathbb{E}\Big[\sup_{(k-N)\vee 0\le i\le k}|z^{\xi,\Delta}(t_i)-z^{\eta,\Delta}(t_i)|^2\Big]$$
$$+2a_2\theta\Delta\mathbb{E}\Big[\sup_{(k-2N)\vee(-N)\le i\le k}|y^{\xi,\Delta}(t_i)-y^{\eta,\Delta}(t_i)|^2\Big]$$
$$\le\mathbb{E}\Big[\sup_{(k-N)\vee 0\le i\le k}|z^{\xi,\Delta}(t_i)-z^{\eta,\Delta}(t_i)|^2\Big]$$
$$+4a_2\theta\tau\sup_{(k-2N)\vee(-N)\le i\le k}\mathbb{E}[|y^{\xi,\Delta}(t_i)-y^{\eta,\Delta}(t_i)|^2]. \tag{2.85}$$

Then the desired argument follows from (2.78) and (2.84). We finish the proof. □

Numerical stability is an important characteristic to measure the reliability of numerical methods in longtime computation. Below, we present the exponential stability of the trivial solution of (1.1). We propose the following condition to ensure the existence of the trivial solution.

Assumption 2.6 *The drift and diffusion coefficients satisfy* $b(\mathbf{0}) = \sigma(\mathbf{0}) = 0$.

It follows from Assumption 2.6 that $x^{\mathbf{0}}(t) \equiv 0$, which implies $x_t^{\mathbf{0}} \equiv \mathbf{0}$. By virtue of (2.1), it is straightforward to obtain the exponential stability of the trivial functional solution of (1.1).

Proposition 2.11 *Let Assumptions 2.1, 2.2, and 2.6 hold. Then the trivial functional solution is mean-square exponentially stable and satisfies*

$$\mathbb{E}[\|x_t^{\xi}\|^2] \le K(1 + \|\xi\|^2)e^{-\lambda t}$$

for any initial data $\xi \in C^d$, *where the positive constant* λ *satisfies* (2.2).

Similarly, by virtue of Proposition 2.10, we can deduce that the θ-EM method inherits the mean-square exponential stability.

Theorem 2.12 *Let Assumptions 2.1, 2.2, and 2.6 hold. Then for any* $\Delta \in (0, 1]$, *the numerical functional solution satisfies*

$$\mathbb{E}[\|y_{t_k}^{\xi,\Delta}\|^2] \le K(1 + |b(\xi)|^2 + \|\xi\|^2)e^{-\lambda_0 t_k}$$

for any initial data $\xi \in C^d$, *where the constant* λ_0 *is defined in Proposition 2.10.*

Remark 2.1 In fact, if the coefficients b and σ are locally Lipschitz continuous, and Assumptions 2.1 and 2.2 are relaxed as

$$2\langle\phi(0), b(\phi)\rangle + |\sigma(\phi)|^2 \le -d_1|\phi(0)|^2 + d_2\int_{-\tau}^{0} |\phi(r)|^2 \mathrm{d}\nu(r),$$

we can also deduce the above stability results of the trivial functional solution and the inheritance of this property by the θ-EM method. Here, d_1 and d_2 are positive constants with $d_1 > d_2$, and ν is a probability measure on $[-\tau, 0]$. The proof is similar to that of Proposition 2.10.

2.3 Mean-Square Convergence Rate

In this subsection, we aim to establish the mean-square convergence rate of the θ-EM method on the infinite time horizon, under the one-sided Lipschitz condition on the drift coefficient (Assumption 2.2) and the globally Lipschitz condition on the diffusion coefficient (Assumption 2.1). The proof is based on the estimates of high-order moments of both the numerical solution and the exact solution, and the estimates between the functional solution and the continuous process of the split-step θ-EM solution (2.4).

Lemma 2.13 *Let Assumptions 2.1–2.5 hold. Then*

$$\sup_{t\geq 0}\mathbb{E}[\|x_t^{\xi}-z_t^{\xi,\Delta}\|^2]\leq K(1+\|\xi\|^{4\beta(1+\beta)^3})\Delta\quad \forall \xi\in C^d.$$

Proof It follows from (1.1) and (2.4) that

$$\begin{aligned}x^{\xi}(t)-z^{\xi,\Delta}(t)&=\theta b(\xi)\Delta+\int_0^t\Big(b(x_s^{\xi})-b(y_s^{\xi,\Delta})\Big)\mathrm{d}s\\&\quad+\int_0^t\Big(\sigma(x_s^{\xi})-\sigma(y_s^{\xi,\Delta})\Big)\mathrm{d}W(s).\end{aligned}$$

By the Itô formula, Young inequality, and Assumption 2.4, we obtain that for any $\varepsilon\in(0,1)$,

$$\begin{aligned}&e^{\varepsilon t}|x^{\xi}(t)-z^{\xi,\Delta}(t)|^2\\&=\int_0^t e^{\varepsilon s}\Big(\varepsilon|x^{\xi}(s)-z^{\xi,\Delta}(s)|^2+2\langle x^{\xi}(s)-z^{\xi,\Delta}(s),b(x_s^{\xi})-b(y_s^{\xi,\Delta})\rangle\\&\quad+|\sigma(x_s^{\xi})-\sigma(y_s^{\xi,\Delta})|^2\Big)\mathrm{d}s+\mathcal{N}_t+\theta^2|b(\xi)|^2\Delta^2\\&\leq K(1+\|\xi\|^{2(\beta+1)})\Delta^2+\int_0^t e^{\varepsilon s}\Big(\varepsilon|x^{\xi}(s)-z^{\xi,\Delta}(s)|^2+2\langle x^{\xi}(s)-z^{\xi,\Delta}(s),\\&\quad b(x_s^{\xi})-b(z_s^{\xi,\Delta})\rangle+(1+\varepsilon)|\sigma(x_s^{\xi})-\sigma(z_s^{\xi,\Delta})|^2\Big)\mathrm{d}s\\&\quad+\int_0^t\varepsilon e^{\varepsilon s}|x^{\xi}(s)-z^{\xi,\Delta}(s)|^2\mathrm{d}s+\mathcal{E}_t+\mathcal{N}_t,\end{aligned}$$

where

$$\mathcal{E}_t:=\int_0^t e^{\varepsilon s}\Big(\frac{1}{\varepsilon}|b(z_s^{\xi,\Delta})-b(y_s^{\xi,\Delta})|^2+(1+\frac{1}{\varepsilon})|\sigma(z_s^{\xi,\Delta})-\sigma(y_s^{\xi,\Delta})|^2\Big)\mathrm{d}s,$$

and $\mathcal{N}_t$ is a martingale defined by

$$\mathcal{N}_t:=2\int_0^t e^{\varepsilon s}\langle x^{\xi}(s)-z^{\xi,\Delta}(s),(\sigma(x_s^{\xi})-\sigma(y_s^{\xi,\Delta}))\mathrm{d}W(s)\rangle.$$

According to Assumptions 2.1 and 2.2, we derive

$$
\begin{aligned}
& e^{\varepsilon t}|x^{\xi}(t)-z^{\xi,\Delta}(t)|^2 \\
&\quad \le K(1+\|\xi\|^{2(\beta+1)})\Delta^2+\int_0^t \Big(-\big(2a_1-2\varepsilon-(1+\varepsilon)L\big)e^{\varepsilon s}|x^{\xi}(s)-z^{\xi,\Delta}(s)|^2 \\
&\qquad +2a_2\int_{-\tau}^{0} e^{\varepsilon s}|x_s^{\xi}(r)-z_s^{\xi,\Delta}(r)|^2 \mathrm{d}\nu_2(r) \\
&\qquad +(1+\varepsilon)L\int_{-\tau}^{0} e^{\varepsilon s}|x_s^{\xi}(r)-z_s^{\xi,\Delta}(r)|^2 \mathrm{d}\nu_1(r)\Big)\mathrm{d}s+\mathcal{E}_t+\mathcal{N}_t \\
&\quad \le K(1+\|\xi\|^{2(\beta+1)})\Delta^2-\Big(2a_1-2\varepsilon-(1+\varepsilon)L-(2a_2+(1+\varepsilon)L)e^{\varepsilon\tau}\Big) \\
&\qquad \times\int_0^t e^{\varepsilon s}|x^{\xi}(s)-z^{\xi,\Delta}(s)|^2 \mathrm{d}s+(2a_2+(1+\varepsilon)L)e^{\varepsilon\tau} \\
&\qquad \times\int_{-\tau}^{0} e^{\varepsilon s}|x^{\xi}(s)-z^{\xi,\Delta}(s)|^2 \mathrm{d}s+\mathcal{E}_t+\mathcal{N}_t,
\end{aligned}
$$

where in the last step we used

$$
\begin{aligned}
& \int_0^t\int_{-\tau}^{0} e^{\varepsilon s}|x_s^{\xi}(r)-z_s^{\xi,\Delta}(r)|^2 \mathrm{d}\nu_l(r)\mathrm{d}s \\
&\quad \le e^{\varepsilon\tau}\int_{-\tau}^{0}\int_0^t e^{\varepsilon(s+r)}|x^{\xi}(s+r)-z^{\xi,\Delta}(s+r)|^2 \mathrm{d}s\mathrm{d}\nu_l(r) \\
&\quad \le e^{\varepsilon\tau}\int_{-\tau}^{0} e^{\varepsilon s}|x^{\xi}(s)-z^{\xi,\Delta}(s)|^2 \mathrm{d}s+e^{\varepsilon\tau}\int_0^t e^{\varepsilon s}|x^{\xi}(s)-z^{\xi,\Delta}(s)|^2 \mathrm{d}s,\quad l=1,2
\end{aligned}
$$

due to the Fubini theorem. In view of $a_1>a_2+L$, choose a sufficiently small number $\varepsilon\in(0,1)$ such that

$$
2a_1-2\varepsilon-(1+\varepsilon)L-(2a_2+(1+\varepsilon)L)e^{\varepsilon\tau}\ge 0.
$$

By Assumption 2.5, we have

$$
(2a_2+(1+\varepsilon)L)e^{\varepsilon\tau}\int_{-\tau}^{0} e^{\varepsilon s}|x^{\xi}(s)-z^{\xi,\Delta}(s)|^2 \mathrm{d}s\le K(1+\|\xi\|^{2(\beta+1)})\Delta.
$$

Thus,

$$
e^{\varepsilon t}|x^{\xi}(t)-z^{\xi,\Delta}(t)|^2\le K(1+\|\xi\|^{2(\beta+1)})\Delta+\mathcal{E}_t+\mathcal{N}_t. \tag{2.86}
$$

By virtue of Assumptions 2.1 and 2.4, we apply the Hölder inequality to deduce

$$\begin{aligned}\mathbb{E}[\mathcal{E}_t] &\le K\mathbb{E}\Big[\int_0^t e^{\varepsilon s}\|z_s^{\xi,\Delta}-y_s^{\xi,\Delta}\|^2(1+\|z_s^{\xi,\Delta}\|^{2\beta}+\|y_s^{\xi,\Delta}\|^{2\beta})\mathrm{d}s\Big]\\ &\le K\int_0^t e^{\varepsilon s}\big(\mathbb{E}[\|z_s^{\xi,\Delta}-y_s^{\xi,\Delta}\|^4]\big)^{\frac{1}{2}}\big(\mathbb{E}[1+\|z_s^{\xi,\Delta}\|^{4\beta}+\|y_s^{\xi,\Delta}\|^{4\beta}]\big)^{\frac{1}{2}}\mathrm{d}s.\end{aligned}$$

Making use of (2.58) and Lemma 2.9 yields

$$\begin{aligned}\mathbb{E}[\mathcal{E}_t] &\le K(1+\|\xi\|^{2(1+\beta)^2})(1+\|\xi\|^{2\beta(1+\beta)})\Delta\int_0^t e^{\varepsilon s}\mathrm{d}s\\ &\le K(1+\|\xi\|^{4\beta(1+\beta)^3})e^{\varepsilon t}\Delta.\end{aligned}$$

This, along with (2.86) implies that

$$\begin{aligned}\mathbb{E}[|x^\xi(t)-z^{\xi,\Delta}(t)|^2] &\le e^{-\varepsilon t}\big(K(1+\|\xi\|^{2(1+\beta)})\Delta+\mathbb{E}[\mathcal{E}_t]\big)\\ &\le K(1+\|\xi\|^{4\beta(1+\beta)^3})\Delta,\end{aligned}\tag{2.87}$$

where the positive constant K is independent of Δ and t. Furthermore, according to (2.86), we apply the Burkholder–Davis–Gundy inequality to derive

$$\begin{aligned}&\mathbb{E}\Big[\sup_{(t-\tau)\vee 0\le u\le t} e^{\varepsilon u}|x^\xi(u)-z^{\xi,\Delta}(u)|^2\Big]\\ &\le K(1+\|\xi\|^{2(1+\beta)})\Delta+\mathbb{E}[\mathcal{E}_t]+\mathbb{E}\Big[\sup_{(t-\tau)\vee 0\le u\le t}\mathcal{N}_u\Big]\\ &\le K(1+\|\xi\|^{4\beta(1+\beta)^3})e^{\varepsilon t}\Delta\\ &\quad+K\mathbb{E}\Big[\Big(\int_{(t-\tau)\vee 0}^t e^{2\varepsilon s}|x^\xi(s)-z^{\xi,\Delta}(s)|^2|\sigma(x_s^\xi)-\sigma(y_s^{\xi,\Delta})|^2\mathrm{d}s\Big)^{\frac{1}{2}}\Big]\\ &\le K(1+\|\xi\|^{4\beta(1+\beta)^3})e^{\varepsilon t}\Delta+K\mathbb{E}\Big[\Big(\Big(\sup_{(t-\tau)\vee 0\le u\le t} e^{\varepsilon u}|x^\xi(u)-z^{\xi,\Delta}(u)|^2\Big)\\ &\quad\times\int_{(t-\tau)\vee 0}^t e^{\varepsilon s}|\sigma(x_s^\xi)-\sigma(y_s^{\xi,\Delta})|^2\mathrm{d}s\Big)^{\frac{1}{2}}\Big].\end{aligned}$$

By virtue of the Young inequality and Assumption 2.1, we arrive at

$$\begin{aligned}&\mathbb{E}\Big[\sup_{(t-\tau)\vee 0\le u\le t} e^{\varepsilon u}|x^\xi(u)-z^{\xi,\Delta}(u)|^2\Big]\\ &\le K(1+\|\xi\|^{4\beta(1+\beta)^3})e^{\varepsilon t}\Delta+\frac{1}{2}\mathbb{E}\Big[\sup_{(t-\tau)\vee 0\le u\le t} e^{\varepsilon u}|x^\xi(u)-z^{\xi,\Delta}(u)|^2\Big]\end{aligned}$$

$$
\begin{aligned}
&+ K\mathbb{E}\Big[\int_{(t-\tau)\vee 0}^{t} e^{\varepsilon s}|\sigma(x_s^{\xi}) - \sigma(y_s^{\xi,\Delta})|^2 \mathrm{d}s\Big] \\
&\leq K(1+\|\xi\|^{4\beta(1+\beta)^3})e^{\varepsilon t}\Delta + \frac{1}{2}\mathbb{E}\Big[\sup_{(t-\tau)\vee 0\leq u\leq t} e^{\varepsilon u}|x^{\xi}(u) - z^{\xi,\Delta}(u)|^2\Big] \\
&+ K\mathbb{E}\Big[\int_{(t-\tau)\vee 0}^{t} e^{\varepsilon s}|\sigma(x_s^{\xi}) - \sigma(z_s^{\xi,\Delta})|^2 \mathrm{d}s\Big] \\
&+ K\mathbb{E}\Big[\int_{(t-\tau)\vee 0}^{t} e^{\varepsilon s}|\sigma(z_s^{\xi,\Delta}) - \sigma(y_s^{\xi,\Delta})|^2 \mathrm{d}s\Big] \\
&\leq K(1+\|\xi\|^{4\beta(1+\beta)^3})e^{\varepsilon t}\Delta + \frac{1}{2}\mathbb{E}\Big[\sup_{(t-\tau)\vee 0\leq u\leq t} e^{\varepsilon u}|x^{\xi}(u) - z^{\xi,\Delta}(u)|^2\Big] \\
&+ K\int_{(t-\tau)\vee 0}^{t} e^{\varepsilon s}\mathbb{E}[|x^{\xi}(s) - z^{\xi,\Delta}(s)|^2]\mathrm{d}s \\
&+ K\int_{(t-\tau)\vee 0}^{t} e^{\varepsilon s}\mathbb{E}[\|z_s^{\xi,\Delta} - y_s^{\xi,\Delta}\|^2]\mathrm{d}s \\
&+ K\int_{(t-\tau)\vee 0}^{t} e^{\varepsilon s}\int_{-\tau}^{0} \mathbb{E}[|x_s^{\xi}(r) - z_s^{\xi,\Delta}(r)|^2]\mathrm{d}\nu_1(r)\mathrm{d}s.
\end{aligned}
$$

It follows from (2.87) and Lemma 2.9 that

$$
\begin{aligned}
&\frac{1}{2}e^{\varepsilon(t-\tau)\vee 0}\mathbb{E}\Big[\sup_{(t-\tau)\vee 0\leq u\leq t} |x^{\xi}(u) - z^{\xi,\Delta}(u)|^2\Big] \\
&\leq \frac{1}{2}\mathbb{E}\Big[\sup_{(t-\tau)\vee 0\leq u\leq t} e^{\varepsilon u}|x^{\xi}(u) - z^{\xi,\Delta}(u)|^2\Big] \\
&\leq K(1+\|\xi\|^{4\beta(1+\beta)^3})e^{\varepsilon t}\Delta + K(1+\|\xi\|^{4\beta(1+\beta)^3})\Delta\int_{(t-\tau)\vee 0}^{t} e^{\varepsilon s}\mathrm{d}s \\
&\leq K(1+\|\xi\|^{4\beta(1+\beta)^3})e^{\varepsilon t}\Delta,
\end{aligned}
$$

which implies

$$
\mathbb{E}\Big[\sup_{(t-\tau)\vee 0\leq u\leq t} |x^{\xi}(u) - z^{\xi,\Delta}(u)|^2\Big] \leq K(1+\|\xi\|^{4\beta(1+\beta)^3})\Delta.
$$

The proof is completed. □

Utilizing Lemmas 2.9 and 2.13, we conclude the following mean-square convergence rate of the θ-EM functional solution.

Theorem 2.14 *Let Assumptions 2.1–2.5 hold. Then for* $\Delta \in (0, 1]$ *and* $\xi \in C^d$,

$$\sup_{k \geq 0} \mathbb{E}[\|x_{t_k}^{\xi} - y_{t_k}^{\xi,\Delta}\|^2] \leq K(1 + \|\xi\|^{4\beta(1+\beta)^3})\Delta.$$

Remark 2.2 In particular, for the SFDE (1.1) with additive noise, the mean-square convergence rate established in Theorem 2.14 can be increased to 1.

Remark 2.3 Building on the longtime mean-square convergence result, we can derive several practical benefits for applications. Specifically, we can obtain the convergence rate of $1/2$ of the numerical invariant measure to the underlying exact one. Nonetheless, for a higher convergence rate, we turn to Chap. 3, addressing the rigorous analysis of the convergence rate for the numerical invariant measure. To this end, we will employ the method of weak convergence analysis, which offers a more refined approach for this investigation. Additionally, as another application, we present in Corollary 4.5 the SLLN and the CLT of the time-average of the θ-EM functional solution.

2.4 Numerical Experiments

In this section, we present numerical algorithms and numerical experiments to verify the mean-square convergence rate of the θ-EM method applied to SFDEs. Two one-dimensional concrete numerical examples are provided to illustrate the theoretical convergence results.

Algorithm 1 illustrates the numerical procedure used to verify the mean-square convergence rate of the θ-EM method for SFDEs. A reference solution computed with a sufficiently small stepsize is used as a proxy for the exact solution, and all approximations are driven by the same underlying Brownian motion path.

Linear Stochastic Delay Differential Equation We first consider the linear stochastic delay differential equation:

$$\mathrm{d}x^{\xi}(t) = (-\alpha x^{\xi}(t) + \beta x^{\xi}(t-\tau))\mathrm{d}t + \sigma \mathrm{d}W(t), \quad t > 0, \tag{2.88}$$

where in the numerical experiment, constants $\alpha = 1.2, \beta = 0.8, \sigma = 0.5, \tau = 1$, and initial datum $\xi(t) = 1$ for $t \in [-\tau, 0]$. We employ $M = 2000$ independent trajectories to estimate expectations of errors. Figure 2.1a illustrates the mean-square convergence rate of the EM method for this model. The mean-square error at the terminal time is plotted against the stepsize on a log-log scale. The dashed line indicates the reference slope of rate 1. The numerical errors exhibit a clear linear decay with respect to the stepsizes, which is consistent with the theoretical mean-square convergence rate one. Figure 2.1b depicts the longtime evolution of the mean-square convergence error between the EM solution and the reference solution.

Algorithm 1 Mean-square convergence rate for θ-EM method applied to SFDEs

Require: SFDE parameters (b, σ, ξ, τ), parameter θ, final time T, stepsizes $\{\Delta_j\}_{j=1}^J$, reference stepsize Δ_{ref}, Monte Carlo size M.

Output: Mean-square errors $\{e_j\}_{j=1}^J$.

Generate M independent Brownian motion increments $\delta W_{\text{ref}}^{(\tilde{m})}(k) \sim \mathcal{N}(0, \Delta_{\text{ref}})$, $k = 0, \ldots, N_{\text{ref}} - 1$, with $N_{\text{ref}} = T/\Delta_{\text{ref}}$.

Set $\tilde{d}_{\text{ref}} = \tau/\Delta_{\text{ref}}$.

Initialize $x_{\text{ref}}^{(\tilde{m})}(t_k)$, $\quad k = -\tilde{d}_{\text{ref}}, \ldots, 0$, $\tilde{m} = 1, \ldots, M$.

for $\tilde{m} = 1, \ldots, M$ **do**

  for $k = 0, \ldots, N_{\text{ref}} - 1$ **do**

$$x_{\text{ref}}^{(\tilde{m})}(t_{k+1}) = x_{\text{ref}}^{(\tilde{m})}(t_k) + \big(\theta b(x_{\text{ref},t_{k+1}}^{(\tilde{m})}) + (1-\theta) b(x_{\text{ref},t_k}^{(\tilde{m})})\big)\Delta_{\text{ref}} + \sigma(x_{\text{ref},t_k}^{(\tilde{m})})\delta W_{\text{ref}}^{(\tilde{m})}(k).$$

  end for

end for

for $j = 1, \ldots, J$ **do**

  Set $\Delta = \Delta_j$, $N = T/\Delta$, $\tilde{d} = \tau/\Delta$, $r = \Delta/\Delta_{\text{ref}}$.

  Construct coarse Brownian motion increments

$$\delta W^{(\tilde{m})}(k) = \sum_{i=kr}^{(k+1)r-1} \delta W_{\text{ref}}^{(\tilde{m})}(i), \quad k = 0, \ldots, N-1.$$

  Initialize $y^{(\tilde{m})}(t_k)$, $\quad k = -\tilde{d}, \ldots, 0$.

  for $\tilde{m} = 1, \ldots, M$ **do**

    for $k = 0, \ldots, N-1$ **do**

$$y^{(\tilde{m})}(t_{k+1}) = y^{(\tilde{m})}(t_k) + \big(\theta b(y_{t_{k+1}}^{(\tilde{m})}) + (1-\theta) b(y_{t_k}^{(\tilde{m})})\big)\Delta + \sigma(y_{t_k}^{(\tilde{m})})\delta W^{(\tilde{m})}(k).$$

    end for

  end for

  Compute the mean-square error $e_j = \Big(\frac{1}{M}\sum_{\tilde{m}=1}^M \big\| y_T^{(\tilde{m})} - x_{\text{ref},T}^{(\tilde{m})} \big\|^2\Big)^{1/2}$.

end for

It can be observed that the mean-square error remains uniformly bounded over long time intervals [0, 500], confirming our longtime mean-square convergence result.

Nonlinear Stochastic Functional Differential Equation We then investigate the mean-square convergence of the backward EM method applied to the following nonlinear SFDE:

$$\begin{aligned} \mathrm{d}x^{\xi}(t) &= \Big(\frac{1}{\tau}\int_{-\tau}^{0} x^{\xi}(t+s)\mathrm{d}s - (x^{\xi}(t))^3 - 2x^{\xi}(t)\Big)\mathrm{d}t \\ &\quad + \Big(\frac{a}{\tau}\int_{-\tau}^{0} x^{\xi}(t+s)\mathrm{d}s + \frac{1}{2}x^{\xi}(t) + \frac{1}{2}\Big)\mathrm{d}W(t), \end{aligned} \tag{2.89}$$

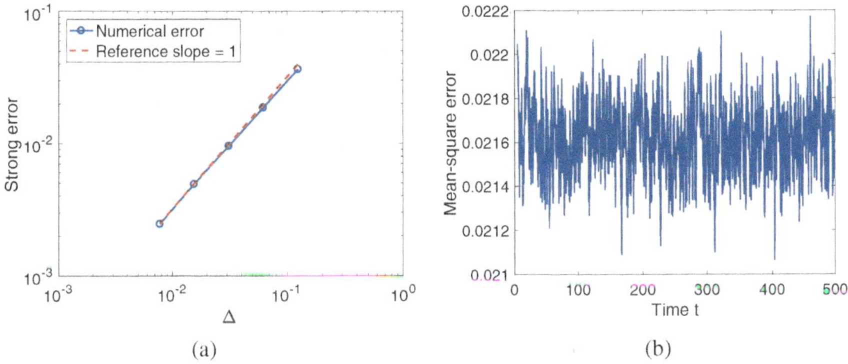

Fig. 2.1 Mean-square convergence for EM method of (2.88); $\alpha = 1.2, \beta = 0.8, \sigma = 0.5, \tau = 1, M = 2000, \xi \equiv 1$. (**a**) $\Delta = 2^{-3}, 2^{-4}, \ldots, 2^{-7}, \Delta_{\text{ref}} = 2^{-12}, T = 1$, and (**b**) $\Delta = 2^{-4}, \Delta_{\text{ref}} = 2^{-8}, T = 500$. (**a**) Mean-square convergence rate. (**b**) Longtime evolution of mean-square error

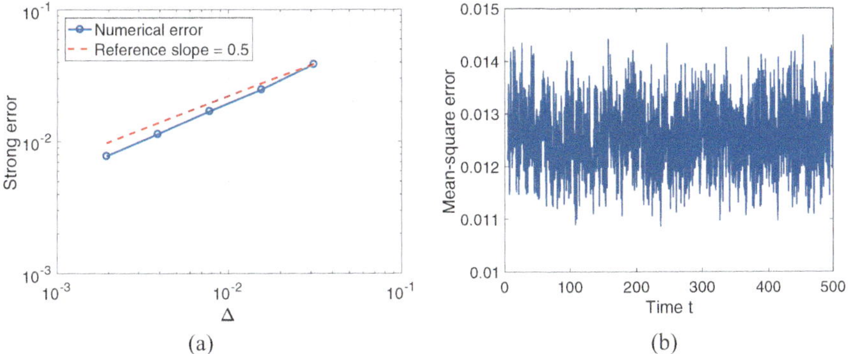

Fig. 2.2 Mean-square convergence for backward EM method of (2.89); $\tau = 0.125, a = 0.1, M = 400, \xi \equiv 1$. (**a**) $\Delta = 2^{-5}, 2^{-6}, \ldots, 2^{-9}, \Delta_{\text{ref}} = 2^{-13}, T = 1$, and (**b**) $\Delta = 2^{-7}, \Delta_{\text{ref}} = 2^{-10}, T = 500$. (**a**) Mean-square convergence rate. (**b**) Longtime evolution of mean-square error

where $\tau = 0.125, a = 0.1$, and the initial datum $\xi(t) = 1$ for $t \in [-\tau, 0]$. We employ $M = 400$ independent trajectories to estimate expectations of errors. As shown in Fig. 2.2a, the numerical errors decrease linearly with respect to $\Delta^{\frac{1}{2}}$, and the slope of the error curve agrees well with the reference slope $\frac{1}{2}$. This confirms the theoretical mean-square convergence rate of the backward EM method for the nonlinear SFDE. Figure 2.2b illustrates the longtime evolution of the mean-square error between the solution of the backward EM method and the reference solution. The mean-square error remains uniformly bounded over long time intervals $[0, 500]$, validating the longtime mean-square convergence result for the nonlinear SFDE with multiplicative noise.

2.5 Summary and Outlook

This chapter has studied the longtime mean-square convergence of the θ-EM method for the SFDE (1.1) with a superlinearly growing drift coefficient, and established a mean-square convergence rate of $\frac{1}{2}$. Existing studies on the mean-square convergence analysis over the finite time horizon include the analysis of the Milstein method in [22], the Wong–Zakai type approximation in [2], EM-type methods in [13, 34, 35], and drift-implicit one-step numerical methods in [1]. For the infinite time horizon setting, we highlight a recent paper [7], in which the mean-square convergence rate of the EM method for the SFDE on the infinite time horizon is proved to be $\frac{1}{2}$ under the globally Lipschitz condition.

In this chapter, we have presented some sufficient conditions for analyzing the problem under consideration. Nevertheless, there is still some flexibility in relaxing these conditions. For example, if Assumption 2.3 is weakened to

$$\langle b(\phi), \phi(0)\rangle \leq K - a_3|\phi(0)|^{\alpha+\epsilon} + a_4 \int_{-\tau}^{0} |\phi(r)|^{\alpha} \mathrm{d}\nu_2(r)$$

with $\epsilon > 0, \alpha \geq 2, a_3, a_4 > 0$, and $\phi \in C^d$, the mean-square convergence result established in this chapter still holds by essentially the same approach. When both the drift and the diffusion coefficients exhibit superlinear growth, some longtime mean-square convergence results for numerical methods applied to stochastic ordinary differential equations have been established; see e.g. [33] for the analysis of the BEM method. However, knowledge regarding the SFDE case remains relatively scarce.

While this chapter focuses on SFDEs driven by Brownian motion, SFDEs driven by other stochastic processes, such as fractional Brownian motion and the Lévy noise, are also of substantial theoretical and practical interest. The numerical analysis on stochastic ordinary and partial differential equations driven by such noises has been extensively developed (see e.g., [5, 6, 8, 17, 28, 29, 37] for Lévy noise and see [15, 19–21, 23–25, 32] for fractional Brownian motion). By contrast, corresponding research on SFDEs remains rather limited. Existing contributions mainly address the convergence analysis in the finite time horizon and stability properties for stochastic delay differential equations driven by Lévy noise or fractional Brownian motion (see [13, 14, 18, 30]), whereas the convergence analysis in the infinite time horizon of numerical methods for general SFDEs in this setting appears to be far less explored. This points to the need for developing appropriate numerical schemes and rigorously analyzing their longtime behavior.

The study of SFDEs presented here motivates consideration of stochastic partial differential equations with memory effects. In this setting, the introduction of time delays leads naturally to stochastic partial differential equations with delay, for which a number of numerical methods have been developed; see e.g., [16, 31, 38, 40]. In addition to the Euclidean spatial domain, stochastic partial differential equations with delay have been considered on manifolds and other geometric settings. We refer to [39] and the references therein for related theoretical analysis and numerical investigations. Building on recent progress in graph theory and stochastic partial differential equations on graphs (see e.g., [3, 4, 9–12, 26, 27]), stochastic delay equations on graphs remain largely unexplored, and the study of their well-posedness together with the construction and analysis of suitable numerical methods for SFDE-type models in this setting constitutes a promising direction for future research.

References

1. E. Buckwar, One-step approximations for stochastic functional differential equations. Appl. Numer. Math. **56**, 667–681 (2006)
2. W. Cao, Z. Zhang, G.E. Karniadakis, Numerical methods for stochastic delay differential equations via the Wong–Zakai approximation. SIAM J. Sci. Comput. **37**, A295–A318 (2015)
3. S. Cerrai, M. Freidlin, SPDEs on narrow domains and on graphs: an asymptotic approach. Ann. Inst. Henri Poincaré Probab. Stat. **53**, 865–899 (2017)
4. S. Cerrai, M. Freidlin, Fast flow asymptotics for stochastic incompressible viscous fluids in $\mathbb{R}^2$ and SPDEs on graphs. Probab. Theory Relat. Fields **173**, 491–535 (2019)
5. Z. Chen, S. Gan, X. Wang, Mean-square approximations of Lévy noise driven SDEs with super-linearly growing diffusion and jump coefficients. Discrete Contin. Dyn. Syst. Ser. B **24**, 4513–4545 (2019)
6. C. Chen, T. Dang, J. Hong, On structure preservation for fully discrete finite difference schemes of stochastic heat equation with Lévy space-time white noise (2024). https://arxiv.org/abs/2409.14064
7. C. Chen, T. Dang, J. Hong, G. Song, On numerical discretizations that preserve probabilistic limit behaviors for time-homogeneous Markov processes. Bernoulli **31**, 3139–3164 (2025)
8. C. Chen, T. Dang, J. Hong, Z. Lei, Fully discrete schemes and L^p-strong convergence orders for the SPDE driven by Lévy noise. J. Comput. Phys. **545** (2026). Paper No. 114475
9. J. Cui, T. Dang, Hamilton–Jacobi–Bellman equation for optimal control of stochastic Wasserstein–Hamiltonian system on graphs (2025). https://arxiv.org/abs/2509.25965
10. J. Cui, D. Sheng, Asymptotic-preserving approximations for stochastic incompressible viscous fluids and spdes on graph (2024). https://arxiv.org/abs/2404.09168
11. J. Cui, D. Sheng, Large and moderate deviation principles for stochastic partial differential equation on graph (2025). https://arxiv.org/abs/2509.05622
12. J. Cui, S. Liu, H. Zhou, Optimal control for stochastic nonlinear Schrödinger equation on graph. SIAM J. Control Optim. **61**, 2021–2042 (2023)
13. K. Dareiotis, C. Kumar, S. Sabanis, On tamed Euler approximations of SDEs driven by Lévy noise with applications to delay equations. SIAM J. Numer. Anal. **54**, 1840–1872 (2016)
14. S. Deng, C. Fei, W. Fei, X. Mao, The truncated EM method for jump-diffusion SDDEs with super-linearly growing diffusion and jump coefficients. J. Comput. Math. **42**, 178–216 (2024)
15. X.-L. Ding, D. Wang, Regularity analysis for SEEs with multiplicative fBms and strong convergence for a fully discrete scheme. IMA J. Numer. Anal. **44**, 1435–1463 (2024)
16. X.-L. Ding, L. Guo, J.J. Nieto, Y. Gao, Regularity analysis for SFDEs driven by mixed fractional Brownian motion and strong convergence of a numerical scheme. Bull. Iranian Math. Soc. **51** (2025). Paper No. 53
17. T. Dunst, E. Hausenblas, A. Prohl, Approximate Euler method for parabolic stochastic partial differential equations driven by space-time Lévy noise. SIAM J. Numer. Anal. **50**, 2873–2896 (2012)
18. S. Gao, J. Hu, L. Tan, C. Yuan, Strong convergence rate of truncated Euler–Maruyama method for stochastic differential delay equations with Poisson jumps. Front. Math. China **16**, 395–423 (2021)

19. J. Hong, C. Huang, Super-convergence analysis on exponential integrator for stochastic heat equation driven by additive fractional brownian motion (2020). https://arxiv.org/abs/2007.02223
20. J. Hong, C. Huang, X. Wang, Optimal rate of convergence for two classes of schemes to stochastic differential equations driven by fractional Brownian motions. IMA J. Numer. Anal. **41**, 1608–1638 (2021)
21. Y. Hu, Y. Liu, D. Nualart, Rate of convergence and asymptotic error distribution of Euler approximation schemes for fractional diffusions. Ann. Appl. Probab. **26**, 1147–1207 (2016)
22. Y. Hu, S.-E. A. Mohammed, F. Yan, Discrete-time approximations of stochastic delay equations: the Milstein scheme. Ann. Probab. **32**, 265–314 (2004)
23. C. Huang, Optimal convergence rate of modified Milstein scheme for SDEs with rough fractional diffusions. J. Differ. Equ. **344**, 325–351 (2023)
24. C. Huang, X. Wang, Strong convergence rate of the Euler scheme for SDEs driven by additive rough fractional noises. Stat. Probab. Lett. **194** (2023). Paper No. 109742
25. M. Kamrani, N. Jamshidi, Implicit Euler approximation of stochastic evolution equations with fractional Brownian motion. Commun. Nonlinear Sci. Numer. Simul. **44**, 1–10 (2017)
26. M. Kovács, E. Sikolya, On the stochastic Allen–Cahn equation on networks with multiplicative noise. Electron. J. Qual. Theory Differ. Equ. **2021**, 24 (2021). Paper No. 7
27. M. Kovács, E. Sikolya, Stochastic reaction-diffusion equations on networks. J. Evol. Equ. **21**, 4213–4260 (2021)
28. C. Kumar, S. Sabanis, On explicit approximations for Lévy driven SDEs with super-linear diffusion coefficients. Electron. J. Probab. **22**, 19 (2017). Paper No. 73
29. C. Kumar, S. Sabanis, On tamed Milstein schemes of SDEs driven by Lévy noise. Discrete Contin. Dyn. Syst. Ser. B **22**, 421–463 (2017)
30. X. Li, W. Cao, Numerical approximation of stochastic delay differential equations driven by fractional Brownian motion. Numer. Math. J. Chin. Univ. **39**, 289–315 (2017)
31. F. Li, M. M. Freitas, Convergence of random attractors for stochastic delay p-Laplacian equation driven by nonlinear colored noise on unbounded thin domains. Appl. Math. Optim. **91**, 34 (2025). Paper No. 47
32. M. Li, Y. Hu, C. Huang, X. Wang, Mean square stability of stochastic theta method for stochastic differential equations driven by fractional Brownian motion. J. Comput. Appl. Math. **420** (2023). Paper No. 114804
33. W. Liu, X. Mao, Y. Wu, The backward Euler–Maruyama method for invariant measures of stochastic differential equations with super-linear coefficients. Appl. Numer. Math. **184**, 137–150 (2023)
34. X. Li, X. Mao, G. Song, An explicit approximation for super-linear stochastic functional differential equations. Stoch. Process. Appl. **169** (2024). Paper No. 104275
35. X. Mao, Numerical solutions of stochastic functional differential equations. LMS J. Comput. Math. **6**, 141–161 (2003)
36. X. Mao, *Stochastic Differential Equations and Applications*, 2nd ed. (Horwood Publishing Limited, Chichester, 2008)
37. P. Przybylowicz, M. Sobieraj, L. Stcepień, Efficient approximation of SDEs driven by countably dimensional Wiener process and Poisson random measure. SIAM J. Numer. Anal. **60**, 824–855 (2022)
38. S. Shi, J. Xu, W. Chen, Strong rate of convergence for an Euler–Galerkin discretization of the stochastic nonlocal partial differential equations with delay. J. Sci. Comput. **105** (2025). Paper No. 10
39. L. Su, W. Michiels, E. Steur, H. Nijmeijer, A method for computation and analysis of partial synchronization manifolds of delay coupled systems, in *Accounting for Constraints in Delay Systems*, vol. 12. Adv. Delays Dyn. (Springer, Cham, 2022), pp. 209–230
40. Y. Wang, G. Cao, P.E. Kloeden, Mean-square stability analysis of stochastic delay evolution equations driven by fractional Brownian motion with Hurst index $H \in (0, 1)$. Discrete Contin. Dyn. Syst. Ser. S **17**, 1073–1100 (2024)

Chapter 3
Numerical Invariant Measure and Weak Convergence Analysis in the Infinite Time Horizon

For the SFDE with either discrete or distributed delay arguments, it is known that the solution is non-Markovian due to the dependence on the past. Hence, the functional solution, which is proved to be Markovian, becomes the focus for studying the existence and uniqueness of the invariant measure of the underlying SFDE. This chapter investigates the existence, uniqueness, and convergence rate of the invariant measure of the numerical functional solution of the θ-EM method. Recall that in Chap. 2, we obtained the time-independent moment boundedness and the exponential attractiveness of the θ-EM functional solution. Building on these properties, in Sect. 3.1, we prove the existence and uniqueness of the numerical invariant measure by using the Krylov–Bogoliubov theorem. In Sect. 3.2, we analyze the longtime weak convergence rate of the θ-EM method, which provides a quantitative estimate between the numerical invariant measure and the exact one. The analysis relies on the Malliavin calculus. We first derive the time-independent estimates for Malliavin derivatives and Gâteaux derivatives of both the exact and numerical functional solutions. Then, we show that the weak convergence rate of the θ-EM method is 1, which is twice the mean-square convergence rate.

3.1 Numerical Invariant Measure

In this section, we show that the θ-EM functional solution admits a unique invariant measure. Recall

$$y_t^{\xi,\Delta} := \sum_{k=0}^{\infty} y_{t_k}^{\xi,\Delta} \mathbf{1}_{[t_k,t_{k+1})}(t)$$

C. Chen et al., *Numerical Analysis of Stochastic Functional Differential Equations*, Lecture Notes in Mathematics 2399, https://doi.org/10.1007/978-981-92-1592-8_3

and define the measure $\mu_t^{\xi,\Delta}$ by

$$\mu_t^{\xi,\Delta}(A) := \mathbb{P}\{\omega \in \Omega : y_t^{\xi,\Delta} \in A\} \quad \forall A \in \mathcal{B}(C^d).$$

Denote by $P_t^{\xi,\Delta}$ the Markov semigroup associated to $\mu_t^{\xi,\Delta}$. According to Propositions 2.4 and 2.10, we can obtain the tightness of the sequence $\{\mu_{t_k}^{\xi,\Delta}\}_{k\in\mathbb{N}}$.

Lemma 3.1 *Let Assumptions 2.1 and 2.2 hold. Then for $\xi \in C^d$ and $\Delta \in (0, 1]$, the family $\{\mu_{t_k}^{\xi,\Delta}\}_{k\in\mathbb{N}}$ is tight.*

Proof Since C^d is a Polish space and

$$\sup_{k\in\mathbb{N}} \mathbb{E}[\|y_{t_k}^{\xi,\Delta}\|^2] \leq K(1 + \|\xi\|^2 + |b(\xi)|^2\Delta^2),$$

by virtue of [14, Lemma 6.14], it suffices to prove that the family $\{\mu_{t_k}^{\xi,\Delta}\}_{k\in\mathbb{N}}$ is a Cauchy sequence in the space $(\mathcal{P}, \mathbb{W}_1)$. In fact, for any $k, i \in \mathbb{N}$, applying [14, Remark 6.5] leads to

$$\mathbb{W}_1(\mu_{t_{k+i}}^{\xi,\Delta}, \mu_{t_k}^{\xi,\Delta}) = \sup_{\|\Psi\|_{Lip}\vee\|\Psi\|_\infty\leq 1} \left| \int_{C^d} \Psi(\phi)\mathrm{d}\mu_{t_{k+i}}^{\xi,\Delta}(\phi) - \int_{C^d} \Psi(\phi)\mathrm{d}\mu_{t_k}^{\xi,\Delta}(\phi) \right|,$$

where $\Psi : C^d \to \mathbb{R}$ satisfies the Lipschitz condition, and norms $\|\cdot\|_{Lip}$ and $\|\cdot\|_\infty$ are defined, respectively, by

$$\|\Psi\|_{Lip} = \sup_{\substack{\phi_1,\phi_2\in C^d\\ \phi_1\neq\phi_2}} \frac{|\Psi(\phi_1) - \Psi(\phi_2)|}{\|\phi_1 - \phi_2\|} \quad \text{and} \quad \|\Psi\|_\infty = \sup_{\phi\in C^d} |\Psi(\phi)|.$$

By the definition of $\mu_\cdot^{\xi,\Delta}$ and the time-homogeneous Markov property of $\{y_{t_k}^{\xi,\Delta}\}_{k\in\mathbb{N}}$, we deduce

$$\begin{aligned}
\mathbb{W}_1(\mu_{t_{k+i}}^{\xi,\Delta}, \mu_{t_k}^{\xi,\Delta}) &= \sup_{\|\Psi\|_{Lip}\vee\|\Psi\|_\infty\leq 1} |\mathbb{E}[\Psi(y_{t_{k+i}}^{\xi,\Delta})] - \mathbb{E}[\Psi(y_{t_k}^{\xi,\Delta})]| \\
&= \sup_{\|\Psi\|_{Lip}\vee\|\Psi\|_\infty\leq 1} \left|\mathbb{E}\Big[\mathbb{E}[\Psi(y_{t_{k+i}}^{\xi,\Delta})|\mathcal{F}_{t_i}]\Big] - \mathbb{E}[\Psi(y_{t_k}^{\xi,\Delta})]\right| \\
&= \sup_{\|\Psi\|_{Lip}\vee\|\Psi\|_\infty\leq 1} \left|\mathbb{E}\Big[\mathbb{E}[\Psi(y_{t_k}^{\zeta,\Delta})]|_{\zeta=y_{t_i}^{\xi,\Delta}}\Big] - \mathbb{E}[\Psi(y_{t_k}^{\xi,\Delta})]\right| \\
&= \sup_{\|\Psi\|_{Lip}\vee\|\Psi\|_\infty\leq 1} \left|\mathbb{E}\Big[\mathbb{E}[\Psi(y_{t_k}^{\zeta,\Delta}) - \Psi(y_{t_k}^{\xi,\Delta})]|_{\zeta=y_{t_i}^{\xi,\Delta}}\Big]\right|.
\end{aligned}$$

It follows from the Lipschitz property of Ψ and Proposition 2.10 that for any constant $M > \|\xi\|$,

$$\begin{aligned}\mathbb{W}_1(\mu_{t_{k+i}}^{\xi,\Delta}, \mu_{t_k}^{\xi,\Delta}) &\leq \mathbb{E}\Big[\mathbb{E}[2 \wedge \|y_{t_k}^{\zeta,\Delta} - y_{t_k}^{\xi,\Delta}\|]\big|_{\zeta = y_{t_i}^{\xi,\Delta}}\Big] \\ &\leq K\mathbb{E}\Big[\big(|b(y_{t_i}^{\xi,\Delta}) - b(\xi)| + \|y_{t_i}^{\xi,\Delta} - \xi\|\big)\mathbf{1}_{\{\|y_{t_i}^{\xi,\Delta}\| \leq M\}}\Big]e^{-\frac{\lambda_0 t_k}{2}} \\ &\quad + 2\mathbb{P}\{\|y_{t_i}^{\xi,\Delta}\| > M\}.\end{aligned}$$

According to the Chebyshev inequality and Proposition 2.4, we have

$$\begin{aligned}\mathbb{W}_1(\mu_{t_{k+i}}^{\xi,\Delta}, \mu_{t_k}^{\xi,\Delta}) &\leq K \sup_{\|\phi\| \leq M} \big(|b(\phi)| + \|\phi\|\big)e^{-\frac{\lambda_0 t_k}{2}} + 2\frac{\mathbb{E}[\|y_{t_i}^{\xi,\Delta}\|^2]}{M^2} \\ &\leq K \sup_{\|\phi\| \leq M} \big(|b(\phi)| + \|\phi\|\big)e^{-\frac{\lambda_0 t_k}{2}} + \frac{K(1 + \|\xi\|^2 + |b(\xi)|^2\Delta^2)}{M^2}.\end{aligned}$$

For any $\varepsilon > 0$, choose a sufficiently large number $M > 0$ such that

$$\frac{K(1 + \|\xi\|^2 + |b(\xi)|^2\Delta^2)}{M^2} < \frac{\varepsilon}{2}.$$

By virtue of the continuity of $b(\cdot)$, choose a sufficiently large number $k \in \mathbb{N}$ such that

$$K \sup_{\|\phi\| \leq M} \big(|b(\phi)| + \|\phi\|\big)e^{-\frac{\lambda_0 t_k}{2}} < \frac{\varepsilon}{2}.$$

Hence, the family $\{\mu_{t_k}^{\xi,\Delta}\}_{k \in \mathbb{N}}$ is a Cauchy sequence in $(\mathcal{P}, \mathbb{W}_1)$. The proof is finished. □

Taking advantage of Proposition 2.10 and Lemma 3.1, we give the existence and uniqueness of the numerical invariant measure of the θ-EM functional solution.

Theorem 3.2 *Let Assumptions 2.1 and 2.2 hold. Then for $\Delta \in (0, 1]$, the θ-EM functional solution admits a unique invariant measure μ^Δ, which satisfies*

$$\sup_{\Delta \in (0,1]} \mu^\Delta(\|\cdot\|^2) := \sup_{\Delta \in (0,1]} \int_{C^d} \|\phi\|^2 \mathrm{d}\mu^\Delta(\phi) \leq K. \tag{3.1}$$

Proof We divide the proof of the existence and uniqueness of the numerical invariant measure into two steps.

Step 1. We present the existence of numerical invariant measure based on the Krylov–Bogoliubov theorem [10, Theorem 3.1.1]. For any $\xi \in C^d$, $\Delta \in (0, 1]$, and $k \in \mathbb{N}$, it follows from Lemma 3.1 that the family $\{\mu_{t_k}^{\xi,\Delta}\}_{k \in \mathbb{N}}$ is tight. Hence we only need to prove that $P_{t_k}^{\xi,\Delta}$ is Feller, i.e., to show that the mapping

$$\xi \to P_{t_k}^{\xi,\Delta}(\Phi) = \mathbb{E}[\Phi(y_{t_k}^{\xi,\Delta})]$$

is continuous for any bounded continuous functional $\Phi : C^d \to \mathbb{R}$. In fact, for $\xi, \eta \in C^d$, by Proposition 2.10, we have

$$y_{t_k}^{\xi,\Delta} \to y_{t_k}^{\eta,\Delta} \quad \text{in } L^2(\Omega; C^d)$$

as $\xi \to \eta$. By [13, p.256, Theorem 2], we derive

$$y_{t_k}^{\xi,\Delta} \to y_{t_k}^{\eta,\Delta} \quad \text{in distribution}$$

as $\xi \to \eta$, which implies the Feller property. Hence, for any $\xi \in C^d$ and $\Delta \in (0, 1]$, the θ-EM functional solution admits an invariant measure μ^Δ.

Step 2. We shall show that for any $\Delta \in (0, 1]$, the numerical invariant measure is unique. Assume that there exist two numerical invariant measures μ_1^Δ and μ_2^Δ. Then in view of the convex property of $\mathbb{W}_2$ and Proposition 2.10, we obtain that for any $k \in \mathbb{N}$,

$$\begin{aligned}
&(\mathbb{W}_2(\mu_1^\Delta, \mu_2^\Delta))^2 \\
&\leq \int_{C^d \times C^d} (\mathbb{W}_2(\mu_{t_k}^{\xi,\Delta}, \mu_{t_k}^{\eta,\Delta}))^2 \mathrm{d}\mu_1^\Delta(\xi)\mathrm{d}\mu_2^\Delta(\eta) \\
&\leq \int_{C^d \times C^d} 1 \wedge \Big(K\big(|b(\xi) - b(\eta)|^2 + \|\xi - \eta\|^2\big)e^{-\lambda_0 t_k}\Big)\mathrm{d}\mu_1^\Delta(\xi)\mathrm{d}\mu_2^\Delta(\eta).
\end{aligned} \tag{3.2}$$

Letting $k \to \infty$ and applying the dominated convergence theorem, we derive

$$\mathbb{W}_2(\mu_1^\Delta, \mu_2^\Delta) = 0,$$

which finishes the proof of Step 2.

In addition, it follows from

$$\lim_{k\to\infty} \mathbb{W}_2(\mu_{t_k}^{\mathbf{0},\Delta}, \mu^\Delta) = 0$$

that for any bounded continuous functional $\Phi : C^d \to \mathbb{R}$,

$$\lim_{k\to\infty} \int_{C^d} \Phi(u)\mathrm{d}\mu_{t_k}^{\mathbf{0},\Delta}(u) = \int_{C^d} \Phi(u)\mathrm{d}\mu^\Delta(u).$$

This, along with Proposition 2.4 implies that

$$\begin{aligned}\int_{C^d}\|u\|^2\mathrm{d}\mu^{\Delta}(u)&=\lim_{M\to\infty}\int_{C^d}(M\wedge\|u\|^2)\mathrm{d}\mu^{\Delta}(u)\\&=\lim_{M\to\infty}\lim_{k\to\infty}\mathbb{E}\big[M\wedge\|y^{\mathbf{0},\Delta}_{t_k}\|^2\big]\\&\le\sup_{k\in\mathbb{N}}\mathbb{E}[\|y^{\mathbf{0},\Delta}_{t_k}\|^2]\le K.\end{aligned}$$

The proof is therefore finished. □

Taking advantage of Theorem 2.14, we can obtain a rough estimate for the error between the numerical invariant measure and the exact one.

Theorem 3.3 *Let Assumptions 2.1to 2.5 hold. Then for any* $\Delta\in(0,1]$,

$$\mathbb{W}_2(\mu,\mu^{\Delta})\le K\Delta^{\frac{1}{2}}.$$

Proof For any $\Delta\in(0,1]$ and $k\in\mathbb{N}$, by virtue of (2.3) and Theorem 2.14, we obtain

$$\begin{aligned}\mathbb{W}_2(\mu,\mu^{\Delta})&\le\mathbb{W}_2(\mu,\mu^{\mathbf{0}}_{t_k})+\mathbb{W}_2(\mu^{\mathbf{0}}_{t_k},\mu^{\mathbf{0},\Delta}_{t_k})+\mathbb{W}_2(\mu^{\mathbf{0},\Delta}_{t_k},\mu^{\Delta})\\&\le Ke^{-\frac{\lambda}{2}t_k}+\big(\mathbb{E}[\|x^{\mathbf{0}}_{t_k}-y^{\mathbf{0},\Delta}_{t_k}\|^2]\big)^{\frac{1}{2}}+\mathbb{W}_2(\mu^{\mathbf{0},\Delta}_{t_k},\mu^{\Delta})\\&\le Ke^{-\frac{\lambda}{2}t_k}+K\Delta^{\frac{1}{2}}+\mathbb{W}_2(\mu^{\mathbf{0},\Delta}_{t_k},\mu^{\Delta}),\end{aligned}\tag{3.3}$$

where the positive constant λ satisfies (2.2). Similar to the proof of (3.2), and by Assumption 2.4, we have

$$\begin{aligned}\mathbb{W}_2(\mu^{\mathbf{0},\Delta}_{t_k},\mu^{\Delta})&\le\Big(\int_{C^d}K\big(|b(\mathbf{0})-b(\eta)|^2+\|\eta\|^2\big)e^{-\lambda_0 t_k}\mathrm{d}\mu^{\Delta}(\eta)\Big)^{\frac{1}{2}}\\&\le Ke^{-\frac{\lambda_0}{2}t_k}\Big(\int_{C^d}\big(1+\|\eta\|^{2(\beta+1)}\big)\mathrm{d}\mu^{\Delta}(\eta)\Big)^{\frac{1}{2}}\\&\le Ke^{-\frac{\lambda_0}{2}t_k}.\end{aligned}$$

This, along with (3.3) implies that

$$\mathbb{W}_2(\mu,\mu^{\Delta})\le Ke^{-\frac{\lambda}{2}t_k}+K\Delta^{\frac{1}{2}}+Ke^{-\frac{\lambda_0}{2}t_k}.$$

Letting $k\to\infty$, we obtain the desired argument. □

3.2 Weak Convergence Analysis

In this section, we investigate the convergence rate of the numerical invariant measure based on the longtime weak convergence analysis of the θ-EM method.

We first introduce some notation. Without loss of generality, let T be a multiple of τ, i.e., there exists a number $N^{\Delta} \in \mathbb{N}_{+}$ such that $T = t_{N^{\Delta}}$. Let $\varphi(t; t_i, \eta)$ and $\varphi_t(t_i, \eta)$ denote the solution and functional solution of the SFDE (1.1) at time t with the initial value $\eta \in C^d$ at t_i, $\Phi(t_k; t_i, \eta)$ and $\Phi_{t_k}(t_i, \eta)$ denote the θ-EM solution and θ-EM functional solution at time t_k with initial value η at t_i. Let $(C([-\tau, T]; \mathbb{R}^d), |||\cdot|||)$ denote the family of continuous functions $\phi : [-\tau, T] \to \mathbb{R}^d$ equipped with the norm $|||\phi||| := \sup_{-\tau \le s \le T} |\phi(s)|$. For $Y \in C([-\tau, T]; \mathbb{R}^d)$, denote its segment process by $\{Y_r(\cdot)\}_{r\ge 0}$ satisfying $Y_r(\cdot) \in C^d$ and $Y_r(s) = Y(r+s)$ for $s \in [-\tau, 0]$.

The error between the numerical invariant measure of the θ-EM method and the invariant measure of (1.1) is split as follows: for the test functional f which belongs to some suitable space,

$$
\begin{aligned}
&|\mu(f) - \mu^{\Delta}(f)| \\
&= |\mathbb{E}[f(x^{\hat{\xi}}(T))] - \mathbb{E}[f(y^{\bar{\xi},\Delta}(T))]| \\
&\le |\mathbb{E}[f(x^{\hat{\xi}}(T))] - \mathbb{E}[f(y^{\hat{\xi},\Delta}(T))]| + |\mathbb{E}[f(y^{\hat{\xi},\Delta}(T))] - \mathbb{E}[f(y^{\bar{\xi},\Delta}(T))]|,
\end{aligned} \tag{3.4}
$$

where $\hat{\xi}, \bar{\xi}$ are initial data with distributions being μ, μ^{Δ}, respectively. When f has the uniformly bounded first-order derivative, according to the mean value theorem and Proposition 2.10, the second term in the right hand of (3.4) decays exponentially in time, namely, $|\mathbb{E}[f(y^{\hat{\xi},\Delta}(T))] - \mathbb{E}[f(y^{\bar{\xi},\Delta}(T))]| \le Ke^{-\frac{\lambda_0}{2}T}$. Hence, it suffices to estimate the term $|\mathbb{E}[f(x^{\hat{\xi}}(T))] - \mathbb{E}[f(y^{\hat{\xi},\Delta}(T))]|$, which is commonly called the weak error of the numerical method.

The decomposition of the weak error is mainly based on a dimensionality reduction argument, which projects the infinite-dimensional spaces C^d and $C([-\tau, T]; \mathbb{R}^d)$ into finite-dimensional ones via linear interpolation. The resulting projection spaces are denoted by C^{Int} and $C^{Int}([-\tau, T]; \mathbb{R}^d)$, respectively. Precisely, for $\xi \in C^d$, let $\xi^{Int} \in C^{Int}$ denote the linear interpolation of ξ at grids $\{(t_k, \xi(t_k))\}_{k=-N}^{0}$. For $Y(\cdot) \in C([-\tau, T]; \mathbb{R}^d)$, the process $Y^{Int}(\cdot) \in C^{Int}([-\tau, T]; \mathbb{R}^d)$ denotes the linear interpolation with respect to $\{(t_k, Y(t_k))\}_{k=-N}^{N^{\Delta}}$, whose segment process $\{Y_r^{Int}(\cdot)\}_{r\ge 0}$ satisfies $Y_r^{Int}(\cdot) \in C^d$ and $Y_r^{Int}(s) = Y^{Int}(r+s)$, $s \in [-\tau, 0]$. It is worth pointing out that $C^{Int} = \tilde{C}^{Int} \otimes \mathbb{R}^d$, where $\tilde{C}^{Int}$ denotes the $(N+1)$-dimensional linear interpolation space of piecewise linear functions from $[-\tau, 0]$ to $\mathbb{R}$ with the following basis functions:

$$
\begin{aligned}
I^{[-N]}(s) &:= \frac{N}{\tau}(t_{-N+1} - s)\mathbf{1}_{\Delta_{-N}}(s), \\
I^{[j]}(s) &:= \frac{N}{\tau}(s - t_{j-1})\mathbf{1}_{\Delta_{j-1}}(s) + \frac{N}{\tau}(t_{j+1} - s)\mathbf{1}_{\Delta_j}(s), \ j = -N+1, \ldots, -1, \\
I^{[0]}(s) &:= \frac{N}{\tau}(s - t_{-1})\mathbf{1}_{\Delta_{-1}}(s),
\end{aligned}
\tag{3.5}
$$

where $\Delta_j := [t_j, t_{j+1})$ for $j = -N, \ldots, -2$, and $\Delta_{-1} := [t_{-1}, 0]$. Thus, any element $\mathrm{u} : [-\tau, 0] \to \mathbb{R}^d$ in C^{Int} can be expressed as $\sum_{j=-N}^{0} I^{[j]}(\cdot)\mathrm{u}_j$, where $I^{[j]} \in \tilde{C}^{Int}$ and $\mathrm{u}_j \in \mathbb{R}^d$.

Denote by $C_b^3(\mathbb{R}^d; \mathbb{R})$ the family of all continuous functions with bounded partial derivatives up to order 3. We always use the relation $\varphi^{Int}(t_l; t_j, \varphi_{t_j}(0, \xi^{Int})) = \varphi^{Int}(t_l; 0, \xi^{Int}) = \varphi(t_l; 0, \xi^{Int})$ for $l \geq j \geq 0$ in the decomposition of the weak error. Let $f \in C_b^3(\mathbb{R}^d; \mathbb{R})$. We split the weak error as

$$
\begin{aligned}
&\mathbb{E}[f(x^{\xi}(T))] - \mathbb{E}[f(y^{\xi,\Delta}(T))] \\
&= \mathbb{E}[f(\varphi(T; 0, \xi))] - \mathbb{E}[f(\Phi(T; 0, \xi^{Int}))] \\
&= \mathbb{E}[f(\Phi(T; t_{N^\Delta}, \varphi^{Int}_{t_{N^\Delta}}(0, \xi)))] - \mathbb{E}[f(\Phi(T; 0, \xi^{Int}))] \\
&= \mathbb{E}[f(\Phi(T; t_{N^\Delta}, \varphi^{Int}_{t_{N^\Delta}}(0, \xi)))] - \mathbb{E}[f(\Phi(T; t_{N^\Delta}, \varphi^{Int}_{t_{N^\Delta}}(0, \xi^{Int})))] \\
&\quad + \mathbb{E}[f(\Phi(T; t_{N^\Delta}, \varphi^{Int}_{t_{N^\Delta}}(0, \xi^{Int})))] - \mathbb{E}[f(\Phi(T; 0, \xi^{Int}))],
\end{aligned}
\tag{3.6}
$$

where we used the relation $\varphi(T; 0, \xi) = \varphi^{Int}_{t_{N^\Delta}}(0, \xi)(0) = \Phi(T; t_{N^\Delta}, \varphi^{Int}_{t_{N^\Delta}}(0, \xi))$.

We split the term $\mathbb{E}[f(\Phi(T; t_{N^\Delta}, \varphi^{Int}_{t_{N^\Delta}}(0, \xi^{Int})))] - \mathbb{E}[f(\Phi(T; 0, \xi^{Int}))]$ further as

$$
\begin{aligned}
&\mathbb{E}[f(\Phi(T; t_{N^\Delta}, \varphi^{Int}_{t_{N^\Delta}}(0, \xi^{Int})))] - \mathbb{E}[f(\Phi(T; 0, \xi^{Int}))] \\
&= \sum_{i=1}^{N^\Delta} \Big\{ \mathbb{E}[f(\Phi(T; t_i, \varphi^{Int}_{t_i}(0, \xi^{Int})))] - \mathbb{E}[f(\Phi(T; t_{i-1}, \varphi^{Int}_{t_{i-1}}(0, \xi^{Int})))] \Big\} \\
&= \sum_{i=1}^{N^\Delta} \mathbb{E}\Big[\mathbb{E}\Big[f(\Phi(T; t_i, \varphi^{Int}_{t_i}(t_{i-1}, \varphi_{t_{i-1}}(0, \xi^{Int})))) \\
&\quad - f(\Phi(T; t_i, \Phi_{t_i}(t_{i-1}, \varphi^{Int}_{t_{i-1}}(0, \xi^{Int})))) \Big| \mathcal{F}_{t_i} \Big]\Big]
\end{aligned}
$$

$$
\begin{aligned}
&= \sum_{i=1}^{N^{\Delta}} \int_0^1 \mathbb{E}\Big[f'(\Phi(T; t_i, Y_i^{\varsigma}))\mathcal{D}\Phi(T; t_i, Y_i^{\varsigma}) \\
&\quad \Big(\varphi_{t_i}^{Int}(t_{i-1}, \varphi_{t_{i-1}}(0, \xi^{Int})) - \Phi_{t_i}(t_{i-1}, \varphi_{t_{i-1}}^{Int}(0, \xi^{Int}))\Big)\Big] \mathrm{d}\varsigma, \qquad (3.7)
\end{aligned}
$$

where we used $\varphi_{t_i}^{Int}(t_{i-1}, \varphi_{t_{i-1}}(0, \xi^{Int})) = \varphi_{t_i}^{Int}(0, \xi^{Int})$, $\Phi(T; t_{i-1}, \varphi_{t_{i-1}}^{Int}(0, \xi^{Int})) = \Phi(T; t_i, \Phi_{t_i}(t_{i-1}, \varphi_{t_{i-1}}^{Int}(0, \xi^{Int})))$, and

$$
Y_i^{\varsigma} := \varsigma \varphi_{t_i}^{Int}(t_{i-1}, \varphi_{t_{i-1}}(0, \xi^{Int})) + (1-\varsigma)\Phi_{t_i}(t_{i-1}, \varphi_{t_{i-1}}^{Int}(0, \xi^{Int})). \qquad (3.8)
$$

We further deal with the C^{Int}-valued random variable

$$
\varphi_{t_i}^{Int}(t_{i-1}, \varphi_{t_{i-1}}(0, \xi^{Int})) - \Phi_{t_i}(t_{i-1}, \varphi_{t_{i-1}}^{Int}(0, \xi^{Int}))
$$

in (3.7). It follows from the expansion $\eta(\cdot) = \sum_{j=-N}^{0} I^{[j]}(\cdot)\eta(t_j)$ for $\eta \in C^{Int}$ and the relation

$$
\Phi(t_i + t_j; t_{i-1}, \varphi_{t_{i-1}}^{Int}(0, \xi^{Int})) = \varphi^{Int}(t_{i+j}; 0, \xi^{Int}), \quad j = -N, \ldots, -1
$$

that

$$
\begin{aligned}
&\varphi_{t_i}^{Int}(t_{i-1}, \varphi_{t_{i-1}}(0, \xi^{Int})) - \Phi_{t_i}(t_{i-1}, \varphi_{t_{i-1}}^{Int}(0, \xi^{Int})) \\
&\quad = \sum_{j=-N}^{0} I^{[j]}\Big(\varphi_{t_i}^{Int}(t_{i-1}, \varphi_{t_{i-1}}(0, \xi^{Int}))(t_j) - \Phi_{t_i}(t_{i-1}, \varphi_{t_{i-1}}^{Int}(0, \xi^{Int}))(t_j)\Big) \\
&\quad = \sum_{j=-N}^{0} I^{[j]}\Big(\varphi^{Int}(t_i + t_j; t_{i-1}, \varphi_{t_{i-1}}(0, \xi^{Int})) - \Phi(t_i + t_j; t_{i-1}, \varphi_{t_{i-1}}^{Int}(0, \xi^{Int}))\Big) \\
&\quad = \sum_{j=-N}^{-1} I^{[j]}\Big(\varphi(t_{i+j}; 0, \xi^{Int}) - \varphi^{Int}(t_{i+j}; 0, \xi^{Int})\Big) \\
&\qquad + I^{[0]}\Big(\varphi(t_i; t_{i-1}, \varphi_{t_{i-1}}(0, \xi^{Int})) - \Phi(t_i; t_{i-1}, \varphi_{t_{i-1}}^{Int}(0, \xi^{Int}))\Big),
\end{aligned}
$$

which together with the relation $\varphi^{Int}(t_l; 0, \xi^{Int}) = \varphi(t_l; 0, \xi^{Int})$ leads to

$$
\begin{aligned}
&\varphi_{t_i}^{Int}(t_{i-1}, \varphi_{t_{i-1}}(0, \xi^{Int})) - \Phi_{t_i}(t_{i-1}, \varphi_{t_{i-1}}^{Int}(0, \xi^{Int})) \\
&\quad = I^{[0]}\Big(\varphi(t_i; t_{i-1}, \varphi_{t_{i-1}}(0, \xi^{Int})) - \Phi(t_i; t_{i-1}, \varphi_{t_{i-1}}^{Int}(0, \xi^{Int}))\Big). \qquad (3.9)
\end{aligned}
$$

Note that for any $i \in \mathbb{N}_+$, by the expression of the solution, we have the following decomposition of the difference in (3.9)

$$\begin{aligned}
&\varphi(t_i; t_{i-1}, \varphi_{t_{i-1}}(0, \xi^{Int})) - \Phi(t_i; t_{i-1}, \varphi_{t_{i-1}}^{Int}(0, \xi^{Int})) \\
&= \int_{t_{i-1}}^{t_i} b(\varphi_r(t_{i-1}, \varphi_{t_{i-1}}(0, \xi^{Int})))\mathrm{d}r + \int_{t_{i-1}}^{t_i} \sigma(\varphi_r(t_{i-1}, \varphi_{t_{i-1}}(0, \xi^{Int})))\mathrm{d}W(r) \\
&\quad - \int_{t_{i-1}}^{t_i} \big[(1-\theta)b(\varphi_{t_{i-1}}^{Int}(0, \xi^{Int})) + \theta b(\Phi_{t_i}(t_{i-1}, \varphi_{t_{i-1}}^{Int}(0, \xi^{Int})))\big]\mathrm{d}r \\
&\quad - \int_{t_{i-1}}^{t_i} \sigma(\varphi_{t_{i-1}}^{Int}(0, \xi^{Int}))\mathrm{d}W(r) = \mathcal{I}_b^i + \mathcal{I}_{b,\theta}^i + \mathcal{I}_\sigma^i,
\end{aligned} \tag{3.10}$$

where

$$\begin{aligned}
\mathcal{I}_b^i &:= \int_{t_{i-1}}^{t_i} \big[b(\varphi_r(t_{i-1}, \varphi_{t_{i-1}}(0, \xi^{Int}))) - b(\varphi_{t_{i-1}}^{Int}(0, \xi^{Int}))\big]\mathrm{d}r, \\
\mathcal{I}_{b,\theta}^i &:= \theta \int_{t_{i-1}}^{t_i} \big[b(\varphi_{t_{i-1}}^{Int}(0, \xi^{Int})) - b(\Phi_{t_i}(t_{i-1}, \varphi_{t_{i-1}}^{Int}(0, \xi^{Int})))\big]\mathrm{d}r, \\
\mathcal{I}_\sigma^i &:= \int_{t_{i-1}}^{t_i} \big[\sigma(\varphi_r(t_{i-1}, \varphi_{t_{i-1}}(0, \xi^{Int}))) - \sigma(\varphi_{t_{i-1}}^{Int}(0, \xi^{Int}))\big]\mathrm{d}W(r).
\end{aligned}$$

According to (3.6)–(3.10), we derive the following decomposition of the weak error

$$\begin{aligned}
&\mathbb{E}[f(x^\xi(T))] - \mathbb{E}[f(y^{\xi,\Delta}(T))] \\
&= \mathbb{E}[f(\Phi(T; t_{N^\Delta}, \varphi_{t_{N^\Delta}}^{Int}(0, \xi)))] - \mathbb{E}[f(\Phi(T; t_{N^\Delta}, \varphi_{t_{N^\Delta}}^{Int}(0, \xi^{Int})))] \\
&\quad + \sum_{i=1}^{N^\Delta} \int_0^1 \mathbb{E}\Big[\Big\langle f'(\Phi(T; t_i, Y_i^\varsigma))\mathcal{D}\Phi(T; t_i, Y_i^\varsigma), 1^{[0]}(\mathcal{I}_b^i + \mathcal{I}_{b,\theta}^i + \mathcal{I}_\sigma^i)\Big\rangle\Big]\mathrm{d}\varsigma \\
&= \mathcal{I}_0 + \mathcal{I}_b + \mathcal{I}_{b,\theta} + \mathcal{I}_\sigma,
\end{aligned} \tag{3.11}$$

where

$$\begin{aligned}
\mathcal{I}_0 &:= \mathbb{E}[f(\Phi(T; t_{N^\Delta}, \varphi_{t_{N^\Delta}}^{Int}(0, \xi)))] - \mathbb{E}[f(\Phi(T; t_{N^\Delta}, \varphi_{t_{N^\Delta}}^{Int}(0, \xi^{Int})))], \\
\mathcal{I}_b &:= \sum_{i=1}^{N^\Delta} \int_0^1 \mathbb{E}\Big[\Big\langle f'(\Phi(T; t_i, Y_i^\varsigma))\mathcal{D}\Phi(T; t_i, Y_i^\varsigma), 1^{[0]}\mathcal{I}_b^i\Big\rangle\Big]\mathrm{d}\varsigma,
\end{aligned}$$

$$\mathcal{I}_{b,\theta} := \sum_{i=1}^{N^\Delta} \int_0^1 \mathbb{E}\Big[\Big\langle f'(\Phi(T;t_i,Y_i^\varsigma))\mathcal{D}\Phi(T;t_i,Y_i^\varsigma), I^{[0]}\mathcal{I}_{b,\theta}^i\Big\rangle\Big]\mathrm{d}\varsigma,$$

$$\mathcal{I}_\sigma := \sum_{i=1}^{N^\Delta} \int_0^1 \mathbb{E}\Big[\Big\langle f'(\Phi(T;t_i,Y_i^\varsigma))\mathcal{D}\Phi(T;t_i,Y_i^\varsigma), I^{[0]}\mathcal{I}_\sigma^i\Big\rangle\Big]\mathrm{d}\varsigma.$$

3.2.1 A Priori Estimates

In this subsection, we present some a priori estimates for the exact functional solution and the θ-EM functional solution. These estimates encompass the time-independent boundedness of high-order moments for the Malliavin derivatives and Gâteaux derivatives of both exact and numerical solutions, ensuring uniform control of the growth of solutions over a long time. These fundamental estimates play a crucial role in the longtime weak convergence analysis of the numerical method. We introduce the following assumptions on the coefficients σ and b.

Assumption 3.1 *Assume that coefficients b and σ have continuous first-order Gâteaux derivatives.*

For any $\phi, \phi_1 \in C^d$, and $t \in \mathbb{R}$, it follows from Assumption 3.1 and the definition of the Gâteaux derivative that $b(\phi_1 + t\phi) - b(\phi_1) = t\mathcal{D}b(\phi_1)\phi + o(t)$, where the symbol $o(\cdot)$ represents a higher-order infinitesimal quantity. Then based on Assumptions 2.1 and 2.2, we have

$$\langle \mathcal{D}b(\phi_1)\phi, \phi(0)\rangle \le -a_1|\phi(0)|^2 + a_2\int_{-\tau}^0 |\phi(r)|^2 \mathrm{d}\nu_2(r), \tag{3.12}$$

$$|\mathcal{D}\sigma(\phi_1)\phi|^2 \le L\Big(|\phi(0)|^2 + \int_{-\tau}^0 |\phi(r)|^2\mathrm{d}\nu_1(r)\Big), \tag{3.13}$$

where $\phi, \phi_1 \in C^d$; see [6, Eq. (2.4)] for a similar proof.

The following lemmas give some regularity estimates of both the exact solution and the numerical solution.

Lemma 3.4 *Let Assumptions 2.1and 2.2 with $a_1 - a_2 - (2p-1)L > 0$ hold for some $p \ge 1$. And let Assumption 3.1 hold. Then for any $t \ge 0$,*

$$\mathbb{E}[|\mathcal{D}x^{\xi}(t)\cdot\eta|^{2p}] \le K\|\eta\|^{2p}e^{-\hat{\lambda}_1 pt},$$

$$\mathbb{E}[\|\mathcal{D}x_t^{\xi}\cdot\eta\|^{2p}] \le K\|\eta\|^{2p}e^{-\hat{\lambda}_1 pt},$$

where $\xi \in L^{2p}(\Omega; C^d)$, $\eta \in C^d$, and constants $K, \hat{\lambda}_1 > 0$.

Proof For $\eta \in \mathcal{C}^d$, $\mathcal{D}x^{\xi}(t)\cdot\eta$ satisfies

$$\mathrm{d}\mathcal{D}x^{\xi}(t)\cdot\eta = \mathcal{D}b(x_t^{\xi})\mathcal{D}x_t^{\xi}\cdot\eta\mathrm{d}t + \mathcal{D}\sigma(x_t^{\xi})\mathcal{D}x_t^{\xi}\cdot\eta\mathrm{d}W(t), \quad t>0$$

with the initial datum $\mathcal{D}x^{\xi}(r)\cdot\eta = \eta(r)$ for $r\in[-\tau,0]$. Let $\hat{\lambda}_1\in(0,1)$ be determined later. Applying the Itô formula, we obtain

$$\begin{aligned}
&\mathrm{d}(e^{\hat{\lambda}_1 pt}|\mathcal{D}x^{\xi}(t)\cdot\eta|^{2p})\\
&\quad=\hat{\lambda}_1 pe^{\hat{\lambda}_1 pt}|\mathcal{D}x^{\xi}(t)\cdot\eta|^{2p}\mathrm{d}t + 2pe^{\hat{\lambda}_1 pt}|\mathcal{D}x^{\xi}(t)\cdot\eta|^{2p-2}\\
&\qquad\times\Big(\langle\mathcal{D}x^{\xi}(t)\cdot\eta, \mathcal{D}b(x_t^{\xi})\mathcal{D}x_t^{\xi}\cdot\eta\rangle\mathrm{d}t + \langle\mathcal{D}x^{\xi}(t)\cdot\eta, \mathcal{D}\sigma(x_t^{\xi})\mathcal{D}x_t^{\xi}\cdot\eta\mathrm{d}W(t)\rangle\Big)\\
&\qquad+2p(p-1)e^{\hat{\lambda}_1 pt}|\mathcal{D}x^{\xi}(t)\cdot\eta|^{2p-4}|(\mathcal{D}\sigma(x_t^{\xi})\mathcal{D}x_t^{\xi}\cdot\eta)^{\top}\mathcal{D}x^{\xi}(t)\cdot\eta|^2\mathrm{d}t\\
&\qquad+pe^{\hat{\lambda}_1 pt}|\mathcal{D}x^{\xi}(t)\cdot\eta|^{2p-2}|\mathcal{D}\sigma(x_t^{\xi})\mathcal{D}x_t^{\xi}\cdot\eta|^2\mathrm{d}t.
\end{aligned}$$

Combining (3.12) and (3.13), we deduce

$$\begin{aligned}
&e^{\hat{\lambda}_1 pt}|\mathcal{D}x^{\xi}(t)\cdot\eta|^{2p}\\
&\quad\le|\eta(0)|^{2p}+\int_0^t\hat{\lambda}_1 pe^{\hat{\lambda}_1 ps}|\mathcal{D}x^{\xi}(s)\cdot\eta|^{2p}\mathrm{d}s+\int_0^t 2pe^{\hat{\lambda}_1 ps}|\mathcal{D}x^{\xi}(s)\cdot\eta|^{2p-2}\\
&\qquad\times\Big[-a_1|\mathcal{D}x^{\xi}(s)\cdot\eta|^2+a_2\int_{-\tau}^0|\mathcal{D}x^{\xi}(s+r)\cdot\eta|^2\mathrm{d}\nu_2(r)\Big]\mathrm{d}s+\int_0^t 2pe^{\hat{\lambda}_1 ps}\\
&\qquad\times|\mathcal{D}x^{\xi}(s)\cdot\eta|^{2p-2}\big\langle\mathcal{D}x^{\xi}(s)\cdot\eta, \mathcal{D}\sigma(x_s^{\xi})\mathcal{D}x_s^{\xi}\cdot\eta\mathrm{d}W(s)\big\rangle+\int_0^t p(2p-1)e^{\hat{\lambda}_1 ps}\\
&\qquad\times|\mathcal{D}x^{\xi}(s)\cdot\eta|^{2p-2}L\Big[|\mathcal{D}x^{\xi}(s)\cdot\eta|^2+\int_{-\tau}^0|\mathcal{D}x^{\xi}(s+r)\cdot\eta|^2\mathrm{d}\nu_1(r)\Big]\mathrm{d}s.
\end{aligned}$$

For the term concerning the double integrals, it follows from the Hölder inequality and the Fubini theorem that

$$\begin{aligned}
&\int_0^t e^{\hat{\lambda}_1 ps}|\mathcal{D}x^{\xi}(s)\cdot\eta|^{2p-2}\int_{-\tau}^0|\mathcal{D}x^{\xi}(s+r)\cdot\eta|^2\mathrm{d}\nu_i(r)\mathrm{d}s\\
&\quad\le\frac{p-1}{p}\int_0^t e^{\hat{\lambda}_1 ps}|\mathcal{D}x^{\xi}(s)\cdot\eta|^{2p}\mathrm{d}s+\frac{1}{p}\int_0^t e^{\hat{\lambda}_1 ps}\int_{-\tau}^0|\mathcal{D}x^{\xi}(s+r)\cdot\eta|^{2p}\mathrm{d}\nu_i(r)\mathrm{d}s\\
&\quad\le\frac{p-1}{p}\int_0^t e^{\hat{\lambda}_1 ps}|\mathcal{D}x^{\xi}(s)\cdot\eta|^{2p}\mathrm{d}s+\frac{1}{p}e^{\hat{\lambda}_1 p\tau}\int_{-\tau}^t e^{\hat{\lambda}_1 pz}|\mathcal{D}x^{\xi}(z)\cdot\eta|^{2p}\mathrm{d}z\\
&\quad\le\frac{1}{p}\tau e^{\hat{\lambda}_1 p\tau}\|\eta\|^{2p}+e^{\hat{\lambda}_1 p\tau}\int_0^t e^{\hat{\lambda}_1 ps}|\mathcal{D}x^{\xi}(s)\cdot\eta|^{2p}\mathrm{d}s, \quad i=1,2.
\end{aligned}$$

Hence, we derive

$$
\begin{aligned}
&\mathbb{E}[e^{\hat{\lambda}_1 pt}|\mathcal{D}x^{\xi}(t)\cdot\eta|^{2p}] \\
&\quad\le|\eta(0)|^{2p}+p\Big(\hat{\lambda}_1-2a_1+2a_2e^{\hat{\lambda}_1 p\tau}+(2p-1)L(1+e^{\hat{\lambda}_1 p\tau})\Big) \\
&\qquad\times\int_0^t e^{\hat{\lambda}_1 ps}\mathbb{E}[|\mathcal{D}x^{\xi}(s)\cdot\eta|^{2p}]\mathrm{d}s+(2a_2+(2p-1)L)\tau e^{\hat{\lambda}_1 p\tau}\|\eta\|^{2p}.
\end{aligned}
$$

Owing to $a_1-a_2-(2p-1)L>0$, there exists a sufficiently small number $\hat{\lambda}_1>0$ such that

$$
\hat{\lambda}_1-2a_1+2a_2e^{\hat{\lambda}_1 p\tau}+(2p-1)L(1+e^{\hat{\lambda}_1 p\tau})\le 0
$$

This gives $\mathbb{E}[e^{\hat{\lambda}_1 pt}|\mathcal{D}x^{\xi}(t)\cdot\eta|^{2p}]\le K\|\eta\|^{2p}$. Moreover, applying the Burkholder–Davis–Gundy inequality, we have

$$
\begin{aligned}
&\mathbb{E}\Big[\sup_{0\vee(t-\tau)\le r\le t}e^{\hat{\lambda}_1 pr}|\mathcal{D}x^{\xi}(r)\cdot\eta|^{2p}\Big] \\
&\quad\le|\eta(0)|^{2p}+(2a_2+(2p-1)L)\tau e^{\hat{\lambda}_1 p\tau}\|\eta\|^{2p}+\mathbb{E}\Big[\sup_{0\vee(t-\tau)\le r\le t}\int_0^r 2pe^{\hat{\lambda}_1 ps} \\
&\qquad\times|\mathcal{D}x^{\xi}(s)\cdot\eta|^{2p-2}\Big\langle\mathcal{D}x^{\xi}(s)\cdot\eta,\mathcal{D}\sigma(x_s^{\xi})\mathcal{D}x_s^{\xi}\cdot\eta\mathrm{d}W(s)\Big\rangle\Big] \\
&\quad\le K\|\eta\|^{2p}+\mathbb{E}\Big[\sup_{0\vee(t-\tau)\le r\le t}\int_{0\vee(t-\tau)}^r 2pe^{\hat{\lambda}_1 ps}|\mathcal{D}x^{\xi}(s)\cdot\eta|^{2p-2}\Big\langle\mathcal{D}x^{\xi}(s)\cdot\eta, \\
&\qquad\mathcal{D}\sigma(x_s^{\xi})\mathcal{D}x_s^{\xi}\cdot\eta\mathrm{d}W(s)\Big\rangle\Big] \\
&\quad\le K\|\eta\|^{2p}+K\mathbb{E}\Big[\Big(\int_{0\vee(t-\tau)}^t e^{2\hat{\lambda}_1 ps}|\mathcal{D}x^{\xi}(s)\cdot\eta|^{4p-2}|\mathcal{D}\sigma(x_s^{\xi})\mathcal{D}x_s^{\xi}\cdot\eta|^2\mathrm{d}s\Big)^{\frac{1}{2}}\Big].
\end{aligned}
$$

Then, we arrive at

$$
\begin{aligned}
&\mathbb{E}\Big[\sup_{0\vee(t-\tau)\le r\le t}e^{\hat{\lambda}_1 pr}|\mathcal{D}x^{\xi}(r)\cdot\eta|^{2p}\Big] \\
&\quad\le K\|\eta\|^{2p}+K\mathbb{E}\Big[\sup_{0\vee(t-\tau)\le s\le t}e^{\hat{\lambda}_1(p-\frac{1}{2})s}|\mathcal{D}x^{\xi}(s)\cdot\eta|^{2p-1} \\
&\qquad\times\Big(\int_{0\vee(t-\tau)}^t e^{\hat{\lambda}_1 s}|\mathcal{D}\sigma(x_s^{\xi})\mathcal{D}x_s^{\xi}\cdot\eta|^2\mathrm{d}s\Big)^{\frac{1}{2}}\Big].
\end{aligned}
$$

According to the Young inequality, (3.13), and the Hölder inequality, we obtain

$$
\begin{aligned}
&\mathbb{E}\Big[\sup_{0\vee(t-\tau)\le r\le t} e^{\hat{\lambda}_1 pr}|\mathcal{D}x^{\tilde{\xi}}(r)\cdot\eta|^{2p}\Big]\\
&\le K\|\eta\|^{2p}+\frac{1}{2}\mathbb{E}\Big[\sup_{0\vee(t-\tau)\le s\le t} e^{\hat{\lambda}_1 ps}|\mathcal{D}x^{\tilde{\xi}}(s)\cdot\eta|^{2p}\Big]\\
&\quad+K\mathbb{E}\Big[\Big(\int_{0\vee(t-\tau)}^{t} e^{\hat{\lambda}_1 s}\big(|\mathcal{D}x^{\tilde{\xi}}(s)\cdot\eta|^2+\int_{-\tau}^{0}|\mathcal{D}x^{\tilde{\xi}}(s+r)\cdot\eta|^2\mathrm{d}\nu_1(r)\big)\mathrm{d}s\Big)^p\Big]\\
&\le K\|\eta\|^{2p}+\frac{1}{2}\mathbb{E}\Big[\sup_{0\vee(t-\tau)\le s\le t} e^{\hat{\lambda}_1 ps}|\mathcal{D}x^{\tilde{\xi}}(s)\cdot\eta|^{2p}\Big]\\
&\quad+\tau^{p-1}K\int_{0\vee(t-\tau)}^{t}\sup_{-\tau\le r\le s}\mathbb{E}\big[e^{\hat{\lambda}_1 pr}|\mathcal{D}x^{\tilde{\xi}}(r)\cdot\eta|^{2p}\big]\mathrm{d}s.
\end{aligned}
$$

Making use of

$$
\mathbb{E}[e^{\hat{\lambda}_1 pt}|\mathcal{D}x^{\tilde{\xi}}(t)\cdot\eta|^{2p}]\le K\|\eta\|^{2p},
$$

we derive

$$
e^{\hat{\lambda}_1 p(0\vee(t-\tau))}\mathbb{E}[\|\mathcal{D}x_t^{\tilde{\xi}}\cdot\eta\|^{2p}]\le K\|\eta\|^{2p},
$$

which gives

$$
\mathbb{E}[\|\mathcal{D}x_t^{\tilde{\xi}}\cdot\eta\|^{2p}]\le K\|\eta\|^{2p}e^{-\hat{\lambda}_1 pt}.
$$

The proof is finished. □

Lemma 3.5 *Let Assumptions 2.1 and 2.2 with $a_1-a_2-(2p-1)L>0$ hold for some $p\ge 1$. Assume further that Assumptions 2.3 and 3.1 hold. Then for any $t\ge 0$ and $u\ge -\tau$,*

$$
\mathbb{E}[\|D_u x_t^{\tilde{\xi}}\|^{2p}]\le Ke^{-\hat{\lambda}_1 p(t-u)}\Big(1+\mathbb{E}[\|\tilde{\xi}\|^{2p}]+\mathbb{E}[\|D_u\tilde{\xi}\|^{2p}]\Big),
$$

where $\tilde{\xi}\in L^{2p}(\Omega;C^d)$ and $D_u\tilde{\xi}\in L^{2p}(\Omega;C^d\otimes\mathbb{R}^m)$, $K>0$, and $\hat{\lambda}_1>0$ is given in Lemma 3.4.

Proof When $0\le u\le t$, $D_u x^{\tilde{\xi}}(t)$ satisfies

$$
\begin{aligned}
D_u x^{\tilde{\xi}}(t)&=\int_u^t \mathcal{D}b(x_s^{\tilde{\xi}})D_u x_s^{\tilde{\xi}}\mathrm{d}s+\int_u^t \mathcal{D}\sigma(x_s^{\tilde{\xi}})D_u x_s^{\tilde{\xi}}\mathrm{d}W(s)\\
&\quad+\sigma(x_u^{\tilde{\xi}})\mathbf{1}_{[0,t]}(u)+D_u\tilde{\xi}(0).
\end{aligned}
$$

By the Itô formula, we derive

$$
\begin{aligned}
&\mathrm{d}(e^{\hat{\lambda}_1 pt}|D_u x^{\tilde{\xi}}(t)|^{2p}) \\
&\quad \le \hat{\lambda}_1 p e^{\hat{\lambda}_1 pt}|D_u x^{\tilde{\xi}}(t)|^{2p}\mathrm{d}t + 2pe^{\hat{\lambda}_1 pt}|D_u x^{\tilde{\xi}}(t)|^{2p-2} \\
&\qquad \times \Big(\langle D_u x^{\tilde{\xi}}(t), \mathcal{D}b(x_t^{\tilde{\xi}})D_u x_t^{\tilde{\xi}}\rangle \mathrm{d}t + \langle D_u x^{\tilde{\xi}}(t), \mathcal{D}\sigma(x_t^{\tilde{\xi}})D_u x_t^{\tilde{\xi}}\mathrm{d}W(t)\rangle\Big) \\
&\qquad + p(2p-1)e^{\hat{\lambda}_1 pt}|D_u x^{\tilde{\xi}}(t)|^{2p-2}|\mathcal{D}\sigma(x_t^{\tilde{\xi}})D_u x_t^{\tilde{\xi}}|^2\mathrm{d}t.
\end{aligned}
$$

Similar to the proof of Lemma 3.4, we have

$$
\begin{aligned}
&\mathbb{E}[e^{\hat{\lambda}_1 pt}|D_u x^{\tilde{\xi}}(t)|^{2p}] \\
&\quad \le K\mathbb{E}[e^{\hat{\lambda}_1 pu}(|\sigma(x_u^{\tilde{\xi}})|^{2p} + |D_u\tilde{\xi}(0)|^{2p})] \\
&\qquad + p\big(\hat{\lambda}_1 - 2a_1 + 2a_2 e^{\hat{\lambda}_1 p\tau} + (2p-1)L(1+e^{\hat{\lambda}_1 p\tau})\big) \\
&\qquad \times \int_u^t e^{\hat{\lambda}_1 ps}\mathbb{E}[|D_u x^{\tilde{\xi}}(s)|^{2p}]\mathrm{d}s \\
&\qquad + (2a_2 + (2p-1)L)\tau e^{\hat{\lambda}_1 p\tau}\|D_u\tilde{\xi}\|^{2p} \\
&\quad \le K\mathbb{E}[e^{\hat{\lambda}_1 pu}(|\sigma(x_u^{\tilde{\xi}})|^{2p} + |D_u\tilde{\xi}(0)|^{2p}] + (2a_2 + (2p-1)L)\tau e^{\hat{\lambda}_1 p\tau}\|D_u\tilde{\xi}\|^{2p},
\end{aligned}
$$

where we used the facts $D_u x^{\tilde{\xi}}(s) = 0$ for $u > s$ and

$$
\begin{aligned}
&\int_u^t \int_{-\tau}^0 e^{\hat{\lambda}_1 ps}|D_u x_s^{\tilde{\xi}}(r)|^{2p}\mathrm{d}\nu_i(r)\mathrm{d}s \\
&\quad = \int_0^t \int_{-\tau}^0 e^{\hat{\lambda}_1 ps}|D_u x_s^{\tilde{\xi}}(r)|^{2p}\mathrm{d}\nu_i(r)\mathrm{d}s \\
&\quad \le \tau e^{\hat{\lambda}_1 p\tau}\|D_u\tilde{\xi}\|^{2p} + e^{\hat{\lambda}_1 p\tau}\int_u^t e^{\hat{\lambda}_1 ps}|D_u x^{\tilde{\xi}}(s)|^{2p}\mathrm{d}s, \quad i = 1, 2.
\end{aligned}
$$

This, along with Assumption 2.1 and (2.35) implies that

$$
\mathbb{E}[|D_u x^{\tilde{\xi}}(t)|^{2p}] \le Ke^{-\hat{\lambda}_1 p(t-u)}\Big(1 + \mathbb{E}[\|\tilde{\xi}\|^{2p}] + \mathbb{E}[\|D_u\tilde{\xi}\|^{2p}]\Big).
$$

Moreover, to obtain the estimates of the functional solution, we apply the chain rule of the Malliavin derivatives, and then use the Burkholder–Davis–Gundy inequality to derive that

$$
\begin{aligned}
&\mathbb{E}\Big[\sup_{u\vee(t-\tau)\le r\le t} e^{\hat{\lambda}_1 pr}|D_u x^{\tilde{\xi}}(r)|^{2p}\Big]\\
&\le Ke^{\hat{\lambda}_1 pu}\mathbb{E}[|\sigma(x_u^{\tilde{\xi}})|^{2p}+|D_u\tilde{\xi}(0)|^{2p}\mathbf{1}_{[-\tau,0]}(u)]+(2a_2+(2p-1)L)\tau e^{\hat{\lambda}_1 p\tau}\\
&\quad\times\|D_u\tilde{\xi}\|^{2p}+\mathbb{E}\Big[\sup_{u\vee(t-\tau)\le r\le t}\int_u^r 2pe^{\hat{\lambda}_1 ps}\Big\langle|D_u x^{\tilde{\xi}}(s)|^{2p-2}D_u x^{\tilde{\xi}}(s),\\
&\quad\mathcal{D}\sigma(x_s^{\tilde{\xi}})D_u x_s^{\tilde{\xi}}\mathrm{d}W(s)\Big\rangle\Big]\\
&\le Ke^{\hat{\lambda}_1 pu}\Big(1+\mathbb{E}[\|\tilde{\xi}\|^{2p}]+\mathbb{E}[\|D_u\tilde{\xi}\|^{2p}]\Big)\\
&\quad+K\mathbb{E}\Big[\Big(\int_{u\vee(t-\tau)}^t e^{2\hat{\lambda}_1 ps}|D_u x^{\tilde{\xi}}(s)|^{4p-2}|\mathcal{D}\sigma(x_s^{\tilde{\xi}})D_u x_s^{\tilde{\xi}}|^2\mathrm{d}s\Big)^{\frac{1}{2}}\Big]\\
&\le Ke^{\hat{\lambda}_1 pu}\Big(1+\mathbb{E}[\|\tilde{\xi}\|^{2p}]+\mathbb{E}[\|D_u\tilde{\xi}\|^{2p}]\Big)+K\mathbb{E}\Big[\Big(\sup_{u\vee(t-\tau)\le s\le t} e^{\hat{\lambda}_1(p-\frac{1}{2})s}\\
&\quad\times|D_u x^{\tilde{\xi}}(s)|^{2p-1}\Big)\Big(\int_{u\vee(t-\tau)}^t e^{\hat{\lambda}_1 s}|\mathcal{D}\sigma(x_s^{\tilde{\xi}})D_u x_s^{\tilde{\xi}}|^2\mathrm{d}s\Big)^{\frac{1}{2}}\Big].
\end{aligned}
$$

Applying the Young inequality to the above inequality, we derive

$$
\begin{aligned}
&\mathbb{E}\Big[\sup_{u\vee(t-\tau)\le r\le t} e^{\hat{\lambda}_1 pr}|D_u x^{\tilde{\xi}}(r)|^{2p}\Big]\\
&\le Ke^{\hat{\lambda}_1 pu}\Big(1+\mathbb{E}[\|\tilde{\xi}\|^{2p}]+\mathbb{E}[\|D_u\tilde{\xi}\|^{2p}]\Big)\\
&\quad+\frac{1}{2}\mathbb{E}\Big[\sup_{u\vee(t-\tau)\le s\le t} e^{\hat{\lambda}_1 ps}|D_u x^{\tilde{\xi}}(s)|^{2p}\Big]\\
&\quad+K\mathbb{E}\Big[\Big(\int_{u\vee(t-\tau)}^t e^{\hat{\lambda}_1 s}|\mathcal{D}\sigma(x_s^{\tilde{\xi}})D_u x_s^{\tilde{\xi}}|^2\mathrm{d}s\Big)^p\Big]\\
&\le Ke^{\hat{\lambda}_1 pu}\Big(1+\mathbb{E}[\|\tilde{\xi}\|^{2p}]+\mathbb{E}[\|D_u\tilde{\xi}\|^{2p}]\Big)\\
&\quad+\frac{1}{2}\mathbb{E}\Big[\sup_{u\vee(t-\tau)\le s\le t} e^{\hat{\lambda}_1 ps}|D_u x^{\tilde{\xi}}(s)|^{2p}\Big]\\
&\quad+\tau^{p-1}K\int_{u\vee(t-\tau)}^t e^{\hat{\lambda}_1 ps}\Big(\sup_{s-\tau\le r\le s}\mathbb{E}\big[|D_u x^{\tilde{\xi}}(r)|^{2p}\big]\Big)\mathrm{d}s\\
&\le K(1+\tau^p)e^{\hat{\lambda}_1 pu}\Big(1+\mathbb{E}[\|\tilde{\xi}\|^{2p}]+\mathbb{E}[\|D_u\tilde{\xi}\|^{2p}]\Big).
\end{aligned}
$$

Hence,

$$
e^{\hat{\lambda}_1 p(t-\tau)}\mathbb{E}[\|D_u x_t^{\tilde{\xi}}\|^{2p}]\le Ke^{\hat{\lambda}_1 pu}\Big(1+\mathbb{E}[\|\tilde{\xi}\|^{2p}]+\mathbb{E}[\|D_u\tilde{\xi}\|^{2p}]\Big),
$$

which implies

$$\mathbb{E}[\|D_u x_t^{\tilde{\xi}}\|^{2p}] \le K e^{\hat{\lambda}_1 p(u-t)}\Big(1+\mathbb{E}[\|\tilde{\xi}\|^{2p}]+\mathbb{E}[\|D_u\tilde{\xi}\|^{2p}]\Big).$$

We complete the proof. □

The pth moment of $\{\mathcal{D}y_{t_k}^{\tilde{\xi},\Delta}\cdot\eta\}_{k\in\mathbb{N}}$ is estimated based on the auxiliary process $\{\mathcal{D}z_{t_k}^{\tilde{\xi},\Delta}\cdot\eta\}_{k\in\mathbb{N}}$. Precisely, by (1.43) and (1.44), taking the Gâteaux derivatives on $z_{t_k}^{\tilde{\xi},\Delta}$, one has that for $k\in\mathbb{N}$,

$$\begin{aligned}
&\mathcal{D}z^{\tilde{\xi},\Delta}(0)\cdot\eta=\eta(0)-\theta\Delta\mathcal{D}b(\tilde{\xi})\cdot\eta,\\
&\mathcal{D}z^{\tilde{\xi},\Delta}(t_k)\cdot\eta=\mathcal{D}y^{\tilde{\xi},\Delta}(t_k)\cdot\eta-\theta\Delta\mathcal{D}b(y_{t_k}^{\tilde{\xi},\Delta})\mathcal{D}y_{t_k}^{\tilde{\xi},\Delta}\cdot\eta,
\end{aligned}\tag{3.14}$$

and

$$\begin{aligned}
\mathcal{D}z^{\tilde{\xi},\Delta}(t_{k+1})\cdot\eta=\;&\mathcal{D}z^{\tilde{\xi},\Delta}(t_k)\cdot\eta+\mathcal{D}b(y_{t_k}^{\tilde{\xi},\Delta})\mathcal{D}y_{t_k}^{\tilde{\xi},\Delta}\cdot\eta\Delta\\
&+\mathcal{D}\sigma(y_{t_k}^{\tilde{\xi},\Delta})\mathcal{D}y_{t_k}^{\tilde{\xi},\Delta}\cdot\eta\delta W_k.
\end{aligned}\tag{3.15}$$

Using (3.12) and the relation (3.14), we derive

$$\begin{aligned}
&|\mathcal{D}z^{\tilde{\xi},\Delta}(t_k)\cdot\eta|^2\\
&\quad\ge|\mathcal{D}y^{\tilde{\xi},\Delta}(t_k)\cdot\eta|^2-2\theta\Delta\langle\mathcal{D}y^{\tilde{\xi},\Delta}(t_k)\cdot\eta,\mathcal{D}b(y_{t_k}^{\tilde{\xi},\Delta})\mathcal{D}y_{t_k}^{\tilde{\xi},\Delta}\cdot\eta\rangle\\
&\quad\ge(1+2a_1\theta\Delta)|\mathcal{D}y^{\tilde{\xi},\Delta}(t_k)\cdot\eta|^2-2a_2\theta\Delta\int_{-\tau}^0|\mathcal{D}y_{t_k}^{\tilde{\xi},\Delta}(r)\cdot\eta|^2\mathrm{d}\nu_2(r),
\end{aligned}$$

which deduces

$$\begin{aligned}
&(1+2a_1\theta\Delta)|\mathcal{D}y^{\tilde{\xi},\Delta}(t_k)\cdot\eta|^2\\
&\quad\le|\mathcal{D}z^{\tilde{\xi},\Delta}(t_k)\cdot\eta|^2+2a_2\theta\Delta\int_{-\tau}^0|\mathcal{D}y_{t_k}^{\tilde{\xi},\Delta}(r)\cdot\eta|^2\mathrm{d}\nu_2(r).
\end{aligned}\tag{3.16}$$

In the following, we give the estimate of $\{\mathcal{D}z^{\tilde{\xi},\Delta}(t_k)\cdot\eta\}_{k\in\mathbb{N}}$ by considering two distinct cases: $\Delta\in(0,\tilde{\Delta}]$ and $\Delta\in(\tilde{\Delta},1]$ for some $\tilde{\Delta}\in(0,1)$.

Lemma 3.6 *Let Assumptions 2.1, 2.2, 2.4, and 3.1 hold. Then for $p \in \mathbb{N}_+$, there exist constants $\tilde{\mathfrak{a}}_1 := \tilde{\mathfrak{a}}_1(L, a_2, \theta, p) > 0$, $\tilde{\lambda}_2 > 0$, and $\tilde{\Delta} \in (0, 1)$ such that for $a_1 > \tilde{\mathfrak{a}}_1$ and $k \in \mathbb{N}$,*

$$\sup_{\Delta \in (0, \tilde{\Delta}]} \mathbb{E}[|\mathcal{D}z^{\tilde{\xi}, \Delta}(t_{k+1}) \cdot \eta|^{2p}] \leq K e^{-\tilde{\lambda}_2 p t_{k+1}} \|\eta\|^{2p} (1 + \mathbb{E}[\|\tilde{\xi}\|^{2p\beta}]).$$

where $\tilde{\xi} \in L^{2p\beta}(\Omega; C^d)$, $\eta \in C^d$, and $K > 0$.

Proof For the case of $p = 1$, one can obtain the result by using the similar technique to that in Lemma 2.3. We are in the position to show the proof of $p > 1$.

Squaring both sides of (3.15), we obtain

$$\begin{aligned}
&|\mathcal{D}z^{\tilde{\xi}, \Delta}(t_{k+1}) \cdot \eta|^2 \\
&= |\mathcal{D}z^{\tilde{\xi}, \Delta}(t_k) \cdot \eta|^2 + |\mathcal{D}b(y_{t_k}^{\tilde{\xi}, \Delta})\mathcal{D}y_{t_k}^{\tilde{\xi}, \Delta} \cdot \eta|^2 \Delta^2 + |\mathcal{D}\sigma(y_{t_k}^{\tilde{\xi}, \Delta})\mathcal{D}y_{t_k}^{\tilde{\xi}, \Delta} \cdot \eta \delta W_k|^2 \\
&\quad + \mathcal{M}_k + 2\langle \mathcal{D}z^{\tilde{\xi}, \Delta}(t_k) \cdot \eta, \mathcal{D}b(y_{t_k}^{\tilde{\xi}, \Delta})\mathcal{D}y_{t_k}^{\tilde{\xi}, \Delta} \cdot \eta \rangle \Delta,
\end{aligned}$$

where $\mathcal{M}_k := 2\langle \mathcal{D}z^{\tilde{\xi}, \Delta}(t_k) \cdot \eta + \mathcal{D}b(y_{t_k}^{\tilde{\xi}, \Delta})\mathcal{D}y_{t_k}^{\tilde{\xi}, \Delta} \cdot \eta\Delta, \mathcal{D}\sigma(y_{t_k}^{\tilde{\xi}, \Delta})\mathcal{D}y_{t_k}^{\tilde{\xi}, \Delta} \cdot \eta \delta W_k \rangle$. It follows from (3.12) and (3.14) that

$$\begin{aligned}
&|\mathcal{D}z^{\tilde{\xi}, \Delta}(t_{k+1}) \cdot \eta|^2 \\
&= \frac{(1-\theta)^2}{\theta^2} |\mathcal{D}z^{\tilde{\xi}, \Delta}(t_k) \cdot \eta|^2 + \frac{1-2\theta}{\theta^2} |\mathcal{D}y^{\tilde{\xi}, \Delta}(t_k) \cdot \eta|^2 \\
&\quad + \frac{2(2\theta - 1)}{\theta^2} \langle \mathcal{D}y^{\tilde{\xi}, \Delta}(t_k) \cdot \eta, \mathcal{D}z^{\tilde{\xi}, \Delta}(t_k) \cdot \eta \rangle + 2\langle \mathcal{D}y^{\tilde{\xi}, \Delta}(t_k) \cdot \eta, \\
&\quad \mathcal{D}b(y_{t_k}^{\tilde{\xi}, \Delta})\mathcal{D}y_{t_k}^{\tilde{\xi}, \Delta} \cdot \eta \rangle \Delta + |\mathcal{D}\sigma(y_{t_k}^{\tilde{\xi}, \Delta})\mathcal{D}y_{t_k}^{\tilde{\xi}, \Delta} \cdot \eta \delta W_k|^2 + \mathcal{M}_k \\
&\leq A_{\theta, \Delta} |\mathcal{D}z^{\tilde{\xi}, \Delta}(t_k) \cdot \eta|^2 + \frac{2\theta - 1}{\theta^2} \Delta |\mathcal{D}y^{\tilde{\xi}, \Delta}(t_k) \cdot \eta|^2 \\
&\quad + 2\langle \mathcal{D}y^{\tilde{\xi}, \Delta}(t_k) \cdot \eta, \mathcal{D}b(y_{t_k}^{\tilde{\xi}, \Delta})\mathcal{D}y_{t_k}^{\tilde{\xi}, \Delta} \cdot \eta \rangle \Delta \\
&\quad + |\mathcal{D}\sigma(y_{t_k}^{\tilde{\xi}, \Delta})\mathcal{D}y_{t_k}^{\tilde{\xi}, \Delta} \cdot \eta \delta W_k|^2 + \mathcal{M}_k \\
&\leq A_{\theta, \Delta} |\mathcal{D}z^{\tilde{\xi}, \Delta}(t_k) \cdot \eta|^2 + \frac{2\theta - 1}{\theta^2} \Delta |\mathcal{D}y^{\tilde{\xi}, \Delta}(t_k) \cdot \eta|^2 - 2\Delta a_1 |\mathcal{D}y^{\tilde{\xi}, \Delta}(t_k) \cdot \eta|^2 \\
&\quad + 2a_2 \Delta \int_{-\tau}^{0} |\mathcal{D}y_{t_k}^{\tilde{\xi}, \Delta}(r) \cdot \eta|^2 \mathrm{d}\nu_2(r) + |\mathcal{D}\sigma(y_{t_k}^{\tilde{\xi}, \Delta})\mathcal{D}y_{t_k}^{\tilde{\xi}, \Delta} \cdot \eta \delta W_k|^2 + \mathcal{M}_k,
\end{aligned} \tag{3.17}$$

where $A_{\theta,\Delta} = \frac{(1-\theta)^2}{\theta^2} + \frac{2\theta-1}{\theta^2(1+\Delta)}$. Hence, taking the pth power on both sides of the above inequality yields

$$\mathbb{E}[|\mathcal{D}z^{\tilde{\xi},\Delta}(t_{k+1}) \cdot \eta|^{2p}] \le A^p_{\theta,\Delta}\mathbb{E}[|\mathcal{D}z^{\tilde{\xi},\Delta}(t_k) \cdot \eta|^{2p}] + \sum_{i=1}^{p} C^i_p \mathbb{E}[\mathcal{I}_i], \tag{3.18}$$

where

$$\mathcal{I}_i := A^{p-i}_{\theta,\Delta}|\mathcal{D}z^{\tilde{\xi},\Delta}(t_k) \cdot \eta|^{2(p-i)}\Big(- \Delta(2a_1 - \frac{2\theta-1}{\theta^2})|\mathcal{D}y^{\tilde{\xi},\Delta}(t_k) \cdot \eta|^2$$
$$+ 2a_2\Delta \int_{-\tau}^{0} |\mathcal{D}y^{\tilde{\xi},\Delta}_{t_k}(r) \cdot \eta|^2 \mathrm{d}\nu_2(r) + |\mathcal{D}\sigma(y^{\tilde{\xi},\Delta}_{t_k})\mathcal{D}y^{\tilde{\xi},\Delta}_{t_k} \cdot \eta\delta W_k|^2 + \mathcal{M}_k\Big)^i.$$

For the term $\mathcal{I}_1$ of (3.18), by (3.13) and the property of the conditional expectation, we obtain that for $\epsilon_1 \in (0, 1)$,

$$\begin{aligned}
\mathbb{E}[\mathcal{I}_1] \le\, & \mathbb{E}\Big[A^{p-1}_{\theta,\Delta}|\mathcal{D}z^{\tilde{\xi},\Delta}(t_k) \cdot \eta|^{2(p-1)}\Big(- \Delta(2a_1 - \frac{2\theta-1}{\theta^2} - L)|\mathcal{D}y^{\tilde{\xi},\Delta}(t_k) \cdot \eta|^2 \\
& + 2a_2\Delta \int_{-\tau}^{0} |\mathcal{D}y^{\tilde{\xi},\Delta}_{t_k}(r) \cdot \eta|^2 \mathrm{d}\nu_2(r) + L\Delta \int_{-\tau}^{0} |\mathcal{D}y^{\tilde{\xi},\Delta}_{t_k}(r) \cdot \eta|^2 \mathrm{d}\nu_1(r)\Big)\Big] \\
\le & -\Delta(2a_1 - \frac{2\theta-1}{\theta^2} - L)A^{p-1}_{\theta,\Delta}\mathbb{E}[|\mathcal{D}z^{\tilde{\xi},\Delta}(t_k) \cdot \eta|^{2(p-1)}|\mathcal{D}y^{\tilde{\xi},\Delta}(t_k) \cdot \eta|^2] \\
& + \frac{L\Delta}{\epsilon_1^{p-1}} \int_{-\tau}^{0} \mathbb{E}[|\mathcal{D}y^{\tilde{\xi},\Delta}_{t_k}(r) \cdot \eta|^{2p}]\mathrm{d}\nu_1(r) \\
& + \frac{2a_2\Delta}{\epsilon_1^{p-1}} \int_{-\tau}^{0} \mathbb{E}[|\mathcal{D}y^{\tilde{\xi},\Delta}_{t_k}(r) \cdot \eta|^{2p}]\mathrm{d}\nu_2(r) \\
& + (2a_2 + L)\epsilon_1\Delta A^p_{\theta,\Delta}\mathbb{E}[|\mathcal{D}z^{\tilde{\xi},\Delta}(t_k) \cdot \eta|^{2p}],
\end{aligned} \tag{3.19}$$

where in the last step we used the Young inequality. To estimate the term $|\mathcal{D}z^{\tilde{\xi},\Delta}(t_k)\cdot \eta|^{2(p-1)}|\mathcal{D}y^{\tilde{\xi},\Delta}(t_k)\cdot\eta|^2$, taking the pth power on both sides of (3.16) and combining the Young inequality, we derive

$$\begin{aligned}
& |\mathcal{D}y^{\tilde{\xi},\Delta}(t_k) \cdot \eta|^{2p} \\
& \le \Big(|\mathcal{D}z^{\tilde{\xi},\Delta}(t_k) \cdot \eta|^2 + 2\theta a_2\Delta \int_{-\tau}^{0} |\mathcal{D}y^{\tilde{\xi},\Delta}_{t_k}(r) \cdot \eta|^2 \mathrm{d}\nu_2(r)\Big)^{2(p-1)}|\mathcal{D}y^{\tilde{\xi},\Delta}(t_k) \cdot \eta|^2 \\
& \le |\mathcal{D}z^{\tilde{\xi},\Delta}(t_k) \cdot \eta|^{2(p-1)}|\mathcal{D}y^{\tilde{\xi},\Delta}(t_k) \cdot \eta|^2 + |\mathcal{D}y^{\tilde{\xi},\Delta}(t_k) \cdot \eta|^2
\end{aligned}$$

$$\times \sum_{j=1}^{p-1} C_{p-1}^{j} |\mathcal{D}z^{\tilde{\xi},\Delta}(t_k)\cdot\eta|^{2(p-1-j)}\Big(2\theta a_2\Delta\int_{-\tau}^{0}|\mathcal{D}y_{t_k}^{\tilde{\xi},\Delta}(r)\cdot\eta|^2\mathrm{d}\nu_2(r)\Big)^{j}$$

$$\leq (1+\epsilon_1)|\mathcal{D}z^{\tilde{\xi},\Delta}(t_k)\cdot\eta|^{2(p-1)}|\mathcal{D}y^{\tilde{\xi},\Delta}(t_k)\cdot\eta|^2 + K_{\epsilon_1}|\mathcal{D}y^{\tilde{\xi},\Delta}(t_k)\cdot\eta|^2$$

$$\times\Big(2\theta a_2\Delta\int_{-\tau}^{0}|\mathcal{D}y_{t_k}^{\tilde{\xi},\Delta}(r)\cdot\eta|^2\mathrm{d}\nu_2(r)\Big)^{p-1}.$$

This yields

$$\begin{aligned}
&-|\mathcal{D}z^{\tilde{\xi},\Delta}(t_k)\cdot\eta|^{2(p-1)}|\mathcal{D}y^{\tilde{\xi},\Delta}(t_k)\cdot\eta|^2\\
&\quad\leq -(1+\epsilon_1)^{-1}|\mathcal{D}y^{\tilde{\xi},\Delta}(t_k)\cdot\eta|^{2p}\\
&\qquad + K_{\epsilon_1}(1+\epsilon_1)^{-1}|\mathcal{D}y^{\tilde{\xi},\Delta}(t_k)\cdot\eta|^2\Big(2\theta a_2\Delta\int_{-\tau}^{0}|\mathcal{D}y_{t_k}^{\tilde{\xi},\Delta}(r)\cdot\eta|^2\mathrm{d}\nu_2(r)\Big)^{p-1}.
\end{aligned} \tag{3.20}$$

Inserting (3.20) into (3.19), we obtain

$$\begin{aligned}
\mathbb{E}[\mathcal{I}_1] \leq\ & \Delta(2a_1-\frac{2\theta-1}{\theta^2}-L)A_{\theta,\Delta}^{p-1}\mathbb{E}\Big[-(1+\epsilon_1)^{-1}|\mathcal{D}y^{\tilde{\xi},\Delta}(t_k)\cdot\eta|^{2p}\\
&+K_{\epsilon_1}(1+\epsilon_1)^{-1}|\mathcal{D}y^{\tilde{\xi},\Delta}(t_k)\cdot\eta|^2\Big(2\theta a_2\Delta\int_{-\tau}^{0}|\mathcal{D}y_{t_k}^{\tilde{\xi},\Delta}(r)\cdot\eta|^2\mathrm{d}\nu_2(r)\Big)^{p-1}\Big]\\
&+(2a_2+L)\epsilon_1\Delta A_{\theta,\Delta}^{p}\mathbb{E}[|\mathcal{D}z^{\tilde{\xi},\Delta}(t_k)\cdot\eta|^{2p}]\\
&+\int_{-\tau}^{0}\mathbb{E}[|\mathcal{D}y_{t_k}^{\tilde{\xi},\Delta}(r)\cdot\eta|^{2p}]\mathrm{d}\nu_2(r)\\
&\times\frac{2a_2}{\epsilon_1^{p-1}}\Delta+\frac{L}{\epsilon_1^{p-1}}\Delta\int_{-\tau}^{0}\mathbb{E}[|\mathcal{D}y_{t_k}^{\tilde{\xi},\Delta}(r)\cdot\eta|^{2p}]\mathrm{d}\nu_1(r)\\
\leq & -\Delta(1+\epsilon_1)^{-1}(2a_1-\frac{2\theta-1}{\theta^2}-L)\\
&\times A_{\theta,\Delta}^{p-1}\mathbb{E}[|\mathcal{D}y^{\tilde{\xi},\Delta}(t_k)\cdot\eta|^{2p}]+(2a_2+L)\\
&\times\epsilon_1\Delta A_{\theta,\Delta}^{p}\mathbb{E}[|\mathcal{D}z^{\tilde{\xi},\Delta}(t_k)\cdot\eta|^{2p}]+K_{\epsilon_1}\Delta^p\int_{-\tau}^{0}\mathbb{E}[|\mathcal{D}y_{t_k}^{\tilde{\xi},\Delta}\cdot\eta|^{2p}]\mathrm{d}\nu_2(r)\\
&+K_{\epsilon_1}\Delta^p\mathbb{E}[|\mathcal{D}y^{\tilde{\xi},\Delta}(t_k)\cdot\eta|^{2p}]+\frac{2a_2}{\epsilon_1^{p-1}}\Delta\int_{-\tau}^{0}\mathbb{E}[|\mathcal{D}y_{t_k}^{\tilde{\xi},\Delta}(r)\cdot\eta|^{2p}]\mathrm{d}\nu_2(r)\\
&+\frac{L}{\epsilon_1^{p-1}}\Delta\int_{-\tau}^{0}\mathbb{E}[|\mathcal{D}y_{t_k}^{\tilde{\xi},\Delta}(r)\cdot\eta|^{2p}]\mathrm{d}\nu_1(r).
\end{aligned}$$

For the term $\mathcal{I}_2$, by the Young inequality and (3.13), we have

$$
\begin{aligned}
\mathbb{E}[\mathcal{I}_2] = \mathbb{E}\Big[& A_{\theta,\Delta}^{p-2}|\mathcal{D}z^{\tilde{\xi},\Delta}(t_k)\cdot\eta|^{2(p-2)}\Big(-\Delta(2a_1-\frac{2\theta-1}{\theta^2})|\mathcal{D}y^{\tilde{\xi},\Delta}(t_k)\cdot\eta|^2 \\
& + 2a_2\Delta\int_{-\tau}^{0}|\mathcal{D}y_{t_k}^{\tilde{\xi},\Delta}(r)\cdot\eta|^2\mathrm{d}\nu_2(r) + |\mathcal{D}\sigma(y_{t_k}^{\tilde{\xi},\Delta})\mathcal{D}y_{t_k}^{\tilde{\xi},\Delta}\cdot\eta\delta W_k|^2 \\
& + 2\Big\langle\mathcal{D}z^{\tilde{\xi},\Delta}(t_k)\cdot\eta+\frac{1}{\theta}(\mathcal{D}y^{\tilde{\xi},\Delta}(t_k)\cdot\eta-\mathcal{D}z^{\tilde{\xi},\Delta}(t_k)\cdot\eta), \\
& \mathcal{D}\sigma(y_{t_k}^{\tilde{\xi},\Delta})\mathcal{D}y_{t_k}^{\tilde{\xi},\Delta}\cdot\eta\delta W_k\Big\rangle\Big)^2\Big] \\
\le \mathbb{E}\Big[& A_{\theta,\Delta}^{p-2}|\mathcal{D}z^{\tilde{\xi},\Delta}(t_k)\cdot\eta|^{2(p-2)} \\
& \Big(\Big(8A_{\theta,\Delta}|\mathcal{D}z^{\tilde{\xi},\Delta}(t_k)\cdot\eta|^2|\mathcal{D}\sigma(y_{t_k}^{\tilde{\xi},\Delta})\mathcal{D}y_{t_k}^{\tilde{\xi},\Delta}\cdot\eta|^2 \\
& + \frac{8}{\theta^2}|\mathcal{D}y^{\tilde{\xi},\Delta}(t_k)\cdot\eta\mathcal{D}\sigma(y_{t_k}^{\tilde{\xi},\Delta})\mathcal{D}y_{t_k}^{\tilde{\xi},\Delta}\cdot\eta|^2\Big)\Delta+\Big(-\Delta(2a_1-\frac{2\theta-1}{\theta^2}) \\
& \times|\mathcal{D}y^{\tilde{\xi},\Delta}(t_k)\cdot\eta|^2+2a_2\Delta\int_{-\tau}^{0}|\mathcal{D}y_{t_k}^{\tilde{\xi},\Delta}(r)\cdot\eta|^2\mathrm{d}\nu_2(r) \\
& + |\mathcal{D}\sigma(y_{t_k}^{\tilde{\xi},\Delta})\mathcal{D}y_{t_k}^{\tilde{\xi},\Delta}\cdot\eta\delta W_k|^2\Big)^2\Big)\Big] \\
\le 8\Delta\Big(& 2\epsilon_1 A_{\theta,\Delta}^{p}\mathbb{E}[|\mathcal{D}z^{\tilde{\xi},\Delta}(t_k)\cdot\eta|^{2p}]+\frac{1}{\epsilon_1^{p-1}}|\mathcal{D}\sigma(y_{t_k}^{\tilde{\xi},\Delta})\mathcal{D}y_{t_k}^{\tilde{\xi},\Delta}\cdot\eta|^{2p} \\
& + \frac{1}{\epsilon_1^{\frac{p-2}{2}}}\frac{1}{\theta^p}|\mathcal{D}y^{\tilde{\xi},\Delta}(t_k)\cdot\eta\mathcal{D}\sigma(y_{t_k}^{\tilde{\xi},\Delta})\mathcal{D}y_{t_k}^{\tilde{\xi},\Delta}\cdot\eta|^{p}\Big) \\
& + K\Delta^2A_{\theta,\Delta}^{p}\mathbb{E}[|\mathcal{D}z^{\tilde{\xi},\Delta}(t_k)\cdot\eta|^{2p}]+K\Delta^2\mathbb{E}[|\mathcal{D}y^{\tilde{\xi},\Delta}(t_k)\cdot\eta|^{2p}] \\
& + K\Delta^2\mathbb{E}\Big[\int_{-\tau}^{0}|\mathcal{D}y_{t_k}^{\tilde{\xi},\Delta}(r)\cdot\eta|^{2p}\mathrm{d}\nu_2(r)\Big] \\
& + K\Delta^2\mathbb{E}\Big[\mathcal{D}\sigma(y_{t_k}^{\tilde{\xi},\Delta})\mathcal{D}y_{t_k}^{\tilde{\xi},\Delta}\cdot\eta|^{2p}\Big] \\
\le 16 & \epsilon_1\Delta A_{\theta,\Delta}^{p}\mathbb{E}[|\mathcal{D}z^{\tilde{\xi},\Delta}(t_k)\cdot\eta|^{2p}]+K\Delta^2A_{\theta,\Delta}^{p}\mathbb{E}[|\mathcal{D}z^{\tilde{\xi},\Delta}(t_k)\cdot\eta|^{2p}] \\
& + (\frac{8}{\epsilon_1^{p-1}}+\frac{4}{\epsilon_1^{\frac{p-2}{2}}}\frac{1}{\theta^p}) \\
& L\Delta\mathbb{E}\Big[|\mathcal{D}y^{\tilde{\xi},\Delta}(t_k)\cdot\eta|^{2p}+\int_{-\tau}^{0}|\mathcal{D}y_{t_k}^{\tilde{\xi},\Delta}(r)\cdot\eta|^{2p}\mathrm{d}\nu_1(r)\Big]
\end{aligned}
$$

$$+\frac{4}{\epsilon_1^{\frac{p-2}{2}}}\frac{1}{\theta^p}\Delta\mathbb{E}[|\mathcal{D}y^{\tilde{\xi},\Delta}(t_k)\cdot\eta|^{2p}]+K\Delta^2\mathbb{E}[|\mathcal{D}y^{\tilde{\xi},\Delta}(t_k)\cdot\eta|^{2p}]$$

$$+K\Delta^2\mathbb{E}\Big[\int_{-\tau}^0|\mathcal{D}y_{t_k}^{\tilde{\xi},\Delta}(r)\cdot\eta|^{2p}\mathrm{d}\nu_1(r)\Big]$$

$$+K\Delta^2\mathbb{E}\Big[\int_{-\tau}^0|\mathcal{D}y_{t_k}^{\tilde{\xi},\Delta}(r)\cdot\eta|^{2p}\mathrm{d}\nu_2(r)\Big].$$

Similarly, for the term $\mathcal{I}_i$ with $i\in\{3,\ldots,p\}$, we obtain that

$$\mathbb{E}[\mathcal{I}_i]\le K\Delta^2A_{\theta,\Delta}^p\mathbb{E}[|\mathcal{D}z^{\tilde{\xi},\Delta}(t_k)\cdot\eta|^{2p}]+K\Delta^2\mathbb{E}[|\mathcal{D}y^{\tilde{\xi},\Delta}(t_k)\cdot\eta|^{2p}]$$

$$+K\Delta^2\mathbb{E}\Big[\int_{-\tau}^0|\mathcal{D}y_{t_k}^{\tilde{\xi},\Delta}(r)\cdot\eta|^{2p}\mathrm{d}\nu_1(r)\Big]$$

$$+K\Delta^2\mathbb{E}\Big[\int_{-\tau}^0|\mathcal{D}y_{t_k}^{\tilde{\xi},\Delta}(r)\cdot\eta|^{2p}\mathrm{d}\nu_2(r)\Big].$$

Therefore, inserting estimates of $\mathcal{I}_i$ for $i\in\{1,\ldots,p\}$ into (3.18), we arrive at

$$\mathbb{E}[|\mathcal{D}z^{\tilde{\xi},\Delta}(t_{k+1})\cdot\eta|^{2p}]$$

$$\le A_{\theta,\Delta}^p\mathbb{E}[|\mathcal{D}z^{\tilde{\xi},\Delta}(t_k)\cdot\eta|^{2p}]+p\mathbb{E}[\mathcal{I}_1]+\frac{p(p-1)}{2}\mathbb{E}[\mathcal{I}_2]+\sum_{i=3}^pC_p^i\mathbb{E}[\mathcal{I}_i]$$

$$\le(1+p(2a_2+L)\epsilon_1\Delta+\frac{16p(p-1)}{2}\epsilon_1\Delta+K\Delta^2)A_{\theta,\Delta}^p\mathbb{E}[|\mathcal{D}z^{\tilde{\xi},\Delta}(t_k)\cdot\eta|^{2p}]$$

$$-\Delta p(1+\epsilon_1)^{-1}(2a_1-\frac{2\theta-1}{\theta^2}-L)A_{\theta,\Delta}^{p-1}\mathbb{E}[|\mathcal{D}y^{\tilde{\xi},\Delta}(t_k)\cdot\eta|^{2p}]+\frac{2pa_2}{\epsilon_1^{p-1}}\Delta$$

$$\times\int_{-\tau}^0\mathbb{E}[|\mathcal{D}y_{t_k}^{\tilde{\xi},\Delta}(r)\cdot\eta|^{2p}]\mathrm{d}\nu_2(r)+\frac{pL}{\epsilon_1^{p-1}}\Delta\int_{-\tau}^0\mathbb{E}[|\mathcal{D}y_{t_k}^{\tilde{\xi},\Delta}(r)\cdot\eta|^{2p}]\mathrm{d}\nu_1(r)$$

$$+\frac{p(p-1)L}{2}(\frac{8}{c_1^{p-1}}+\frac{4}{\epsilon_1^{\frac{p-2}{2}}}\frac{1}{\theta^p})\Delta\mathbb{E}\Big[|\mathcal{D}y^{\tilde{\xi},\Delta}(t_k)\cdot\eta|^{2p}$$

$$+\int_{-\tau}^0|\mathcal{D}y_{t_k}^{\tilde{\xi},\Delta}(r)\cdot\eta|^{2p}\mathrm{d}\nu_1(r)\Big]+\frac{2p(p-1)}{\theta^p\epsilon_1^{\frac{p-2}{2}}}\Delta\mathbb{E}[|\mathcal{D}y^{\tilde{\xi},\Delta}(t_k)\cdot\eta|^{2p}]$$

$$+(K\Delta^2+K_{\epsilon_1}\Delta^p)\mathbb{E}[|\mathcal{D}y^{\tilde{\xi},\Delta}(t_k)\cdot\eta|^{2p}]$$

$$+K\Delta^2\mathbb{E}\Big[\int_{-\tau}^0|\mathcal{D}y_{t_k}^{\tilde{\xi},\Delta}(r)\cdot\eta|^{2p}\mathrm{d}\nu_1(r)\Big]$$

$$
\begin{aligned}
&+ (K\Delta^2 + K_{\epsilon_1}\Delta^p)\mathbb{E}\Big[\int_{-\tau}^{0} |\mathcal{D}y_{t_k}^{\tilde{\xi},\Delta}(r)\cdot\eta|^{2p}\mathrm{d}\nu_2(r)\Big] \\
&\leq (1 + K\Delta(\epsilon_1+\Delta))A_{\theta,\Delta}^{p}\mathbb{E}[|\mathcal{D}z^{\tilde{\xi},\Delta}(t_k)\cdot\eta|^{2p}] + \Delta J_{1,k} + \Delta^2 J_{2,k},
\end{aligned}
$$

where

$$
\begin{aligned}
J_{1,k} &:= -p(1+\epsilon_1)^{-1}(2a_1 - \frac{2\theta-1}{\theta^2} - L)A_{\theta,\Delta}^{p-1}\mathbb{E}[|\mathcal{D}y^{\tilde{\xi},\Delta}(t_k)\cdot\eta|^{2p}] + \frac{2pa_2}{\epsilon_1^{p-1}} \\
&\quad\times\int_{-\tau}^{0}\mathbb{E}[|\mathcal{D}y_{t_k}^{\tilde{\xi},\Delta}(r)\cdot\eta|^{2p}]\mathrm{d}\nu_2(r) + \frac{pL}{\epsilon_1^{p-1}}\int_{-\tau}^{0}\mathbb{E}[|\mathcal{D}y_{t_k}^{\tilde{\xi},\Delta}(r)\cdot\eta|^{2p}]\mathrm{d}\nu_1(r) \\
&\quad+\frac{p(p-1)L}{2}(\frac{8}{\epsilon_1^{p-1}} + \frac{4}{\epsilon_1^{\frac{p-2}{2}}}\frac{1}{\theta^p})\mathbb{E}\Big[|\mathcal{D}y^{\tilde{\xi},\Delta}(t_k)\cdot\eta|^{2p} \\
&\quad+\int_{-\tau}^{0}|\mathcal{D}y_{t_k}^{\tilde{\xi},\Delta}(r)\cdot\eta|^{2p}\mathrm{d}\nu_1(r)\Big] + \frac{2p(p-1)}{\theta^p\epsilon_1^{\frac{p-2}{2}}}\mathbb{E}[|\mathcal{D}y^{\tilde{\xi},\Delta}(t_k)\cdot\eta|^{2p}], \\
J_{2,k} &:= (K + K_{\epsilon_1}\Delta^{p-2})\mathbb{E}[|\mathcal{D}y^{\tilde{\xi},\Delta}(t_k)\cdot\eta|^{2p}] + K\mathbb{E}\Big[\int_{-\tau}^{0}|\mathcal{D}y_{t_k}^{\tilde{\xi},\Delta}(r)\cdot\eta|^{2p}\mathrm{d}\nu_1(r)\Big] \\
&\quad+(K + K_{\epsilon_1}\Delta^{p-2})\mathbb{E}\Big[\int_{-\tau}^{0}|\mathcal{D}y_{t_k}^{\tilde{\xi},\Delta}(r)\cdot\eta|^{2p}\mathrm{d}\nu_2(r)\Big].
\end{aligned}
$$

Then we derive that for $\tilde{\lambda}_2 > 0$,

$$
\begin{aligned}
&e^{\tilde{\lambda}_2 p t_{k+1}}\mathbb{E}[|\mathcal{D}z^{\tilde{\xi},\Delta}(t_{k+1})\cdot\eta|^{2p}] \\
&\quad= \sum_{l=0}^{k}\Big(e^{\tilde{\lambda}_2 p t_{l+1}}\mathbb{E}[|\mathcal{D}z^{\tilde{\xi},\Delta}(t_{l+1})\cdot\eta|^{2p}] - e^{\tilde{\lambda}_2 p t_l}\mathbb{E}[|\mathcal{D}z^{\tilde{\xi},\Delta}(t_l)\cdot\eta|^{2p}]\Big) \\
&\qquad+\mathbb{E}[|\mathcal{D}z^{\tilde{\xi},\Delta}(0)\cdot\eta|^{2p}] \\
&\quad\leq \mathbb{E}[|\mathcal{D}z^{\tilde{\xi},\Delta}(0)\cdot\eta|^{2p}] + \sum_{l=0}^{k} e^{\tilde{\lambda}_2 p t_{l+1}}(\Delta J_{1,l} + \Delta^2 J_{2,l}) \\
&\qquad+\Big((1 + K\Delta(\epsilon_1+\Delta))A_{\theta,\Delta}^{p}e^{\tilde{\lambda}_2 p\Delta} - 1\Big)\sum_{l=0}^{k} e^{\tilde{\lambda}_2 p t_l}\mathbb{E}[|\mathcal{D}z^{\tilde{\xi},\Delta}(t_l)\cdot\eta|^{2p}].
\end{aligned}
\tag{3.21}
$$

Similar to (2.27), we deduce

$$\Delta\sum_{l=0}^{k} e^{\tilde{\lambda}_2 p t_{l+1}} J_{1,l} \le -R(\epsilon_1, a_1, \tilde{\lambda}_2)\Delta\sum_{l=0}^{k} e^{\tilde{\lambda}_2 p t_{l+1}}\mathbb{E}[|\mathcal{D}y^{\xi,\Delta}(t_l)\cdot\eta|^{2p}]$$
$$+ K_{\epsilon_1} e^{\tilde{\lambda}_2\tau}\tau\mathbb{E}[\|\mathcal{D}y_0^{\xi,\Delta}\cdot\eta\|^{2p}]$$

and

$$\Delta^2\sum_{l=0}^{k} e^{\tilde{\lambda}_2 p t_{l+1}} J_{2,l} \le \Delta^2\big(K + K_{\epsilon_1}\Delta^{p-2}\big)\sum_{l=0}^{k} e^{\tilde{\lambda}_2 p t_{l+1}}\mathbb{E}[|\mathcal{D}y^{\xi,\Delta}(t_l)\cdot\eta|^{2p}]$$
$$+ K_{\epsilon_1} e^{\tilde{\lambda}_2\tau}\tau\mathbb{E}[\|\mathcal{D}y_0^{\xi,\Delta}\cdot\eta\|^{2p}],$$

where

$$R(\epsilon_1, a_1, \tilde{\lambda}_2) := p(1+\epsilon_1)^{-1}(2a_1 - \frac{2\theta-1}{\theta^2} - L)A_{\theta,\Delta}^{p-1} - \frac{2pa_2 e^{\tilde{\lambda}_2\tau}}{\epsilon_1^{p-1}} - \frac{pLe^{\tilde{\lambda}_2\tau}}{\epsilon_1^{p-1}}$$
$$- \frac{p(p-1)L(1+e^{\tilde{\lambda}_2\tau})}{2}\Big(\frac{8}{\epsilon_1^{p-1}} + \frac{4}{\epsilon_1^{\frac{p-2}{2}}}\frac{1}{\theta^p}\Big) - \frac{2p(p-1)}{\theta^p\epsilon_1^{\frac{p-2}{2}}}.$$

Note that $A_{\theta,\Delta} = 1 - \frac{(2\theta-1)\Delta}{\theta^2(1+\Delta)} \le e^{-\frac{(2\theta-1)}{\theta^2(1+\Delta)}\Delta}$ implies $(1+K\Delta(\epsilon_1+\Delta))A_{\theta,\Delta}^{p}e^{\tilde{\lambda}_2 p\Delta} - 1 \le e^{(K(\epsilon_1+\Delta)+\tilde{\lambda}_2 p - \frac{p(2\theta-1)}{\theta^2(1+\Delta)})\Delta} - 1$. Let $\epsilon_1 \in (0, \frac{1}{K}\frac{p(2\theta-1)}{2\theta^2})$ and denote

$$\mathcal{R}(\epsilon_1, a_1, \tilde{\lambda}_2) := p(1+\epsilon_1)^{-1}(2a_1 - \frac{2\theta-1}{\theta^2} - L)\Big(\frac{(1-\theta)^2}{\theta^2} + \frac{2\theta-1}{2\theta^2}\Big)^{p-1}$$
$$- \frac{2pa_2 e^{\tilde{\lambda}_2\tau}}{\epsilon_1^{p-1}} - \frac{pLe^{\tilde{\lambda}_2\tau}}{\epsilon_1^{p-1}}$$
$$- \frac{p(p-1)L(1+e^{\tilde{\lambda}_2\tau})}{2}\Big(\frac{8}{\epsilon_1^{p-1}} + \frac{4}{\epsilon_1^{\frac{p-2}{2}}}\frac{1}{\theta^p}\Big)$$
$$- \frac{2p(p-1)}{\theta^p\epsilon_1^{\frac{p-2}{2}}}.$$

Take a sufficiently large number $\tilde{\mathfrak{a}}_1$ such that $\mathcal{R}(\epsilon_1, \tilde{\mathfrak{a}}_1, 0) > 0$. Then choose a sufficiently small number $\tilde{\tilde{\lambda}}_2$ such that $K\epsilon_1 + \tilde{\tilde{\lambda}}_2 p < \frac{p(2\theta-1)}{2\theta^2}$ and $\mathcal{R}(\epsilon_1, \tilde{\mathfrak{a}}_1, \tilde{\tilde{\lambda}}_2) > 0$. Furthermore, there exists a number $\tilde{\Delta} \in (0, 1]$ such that for any $\Delta \in (0, \tilde{\Delta}]$,

$K(\epsilon_1 + \Delta) + \tilde{\tilde{\lambda}}_2 p - \frac{p(2\theta-1)}{2\theta^2} < 0$ and $-R(\epsilon_1, \tilde{\mathfrak{a}}_1, \tilde{\lambda}_2) + \Delta(K + K_{\epsilon_1}\Delta^{p-2}) < 0$, which yields

$$(1 + K\Delta(\epsilon_1 + \Delta))A^p_{\theta,\Delta}e^{\tilde{\tilde{\lambda}}_2 p\Delta} - 1 \le e^{(K(\epsilon_1+\Delta)+\tilde{\tilde{\lambda}}_2 p - \frac{p(2\theta-1)}{\theta^2(1+\Delta)})\Delta} - 1 \le 0.$$

Hence, combining (3.21) we have that for $a_1 > \tilde{\mathfrak{a}}_1$, $\tilde{\lambda}_2 \in (0, \tilde{\tilde{\lambda}}_2]$, and $\Delta \in (0, \tilde{\Delta}]$,

$$\begin{aligned} e^{\tilde{\lambda}_2 p t_{k+1}}\mathbb{E}[|\mathcal{D}z^{\tilde{\xi},\Delta}(t_{k+1}) \cdot \eta|^{2p}] &\le \mathbb{E}[|\mathcal{D}z^{\tilde{\xi},\Delta}(0) \cdot \eta|^{2p}] + Ke^{\tilde{\lambda}_2\tau}\tau\|\eta\|^{2p} \\ &\le K\|\eta\|^{2p} + K\mathbb{E}[\|\mathcal{D}b(\tilde{\xi}) \cdot \eta\|^{2p}]. \end{aligned}$$

According to Assumptions 2.4 and 3.1, we have $\|\mathcal{D}b(\tilde{\xi}) \cdot \eta\| \le (1 + \|\tilde{\xi}\|^{\beta})\|\eta\|$, which implies

$$e^{\tilde{\lambda}_2 p t_{k+1}}\mathbb{E}[|\mathcal{D}z^{\tilde{\xi},\Delta}(t_{k+1}) \cdot \eta|^{2p}] \le K\|\eta\|^{2p}(1 + \mathbb{E}[\|\tilde{\xi}\|^{2p\beta}]).$$

The proof is completed. □

Lemma 3.7 *Let Assumptions 2.1, 2.2, 2.4, and 3.1 hold. Then for $p \in \mathbb{N}_+$ with $p \ge 2$, there exist constants $\mathfrak{a}_1 := \mathfrak{a}_1(L, a_2, \theta, p) \ge \tilde{\mathfrak{a}}_1$, $\hat{\lambda}_2 \in (0, \tilde{\lambda}_2]$ such that for $a_1 > \tilde{\mathfrak{a}}_1$, $\Delta \in (0, 1]$, and $k \in \mathbb{N}$,*

$$e^{\hat{\lambda}_2 p t_{k+1}}\mathbb{E}[|\mathcal{D}z^{\tilde{\xi},\Delta}(t_{k+1}) \cdot \eta|^{2p}] \le K\|\eta\|^{2p}(1 + \mathbb{E}[\|\tilde{\xi}\|^{2p\beta}]),$$

where $\tilde{\xi} \in L^{2p\beta}(\Omega; C^d)$, $\eta \in C^d$, $K > 0$, and $\tilde{\mathfrak{a}}_1, \tilde{\lambda}_2$ are given in Lemma 3.6.

Proof In light of Lemma 3.6, we may restrict our attention to the case where $\Delta \in (\tilde{\Delta}, 1]$. Similar to the proof of Lemma 3.6, for any $\epsilon_2 \in (0, 1)$, terms $\mathcal{I}_1$ and $\mathcal{I}_2$ of (3.18) can be estimated as

$$\begin{aligned} \mathbb{E}[\mathcal{I}_1] \le\, & -\Delta(1+\epsilon_2)^{-1}(2a_1 - \frac{2\theta-1}{\theta^2} - L) \\ & \times A^{p-1}_{\theta,\Delta}\mathbb{E}[|\mathcal{D}y^{\tilde{\xi},\Delta}(t_k) \cdot \eta|^{2p}] + (2a_2 + L) \\ & \times \epsilon_2 A^p_{\theta,\Delta}\mathbb{E}[|\mathcal{D}z^{\tilde{\xi},\Delta}(t_k) \cdot \eta|^{2p}] \\ & + K_{\epsilon_2}(1+\epsilon_2)^{-1}\Delta^p A^{p-1}_{\theta,\Delta}\mathbb{E}[|\mathcal{D}y^{\tilde{\xi},\Delta}(t_k) \cdot \eta|^{2p}] \\ & + \big(K_{\epsilon_2}(1+\epsilon_2)^{-1} + \frac{2a_2}{\epsilon_2^{p-1}}\big)\Delta^p \int_{-\tau}^0 \mathbb{E}[|\mathcal{D}y^{\tilde{\xi},\Delta}_{t_k}(r) \cdot \eta|^{2p}]\mathrm{d}\nu_2(r) \\ & + \frac{L}{\epsilon_2^{p-1}}\Delta^p \int_{-\tau}^0 \mathbb{E}[|\mathcal{D}y^{\tilde{\xi},\Delta}_{t_k}(r) \cdot \eta|^{2p}]\mathrm{d}\nu_1(r), \end{aligned}$$

and

$$\mathbb{E}[\mathcal{I}_2] \le 17\epsilon_2 A^p_{\theta,\Delta}\mathbb{E}[|\mathcal{D}z^{\tilde{\xi},\Delta}(t_k)\cdot\eta|^{2p}] + (\frac{8}{\epsilon_2^{p-1}}\Delta^p + \frac{4}{\epsilon_2^{\frac{p-2}{2}}}\frac{1}{\theta^p}\Delta^{\frac{p}{2}})L$$
$$\times\mathbb{E}\Big[|\mathcal{D}y^{\tilde{\xi},\Delta}(t_k)\cdot\eta|^{2p} + \int_{-\tau}^0 |\mathcal{D}y^{\tilde{\xi},\Delta}_{t_k}(r)\cdot\eta|^{2p}\mathrm{d}\nu_1(r)\Big]$$
$$+\frac{4}{\epsilon_2^{\frac{p-2}{2}}}\frac{1}{\theta^p}\Delta^{\frac{p}{2}}|\mathcal{D}y^{\tilde{\xi},\Delta}(t_k)\cdot\eta|^{2p} + K_{\epsilon_2}\Delta^p\mathbb{E}[|\mathcal{D}y^{\tilde{\xi},\Delta}(t_k)\cdot\eta|^{2p}]$$
$$+K_{\epsilon_2}\Delta^p\mathbb{E}\Big[\int_{-\tau}^0 |\mathcal{D}y^{\tilde{\xi},\Delta}_{t_k}(r)\cdot\eta|^{2p}\mathrm{d}\nu_1(r) + \int_{-\tau}^0 |\mathcal{D}y^{\tilde{\xi},\Delta}_{t_k}(r)\cdot\eta|^{2p}\mathrm{d}\nu_2(r)\Big].$$

For terms $\mathcal{I}_i$ with $i\in\{3,\ldots,p\}$, we have

$$\mathbb{E}[\mathcal{I}_i] \le K\epsilon_2 A^p_{\theta,\Delta}\mathbb{E}[|\mathcal{D}z^{\tilde{\xi},\Delta}(t_k)\cdot\eta|^{2p}] + K_{\epsilon_2}\Delta^{\frac{p}{2}}\mathbb{E}[|\mathcal{D}y^{\tilde{\xi},\Delta}(t_k)\cdot\eta|^{2p}]$$
$$+K_{\epsilon_2}\Delta^{\frac{p}{2}}\mathbb{E}\Big[\int_{-\tau}^0 |\mathcal{D}y^{\tilde{\xi},\Delta}_{t_k}(r)\cdot\eta|^{2p}\mathrm{d}\nu_1(r) + \int_{-\tau}^0 |\mathcal{D}y^{\tilde{\xi},\Delta}_{t_k}(r)\cdot\eta|^{2p}\mathrm{d}\nu_2(r)\Big].$$

Combining these estimates, we obtain that for $\hat{\lambda}_2\in(0,\tilde{\lambda}_2]$,

$$\begin{aligned}
&e^{\hat{\lambda}_2 p t_{k+1}}\mathbb{E}[|\mathcal{D}z^{\tilde{\xi},\Delta}(t_{k+1})\cdot\eta|^{2p}]\\
&\quad\le \mathbb{E}[|\mathcal{D}z^{\tilde{\xi},\Delta}(0)\cdot\eta|^{2p}] + [(1+K\epsilon_2)A^p_{\theta,\Delta}e^{\hat{\lambda}_2 p\Delta}-1]\\
&\quad\times\sum_{l=0}^k e^{\hat{\lambda}_2 p t_l}\mathbb{E}[|\mathcal{D}z^{\tilde{\xi},\Delta}(t_l)\cdot\eta|^{2p}]\\
&\quad-\Big(p(1+\epsilon_2)^{-1}(2a_1-\frac{2\theta-1}{\theta^2}-L)A^{p-1}_{\theta,\Delta}-(\frac{2pa_2e^{\hat{\lambda}_2\tau}}{\epsilon_2^{p-1}}+\frac{pLe^{\hat{\lambda}_2\tau}}{\epsilon_2^{p-1}})\Delta^{p-1}\\
&\quad-\frac{p(p-1)L(1+e^{\hat{\lambda}_2\tau})}{2}(\frac{8}{\epsilon_2^{p-1}}\Delta^{p-1}+\frac{4}{\epsilon_2^{\frac{p-2}{2}}}\frac{1}{\theta^p}\Delta^{\frac{p-2}{2}})-\frac{4}{\epsilon_2^{\frac{p-2}{2}}}\frac{1}{\theta^p}\Delta^{\frac{p-2}{2}}\\
&\quad-K_{\epsilon_2}\Delta^{\frac{p-2}{2}}\Big)\Delta\sum_{l=0}^k e^{\hat{\lambda}_2 p t_{l+1}}\mathbb{E}[|\mathcal{D}y^{\tilde{\xi},\Delta}(t_l)\cdot\eta|^{2p}] + Ke^{\hat{\lambda}_2\tau}\tau\|\eta\|^{2p}.
\end{aligned}\tag{3.22}$$

Noting $(1+K\epsilon_2)A^p_{\theta,\Delta}e^{\hat{\lambda}_2 p\Delta}\le e^{K\epsilon_2+\hat{\lambda}_2 p\Delta-\frac{p(2\theta-1)\tilde{\Delta}}{2\theta^2}}$, we choose a sufficiently small number $\epsilon_2>0$ such that $K\epsilon_2<\frac{p(2\theta-1)\tilde{\Delta}}{2\theta^2}$. Let $\mathfrak{a}_1\ge\tilde{\mathfrak{a}}_1$ be sufficiently large such that

$$p(1+\epsilon_2)^{-1}(2\mathfrak{a}_1-\frac{2\theta-1}{\theta^2}-L)(\frac{(1-\theta)^2}{\theta^2}+\frac{2\theta-1}{\theta^2(1+\tilde{\Delta})})^{p-1}-(\frac{2pa_2}{\epsilon_2^{p-1}}+\frac{pL}{\epsilon_2^{p-1}})$$
$$-\frac{p(p-1)L}{2}(\frac{8}{\epsilon_2^{p-1}}+\frac{4}{\epsilon_2^{\frac{p-2}{2}}}\frac{1}{\theta^p})-\frac{4}{\epsilon_2^{\frac{p-2}{2}}}\frac{1}{\theta^p}-K_{\epsilon_2}>0,$$

where $\tilde{\mathfrak{a}}_1$ is given by Lemma 3.6. Letting the positive constant $\hat{\lambda}_2$ sufficiently small, we obtain that $K\epsilon_2+\hat{\lambda}_2 p\tilde{\Delta}-\frac{p(2\theta-1)\tilde{\Delta}}{2\theta^2}\le 0$ and for $a_1\ge\mathfrak{a}_1$ and $\Delta\in(\tilde{\Delta},1]$,

$$p(1+\epsilon_2)^{-1}(2a_1-\frac{2\theta-1}{\theta^2}-L)A_{\theta,\Delta}^{p-1}-(\frac{2pa_2e^{\hat{\lambda}_2\tau}}{\epsilon_2^{p-1}}+\frac{pLe^{\hat{\lambda}_2\tau}}{\epsilon_2^{p-1}})\Delta^{p-1}$$
$$-\frac{p(p-1)L(1+e^{\hat{\lambda}_2\tau})}{2}(\frac{8}{\epsilon_2^{p-1}}\Delta^{p-1}+\frac{4}{\epsilon_2^{\frac{p-2}{2}}}\frac{1}{\theta^p}\Delta^{\frac{p-2}{2}})-\frac{4}{\epsilon_2^{\frac{p-2}{2}}}\frac{1}{\theta^p}\Delta^{\frac{p-2}{2}}$$
$$-K_{\epsilon_2}\Delta^{\frac{p-2}{2}}\ge 0.$$

Combining (3.22), we derive

$$\begin{aligned}e^{\hat{\lambda}_2 pt_{k+1}}\mathbb{E}[|\mathcal{D}z^{\tilde{\xi},\Delta}(t_{k+1})\cdot\eta|^{2p}]&\le K(\|\mathcal{D}z_0^{\tilde{\xi},\Delta}\cdot\eta\|^{2p}+\|\eta\|^{2p})\\&\le K\|\eta\|^{2p}(1+\mathbb{E}[\|\tilde{\xi}\|^{2p\beta}]).\end{aligned}$$

We complete the proof. □

Lemma 3.8 *Let conditions in Lemma 3.7 hold. Then*

$$\mathbb{E}[\|\mathcal{D}y_{t_k}^{\tilde{\xi},\Delta}\cdot\eta\|^{2p}]\le Ke^{-\hat{\lambda}_3 pt_k}\|\eta\|^{2p}(1+\mathbb{E}[\|\tilde{\xi}\|^{2p\beta}]),$$

where $\tilde{\xi}\in L^{2p\beta}(\Omega;\mathcal{C}^d)$, $\eta\in\mathcal{C}^d$, $K>0$, *and* $\hat{\lambda}_3\in(0,\hat{\lambda}_2)$ *with* $\hat{\lambda}_2$ *given by Lemma 3.7.*

Proof The proof is divided into two steps.

Step 1. We show that for $p\ge 2$ and $k\in\mathbb{N}$,

$$e^{\hat{\lambda}_2 pt_{k+1}}\mathbb{E}[|\mathcal{D}y^{\tilde{\xi},\Delta}(t_{k+1})\cdot\eta|^{2p}]\le K\|\eta\|^{2p}(1+\mathbb{E}[\|\tilde{\xi}\|^{2p\beta}]). \quad (3.23)$$

Similar to (2.74), we obtain from (3.16) that for $\tilde{\varepsilon}\in(0,1)$,

$$\begin{aligned}&\big(1+(2a_1\theta\Delta)^p\big)|\mathcal{D}y^{\tilde{\xi},\Delta}(t_k)\cdot\eta|^{2p}\\&\le K_{\tilde{\varepsilon}}|\mathcal{D}z^{\tilde{\xi},\Delta}(t_k)\cdot\eta|^{2p}+(2a_2\theta\Delta)^p(1+\tilde{\varepsilon})\int_{-\tau}^0|\mathcal{D}y_{t_k}^{\tilde{\xi},\Delta}(r)\cdot\eta|^{2p}\mathrm{d}\nu_2(r)\end{aligned}$$

$$
\begin{aligned}
&\le K_{\tilde{\varepsilon}}|\mathcal{D}z^{\tilde{\xi},\Delta}(t_k)\cdot\eta|^{2p}+(2a_2\theta\Delta)^p(1+\tilde{\varepsilon})\int_{-\tau}^{-\Delta}|\mathcal{D}y_{t_k}^{\tilde{\xi},\Delta}(r)\cdot\eta|^{2p}\mathrm{d}\nu_2(r)\\
&\quad+(2a_2\theta\Delta)^p(1+\tilde{\varepsilon})\\
&\quad\times\int_{-\Delta}^{0}\Big(\frac{-r}{\Delta}|\mathcal{D}y^{\tilde{\xi},\Delta}(t_{k-1})\cdot\eta|^{2p}+\frac{r+\Delta}{\Delta}|\mathcal{D}y^{\tilde{\xi},\Delta}(t_k)\cdot\eta|^{2p}\Big)\mathrm{d}\nu_2(r)\\
&\le K_{\tilde{\varepsilon}}|\mathcal{D}z^{\tilde{\xi},\Delta}(t_k)\cdot\eta|^{2p}+(2a_2\theta\Delta)^p(1+\tilde{\varepsilon})\int_{-\Delta}^{0}\frac{r+\Delta}{\Delta}\mathrm{d}\nu_2(r)|\mathcal{D}y^{\tilde{\xi},\Delta}(t_k)\cdot\eta|^{2p}\\
&\quad+(2a_2\theta\Delta)^p(1+\tilde{\varepsilon})\Big(\nu_2([-\tau,-\Delta])+\int_{-\Delta}^{0}\frac{-r}{\Delta}\mathrm{d}\nu_2(r)\Big)\\
&\quad\times\sup_{k-N\le i\le k-1}|\mathcal{D}y^{\tilde{\xi},\Delta}(t_i)\cdot\eta|^{2p}.
\end{aligned}
\tag{3.24}
$$

It follows from $R(\epsilon_1,\tilde{a}_1,\tilde{\lambda}_2)>0$ in the proof of Lemma 3.6, $\hat{\lambda}_2\le\tilde{\lambda}_2$, and $a_1\ge\tilde{a}_1$ that $a_1>a_2e^{\hat{\lambda}_2\tau}$. Choose a sufficiently small $\tilde{\varepsilon}\in(0,1)$ such that $(1+\tilde{\varepsilon})a_2^pe^{\hat{\lambda}_2p\tau}\le a_1^p$. This, along with (3.24) derives that

$$
\begin{aligned}
&\Big(1+(2a_1\theta\Delta)^p\big(1-\int_{-\Delta}^{0}\frac{r+\Delta}{\Delta}\mathrm{d}\nu_2(r)\big)\Big)|\mathcal{D}y^{\tilde{\xi},\Delta}(t_k)\cdot\eta|^{2p}\\
&\quad\le(2a_2\theta\Delta)^p(1+\tilde{\varepsilon})\Big(1-\int_{-\Delta}^{0}\frac{r+\Delta}{\Delta}\mathrm{d}\nu_2(r)\Big)\sup_{k-N\le i\le k-1}|\mathcal{D}y^{\tilde{\xi},\Delta}(t_i)\cdot\eta|^{2p}\\
&\qquad+K_{\tilde{\varepsilon}}|\mathcal{D}z^{\tilde{\xi},\Delta}(t_k)\cdot\eta|^{2p}.
\end{aligned}
\tag{3.25}
$$

Next, we shall show (3.23) by the induction argument. To make the proof clearer, we replace the constant K in Lemma 3.7 with a fixed constant K_z, i.e.,

$$
\mathbb{E}[|\mathcal{D}z^{\tilde{\xi},\Delta}(t_k)\cdot\eta|^{2p}]\le K_z\|\eta\|^{2p}(1+\mathbb{E}[\|\tilde{\xi}\|^{2p\beta}])e^{-\hat{\lambda}_2pt_k}.
$$

For $k=1$, by (3.25), Lemma 3.7, and $\mathcal{D}\xi(t_j)\cdot\eta=\eta(t_j)$, $j\in\{-N,-N+1,\dots,0\}$, we arrive at

$$
\begin{aligned}
&\Big(1+(2a_1\theta\Delta)^p\big(1-\int_{-\Delta}^{0}\frac{r+\Delta}{\Delta}\mathrm{d}\nu_2(r)\big)\Big)\mathbb{E}[|\mathcal{D}y^{\tilde{\xi},\Delta}(t_1)\cdot\eta|^{2p}]\\
&\quad\le(2a_2\theta\Delta)^p(1+\tilde{\varepsilon})\Big(1-\int_{-\Delta}^{0}\frac{r+\Delta}{\Delta}\mathrm{d}\nu_2(r)\Big)\|\eta\|^{2p}+K_{\tilde{\varepsilon}}\mathbb{E}[|\mathcal{D}z^{\tilde{\xi},\Delta}(t_1)\cdot\eta|^{2p}]\\
&\quad\le(1\vee K_{\tilde{\varepsilon}}K_z)\Big(1+(2a_2\theta\Delta)^p(1+\tilde{\varepsilon})e^{\hat{\lambda}_2p\Delta}\big(1-\int_{-\Delta}^{0}\frac{r+\Delta}{\Delta}\mathrm{d}\nu_2(r)\big)\Big)\\
&\qquad\times\|\eta\|^{2p}(1+\mathbb{E}[\|\tilde{\xi}\|^{2p\beta}])e^{-\hat{\lambda}_2pt_1}.
\end{aligned}
$$

Utilizing $(1+\tilde{\varepsilon})a_2^p e^{\hat{\lambda}_2 p\tau} \leq a_1^p$, one has

$$1+(2a_1\theta\Delta)^p\big(1-\int_{-\Delta}^0 \frac{r+\Delta}{\Delta}\mathrm{d}\nu_2(r)\big)$$
$$\geq 1+(2a_2\theta\Delta)^p(1+\tilde{\varepsilon})e^{\hat{\lambda}_2 p\Delta}\big(1-\int_{-\Delta}^0 \frac{r+\Delta}{\Delta}\mathrm{d}\nu_2(r)\big),$$

which derives

$$\mathbb{E}[|\mathcal{D}y^{\tilde{\xi},\Delta}(t_1)\cdot\eta|^{2p}] \leq (1\vee K_{\tilde{\varepsilon}}K_z)\|\eta\|^{2p}(1+\mathbb{E}[\|\tilde{\xi}\|^{2p\beta}])e^{-\hat{\lambda}_2 p t_1}. \tag{3.26}$$

Assume that for some integer $k>1$ and $l\in\{1,2,\ldots,k\}$,

$$\mathbb{E}[|\mathcal{D}y^{\tilde{\xi},\Delta}(t_l)\cdot\eta|^{2p}] \leq (1\vee K_{\tilde{\varepsilon}}K_z)\|\eta\|^{2p}(1+\mathbb{E}[\|\tilde{\xi}\|^{2p\beta}])e^{-\hat{\lambda}_2 p t_l}. \tag{3.27}$$

Then for $k+1$, we arrive at

$$\begin{aligned}
&\Big(1+(2a_1\theta\Delta)^p\big(1-\int_{-\Delta}^0 \frac{r+\Delta}{\Delta}\mathrm{d}\nu_2(r)\big)\Big)\mathbb{E}[|\mathcal{D}y^{\tilde{\xi},\Delta}(t_{k+1})\cdot\eta|^{2p}]\\
&\leq (2a_2\theta\Delta)^p(1+\tilde{\varepsilon})\Big(1-\int_{-\Delta}^0 \frac{r+\Delta}{\Delta}\mathrm{d}\nu_2(r)\Big)(1\vee K_{\tilde{\varepsilon}}K_z)\|\eta\|^{2p}\\
&\quad\times(1+\mathbb{E}[\|\tilde{\xi}\|^{2p\beta}])e^{-\hat{\lambda}_2 p t_{k+1-N}}+K_{\tilde{\varepsilon}}K_z\|\eta\|^{2p}(1+\mathbb{E}[\|\tilde{\xi}\|^{2p\beta}])e^{-\hat{\lambda}_2 p t_{k+1}}\\
&\leq (1\vee K_{\tilde{\varepsilon}}K_z)\Big(1+(2a_2\theta\Delta)^p(1+\tilde{\varepsilon})e^{\hat{\lambda}_2 p\tau}\big(1-\int_{-\Delta}^0 \frac{r+\Delta}{\Delta}\mathrm{d}\nu_2(r)\big)\Big)\\
&\quad\times\|\eta\|^{2p}(1+\mathbb{E}[\|\tilde{\xi}\|^{2p\beta}])e^{-\hat{\lambda}_2 p t_{k+1}},
\end{aligned}$$

which implies

$$\mathbb{E}[|\mathcal{D}y^{\tilde{\xi},\Delta}(t_{k+1})\cdot\eta|^{2p}] \leq (1\vee K_{\tilde{\varepsilon}}K_z)\|\eta\|^{2p}(1+\mathbb{E}[\|\tilde{\xi}\|^{2p\beta}])e^{-\hat{\lambda}_2 p t_{k+1}}. \tag{3.28}$$

Hence, (3.23) follows from (3.26)–(3.28).

Step 2. We prove that for $p\geq 2$, $k\in\mathbb{N}$, and $\hat{\lambda}_3\in(0,\hat{\lambda}_2)$,

$$\mathbb{E}[\|\mathcal{D}y_{t_k}^{\tilde{\xi},\Delta}\cdot\eta\|^{2p}] \leq Ke^{-\hat{\lambda}_3 p t_k}\|\eta\|^{2p}(1+\mathbb{E}[\|\tilde{\xi}\|^{2p\beta}]).$$

In fact, based on (3.17) and $2a_1 > \frac{2\theta-1}{\theta^2}$, which is ensured by the condition $\mathcal{R}(\epsilon_1,\tilde{a}_1,0)>0$ in the proof of Lemma 3.6 together with $a_1\geq\tilde{a}_1$, we obtain

$$
\begin{aligned}
& e^{\hat{\lambda}_2 p t_{k+1}} |\mathcal{D} z^{\tilde{\xi},\Delta}(t_{k+1}) \cdot \eta|^{2p} \\
&= \sum_{l=0}^{k} \Big(e^{\hat{\lambda}_2 p t_{l+1}} |\mathcal{D} z^{\tilde{\xi},\Delta}(t_{l+1}) \cdot \eta|^{2p} - e^{\hat{\lambda}_2 p t_l} |\mathcal{D} z^{\tilde{\xi},\Delta}(t_l) \cdot \eta|^{2p} \Big) \\
&+ |\mathcal{D} z^{\tilde{\xi},\Delta}(0) \cdot \eta|^{2p} \\
&\leq |\mathcal{D} z^{\tilde{\xi},\Delta}(0) \cdot \eta|^{2p} + \sum_{l=0}^{k} (A^p_{\theta,\Delta} e^{\hat{\lambda}_2 p \Delta} - 1) e^{\hat{\lambda}_2 p t_l} |\mathcal{D} z^{\tilde{\xi},\Delta}(t_l) \cdot \eta|^{2p} \\
&\quad + \sum_{i=1}^{p} \sum_{l=0}^{k} C^i_p \hat{\mathcal{I}}_{i,l},
\end{aligned}
$$

where

$$
\begin{aligned}
\hat{\mathcal{I}}_{i,l} := e^{\hat{\lambda}_2 p t_{l+1}} A^{p-i}_{\theta,\Delta} |\mathcal{D} z^{\tilde{\xi},\Delta}(t_l) \cdot \eta|^{2(p-i)} \Big(& 2a_2 \Delta \int_{-\tau}^{0} |\mathcal{D} y^{\tilde{\xi},\Delta}_{t_l}(r) \cdot \eta|^2 \mathrm{d}\nu_2(r) \\
& + |\mathcal{D}\sigma(y^{\tilde{\xi},\Delta}_{t_l}) \mathcal{D} y^{\tilde{\xi},\Delta}_{t_l} \cdot \eta \delta W_l|^2 + \mathcal{M}_l \Big)^i
\end{aligned}
$$

with

$$
\mathcal{M}_l = 2\Big\langle \mathcal{D} z^{\tilde{\xi},\Delta}(t_l) \cdot \eta + \frac{1}{\theta} (\mathcal{D} y^{\tilde{\xi},\Delta}(t_l) \cdot \eta - \mathcal{D} z^{\tilde{\xi},\Delta}(t_l) \cdot \eta), \mathcal{D}\sigma(y^{\tilde{\xi},\Delta}_{t_l}) \mathcal{D} y^{\tilde{\xi},\Delta}_{t_l} \cdot \eta \delta W_l \Big\rangle.
$$

It follows from the proofs of Lemmas 3.6 and 3.7 that $A^p_{\theta,\Delta} e^{\hat{\lambda}_2 p \Delta} - 1 \leq 0$ for $\Delta \in (0, 1]$, which implies

$$
e^{\hat{\lambda}_2 p t_{k+1}} |\mathcal{D} z^{\tilde{\xi},\Delta}(t_{k+1}) \cdot \eta|^{2p} \leq |\mathcal{D} z^{\tilde{\xi},\Delta}(0) \cdot \eta|^{2p} + \sum_{i=1}^{p} \sum_{l=0}^{k} C^i_p \hat{\mathcal{I}}_{i,l}.
$$

Thus,

$$
\begin{aligned}
& \mathbb{E}\Big[\sup_{(k-N)\vee 0 \leq j \leq k} e^{\hat{\lambda}_2 p t_{j+1}} |\mathcal{D} z^{\tilde{\xi},\Delta}(t_{j+1}) \cdot \eta|^{2p} \Big] \\
&\quad \leq \mathbb{E}[|\mathcal{D} z^{\tilde{\xi},\Delta}(0) \cdot \eta|^{2p}] + \sum_{i=1}^{p} C^i_p \mathbb{E}\Big[\sup_{(k-N)\vee 0 \leq j \leq k} \sum_{l=0}^{j} \hat{\mathcal{I}}_{i,l} \Big]. \qquad (3.29)
\end{aligned}
$$

We estimate $\mathbb{E}\Big[\sup_{(k-N)\vee 0 \leq j \leq k} \sum_{l=0}^{j} \hat{\mathcal{I}}_{1,l} \Big]$. Applying the Young inequality and (3.13) yields

$$\mathbb{E}\Big[\sup_{(k-N)\vee 0\le j\le k}\sum_{l=0}^{j}\hat{\mathcal{I}}_{1,l}\Big]$$

$$=\mathbb{E}\Big[A_{\theta,\Delta}^{p-1}\sup_{(k-N)\vee 0\le j\le k}\sum_{l=0}^{j}e^{\hat{\lambda}_2 p t_{l+1}}|\mathcal{D}z^{\tilde{\xi},\Delta}(t_l)\cdot\eta|^{2(p-1)}\mathcal{M}_l\Big]$$

$$+\mathbb{E}\Big[A_{\theta,\Delta}^{p-1}\sum_{l=0}^{k}e^{\hat{\lambda}_2 p t_{l+1}}|\mathcal{D}z^{\tilde{\xi},\Delta}(t_l)\cdot\eta|^{2(p-1)}\Big(2a_2\Delta\int_{-\tau}^{0}|\mathcal{D}y_{t_l}^{\tilde{\xi},\Delta}(r)\cdot\eta|^2\mathrm{d}\nu_2(r)$$

$$+|\mathcal{D}\sigma(y_{t_l}^{\tilde{\xi},\Delta})\mathcal{D}y_{t_l}^{\tilde{\xi},\Delta}\cdot\eta\delta W_l|^2\Big)\Big]$$

$$\le\mathbb{E}\Big[A_{\theta,\Delta}^{p-1}\sup_{(k-N)\vee 0\le j\le k}\sum_{l=0}^{j}e^{\hat{\lambda}_2 p t_{l+1}}|\mathcal{D}z^{\tilde{\xi},\Delta}(t_l)\cdot\eta|^{2(p-1)}\mathcal{M}_l\Big]$$

$$+K\Delta\sum_{l=0}^{k}e^{\hat{\lambda}_2 p t_l}\mathbb{E}\Big[|\mathcal{D}z^{\tilde{\xi},\Delta}(t_l)\cdot\eta|^{2p}+\int_{-\tau}^{0}|\mathcal{D}y_{t_l}^{\tilde{\xi},\Delta}(r)\cdot\eta|^{2p}\mathrm{d}\nu_2(r)$$

$$+|\mathcal{D}\sigma(y_{t_l}^{\tilde{\xi},\Delta})\mathcal{D}y_{t_l}^{\tilde{\xi},\Delta}\cdot\eta|^{2p}\Big]$$

$$\le\mathbb{E}\Big[A_{\theta,\Delta}^{p-1}\sup_{(k-N)\vee 0\le j\le k}\sum_{l=0}^{j}e^{\hat{\lambda}_2 p t_{l+1}}|\mathcal{D}z^{\tilde{\xi},\Delta}(t_l)\cdot\eta|^{2(p-1)}\mathcal{M}_l\Big]$$

$$+K\Delta\sum_{l=0}^{k}e^{\hat{\lambda}_2 p t_l}\mathbb{E}\Big[|\mathcal{D}z^{\tilde{\xi},\Delta}(t_l)\cdot\eta|^{2p}+|\mathcal{D}y^{\tilde{\xi},\Delta}(t_l)\cdot\eta|^{2p}\Big]+K\|\eta\|^{2p}.$$

Using Lemma 3.7 and (3.23), we obtain

$$\mathbb{E}\Big[\sup_{(k-N)\vee 0\le j\le k}\sum_{l=0}^{j}\hat{\mathcal{I}}_{1,l}\Big]$$
$$\le\mathbb{E}\Big[A_{\theta,\Delta}^{p-1}\sup_{(k-N)\vee 0\le j\le k}\sum_{l=0}^{j}e^{\hat{\lambda}_2 p t_{l+1}}|\mathcal{D}z^{\tilde{\xi},\Delta}(t_l)\cdot\eta|^{2(p-1)}\mathcal{M}_l\Big]$$
$$+K\|\eta\|^{2p}(1+\mathbb{E}[\|\tilde{\xi}\|^{2p\beta}])t_k+K\|\eta\|^{2p}. \tag{3.30}$$

Following a similar argument as used in (2.23), (2.24), and (2.32), we deduce

$$\mathbb{E}\Big[A_{\theta,\Delta}^{p-1}\sup_{(k-N)\vee 0\le j\le k}\sum_{l=0}^{j}e^{\hat{\lambda}_2 p t_{l+1}}|\mathcal{D}z^{\tilde{\xi},\Delta}(t_l)\cdot\eta|^{2(p-1)}\mathcal{M}_l\Big]$$

$$
\begin{aligned}
&\le K\mathbb{E}\Big[\Big(\sum_{l=(k-N)\vee 0}^{k} e^{2\hat{\lambda}_2 p t_{l+1}}|\mathcal{D}z^{\tilde{\xi},\Delta}(t_l)\cdot\eta|^{4p-2}|\mathcal{D}\sigma(y_{t_l}^{\tilde{\xi},\Delta})\mathcal{D}y_{t_l}^{\tilde{\xi},\Delta}\cdot\eta|^2\Delta\Big)^{\frac{1}{2}}\Big]\\
&\quad + K\mathbb{E}\Big[\Big(\sum_{l=(k-N)\vee 0}^{k} e^{2\hat{\lambda}_2 p t_{l+1}}|\mathcal{D}z^{\tilde{\xi},\Delta}(t_l)\cdot\eta|^{4p-4}|\mathcal{D}y^{\tilde{\xi},\Delta}(t_l)\cdot\eta|^2\\
&\quad \times|\mathcal{D}\sigma(y_{t_l}^{\tilde{\xi},\Delta})\mathcal{D}y_{t_l}^{\tilde{\xi},\Delta}\cdot\eta|^2\Delta\Big)^{\frac{1}{2}}\Big]\\
&\le \frac{1}{2}\mathbb{E}\Big[\sup_{(k-N)\vee 0\le i\le k} e^{\hat{\lambda}_2 p t_{i+1}}|\mathcal{D}z^{\tilde{\xi},\Delta}(t_{i+1})\cdot\eta|^{2p}\Big]\\
&\quad + \frac{1}{2}\mathbb{E}\Big[e^{\hat{\lambda}_2 p t_{(k-N)\vee 0}}|\mathcal{D}z^{\tilde{\xi},\Delta}(t_{(k-N)\vee 0})\cdot\eta|^{2p}\Big]\\
&\quad + K\Delta\mathbb{E}\Big[\sum_{l=(k-N)\vee 0}^{k} e^{\hat{\lambda}_2 p t_l}|\mathcal{D}y^{\tilde{\xi},\Delta}(t_l)\cdot\eta|^{2p}\Big]\\
&\quad + K\Delta\mathbb{E}\Big[\sum_{l=(k-N)\vee 0}^{k} e^{\hat{\lambda}_2 p t_l}|\mathcal{D}\sigma(y_{t_l}^{\tilde{\xi},\Delta})\mathcal{D}y_{t_l}^{\tilde{\xi},\Delta}\cdot\eta|^{2p}\Big].
\end{aligned}
$$

Making use of (3.13), Lemma 3.7, and (3.23), we arrive at

$$
\begin{aligned}
&\mathbb{E}\Big[A_{\theta,\Delta}^{p-1}\sup_{(k-N)\vee 0\le j\le k}\sum_{l=0}^{j} e^{\hat{\lambda}_2 p t_{l+1}}|\mathcal{D}z^{\tilde{\xi},\Delta}(t_l)\cdot\eta|^{2(p-1)}\mathcal{M}_l\Big]\\
&\quad\le \frac{1}{2}\mathbb{E}\Big[\sup_{(k-N)\vee 0\le i\le k} e^{\hat{\lambda}_2 p t_{i+1}}|\mathcal{D}z^{\tilde{\xi},\Delta}(t_{i+1})\cdot\eta|^{2p}\Big]\\
&\quad\quad + K\|\eta\|^{2p}(1+\mathbb{E}[\|\tilde{\xi}\|^{2p\beta}])+K\|\eta\|^{2p}.
\end{aligned}
\tag{3.31}
$$

Inserting (3.31) into (3.30) derives

$$
\begin{aligned}
&\mathbb{E}\Big[\sup_{(k-N)\vee 0\le j\le k}\sum_{l=0}^{j}\hat{\mathcal{I}}_{1,l}\Big]\\
&\quad\le \frac{1}{2}\mathbb{E}\Big[\sup_{(k-N)\vee 0\le i\le k} e^{\hat{\lambda}_2 p t_{l+1}}|\mathcal{D}z^{\tilde{\xi},\Delta}(t_{i+1})\cdot\eta|^{2p}\Big]\\
&\quad\quad + K\|\eta\|^{2p}(1+\mathbb{E}[\|\tilde{\xi}\|^{2p\beta}])(1+t_k)+K\|\eta\|^{2p}.
\end{aligned}
\tag{3.32}
$$

For $i\in\{2,3,\ldots,p\}$, based on the Young inequality, (3.13), Lemma 3.7, and (3.23), $\mathbb{E}\Big[\sup_{(k-N)\vee 0\le j\le k}\sum_{l=0}^{j}\hat{\mathcal{I}}_{i,l}\Big]$ satisfies

$$
\begin{aligned}
&\mathbb{E}\Big[\sup_{(k-N)\vee 0\le j\le k}\sum_{l=0}^{j}\hat{\mathcal{I}}_{i,l}\Big]\\
&\le K\mathbb{E}\Big[\sum_{l=0}^{k}e^{\hat{\lambda}_2 pt_{l+1}}|\mathcal{D}z^{\tilde{\xi},\Delta}(t_l)\cdot\eta|^{2(p-i)}\Big((K\Delta)^i\int_{-\tau}^{0}|\mathcal{D}y_{t_l}^{\tilde{\xi},\Delta}(r)\cdot\eta|^i\mathrm{d}\nu_2(r)\\
&\quad+|\mathcal{D}\sigma(y_{t_l}^{\tilde{\xi},\Delta})\mathcal{D}y_{t_l}^{\tilde{\xi},\Delta}\cdot\eta\delta W_l|^{2i}+|\mathcal{M}_l|^i\Big)\Big]\\
&\le K\Delta\sum_{l=0}^{k}e^{\hat{\lambda}_2 pt_l}\mathbb{E}\Big[|\mathcal{D}z^{\tilde{\xi},\Delta}(t_l)\cdot\eta|^{2p}+|\mathcal{D}y^{\tilde{\xi},\Delta}(t_l)\cdot\eta|^{2p}\Big]+K\|\eta\|^{2p}\\
&\le K\|\eta\|^{2p}(1+\mathbb{E}[\|\tilde{\xi}\|^{2p\beta}])t_k+K\|\eta\|^{2p}.
\end{aligned}\tag{3.33}
$$

Plugging (3.32) and (3.33) into (3.29) leads to

$$
\begin{aligned}
&\mathbb{E}\Big[\sup_{(k-N)\vee 0\le i\le k}e^{\hat{\lambda}_2 pt_{l+1}}|\mathcal{D}z^{\tilde{\xi},\Delta}(t_{i+1})|^{2p}\Big]\\
&\le K\|\eta\|^{2p}(1+\mathbb{E}[\|\tilde{\xi}\|^{2p\beta}])(1+t_k)+K\|\eta\|^{2p}.
\end{aligned}
$$

Similar to (2.33), we obtain

$$
\begin{aligned}
&\mathbb{E}\Big[\sup_{(k-N)\vee 0\le l\le k}e^{\hat{\lambda}_2 pt_{l+1}}|\mathcal{D}y^{\tilde{\xi},\Delta}(t_{l+1})\cdot\eta|^{2p}\Big]\\
&\le\mathbb{E}\Big[\sup_{(k-N)\vee 0\le l\le k}e^{\hat{\lambda}_2 pt_{l+1}}|\mathcal{D}z^{\tilde{\xi},\Delta}(t_{l+1})\cdot\eta|^{2p}\Big]\\
&\quad+K\Delta e^{\hat{\lambda}_2 p\tau}2N\sup_{(k-2N+1)\vee -N\le l\le k+1}e^{\hat{\lambda}_2 pt_l}\mathbb{E}[|\mathcal{D}y^{\tilde{\xi},\Delta}(t_l)\cdot\eta|^{2p}]\\
&\le\mathbb{E}\Big[\sup_{(k-N)\vee 0\le l\le k}e^{\hat{\lambda}_2 pt_{l+1}}|\mathcal{D}z^{\tilde{\xi},\Delta}(t_{l+1})\cdot\eta|^{2p}\Big]\\
&\quad+K\|\eta\|^{2p}(1+\mathbb{E}[\|\tilde{\xi}\|^{2p\beta}])+K\|\eta\|^{2p}.
\end{aligned}
$$

Above two inequalities imply

$$
\begin{aligned}
&\mathbb{E}\Big[\sup_{(k-N)\vee 0\le k\le k}|\mathcal{D}y^{\tilde{\xi},\Delta}(t_{l+1})\cdot\eta|^{2p}\Big]\\
&\le e^{-\hat{\lambda}_2 pt_{k+1-N}}\Big(K\|\eta\|^{2p}(1+\mathbb{E}[\|\xi\|^{2p\beta}])(t_k+1)+K\|\eta\|^{2p}\Big)\\
&\le K\|\eta\|^{2p}(1+\mathbb{E}[\|\tilde{\xi}\|^{2p\beta}])e^{-\hat{\lambda}_3 pt_k},
\end{aligned}
$$

where $\hat{\lambda}_3\in(0,\hat{\lambda}_2)$. We finish the proof. □

Lemma 3.9 *Let Assumptions 2.1 to 2.3 and 3.1 hold. Then for $p \in \mathbb{N}_+$, $a_1 > \mathfrak{a}_1$, $\Delta \in (0, 1]$, $k \in \mathbb{N}$, and $u \geq -\tau$,*

$$\mathbb{E}[\|D_u y_{t_k}^{\xi,\Delta}\|^{2p}] \\ \leq K e^{-\hat{\lambda}_3 p(t_k-u)}\Big(1 + \mathbb{E}[\|\xi\|^{2p(\beta+1)}] + \mathbb{E}[\|D_u\xi\|^{2p}] + \mathbb{E}[|\mathcal{D}b(\xi)D_u\xi|^{2p}]\Big),$$

where $\xi \in L^{2p(\beta+1)}(\Omega; \mathcal{C}^d)$, $D_u\xi \in L^{2p}(\Omega; \mathcal{C}^d \otimes \mathbb{R}^m)$, $\mathcal{D}b(\xi)D_u\xi \in L^{2p}(\Omega; \mathbb{R}^{d\times m})$, $\eta \in \mathcal{C}^d$, $K > 0$, $\hat{\lambda}_3 \in (0, \hat{\lambda}_2)$ with $\hat{\lambda}_2$ being given by Lemma 3.7, and $\mathfrak{a}_1$ is given in Lemma 3.7.

Proof The proof is similar to that of Lemma 3.8 and is omitted. □

The following assumption on the second-order derivatives of coefficients is used to derive estimates of the second-order Malliavin derivatives and Gâteaux derivatives for the exact functional solution and the numerical one.

Assumption 3.2 *Assume that coefficients b and σ have continuous derivatives up to order 2 satisfying that for any $\phi, \phi_1, \phi_2 \in \mathcal{C}^d$,*

$$|\mathcal{D}^2 b(\phi)(\phi_1, \phi_2)| \leq K(1 + \|\phi\|^{(\beta-1)\vee 0})\|\phi_1\|\|\phi_2\|,$$
$$|\mathcal{D}^2 \sigma(\phi)(\phi_1, \phi_2)| \leq K\|\phi_1\|\|\phi_2\|,$$

where $K > 0$, and the constant $\beta > 0$ is given in Assumption 2.4.

Lemma 3.10 *Let Assumptions 2.1 and 2.2 with $a_1 - a_2 - (2p-1)L > 0$ hold for some $p \geq 1$. And let Assumptions 2.3, 3.1, and 3.2 hold. Then for any $t \geq 0$ and $u \geq -\tau$,*

$$\mathbb{E}[|D_u \mathcal{D}x^{\xi}(t)\cdot\eta|^{2p}] \leq K e^{-\hat{\lambda}_1 pt}\|\eta\|^{2p}\Big(1 + \mathbb{E}[\|\xi\|^{4p((\beta-1)\vee 1)}] + \mathbb{E}[\|D_u\xi\|^{8p}]\Big),$$
$$\mathbb{E}[\|D_u \mathcal{D}x_t^{\xi}\cdot\eta\|^{2p}] \leq K e^{-\hat{\lambda}_1 pt}\|\eta\|^{2p}\Big(1 + \mathbb{E}[\|\xi\|^{4p((\beta-1)\vee 1)}] + \mathbb{E}[\|D_u\xi\|^{8p}]\Big),$$

where $\xi \in L^{4p((\beta-1)\vee 1)}(\Omega; \mathcal{C}^d)$ and $D_u\xi \in L^{8p}(\Omega; \mathcal{C}^d \otimes \mathbb{R}^m)$, $\eta \in \mathcal{C}^d$, $K > 0$, and $\hat{\lambda}_1 > 0$ is given in Lemma 3.4.

Proof For $u \leq t$, by the chain rule, $D_u\mathcal{D}x^{\xi}(t)\cdot\eta$ satisfies

$$\begin{aligned} D_u\mathcal{D}x^{\xi}(t)\cdot\eta &= \int_u^t \Big(\mathcal{D}^2 b(x_s^{\xi})(D_u x_s^{\xi}, \mathcal{D}x_s^{\xi}\cdot\eta) + \mathcal{D}b(x_s^{\xi})D_u\mathcal{D}x_s^{\xi}\cdot\eta\Big)\mathrm{d}s \\ &\quad + \int_u^t \Big(\mathcal{D}^2\sigma(x_s^{\xi})(D_u x_s^{\xi}, \mathcal{D}x_s^{\xi}\cdot\eta) + \mathcal{D}\sigma(x_s^{\xi})D_u\mathcal{D}x_s^{\xi}\cdot\eta\Big)\mathrm{d}W(s) \\ &\quad + \mathcal{D}\sigma(x_u^{\xi})\mathcal{D}x_u^{\xi}\cdot\eta\mathbf{1}_{[0,t]}(u). \end{aligned}$$

Applying the Itô formula to the function $e^{\hat{\lambda}_1 pt}|D_u \mathcal{D}x^{\xi}(t)\cdot\eta|^{2p}$, we have

$$
\begin{aligned}
&\mathrm{d}(e^{\hat{\lambda}_1 pt}|D_u \mathcal{D}x^{\xi}(t)\cdot\eta|^{2p})\\
&\quad\le \hat{\lambda}_1 p e^{\hat{\lambda}_1 pt}|D_u \mathcal{D}x^{\xi}(t)\cdot\eta|^{2p}\mathrm{d}t + 2pe^{\hat{\lambda}_1 pt}|D_u \mathcal{D}x^{\xi}(t)\cdot\eta|^{2p-2}\\
&\qquad\times\Big[\Big\langle D_u\mathcal{D}x^{\xi}(t)\cdot\eta,\ \mathcal{D}^2 b(x_t^{\xi})(D_u x_t^{\xi},\mathcal{D}x_t^{\xi}\cdot\eta)+\mathcal{D}b(x_t^{\xi})D_u\mathcal{D}x_t^{\xi}\cdot\eta\Big\rangle\mathrm{d}t\\
&\qquad+\Big\langle D_u\mathcal{D}x^{\xi}(t)\cdot\eta,\ \Big(\mathcal{D}^2\sigma(x_t^{\xi})(D_u x_t^{\xi},\mathcal{D}x_t^{\xi}\cdot\eta)+\mathcal{D}\sigma(x_t^{\xi})D_u\mathcal{D}x_t^{\xi}\cdot\eta\Big)\mathrm{d}W(t)\Big\rangle\Big]\\
&\qquad+p(2p-1)e^{\hat{\lambda}_1 pt}|D_u\mathcal{D}x^{\xi}(t)\cdot\eta|^{2p-2}|\mathcal{D}^2\sigma(x_t^{\xi})(D_u x_t^{\xi},\mathcal{D}x_t^{\xi}\cdot\eta)\\
&\qquad+\mathcal{D}\sigma(x_t^{\xi})D_u\mathcal{D}x_t^{\xi}\cdot\eta|^2\mathrm{d}t.
\end{aligned}
$$

Taking expectations on both sides, and using Assumptions 3.1 and 3.2 together with the Young inequality, we obtain

$$
\begin{aligned}
&\mathbb{E}[e^{\hat{\lambda}_1 pt}|D_u\mathcal{D}x^{\xi}(t)\cdot\eta|^{2p}]\\
&\quad\le \mathbb{E}[e^{\hat{\lambda}_1 pu}|\mathcal{D}\sigma(x_u^{\xi})\mathcal{D}x_u^{\xi}\cdot\eta|^{2p}]+\int_u^t\mathbb{E}\Big[\hat{\lambda}_1 pe^{\hat{\lambda}_1 ps}|D_u\mathcal{D}x^{\xi}(s)\cdot\eta|^{2p}+2pe^{\hat{\lambda}_1 ps}\\
&\qquad\times|D_u\mathcal{D}x^{\xi}(s)\cdot\eta|^{2p-2}\\
&\quad\times\Big(-a_1|D_u\mathcal{D}x^{\xi}(s)\cdot\eta|^2+a_2\int_{-\tau}^0|D_u\mathcal{D}x^{\xi}(s+r)\cdot\eta|^2\mathrm{d}\nu_2(r)\Big)\\
&\qquad+Ke^{\hat{\lambda}_1 ps}|D_u\mathcal{D}x^{\xi}(s)\cdot\eta|^{2p-1}(1+\|x_s^{\xi}\|^{(\beta-1)\vee 0})\|D_u x_s^{\xi}\|\|\mathcal{D}x_s^{\xi}\cdot\eta\|\\
&\qquad+p(2p-1)e^{\hat{\lambda}_1 ps}|D_u\mathcal{D}x^{\xi}(s)\cdot\eta|^{2p-2}\Big((1+\frac{1}{\gamma_1})|\mathcal{D}^2\sigma(x_s^{\xi})(D_u x_s^{\xi},\mathcal{D}x_s^{\xi}\cdot\eta)|^2\\
&\qquad+(1+\gamma_1)L\big(|D_u\mathcal{D}x^{\xi}(s)\cdot\eta|^2+\int_{-\tau}^0|D_u\mathcal{D}x^{\xi}(s+r)\cdot\eta|^2\mathrm{d}\nu_1(r)\big)\Big)\Big]\mathrm{d}s\\
&\quad\le\mathbb{E}[e^{\hat{\lambda}_1 pu}|\mathcal{D}\sigma(x_u^{\xi})\mathcal{D}x_u^{\xi}\cdot\eta|^{2p}]+pC(\hat{\lambda}_1,\gamma_1)\int_u^t e^{\hat{\lambda}_1 ps}\mathbb{E}[|D_u\mathcal{D}x^{\xi}(s)\cdot\eta|^{2p}]\mathrm{d}s\\
&\qquad+K\int_u^t e^{\hat{\lambda}_1 ps}\mathbb{E}[|D_u\mathcal{D}x^{\xi}(s)\cdot\eta|^{2p-1}(1+\|x_s^{\xi}\|^{(\beta-1)\vee 0})\|D_u x_s^{\xi}\|\|\mathcal{D}x_s^{\xi}\cdot\eta\|]\mathrm{d}s\\
&\qquad+p(2p-1)(1+\frac{1}{\gamma_1})\\
&\qquad\times\int_u^t e^{\hat{\lambda}_1 ps}\mathbb{E}[|D_u\mathcal{D}x^{\xi}(s)\cdot\eta|^{2p-2}|\mathcal{D}^2\sigma(x_s^{\xi})D_u x_s^{\xi}\mathcal{D}x_s^{\xi}\cdot\eta|^2]\mathrm{d}s
\end{aligned}
$$

$$
\begin{aligned}
&\leq \mathbb{E}[e^{\hat{\lambda}_1 pu}|\mathcal{D}\sigma(x_u^{\tilde{\xi}})\mathcal{D}x_u^{\tilde{\xi}}\cdot\eta|^{2p}] + pC(\hat{\lambda}_1,\gamma_1)\int_u^t e^{\hat{\lambda}_1 ps}\mathbb{E}[|D_u\mathcal{D}x^{\tilde{\xi}}(s)\cdot\eta|^{2p}]\mathrm{d}s \\
&\quad + \int_u^t \Big(p\gamma_2 e^{\hat{\lambda}_1 ps}\mathbb{E}[|D_u\mathcal{D}x^{\tilde{\xi}}(s)\cdot\eta|^{2p}] + pK_{\gamma_2}e^{\hat{\lambda}_1 ps}\mathbb{E}[(1+\|x_s^{\tilde{\xi}}\|^{2p((\beta-1)\vee 0)}) \\
&\quad \times \|D_u x_s^{\tilde{\xi}}\|^{2p}\|\mathcal{D}x_s^{\tilde{\xi}}\cdot\eta\|^{2p}]\Big)\mathrm{d}s + p(2p-1)\int_u^t \Big(\gamma_3 e^{\hat{\lambda}_1 ps}\mathbb{E}[|D_u\mathcal{D}x^{\tilde{\xi}}(s)\cdot\eta|^{2p}] \\
&\quad + K_{\gamma_3}e^{\hat{\lambda}_1 ps}\mathbb{E}[\|D_u x_s^{\tilde{\xi}}\|^{2p}\|\mathcal{D}x_s^{\tilde{\xi}}\cdot\eta\|^{2p}]\Big)\mathrm{d}s.
\end{aligned}
$$

where $C(\hat{\lambda}_1,\gamma_1) := \hat{\lambda}_1 - 2a_1 + 2a_2 e^{\hat{\lambda}_1 p\tau} + (2p-1)(1+\gamma_1)L(1+e^{\hat{\lambda}_1 p\tau})$. By taking $\gamma_1,\gamma_2,\gamma_3$ sufficiently small such that $C(\hat{\lambda}_1,\gamma_1)+\gamma_2+(2p-1)\gamma_3 \leq 0$, and combining Lemmas 3.4 and 3.5, we arrive at

$$
\begin{aligned}
&\mathbb{E}[e^{\hat{\lambda}_1 pt}|D_u\mathcal{D}x^{\tilde{\xi}}(t)\cdot\eta|^{2p}] \\
&\quad \leq \mathbb{E}[e^{\hat{\lambda}_1 pu}|\mathcal{D}\sigma(x_u^{\tilde{\xi}})\mathcal{D}x_u^{\tilde{\xi}}\cdot\eta|^{2p}] + K\int_u^t e^{\hat{\lambda}_1 ps}\mathbb{E}\Big[(1 \\
&\qquad + \|x_s^{\tilde{\xi}}\|^{2p((\beta-1)\vee 0)})\|D_u x_s^{\tilde{\xi}}\|^{2p}\|\mathcal{D}x_s^{\tilde{\xi}}\cdot\eta\|^{2p} + \|D_u x_s^{\tilde{\xi}}\|^{2p}\|\mathcal{D}x_s^{\tilde{\xi}}\cdot\eta\|^{2p}\Big]\mathrm{d}s \\
&\quad \leq \mathbb{E}[e^{\hat{\lambda}_1 pu}|\mathcal{D}\sigma(x_u^{\tilde{\xi}})\mathcal{D}x_u^{\tilde{\xi}}\cdot\eta|^{2p}] + K\int_u^t e^{\hat{\lambda}_1 ps}\Big(\mathbb{E}[1+\|x_s^{\tilde{\xi}}\|^{4p((\beta-1)\vee 0)}]\Big)^{\frac{1}{2}} \\
&\qquad \times \Big(\mathbb{E}[\|D_u x_s^{\tilde{\xi}}\|^{8p}]\Big)^{\frac{1}{4}}\Big(\mathbb{E}[\|\mathcal{D}x_s^{\tilde{\xi}}\cdot\eta\|^{8p}]\Big)^{\frac{1}{4}}\mathrm{d}s \\
&\quad \leq \mathbb{E}[e^{\hat{\lambda}_1 pu}|\mathcal{D}\sigma(x_u^{\tilde{\xi}})\mathcal{D}x_u^{\tilde{\xi}}\cdot\eta|^{2p}] + K\int_u^t e^{\hat{\lambda}_1 ps}\Big(1+(\mathbb{E}[\|\tilde{\xi}\|^{4p((\beta-1)\vee 0)}])^{\frac{1}{2}} \\
&\qquad + e^{-\hat{\lambda}_1 ps-\hat{\lambda}_1 p(s-u)}\|\eta\|^{2p}\Big(1+(\mathbb{E}[\|\tilde{\xi}\|^{8p}])^{\frac{1}{4}} + (\mathbb{E}[\|D_u\tilde{\xi}\|^{8p}])^{\frac{1}{4}}\Big)\Big)\mathrm{d}s \\
&\quad \leq Ke^{\hat{\lambda}_1 pu}\mathbb{E}[\|\mathcal{D}x_u^{\tilde{\xi}}\cdot\eta\|^{2p}] + K\|\eta\|^{2p}\Big(1+\mathbb{E}[\|\tilde{\xi}\|^{4p((\beta-1)\vee 1)}] + \mathbb{E}[\|D_u\tilde{\xi}\|^{8p}]\Big) \\
&\quad \leq K\|\eta\|^{2p}\Big(1+\mathbb{E}[\|\tilde{\xi}\|^{4p((\beta-1)\vee 1)}] + \mathbb{E}[\|D_u\tilde{\xi}\|^{8p}]\Big),
\end{aligned}
$$

which implies

$$
\mathbb{E}[|D_u\mathcal{D}x^{\tilde{\xi}}(t)\cdot\eta|^{2p}] \leq Ke^{-\hat{\lambda}_1 pt}\|\eta\|^{2p}(1+\mathbb{E}[\|\tilde{\xi}\|^{4p((\beta-1)\vee 1)}] + \mathbb{E}[\|D_u\tilde{\xi}\|^{8p}]).
$$

Similar to the proof of Lemmas 3.4 and 3.5, by using the Burkholder–Davis–Gundy inequality, one has

$$
\mathbb{E}[\|D_u\mathcal{D}x_t^{\tilde{\xi}}\cdot\eta\|^{2p}] \leq Ke^{-\hat{\lambda}_1 pt}\|\eta\|^{2p}\Big(1+\mathbb{E}[\|\tilde{\xi}\|^{4p((\beta-1)\vee 1)}] + \mathbb{E}[\|D_u\tilde{\xi}\|^{8p}]\Big).
$$

We finish the proof. □

Lemma 3.11 *Let Assumptions 2.1 and 2.2 with $a_1 - a_2 - (2p-1)L > 0$ hold for some $p \geq 1$. And let Assumptions 2.3, 3.1, and 3.2 hold. Then for $t \geq 0$, $u \geq -\tau$, and $w \geq -\tau$,*

$$\mathbb{E}[|D_w D_u x^{\xi}(t)|^{2p}] \leq K e^{-\hat{\lambda}_1 p(t-u\vee w)}\Big(1 + \mathbb{E}[\|\xi\|^{4p((\beta-1)\vee 1)}] + \mathbb{E}[\|D_u\xi\|^{8p}] + \mathbb{E}[\|D_w\xi\|^{8p}] + \mathbb{E}[\|D_w D_u\xi\|^{2p}]\Big),$$

$$\mathbb{E}[\|D_w D_u x_t^{\xi}\|^{2p}] \leq K e^{-\hat{\lambda}_1 p(t-u\vee w)}\Big(1 + \mathbb{E}[\|\xi\|^{4p((\beta-1)\vee 1)}] + \mathbb{E}[\|D_u\xi\|^{8p}] + \mathbb{E}[\|D_w\xi\|^{8p}] + \mathbb{E}[\|D_w D_u\xi\|^{2p}]\Big),$$

where $\xi \in L^{4p((\beta-1)\vee 1)}(\Omega; C^d)$, $D_u\xi \in L^{8p}(\Omega; C^d \otimes \mathbb{R}^m)$, $D_w D_u\xi \in L^{2p}(\Omega; C^d)$, $K > 0$, and $\hat{\lambda}_1$ is given by Lemma 3.4.

Proof The proof is similar to that of Lemma 3.10. We only give the sketch of the proof. $D_w D_u x^{\xi}(t)$ satisfies

$$\begin{aligned} D_w D_u x^{\xi}(t) = &\int_{u\vee w}^{t} \Big(\mathcal{D}^2 b(x_s^{\xi})(D_w x_s^{\xi}, D_u x_s^{\xi}) + \mathcal{D}b(x_s^{\xi}) D_w D_u x_s^{\xi}\Big)\mathrm{d}s \\ &+ \int_{u\vee w}^{t} \Big(\mathcal{D}^2\sigma(x_s^{\xi})(D_w x_s^{\xi}, D_u x_s^{\xi}) + \mathcal{D}\sigma(x_s^{\xi}) D_w D_u x_s^{\xi}\Big)\mathrm{d}W(s) \\ &+ \mathcal{D}\sigma(x_w^{\xi}) D_u x_w^{\xi} \mathbf{1}_{[u,t]}(w) + \mathcal{D}\sigma(x_u^{\xi}) D_w x_u^{\xi} \mathbf{1}_{[w,t]}(u) + D_w D_u\xi(0). \end{aligned}$$

Applying the Itô formula, (3.12) and (3.13), and Assumption 3.2, we deduce

$$\begin{aligned} &\mathbb{E}[e^{\hat{\lambda}_1 pt}|D_w D_u x^{\xi}(t)|^{2p}] \\ &\leq \mathbb{E}\Big[e^{\hat{\lambda}_1 p(u\vee w)}|\mathcal{D}\sigma(x_w^{\xi}) D_u x_w^{\xi} \mathbf{1}_{[u,u\vee w]}(w) \\ &\quad + \mathcal{D}\sigma(x_u^{\xi}) D_w x_u^{\xi} \mathbf{1}_{[w,u\vee w]}(u) + D_w D_u\xi(0)|^{2p}\Big] \\ &\quad + K\int_{u\vee w}^{t} e^{\hat{\lambda}_1 ps}\mathbb{E}[(1 + \|x_s^{\xi}\|^{2p((\beta-1)\vee 0)})\|D_w x_s^{\xi}\|^{2p}\|D_u x_s^{\xi}\|^{2p}]\mathrm{d}s. \end{aligned}$$

It follows from (2.35) and Lemma 3.5 that

$$\mathbb{E}[|D_w D_u x^{\xi}(t)|^{2p}] \leq K e^{-\hat{\lambda}_1 p(t-u\vee w)}\Big(1 + \mathbb{E}[\|\xi\|^{4p((\beta-1)\vee 1)}] + \mathbb{E}[\|D_u\xi\|^{8p}] + \mathbb{E}[\|D_w\xi\|^{8p}] + \mathbb{E}[\|D_w D_u\xi\|^{2p}]\Big).$$

Similarly, one can derive

$$\mathbb{E}[\|D_w D_u x_t^{\tilde{\xi}}\|^{2p}] \le K e^{-\hat{\lambda}_1 p(t-u\vee w)}\Big(1+\mathbb{E}[\|\tilde{\xi}\|^{4p((\beta-1)\vee 1)}]+\mathbb{E}[\|D_u\tilde{\xi}\|^{8p}]$$
$$+\mathbb{E}[\|D_w\xi\|^{8p}]+\mathbb{E}[\|D_w D_u\tilde{\xi}\|^{2p}]\Big).$$

The proof is completed. □

To derive estimates of the third-order derivatives of the numerical functional solution, we give the following assumption on the third order derivatives of coefficients b and σ.

Assumption 3.3 *Assume that coefficients b and σ have continuous derivatives up to order 3 satisfying that for any $\phi, \phi_1, \phi_2, \phi_3 \in C^d$,*

$$|\mathcal{D}^3 b(\phi)(\phi_1,\phi_2,\phi_3)| \le K(1+\|\phi\|^{(\beta-2)\vee 0})\|\phi_1\|\|\phi_2\|\|\phi_3\|,$$
$$|\mathcal{D}^3 \sigma(\phi)(\phi_1,\phi_2,\phi_3)| \le K\|\phi_1\|\|\phi_2\|\|\phi_3\|,$$

where $K>0$, the constant $\beta>0$ is given in Assumption 2.4.

Similarly, we have the following estimates of the numerical solution.

Lemma 3.12 *Let Assumptions 2.1 to 2.3 and 3.1 to 3.3 hold. Then for $p\in\mathbb{N}_+$, there exists a number $\mathfrak{a}_2 := \mathfrak{a}_2(L, a_2, \theta, p) > 0$ such that for $a_1 > \mathfrak{a}_2$, $\Delta\in(0,1]$, $u \ge -\tau$, and $w \ge -\tau$,*

$$\mathbb{E}[\|D_u\mathcal{D}y_{t_k}^{\tilde{\xi},\Delta}\cdot\eta\|^{2p}] \le K e^{-\hat{\lambda}_3 t_k}\|\eta\|^{2p}\big(1+\mathbb{E}[\|\tilde{\xi}\|^{p'p}]+\mathbb{E}[\|D_u\tilde{\xi}\|^{16p}]\big),$$
$$\mathbb{E}[\|D_w D_u y_{t_k}^{\tilde{\xi},\Delta}\|^{2p}] \le K e^{-\hat{\lambda}_3(t_k-u\vee w)}\big(1+\mathbb{E}[\|\tilde{\xi}\|^{p'p}]+\mathbb{E}[\|D_u\tilde{\xi}\|^{16p}]$$
$$+\mathbb{E}[\|D_w\tilde{\xi}\|^{16p}]+\mathbb{E}[\|D_w D_u\tilde{\xi}\|^{2p}]\big),$$
$$\mathbb{E}[\|D_w D_u\mathcal{D}y_{t_k}^{\tilde{\xi},\Delta}\cdot\eta\|^{2p}] \le K e^{-\hat{\lambda}_3 t_k}\|\eta\|^{2p}\big(1+\mathbb{E}[\|\tilde{\xi}\|^{p'p}]+\mathbb{E}[\|D_u\tilde{\xi}\|^{32p}]$$
$$+\mathbb{E}[\|D_w\tilde{\xi}\|^{32p}]+\mathbb{E}[\|D_w D_u\tilde{\xi}\|^{4p}]\big),$$

where $\tilde{\xi}\in L^{p'p}(\Omega; C^d)$ with some $p'>0$, $\eta\in C^d$, $K>0$, and $\hat{\lambda}_3>0$ is the number given in Lemma 3.8.

3.2.2 Convergence Rate via Weak Convergence Analysis

For the SFDE with distributed delay, under the globally Lipschitz continuous condition on coefficients, there have been some works devoted to the weak convergence of numerical methods on the finite time horizon. For instance, authors in [3] investigate the weak convergence of the EM method by using the infinite-dimensional version of the Kolmogorov equation; Authors in [9] obtain the weak convergence rate of the EM method by developing a dual approach. To deepen the study, this subsection is devoted to presenting the weak convergence rate on the infinite time horizon of the θ-EM method for the SFDE with superlinearly growing coefficients. Before that, we propose the following assumption on the coefficients b and σ.

Assumption 3.4 *Assume that there exist constants $n_b, n_\sigma \in \mathbb{N}_+$ and probability measures ν_3^i, ν_4^j on $[-\tau, 0]$ for $i \in \{1, \ldots, n_b\}$, $j \in \{1, \ldots, n_\sigma\}$, such that coefficients b and σ satisfy that for any $\phi_1, \phi_2 \in C^d$,*

$$\mathcal{D}b(\phi_1)\phi_2 = \sum_{i=1}^{n_b} \int_{-\tau}^{0} k_b^i(\phi_1)\phi_2(s)\mathrm{d}\nu_3^i(s),$$

$$\mathcal{D}\sigma(\phi_1)\phi_2 = \sum_{j=1}^{n_\sigma} \int_{-\tau}^{0} k_\sigma^j(\phi_1)\phi_2(s)\mathrm{d}\nu_4^j(s),$$

where $k_b^i : C^d \to \mathbb{R}^{d\times d}$ and $k_\sigma^j : C^d \to \mathcal{L}(\mathbb{R}^d; \mathbb{R}^{d\times m})$ satisfy

$$\sup_{i\in\{1,\ldots,n_b\}} \Big(\big\| \mathcal{D}k_b^i(\phi_1) \big\|_{\mathcal{L}(C^d;\mathbb{R}^{d\times d})} + |k_b^i(\phi_1)| \Big) \le K(1 + \|\phi_1\|^\beta),$$

$$\sup_{j\in\{1,\ldots,n_\sigma\}} \sup_{\phi_1\in C^d} \Big(\big\| \mathcal{D}k_\sigma^j(\phi_1) \big\|_{\mathcal{L}(C^d;\mathcal{L}(\mathbb{R}^d;\mathbb{R}^{d\times m}))} + \|k_\sigma^j(\phi_1)\|_{\mathcal{L}(\mathbb{R}^d;\mathbb{R}^{d\times m})} \Big) \le K.$$

Examples satisfying Assumption 3.4 are given as follows.

Example 3.1 Coefficients b and σ considered in [3, 9] satisfy Assumption 3.4, where the coefficients have the forms: $b(\phi) = \tilde{b}(\int_{-\tau}^0 \phi(r)\mathrm{d}\nu_3(r))$ and $\sigma(\phi) = \tilde{\sigma}(\int_{-\tau}^0 \phi(r)\mathrm{d}\nu_4(r))$ for $\phi \in C^d$ with some functions $\tilde{b} : \mathbb{R}^d \to \mathbb{R}^d$ and $\tilde{\sigma} : \mathbb{R}^d \to \mathbb{R}^{d\times m}$. In fact, it can be calculated that for $\phi_1 \in C^d$,

$$\mathcal{D}b(\phi)\phi_1 = \mathcal{D}\tilde{b}\Big(\int_{-\tau}^0 \phi(r)\mathrm{d}\nu_3(r)\Big)\phi_1 = \int_{-\tau}^0 \mathcal{D}\tilde{b}\Big(\int_{-\tau}^0 \phi(r)\mathrm{d}\nu_3(r)\Big)\phi_1(s)\mathrm{d}\nu_3(s),$$

$$\mathcal{D}\sigma(\phi)\phi_1 = \mathcal{D}\tilde{\sigma}\Big(\int_{-\tau}^0 \phi(r)\mathrm{d}\nu_4(r)\Big)\phi_1 = \int_{-\tau}^0 \mathcal{D}\tilde{\sigma}\Big(\int_{-\tau}^0 \phi(r)\mathrm{d}\nu_4(r)\Big)\phi_1(s)\mathrm{d}\nu_4(s).$$

Example 3.2 Let $b(\phi) = \frac{1}{\tau}\int_{-\tau}^0 \phi(s)\mathrm{d}s - |\phi(0)|^2\phi(0) - 2\phi(0)$, $\phi \in C^d$. Then it can be verified that Assumption 3.4 is satisfied: for $\phi_1 \in C^d$,

$$\mathcal{D}b(\phi)\phi_1 = \frac{1}{\tau}\int_{-\tau}^0 \phi_1(s)\mathrm{d}s - 2\phi(0)^\top\phi_1(0)\phi(0) - |\phi(0)|^2\phi_1(0) - 2\phi_1(0).$$

In this example, $n_b = 2$, $k_b^1(\phi) = \mathrm{Id}_{d\times d}$, $k_b^2(\phi) = -2\phi(0)\phi^\top(0) - (|\phi(0)|^2 + 2)\mathrm{Id}_{d\times d}$, $\nu_3^1(\mathrm{d}s) = \frac{1}{\tau}\mathrm{d}s$, $\nu_3^2(\mathrm{d}s) = \delta_0(\mathrm{d}s)$.

The following theorem establishes the weak convergence of the θ-EM method, with the proof presented in the next subsection.

Theorem 3.13 *Let conditions in Lemma 3.12, Assumption 2.5 with $\rho \geq 1$, and Assumption 3.4 hold. Then for $f \in C_b^3(\mathbb{R}^d;\mathbb{R})$,*

$$|\mathbb{E}[f(x^\xi(T))] - \mathbb{E}[f(y^{\xi,\Delta}(T))]| \leq K\Delta,$$

where K is independent of T.

As a consequence, we obtain the convergence rate of the numerical invariant measure.

Theorem 3.14 *Let conditions in Theorem 3.13 hold. Then for $f \in C_b^3(\mathbb{R}^d;\mathbb{R})$,*

$$|\mu(f) - \mu^\Delta(f)| \leq K\Delta.$$

3.2.3 *Proof of Theorem 3.13*

We are in the position to show the proof of Theorem 3.13. To simplify the proof of Theorem 3.13, we take parameters n_b, n_σ in Assumption 3.4 to be $n_b = n_\sigma = 1$. The case of $n_b > 1$, $n_\sigma > 1$ can be proven similarly. By (3.11), it suffices to provide estimates for the terms $\mathcal{I}_0, \mathcal{I}_b, \mathcal{I}_{b,\theta}$, and $\mathcal{I}_\sigma$.

Lemma 3.15 *Let conditions in Theorem 3.13 hold. Then $|\mathcal{I}_b| \leq K\Delta$.*

Proof By the Taylor formula, we have

$$\mathcal{I}_b = \sum_{i=1}^{N^\Delta} \int_0^1 \mathbb{E}\Big[\Big\langle f'(\Phi(T;t_i,Y_i^\varsigma))\mathcal{D}\Phi(T;t_i,Y_i^\varsigma), I^{[0]} \int_{t_{i-1}}^{t_i} \int_0^1 \mathcal{D}b(Z_{i,r}^{\beta_1})$$
$$\big(\varphi_r(t_{i-1}, \varphi_{t_{i-1}}(0,\xi^{Int})) - \varphi_{t_{i-1}}^{Int}(0,\xi^{Int})\big)\mathrm{d}\beta_1\mathrm{d}r\Big\rangle\Big]\mathrm{d}\varsigma,$$

where $Z_{i,r}^{\beta_1} := \beta_1\varphi_r(t_{i-1}, \varphi_{t_{i-1}}(0,\xi^{Int})) + (1-\beta_1)\varphi_{t_{i-1}}^{Int}(0,\xi^{Int})$. To estimate $\mathcal{I}_b$, we need to split $\varphi_r(t_{i-1}, \varphi_{t_{i-1}}(0,\xi^{Int})) - \varphi_{t_{i-1}}^{Int}(0,\xi^{Int})$ for $r \in [t_{i-1}, t_i)$, based on the definitions of the exact functional solution and its linear interpolation. For $s \in [t_j, t_{j+1}] \subset [-\tau, 0]$, we have

$$\varphi_{t_{i-1}}^{Int}(0,\xi^{Int})(s)=\frac{t_{j+1}-s}{\Delta}\varphi(t_{i+j-1};0,\xi^{Int})+\frac{s-t_j}{\Delta}\varphi(t_{i+j};0,\xi^{Int}). \tag{3.34}$$

We also have $r+s\in[t_{i+j-1},t_{i+j+1})$, which leads to the following two cases. For notational simplicity and to illustrate the main idea of the proof, we suppose $t_{i+j-1}\geq 0$. The case $t_{i+j-1}<0$, which involves contributions from the initial values on $[-\tau,0]$, can be treated similarly.

Case 1: $r+s\in[t_{i+j-1},t_{i+j})$. In this case, by the integral form of the exact solution of (1.1), we have

$$\begin{aligned}
&\varphi_r(t_{i-1},\varphi_{t_{i-1}}(0,\xi^{Int}))(s)\\
&\quad=\varphi(r+s;0,\xi^{Int})\\
&\quad=\varphi(t_{i+j-1};0,\xi^{Int})+\int_{t_{i+j-1}}^{r+s}b(\varphi_u(0,\xi^{Int}))\mathrm{d}u+\int_{t_{i+j-1}}^{r+s}\sigma(\varphi_u(0,\xi^{Int}))\mathrm{d}W(u).
\end{aligned}$$

Hence, combining this with (3.34) yields

$$\begin{aligned}
&\varphi_r(t_{i-1},\varphi_{t_{i-1}}(0,\xi^{Int}))(s)-\varphi_{t_{i-1}}^{Int}(0,\xi^{Int})(s)\\
&\quad=\frac{t_j-s}{\Delta}(\varphi(t_{i+j};0,\xi^{Int})-\varphi(t_{i+j-1};0,\xi^{Int}))\\
&\qquad+\int_{t_{i+j-1}}^{r+s}b(\varphi_u(0,\xi^{Int}))\mathrm{d}u+\int_{t_{i+j-1}}^{r+s}\sigma(\varphi_u(0,\xi^{Int}))\mathrm{d}W(u).
\end{aligned}$$

Case 2: $r+s\in[t_{i+j},t_{i+j+1})$. In this case, by the integral form of the exact solution of (1.1) again, we have

$$\begin{aligned}
\varphi_r(t_{i-1},\varphi_{t_{i-1}}(0,\xi^{Int}))(s)&=\varphi(t_{i+j};0,\xi^{Int})+\int_{t_{i+j}}^{r+s}b(\varphi_u(0,\xi^{Int}))\mathrm{d}u\\
&\quad+\int_{t_{i+j}}^{r+s}\sigma(\varphi_u(0,\xi^{Int}))\mathrm{d}W(u).
\end{aligned}$$

This, together with (3.34) leads to

$$\begin{aligned}
&\varphi_r(t_{i-1},\varphi_{t_{i-1}}(0,\xi^{Int}))(s)-\varphi_{t_{i-1}}^{Int}(0,\xi^{Int})(s)\\
&\quad=\frac{t_{j+1}-s}{\Delta}(\varphi(t_{i+j};0,\xi^{Int})-\varphi(t_{i+j-1};0,\xi^{Int}))\\
&\qquad+\int_{t_{i+j}}^{r+s}b(\varphi_u(0,\xi^{Int}))\mathrm{d}u+\int_{t_{i+j}}^{r+s}\sigma(\varphi_u(0,\xi^{Int}))\mathrm{d}W(u).
\end{aligned}$$

Combining Case 1 and Case 2, we deduce

$$
\begin{aligned}
&\varphi_r(t_{i-1}, \varphi_{t_{i-1}}(0, \xi^{Int})) - \varphi^{Int}_{t_{i-1}}(0, \xi^{Int}) \\
&= \sum_{j=-N}^{-1} \Big\{ \mathbf{1}_{[t_{i+j-1}-r, t_{i+j}-r)}(\cdot) \Big[\frac{t_j - \cdot}{\Delta} \Big(\int_{t_{i+j-1}}^{t_{i+j}} b(\varphi_u(0, \xi^{Int}))\mathrm{d}u \\
&+ \int_{t_{i+j-1}}^{t_{i+j}} \sigma(\varphi_u(0, \xi^{Int}))\mathrm{d}W(u) \Big) + \int_{t_{i+j-1}}^{r+\cdot} b(\varphi_u(0, \xi^{Int}))\mathrm{d}u \\
&+ \int_{t_{i+j-1}}^{r+\cdot} \sigma(\varphi_u(0, \xi^{Int}))\mathrm{d}W(u) \Big] + \mathbf{1}_{[t_{i+j}-r, t_{i+j+1}-r)}(\cdot) \\
&\times \Big[\frac{t_{j+1} - \cdot}{\Delta} \Big(\int_{t_{i+j-1}}^{t_{i+j}} b(\varphi_u(0, \xi^{Int}))\mathrm{d}u + \int_{t_{i+j-1}}^{t_{i+j}} \sigma(\varphi_u(0, \xi^{Int}))\mathrm{d}W(u) \Big) \\
&+ \int_{t_{i+j}}^{r+\cdot} b(\varphi_u(0, \xi^{Int}))\mathrm{d}u + \int_{t_{i+j}}^{r+\cdot} \sigma(\varphi_u(0, \xi^{Int}))\mathrm{d}W(u) \Big] \Big\}.
\end{aligned} \tag{3.35}
$$

Inserting the above equality into $\mathcal{I}_b$, we are in the position to estimate $\mathcal{I}_b$. We only estimate the sub-term

$$
\begin{aligned}
\mathcal{I}_b^0 := \sum_{i=1}^{N^\Delta} \int_0^1 \mathbb{E}\Big[\Big\langle f'(\Phi(T; t_i, Y_i^\varsigma)) \mathcal{D}\Phi(T; t_i, Y_i^\varsigma), I^{[0]} \int_{t_{i-1}}^{t_i} \int_0^1 \mathcal{D}b(Z_{i,r}^{\beta_1}) \\
\sum_{j=-N}^{-1} \mathbf{1}_{[t_{i+j-1}-r, t_{i+j}-r)}(\cdot) \frac{t_j - \cdot}{\Delta} \int_{t_{i+j-1}}^{t_{i+j}} b(\varphi_u(0, \xi^{Int}))\mathrm{d}u\mathrm{d}\beta_1 \mathrm{d}r \Big\rangle \Big] \mathrm{d}\varsigma,
\end{aligned}
$$

the sub-term

$$
\begin{aligned}
\mathcal{I}_b^1 := \sum_{i=1}^{N^\Delta} \int_0^1 \mathbb{E}\Big[\Big\langle f'(\Phi(T; t_i, Y_i^\varsigma)) \mathcal{D}\Phi(T; t_i, Y_i^\varsigma), I^{[0]} \int_{t_{i-1}}^{t_i} \int_0^1 \mathcal{D}b(Z_{i,r}^{\beta_1}) \\
\sum_{j=-N}^{-1} \mathbf{1}_{[t_{i+j-1}-r, t_{i+j}-r)}(\cdot) \frac{t_j - \cdot}{\Delta} \int_{t_{i+j-1}}^{t_{i+j}} \sigma(\varphi_u(0, \xi^{Int}))\mathrm{d}W(u)\mathrm{d}\beta_1 \mathrm{d}r \Big\rangle \Big] \mathrm{d}\varsigma,
\end{aligned}
$$

and the sub-term

$$
\begin{aligned}
\mathcal{I}_b^2 := \sum_{i=1}^{N^\Delta} \int_0^1 \mathbb{E}\Big[\Big\langle f'(\Phi(T; t_i, Y_i^\varsigma)) \mathcal{D}\Phi(T; t_i, Y_i^\varsigma), I^{[0]} \int_{t_{i-1}}^{t_i} \int_0^1 \mathcal{D}b(Z_{i,r}^{\beta_1}) \\
\sum_{j=-N}^{-1} \mathbf{1}_{[t_{i+j-1}-r, t_{i+j}-r)}(\cdot) \int_{t_{i+j-1}}^{r+\cdot} \sigma(\varphi_u(0, \xi^{Int}))\mathrm{d}W(u)\mathrm{d}\beta_1 \mathrm{d}r \Big\rangle \Big] \mathrm{d}\varsigma,
\end{aligned}
$$

since other sub-terms in $\mathcal{I}_b$ can be estimated similarly.

Estimate of term I_b^0. It follows from Assumption 3.4 that $|\mathcal{D}b(x)y| \leq K(1+\|x\|^{\beta})\|y\|$. This, along with $\sup_{u\in\mathbb{R}^d}|f'(u)| \leq K$, Assumption 2.4, and Lemma 3.8 implies that

$$
\begin{aligned}
|I_b^0| &\leq K\sum_{i=1}^{N^\Delta}\int_0^1 \Big\|\mathcal{D}\Phi(T;t_i,Y_i^\varsigma)I^{[0]}\int_{t_{i-1}}^{t_i}\int_0^1 \mathcal{D}b(Z_{i,r}^{\beta_1})\sum_{j=-N}^{-1}\mathbf{1}_{[t_{i+j-1}-r,t_{i+j}-r)}(\cdot)\\
&\quad\times\frac{t_j-\cdot}{\Delta}\int_{t_{i+j-1}}^{t_{i+j}} b(\varphi_u(0,\xi^{Int}))\mathrm{d}u\mathrm{d}\beta_1\mathrm{d}r\Big\|_{L^2(\Omega)}\mathrm{d}\varsigma\\
&\leq K\sum_{i=1}^{N^\Delta}\int_0^1\Big(e^{-\hat\lambda_3(T-t_i)}\mathbb{E}\Big[\Big\|I^{[0]}\int_{t_{i-1}}^{t_i}\int_0^1 \mathcal{D}b(Z_{i,r}^{\beta_1})\sum_{j=-N}^{-1}\mathbf{1}_{[t_{i+j-1}-r,t_{i+j}-r)}(\cdot)\\
&\quad\times\frac{t_j-\cdot}{\Delta}\int_{t_{i+j-1}}^{t_{i+j}} b(\varphi_u(0,\xi^{Int}))\mathrm{d}u\mathrm{d}\beta_1\mathrm{d}r\Big\|^2(1+\|Y_i^\varsigma\|^{2\beta})\Big]\Big)^{\frac12}\mathrm{d}\varsigma\\
&\leq K\sum_{i=1}^{N^\Delta}\int_0^1 e^{-\frac{\hat\lambda_3(T-t_i)}{2}}\Big(\mathbb{E}\Big[\Big(\int_{t_{i-1}}^{t_i}\int_0^1(1+\|Z_{i,r}^{\beta_1}\|^{\beta})\int_{t_{i-2}}^{t_{i-1}}(1+\\
&\quad\|\varphi_u(0,\xi^{Int})\|^{\beta+1})\mathrm{d}u\mathrm{d}\beta_1\mathrm{d}r\Big)^4\Big]\Big(1+\sup_{i\geq0}\mathbb{E}[\|Y_i^\varsigma\|^{4\beta}]\Big)\Big)^{\frac14}\mathrm{d}\varsigma\\
&\leq K\Delta^2\sum_{i=1}^{N^\Delta}e^{-\frac{\hat\lambda_3(T-t_i)}{2}}\\
&\quad\big(1+\sup_{u\geq0}\mathbb{E}[\|\varphi_u(0,\xi^{Int})\|^{4(2\beta+1)}]\big)^{\frac14}\sup_{i\geq0}\big(1+\mathbb{E}[\|Y_i^\varsigma\|^{4\beta}]\big)^{\frac14},
\end{aligned}
$$

where we used the fact $I^{[0]}(s)\phi(s) = \mathbf{0}$ for $\phi \in C^d$, $s \in [-\tau,-\Delta]$ in the penultimate inequality, and applied

$$
\begin{aligned}
&\mathbb{E}\Big[\Big(\int_{t_{i-1}}^{t_i}\int_0^1(1+\|Z_{i,r}^{\beta_1}\|^{\beta})\int_{t_{i-2}}^{t_{i-1}}(1+\|\varphi_u(0,\xi^{Int})\|^{\beta+1})\mathrm{d}u\mathrm{d}\beta_1\mathrm{d}r\Big)^4\Big]\\
&\quad\leq K\mathbb{E}\Big[\Big(1+\sup_{r\in[t_{i-1},t_i]}\|Z_{i,r}^{\beta_1}\|^{4\beta}\Big)\\
&\qquad\times\Big(1+\|\varphi_{t_{i-2}}(0,\xi^{Int})\|^{4(\beta+1)}+\|\varphi_{t_{i-1}}(0,\xi^{Int})\|^{4(\beta+1)}\Big)\Big]\Delta^2\\
&\quad\leq K\Delta^2\Big(1+\sup_{u\geq0}\mathbb{E}\big[\|\varphi_u(0,\xi^{Int})\|^{4(2\beta+1)}\big]\Big)
\end{aligned}
$$

in the last inequality. Making use of (2.35), Proposition 2.8, and

$$\sup_{T>0} \Delta \sum_{i=1}^{N^\Delta} e^{-\frac{\hat{\lambda}_3(T-t_i)}{2}} = \sup_{N^\Delta>0} \Delta \sum_{i=1}^{N^\Delta} e^{-\frac{\hat{\lambda}_3 \Delta(N^\Delta - i)}{2}} < \infty,$$

yields $|\mathcal{I}_b^0| \leq K\Delta$.

Estimate of term $\mathcal{I}_b^1$. By the duality formula (C.1), the term $\mathcal{I}_b^1$ can be estimated as

$$\begin{aligned}
\mathcal{I}_b^1 = &\sum_{i=1}^{N^\Delta} \int_0^1 \int_{t_{i-1}}^{t_i} \int_0^1 \sum_{j=-N}^{-1} \mathbf{1}_{[t_{i+j-1}-r, t_{i+j}-r)}(\cdot) \frac{t_j - \cdot}{\Delta} \mathbb{E}\Big[\Big\langle (I^{[0]}\mathcal{D}b(Z_{i,r}^{\beta_1}))^* \\
& f'(\Phi(T; t_i, Y_i^\varsigma))\mathcal{D}\Phi(T; t_i, Y_i^\varsigma), \int_{t_{i+j-1}}^{t_{i+j}} \sigma(\varphi_u(0, \xi^{Int}))\mathrm{d}W(u)\Big\rangle\Big]\mathrm{d}\beta_1 \mathrm{d}r \mathrm{d}\varsigma \\
= &\sum_{i=1}^{N^\Delta} \int_0^1 \int_{t_{i-1}}^{t_i} \int_0^1 \int_{t_{i+j-1}}^{t_{i+j}} \sum_{j=-N}^{-1} \mathbf{1}_{[t_{i+j-1}-r, t_{i+j}-r)}(\cdot) \frac{t_j - \cdot}{\Delta} \\
& \times \mathbb{E}\Big[\Big\langle D_u\big[(I^{[0]}\mathcal{D}b(Z_{i,r}^{\beta_1}))^* \\
& f'(\Phi(T; t_i, Y_i^\varsigma))\mathcal{D}\Phi(T; t_i, Y_i^\varsigma)\big], \sigma(\varphi_u(0, \xi^{Int}))\Big\rangle\Big]\mathrm{d}u \mathrm{d}\beta_1 \mathrm{d}r \mathrm{d}\varsigma \\
= &\sum_{i=1}^{N^\Delta} \int_0^1 \int_{t_{i-1}}^{t_i} \int_0^1 \int_{t_{i+j-1}}^{t_{i+j}} \Big\{\mathbb{E}\Big[\Big\langle f'(\Phi(T; t_i, Y_i^\varsigma))\mathcal{D}\Phi(T; t_i, Y_i^\varsigma), I^{[0]}\mathcal{D}^2 b(Z_{i,r}^{\beta_1}) \\
& \Big(D_u Z_{i,r}^{\beta_1}, \sum_{j=-N}^{-1} \mathbf{1}_{[t_{i+j-1}-r, t_{i+j}-r)}(\cdot) \frac{t_j - \cdot}{\Delta} \sigma(\varphi_u(0, \xi^{Int}))\Big)\Big\rangle\Big] \\
& + \mathbb{E}\Big[\Big\langle f'(\Phi(T; t_i, Y_i^\varsigma)) D_u \mathcal{D}\Phi(T; t_i, Y_i^\varsigma), I^{[0]}\mathcal{D}b(Z_{i,r}^{\beta_1}) \\
& \sum_{j=-N}^{-1} \mathbf{1}_{[t_{i+j-1}-r, t_{i+j}-r)}(\cdot) \frac{t_j - \cdot}{\Delta} \sigma(\varphi_u(0, \xi^{Int}))\Big\rangle\Big] + \mathbb{E}\Big[\Big\langle f''(\Phi(T; t_i, Y_i^\varsigma)) \\
& D_u \Phi(T; t_i, Y_i^\varsigma), \mathcal{D}\Phi(T; t_i, Y_i^\varsigma) I^{[0]}\mathcal{D}b(Z_{i,r}^{\beta_1}) \\
& \sum_{j=-N}^{-1} \mathbf{1}_{[t_{i+j-1}-r, t_{i+j}-r)}(\cdot) \frac{t_j - \cdot}{\Delta} \sigma(\varphi_u(0, \xi^{Int}))\Big\rangle\Big]\Big\}\mathrm{d}u \mathrm{d}\beta_1 \mathrm{d}r \mathrm{d}\varsigma,
\end{aligned}$$

This, along with $\sup_{u\in\mathbb{R}^d}(|f'(u)| + |f''(u)|) \leq K$ leads to

$$|\mathcal{I}_b^1| \leq \mathcal{I}_{b,1}^1 + \mathcal{I}_{b,2}^1 + \mathcal{I}_{b,3}^1,$$

where

$$\mathcal{I}_{b,1}^1 := K\sum_{i=1}^{N^\Delta}\int_0^1\int_{t_{i-1}}^{t_i}\int_0^1\int_{t_{i+j-1}}^{t_{i+j}}\Big\|\mathcal{D}\Phi(T;t_i,Y_i^\varsigma)I^{[0]}\mathcal{D}^2b\big(Z_{i,r}^{\beta_1}\big)(D_u Z_{i,r}^{\beta_1},$$

$$\sum_{j=-N}^{-1}\mathbf{1}_{[t_{i+j-1}-r,t_{i+j}-r)}(\cdot)\frac{t_j-\cdot}{\Delta}\sigma(\varphi_u(0,\xi^{Int}))\big)\Big\|_{L^2(\Omega)}\mathrm{d}u\mathrm{d}\beta_1\mathrm{d}r\mathrm{d}\varsigma,$$

$$\mathcal{I}_{b,2}^1 := K\sum_{i=1}^{N^\Delta}\int_0^1\int_{t_{i-1}}^{t_i}\int_0^1\int_{t_{i+j-1}}^{t_{i+j}}\Big\|D_u\mathcal{D}\Phi(T;t_i,Y_i^\varsigma)I^{[0]}\mathcal{D}b(Z_{i,r}^{\beta_1})$$

$$\sum_{j=-N}^{-1}\mathbf{1}_{[t_{i+j-1}-r,t_{i+j}-r)}(\cdot)\frac{t_j-\cdot}{\Delta}\sigma(\varphi_u(0,\xi^{Int}))\Big\|_{L^2(\Omega)}\mathrm{d}u\mathrm{d}\beta_1\mathrm{d}r\mathrm{d}\varsigma,$$

$$\mathcal{I}_{b,3}^1 := K\sum_{i=1}^{N^\Delta}\int_0^1\int_{t_{i-1}}^{t_i}\int_0^1\int_{t_{i+j-1}}^{t_{i+j}}\big\|D_u\Phi(T;t_i,Y_i^\varsigma)|\big\|_{L^2(\Omega)}\Big\|\mathcal{D}\Phi(T;t_i,Y_i^\varsigma)I^{[0]}$$

$$\mathcal{D}b(Z_{i,r}^{\beta_1})\sum_{j=-N}^{-1}\mathbf{1}_{[t_{i+j-1}-r,t_{i+j}-r)}(\cdot)\frac{t_j-\cdot}{\Delta}\sigma(\varphi_u(0,\xi^{Int}))\Big\|_{L^2(\Omega)}\mathrm{d}u\mathrm{d}\beta_1\mathrm{d}r\mathrm{d}\varsigma.$$

By (2.35), Proposition 2.8, Lemmas 3.5 and 3.8, and Assumptions 2.1 and 3.2, we have

$$\mathcal{I}_{b,1}^1 \le K\sum_{i=1}^{N^\Delta}\int_0^1\int_{t_{i-1}}^{t_i}\int_0^1\int_{t_{i+j-1}}^{t_{i+j}}e^{-\frac{\hat{\lambda}_3(T-t_i)}{2}}\Big(\mathbb{E}\Big[\big\|I^{[0]}\mathcal{D}^2b(Z_{i,r}^{\beta_1})\big(D_u Z_{i,r}^{\beta_1},$$

$$\sum_{j=-N}^{-1}\mathbf{1}_{[t_{i+j-1}-r,t_{i+j}-r)}(\cdot)\frac{t_j-\cdot}{\Delta}\sigma(\varphi_u(0,\xi^{Int}))\big)\big\|^2(1+\|Y_i^\varsigma\|^{2\beta})\Big]\Big)^{\frac{1}{2}}$$

$$\mathrm{d}u\mathrm{d}\beta_1\mathrm{d}r\mathrm{d}\varsigma \le K\Delta.$$

For the term $\mathcal{I}_{b,2}^1$, by virtue of (2.35), Lemmas 3.5, 3.9, and 3.12, we arrive at

$$\mathcal{I}_{b,2}^1 \le K\sum_{i=1}^{N^\Delta}\int_0^1\int_{t_{i-1}}^{t_i}\int_0^1\int_{t_{i+j-1}}^{t_{i+j}}e^{-\frac{\hat{\lambda}_3(T-t_i)}{2}}\Big(\mathbb{E}\Big[\Big\|I^{[0]}\mathcal{D}b(Z_{i,r}^{\beta_1})$$

$$\sum_{j=-N}^{-1}\mathbf{1}_{[t_{i+j-1}-r,t_{i+j}-r)}(\cdot)\frac{t_j-\cdot}{\Delta}\sigma(\varphi_u(0,\xi^{Int}))\Big\|^2$$

$$\times (1 + \|Y_i^{\varsigma}\|^{p'} + \|D_u Y_i^{\varsigma}\|^{16})\Big]\Big)^{\frac{1}{2}}$$

$$\mathrm{d}u\mathrm{d}\beta_1\mathrm{d}r\mathrm{d}\varsigma$$

$$\leq K\sum_{i=1}^{N^{\Delta}}\int_0^1\int_{t_{i-1}}^{t_i}\int_0^1\int_{t_{i-2}}^{t_{i-1}} e^{-\frac{\hat{\lambda}_3(T-t_i)}{2}}\Big(\mathbb{E}\big[(1+\|Z_{i,r}^{\beta_1}\|^{4\beta})|\sigma(\varphi_u(0,\xi^{Int}))|^4\big]$$

$$+\big(1+\mathbb{E}[\|Y_i^{\varsigma}\|^{2p'}]+\mathbb{E}[\|D_u Y_i^{\varsigma}\|^{32}]\big)\Big)^{\frac{1}{4}}\mathrm{d}u\mathrm{d}\beta_1\mathrm{d}r\mathrm{d}\varsigma \leq K\Delta.$$

Similar to the estimates of $\mathcal{I}_{b,1}^1$ and $\mathcal{I}_{b,2}^1$, for the term $\mathcal{I}_{b,3}^1$, we obtain $\mathcal{I}_{b,3}^1 \leq K\Delta$.

Hence, we can derive that $|\mathcal{I}_b^1| \leq K\Delta$.

Estimate of term $\mathcal{I}_b^2$. Based on Assumption 3.4 and (C.1), we have

$$\mathcal{I}_b^2 = \sum_{i=1}^{N^{\Delta}}\int_0^1 \mathbb{E}\Big[\Big\langle f'(\Phi(T;t_i,Y_i^{\varsigma}))\mathcal{D}\Phi(T;t_i,Y_i^{\varsigma}), I^{[0]}\int_{t_{i-1}}^{t_i}\int_0^1\int_{-\tau}^0 k_b^1(Z_{i,r}^{\beta_1})$$

$$\sum_{j=-N}^{-1}\mathbf{1}_{[t_{i+j-1}-r,t_{i+j}-r)}(s)\int_{t_{i+j-1}}^{r+s}\sigma(\varphi_u(0,\xi^{Int}))\mathrm{d}W(u)\mathrm{d}v_3^1(s)\mathrm{d}\beta_1\mathrm{d}r\Big\rangle\Big]\mathrm{d}\varsigma$$

$$=\sum_{i=1}^{N^{\Delta}}\sum_{j=-N}^{-1}\int_0^1\int_{t_{i-1}}^{t_i}\int_0^1\int_{-\tau\vee(t_{i+j-1}-r)}^{(t_{i+j}-r)\wedge 0}\mathbb{E}\Big[\Big\langle (I^{[0]}k_b^1(Z_{i,r}^{\beta_1}))^* f'(\Phi(T;t_i,Y_i^{\varsigma}))$$

$$\mathcal{D}\Phi(T;t_i,Y_i^{\varsigma}), \int_{t_{i+j-1}}^{r+s}\sigma(\varphi_u(0,\xi^{Int}))\mathrm{d}W(u)\Big\rangle\Big]\mathrm{d}v_3^1(s)\mathrm{d}\beta_1\mathrm{d}r\mathrm{d}\varsigma$$

$$=\sum_{i=1}^{N^{\Delta}}\sum_{j=-N}^{-1}\int_0^1\int_{t_{i-1}}^{t_i}\int_0^1\int_{-\tau\vee(t_{i+j-1}-r)}^{(t_{i+j}-r)\wedge 0}\int_{t_{i+j-1}}^{r+s}\mathbb{E}\Big[\Big\langle D_u[(I^{[0]}k_b^1(Z_{i,r}^{\beta_1}))^*$$

$$f'(\Phi(T;t_i,Y_i^{\varsigma}))\mathcal{D}\Phi(T;t_i,Y_i^{\varsigma})], \sigma(\varphi_u(0,\xi^{Int}))\Big\rangle\Big]\mathrm{d}u\mathrm{d}v_3^1(s)\mathrm{d}\beta_1\mathrm{d}r\mathrm{d}\varsigma$$

$$=\sum_{i=1}^{N^{\Delta}}\sum_{j=-N}^{-1}\int_0^1\int_{t_{i-1}}^{t_i}\int_0^1\int_{-\tau\vee(t_{i+j-1}-r)}^{(t_{i+j}-r)\wedge 0}$$

$$\int_{t_{i+j-1}}^{r+s}\Big\{\mathbb{E}\Big[\Big\langle f'(\Phi(T;t_i,Y_i^{\varsigma}))\mathcal{D}\Phi(T;t_i,Y_i^{\varsigma}),$$

$$I^{[0]}D_u k_b^1(Z_{i,r}^{\beta_1})\sigma(\varphi_u(0,\xi^{Int}))\Big\rangle\Big]+\mathbb{E}\Big[\Big\langle f'(\Phi(T;t_i,Y_i^{\varsigma}))D_u\mathcal{D}\Phi(T;t_i,Y_i^{\varsigma}),$$

$$I^{[0]}k_b^1(Z_{i,r}^{\beta_1})\sigma(\varphi_u(0,\xi^{Int}))\Big\rangle\Big]+\mathbb{E}\Big[\Big\langle f''(\Phi(T;t_i,Y_i^{\varsigma}))D_u\Phi(T;t_i,Y_i^{\varsigma})$$

$$\mathcal{D}\Phi(T;t_i,Y_i^{\varsigma}), I^{[0]}k_b^1(Z_{i,r}^{\beta_1})\sigma(\varphi_u(0,\xi^{Int}))\Big\rangle\Big]\Big\}\mathrm{d}u\mathrm{d}v_3^1(s)\mathrm{d}\beta_1\mathrm{d}r\mathrm{d}\varsigma,$$

which leads to

$$|\mathcal{I}_b^2| \leq K \sum_{i=1}^{N^\Delta} \sum_{j=-N}^{-1} \int_0^1 \int_{t_{i-1}}^{t_i} \int_0^1 \int_{-\tau\vee(t_{i+j-1}-r)}^{(t_{i+j}-r)\wedge 0} \int_{t_{i+j-1}}^{r+s} \Big\{ \big\| \mathcal{D}\Phi(T;t_i,Y_i^\varsigma) I^{[0]} D_u k_b^1(Z_{i,r}^{\beta_1}) \sigma(\varphi_u(0,\xi^{Int})) \big\|_{L^2(\Omega)} + \big\| D_u \mathcal{D}\Phi(T;t_i,Y_i^\varsigma) I^{[0]} k_b^1(Z_{i,r}^{\beta_1}) \sigma(\varphi_u(0,\xi^{Int})) \big\|_{L^2(\Omega)} + \|D_u\Phi(T;t_i,Y_i^\varsigma)\|_{L^2(\Omega)} \big\|\mathcal{D}\Phi(T;t_i,Y_i^\varsigma) I^{[0]} k_b^1(Z_{i,r}^{\beta_1}) \sigma(\varphi_u(0,\xi^{Int})) \big\|_{L^2(\Omega)} \Big\} \mathrm{d}u\mathrm{d}\nu_3^1(s)\mathrm{d}\beta_1\mathrm{d}r\mathrm{d}\varsigma \leq K\Delta.$$

Here we used the property

$$\sup_{u\in[t_{i+j},t_{i+j+1}]} \mathbb{E}[\|D_u k_b^1(Z_{i,r}^{\beta_1})\|^{2p_0}_{\mathcal{L}(C^d;\mathbb{R}^{d\times d})}] + \mathbb{E}[|k_b^1(Z_{i,r}^{\beta_1})|^{2p_0}] \leq K$$

for some $p_0 > 0$, which is an application of Assumption 3.4, (2.35), and Lemma 3.5.

Hence, the term $\mathcal{I}_b^2$ is estimated as $|\mathcal{I}_b^2| \leq K\Delta$.

Combining estimates of $\mathcal{I}_b^0$, $\mathcal{I}_b^1$, and $\mathcal{I}_b^2$, we complete the proof. □

Lemma 3.16 *Let conditions in Theorem 3.13 hold. Then* $|\mathcal{I}_\sigma| \leq K\Delta$.

Proof Recalling the definition of $\mathcal{I}_\sigma$, we have

$$\mathcal{I}_\sigma = \sum_{i=1}^{N^\Delta} \int_0^1 \mathbb{E}\Big[\Big\langle f'(\Phi(T;t_i,Y_i^\varsigma))\mathcal{D}\Phi(T;t_i,Y_i^\varsigma), I^{[0]} \int_{t_{i-1}}^{t_i} \int_0^1 \mathcal{D}\sigma(Z_{i,r}^{\beta_1}) (\varphi_r(t_{i-1},\varphi_{t_{i-1}}(0,\xi^{Int})) - \varphi_{t_{i-1}}^{Int}(0,\xi^{Int}))\mathrm{d}\beta_1\mathrm{d}W(r)\Big\rangle\Big]\mathrm{d}\varsigma. \tag{3.36}$$

Following a similar strategy for estimating $\mathcal{I}_b$, we need to split $\varphi_r(t_{i-1},\varphi_{t_{i-1}}(0,\xi^{Int})) - \varphi_{t_{i-1}}^{Int}(0,\xi^{Int})$ for $r \in [t_{i-1},t_i)$. For $s \in [t_j,t_{j+1}] \subset [-\tau,0]$, one has $r+s \in [t_{i+j-1},t_{i+j+1})$. For notational convenience and to highlight the main idea of the proof, we continue to assume $t_{i+j-1} \geq 0$. The case where $t_{i+j-1} < 0$, which involves contributions from the initial values on $[-\tau,0]$, can be handled analogously. Inserting (3.35) into (3.36), we are in the position to estimate $\mathcal{I}_\sigma$. For the estimate of $\mathcal{I}_\sigma$, we only give the estimates of the sub-term

$$\mathcal{I}_\sigma^1 := \sum_{i=1}^{N^\Delta} \int_0^1 \mathbb{E}\Big[\Big\langle f'(\Phi(T;t_i,Y_i^\varsigma))\mathcal{D}\Phi(T;t_i,Y_i^\varsigma), I^{[0]} \int_{t_{i-1}}^{t_i} \int_0^1 \mathcal{D}\sigma(Z_{i,r}^{\beta_1}) \sum_{j=-N}^{-1} \mathbf{1}_{[t_{i+j-1}-r,t_{i+j}-r)}(\cdot)\frac{t_j-\cdot}{\Delta} \int_{t_{i+j-1}}^{t_{i+j}} \sigma(\varphi_u(0,\xi^{Int}))\mathrm{d}W(u)\mathrm{d}\beta_1\mathrm{d}W(r)\Big\rangle\Big]\mathrm{d}\varsigma,$$

and the sub-term

$$\mathcal{I}_\sigma^2 := \sum_{i=1}^{N^\Delta} \int_0^1 \mathbb{E}\Big[\Big\langle f'(\Phi(T;t_i,Y_i^\varsigma))\mathcal{D}\Phi(T;t_i,Y_i^\varsigma), I^{[0]} \int_{t_{i-1}}^{t_i} \int_0^1 \mathcal{D}\sigma(Z_{i,r}^{\beta_1})$$

$$\sum_{j=-N}^{-1} \mathbf{1}_{[t_{i+j-1}-r,t_{i+j}-r)}(\cdot) \int_{t_{i+j-1}}^{r+\cdot} \sigma(\varphi_u(0,\xi^{Int}))\mathrm{d}W(u)\mathrm{d}\beta_1\mathrm{d}W(r)\Big\rangle\Big]\mathrm{d}\varsigma,$$

since other terms can be estimated similarly.

Estimate of term $\mathcal{I}_\sigma^1$. By (C.1), we derive

$$\mathcal{I}_\sigma^1 = \sum_{i=1}^{N^\Delta} \int_0^1 \int_{t_{i-1}}^{t_i} \mathbb{E}\Big[\Big\langle D_r[f'(\Phi(T;t_i,Y_i^\varsigma))\mathcal{D}\Phi(T;t_i,Y_i^\varsigma)], I^{[0]} \int_0^1 \mathcal{D}\sigma(Z_{i,r}^{\beta_1})$$

$$\sum_{j=-N}^{-1} \mathbf{1}_{[t_{i+j-1}-r,t_{i+j}-r)}(\cdot)\frac{t_j-\cdot}{\Delta} \int_{t_{i+j-1}}^{t_{i+j}} \sigma(\varphi_u(0,\xi^{Int}))\mathrm{d}W(u)\Big\rangle\mathrm{d}\beta_1\Big]\mathrm{d}r\mathrm{d}\varsigma$$

$$= \sum_{i=1}^{N^\Delta} \int_0^1 \int_{t_{i-1}}^{t_i} \int_0^1 \int_{t_{i+j-1}}^{t_{i+j}} \mathbb{E}\Big[\Big\langle D_u\Big((I^{[0]}\mathcal{D}\sigma(Z_{i,r}^{\beta_1})$$

$$\sum_{j=-N}^{-1} \mathbf{1}_{[t_{i+j-1}-r,t_{i+j}-r)}(\cdot)\frac{t_j-\cdot}{\Delta})^*$$

$$D_r[f'(\Phi(T;t_i,Y_i^\varsigma))\mathcal{D}\Phi(T;t_i,Y_i^\varsigma)]\Big), \sigma(\varphi_u(0,\xi^{Int}))\Big\rangle\Big]\mathrm{d}u\mathrm{d}\beta_1\mathrm{d}r\mathrm{d}\varsigma$$

$$= \sum_{i=1}^{N^\Delta} \int_0^1 \int_{t_{i-1}}^{t_i} \int_0^1 \int_{t_{i+j-1}}^{t_{i+j}}$$

$$\mathbb{E}\Big[\Big\langle D_r[f'(\Phi(T;t_i,Y_i^\varsigma))\mathcal{D}\Phi(T;t_i,Y_i^\varsigma)], I^{[0]}\mathcal{D}^2\sigma(Z_{i,r}^{\beta_1})$$

$$\Big(D_u Z_{i,r}^{\beta_1}, \sum_{j=-N}^{-1} \mathbf{1}_{[t_{i+j-1}-r,t_{i+j}-r)}(\cdot)\frac{t_j-\cdot}{\Delta}\sigma(\varphi_u(0,\xi^{Int}))\Big)\Big\rangle\Big]$$

$$+ \mathbb{E}\Big[\Big\langle D_u D_r[f'(\Phi(T;t_i,Y_i^\varsigma))\mathcal{D}\Phi(T;t_i,Y_i^\varsigma)], I^{[0]}\mathcal{D}\sigma(Z_{i,r}^{\beta_1})$$

$$\sum_{j=-N}^{-1} \mathbf{1}_{[t_{i+j-1}-r,t_{i+j}-r)}(\cdot)\frac{t_j-\cdot}{\Delta}\sigma(\varphi_u(0,\xi^{Int}))\Big\rangle\Big]\mathrm{d}u\mathrm{d}\beta_1\mathrm{d}r\mathrm{d}\varsigma.$$

This, along with $f \in C_b^3(\mathbb{R}^d;\mathbb{R})$ leads to

$$
\begin{aligned}
|\mathcal{I}_\sigma^1| \le K \sum_{i=1}^{N^\Delta} \int_0^1 \int_{t_{i-1}}^{t_i} \int_0^1 \int_{t_{i+j-1}}^{t_{i+j}} \Big\{ \Big\| D_r \mathcal{D}\Phi(T;t_i,Y_i^\varsigma) I^{[0]} \mathcal{D}^2\sigma(Z_{i,r}^{\beta_1}) \\
\Big(D_u Z_{i,r}^{\beta_1}, \sum_{j=-N}^{-1} \mathbf{1}_{[t_{i+j-1}-r,t_{i+j}-r)}(\cdot) \frac{t_j-\cdot}{\Delta} \sigma(\varphi_u(0,\xi^{Int})) \Big) \Big\|_{L^2(\Omega)} \\
+ \|D_r \Phi(T;t_i,Y_i^\varsigma)\|_{L^2(\Omega)} \Big\| \mathcal{D}\Phi(T;t_i,Y_i^\varsigma) I^{[0]} \mathcal{D}^2\sigma(Z_{i,r}^{\beta_1}) \big(D_u Z_{i,r}^{\beta_1}, \\
\sum_{j=-N}^{-1} \mathbf{1}_{[t_{i+j-1}-r,t_{i+j}-r)}(\cdot) \frac{t_j-\cdot}{\Delta} \sigma(\varphi_u(0,\xi^{Int})) \big) \Big\|_{L^2(\Omega)} \\
+ \Big\| D_u D_r \mathcal{D}\Phi(T;t_i,Y_i^\varsigma) \\
I^{[0]} \mathcal{D}\sigma(Z_{i,r}^{\beta_1}) \sum_{j=-N}^{-1} \mathbf{1}_{[t_{i+j-1}-r,t_{i+j}-r)}(\cdot) \frac{t_j-\cdot}{\Delta} \sigma(\varphi_u(0,\xi^{Int})) \Big\|_{L^2(\Omega)} \\
+ \|D_u \Phi(T;t_i,Y_i^\varsigma)\|_{L^2(\Omega)} \Big\| D_r \mathcal{D}\Phi(T;t_i,Y_i^\varsigma) I^{[0]} \mathcal{D}\sigma(Z_{i,r}^{\beta_1}) \\
\sum_{j=-N}^{-1} \mathbf{1}_{[t_{i+j-1}-r,t_{i+j}-r)}(\cdot) \frac{t_j-\cdot}{\Delta} \sigma(\varphi_u(0,\xi^{Int})) \Big\|_{L^2(\Omega)} \\
+ \|D_r \Phi(T;t_i,Y_i^\varsigma)\|_{L^2(\Omega)} \\
\times \Big\| D_u \mathcal{D}\Phi(T;t_i,Y_i^\varsigma) I^{[0]} \mathcal{D}\sigma(Z_{i,r}^{\beta_1}) \sum_{j=-N}^{-1} \mathbf{1}_{[t_{i+j-1}-r,t_{i+j}-r)}(\cdot) \frac{t_j-\cdot}{\Delta} \\
\times \sigma(\varphi_u(0,\xi^{Int})) \Big\|_{L^2(\Omega)} + \|D_u D_r \Phi(T;t_i,Y_i^\varsigma)\|_{L^2(\Omega)} \Big\| \mathcal{D}\Phi(T;t_i,Y_i^\varsigma) \\
I^{[0]} \mathcal{D}\sigma(Z_{i,r}^{\beta_1}) \sum_{j=-N}^{-1} \mathbf{1}_{[t_{i+j-1}-r,t_{i+j}-r)}(\cdot) \frac{t_j-\cdot}{\Delta} \sigma(\varphi_u(0,\xi^{Int})) \Big\|_{L^2(\Omega)} \\
+ \|D_u \Phi(T;t_i,Y_i^\varsigma) D_r \Phi(T;t_i,Y_i^\varsigma)\|_{L^2(\Omega)} \Big\| \mathcal{D}\Phi(T;t_i,Y_i^\varsigma) I^{[0]} \mathcal{D}\sigma(Z_{i,r}^{\beta_1}) \\
\sum_{j=-N}^{-1} \mathbf{1}_{\{i+j\ge 1\}} \mathbf{1}_{[t_{i+j-1}-r,t_{i+j}-r)}(\cdot) \frac{t_j-\cdot}{\Delta} \sigma(\varphi_u(0,\xi^{Int})) \Big\|_{L^2(\Omega)} \Big\} \mathrm{d}u \mathrm{d}\beta_1 \mathrm{d}r \mathrm{d}\varsigma \\
\le K\Delta.
\end{aligned}
$$

Estimate of term $\mathcal{I}_\sigma^2$. It follows from Assumption 3.4 and (C.1) that

$$
\begin{aligned}
\mathcal{I}_\sigma^2 &= \sum_{i=1}^{N^\Delta} \int_0^1 \mathbb{E}\Big[\Big\langle f'(\Phi(T;t_i,Y_i^\varsigma))\mathcal{D}\Phi(T;t_i,Y_i^\varsigma), \int_{t_{i-1}}^{t_i}\int_0^1\int_{-\tau}^0 k_\sigma^1(Z_{i,r}^{\beta_1})I^{[0]}(s) \\
&\quad \sum_{j=-N}^{-1} \mathbf{1}_{[t_{i+j-1}-r,t_{i+j}-r)}(s) \\
&\quad \int_{t_{i+j-1}}^{r+s} \sigma(\varphi_u(0,\xi^{Int}))\mathrm{d}W(u)\mathrm{d}\nu_4^1(s)\mathrm{d}\beta_1\mathrm{d}W(r)\Big\rangle\Big]\mathrm{d}\varsigma \\
&= \sum_{i=1}^{N^\Delta} \int_0^1 \int_{t_{i-1}}^{t_i} \mathbb{E}\Big[\Big\langle D_r[f'(\Phi(T;t_i,Y_i^\varsigma))\mathcal{D}\Phi(T;t_i,Y_i^\varsigma)], \\
&\quad \int_0^1 \sum_{j=-N}^{-1} \int_{(t_{i+j-1}-r)\vee-\tau}^{(t_{i+j}-r)\wedge 0} k_\sigma^1(Z_{i,r}^{\beta_1})I^{[0]}(s)\int_{t_{i+j-1}}^{r+s} \sigma(\varphi_u(0,\xi^{Int}))\mathrm{d}W(u)\Big\rangle\Big] \\
&\quad \mathrm{d}\nu_4^1(s)\mathrm{d}\beta_1\mathrm{d}r\mathrm{d}\varsigma \\
&= \sum_{i=1}^{N^\Delta}\sum_{j=-N}^{-1} \int_0^1\int_{t_{i-1}}^{t_i}\int_0^1\int_{(t_{i+j-1}-r)\vee-\tau}^{(t_{i+j}-r)\wedge 0}\int_{t_{i+j-1}}^{r+s} \mathbb{E}\Big[\Big\langle D_u\Big(\big(\sum_{j=-N}^{-1} k_\sigma^1(Z_{i,r}^{\beta_1})I^{[0]}(s)\big)^* \\
&\quad D_r[f'(\Phi(T;t_i,Y_i^\varsigma))\mathcal{D}\Phi(T;t_i,Y_i^\varsigma)]\Big), \sigma(\varphi_u(0,\xi^{Int}))\Big\rangle\Big]\mathrm{d}u\mathrm{d}\nu_4^1(s)\mathrm{d}\beta_1\mathrm{d}r\mathrm{d}\varsigma \\
&= \sum_{i=1}^{N^\Delta}\sum_{j=-N}^{-1} \int_0^1\int_{t_{i-1}}^{t_i}\int_0^1\int_{(t_{i+j-1}-r)\vee-\tau}^{(t_{i+j}-r)\wedge 0}\int_{t_{i+j-1}}^{r+s} \mathbb{E}\Big[\Big\langle D_r[f'(\Phi(T;t_i,Y_i^\varsigma)) \\
&\quad \mathcal{D}\Phi(T;t_i,Y_i^\varsigma)], D_u(k_\sigma^1(Z_{i,r}^{\beta_1}))I^{[0]}(s)\sigma(\varphi_u(0,\xi^{Int}))\Big\rangle\Big] \\
&\quad + \mathbb{E}\Big[\Big\langle D_uD_r[f'(\Phi(T;t_i,Y_i^\varsigma))\mathcal{D}\Phi(T;t_i,Y_i^\varsigma)], \\
&\quad k_\sigma^1(Z_{i,r}^{\beta_1})I^{[0]}(s)\sigma(\varphi_u(0,\xi^{Int}))\Big\rangle\Big] \\
&\quad \mathrm{d}u\mathrm{d}\nu_4^1(s)\mathrm{d}\beta_1\mathrm{d}r\mathrm{d}\varsigma.
\end{aligned}
$$

This leads to

$$
\begin{aligned}
\mathcal{I}_\sigma^2 &\le K\sum_{i=1}^{N^\Delta}\sum_{j=-N}^{-1} \int_0^1\int_{t_{i-1}}^{t_i}\int_0^1\int_{(t_{i+j-1}-r)\vee-\tau}^{(t_{i+j}-r)\wedge 0}\int_{t_{i+j}}^{r+s} \Big\{\Big\| D_r\mathcal{D}\Phi(T;t_i,Y_i^\varsigma) \\
&\quad D_u(k_\sigma^1(Z_{i,r}^{\beta_1}))I^{[0]}(s)\sigma(\varphi_u(0,\xi^{Int}))\Big\|_{L^2(\Omega)} + \|D_r\Phi(T;t_i,Y_i^\varsigma)\|_{L^2(\Omega)}
\end{aligned}
$$

$$\times \left\| \mathcal{D}\Phi(T;t_i,Y_i^{\varsigma})D_u(k_\sigma^1(Z_{i,r}^{\beta_1}))I^{[0]}(s)\sigma(\varphi_u(0,\xi^{Int}))\right\|_{L^2(\Omega)}$$

$$+ \left\| D_u D_r \mathcal{D}\Phi(T;t_i,Y_i^{\varsigma})k_\sigma^1(Z_{i,r}^{\beta_1})I^{[0]}(s)\sigma(\varphi_u(0,\xi^{Int}))\right\|_{L^2(\Omega)}$$

$$+ \|D_u\Phi(T;t_i,Y_i^{\varsigma})\|_{L^2(\Omega)}$$

$$\times \left\| D_r \mathcal{D}\Phi(T;t_i,Y_i^{\varsigma})k_\sigma^1(Z_{i,r}^{\beta_1})I^{[0]}(s)\sigma(\varphi_u(0,\xi^{Int}))\right\|_{L^2(\Omega)}$$

$$+ \|D_r\Phi(T;t_i,Y_i^{\varsigma})\|_{L^2(\Omega)}$$

$$\times \left\| D_u \mathcal{D}\Phi(T;t_i,Y_i^{\varsigma})k_\sigma^1(Z_{i,r}^{\beta_1})I^{[0]}(s)\sigma(\varphi_u(0,\xi^{Int}))\right\|_{L^2(\Omega)}$$

$$+ \|D_u D_r\Phi(T;t_i,Y_i^{\varsigma})\|_{L^2(\Omega)}$$

$$\times \left\| \mathcal{D}\Phi(T;t_i,Y_i^{\varsigma})k_\sigma^1(Z_{i,r}^{\beta_1})I^{[0]}(s)\sigma(\varphi_u(0,\xi^{Int}))\right\|_{L^2(\Omega)}$$

$$+ \|D_u\Phi(T;t_i,Y_i^{\varsigma})D_r\Phi(T;t_i,Y_i^{\varsigma})\|_{L^2(\Omega)} \left\| \mathcal{D}\Phi(T;t_i,Y_i^{\varsigma})k_\sigma^1(Z_{i,r}^{\beta_1})I^{[0]}(s)\right.$$

$$\left.\sigma(\varphi_u(0,\xi^{Int}))\right\|_{L^2(\Omega)}\Big\}\mathrm{d}u\mathrm{d}v_4^1(s)\mathrm{d}\beta_1\mathrm{d}r\mathrm{d}\varsigma \le K\Delta,$$

where we used the property

$$\sup_{i,j}\ \sup_{u\in[t_{i+j},t_{i+j+1}]} \mathbb{E}\Big[\|D_u k_\sigma^1(Z_{i,r}^{\beta_1})\|^{2p_0}_{\mathcal{L}(C^d;\mathcal{L}(\mathbb{R}^d;\mathbb{R}^{d\times m}))}\Big]$$

$$+\mathbb{E}\Big[\|k_\sigma^1(Z_{i,r}^{\beta_1})\|^{2p_0}_{\mathcal{L}(\mathbb{R}^d;\mathbb{R}^{d\times m})}\Big] \le K$$

for some $p_0 > 0$. We complete the proof. □

Proof of Theorem 3.13 By (3.11), Lemma 3.15, and Lemma 3.16, it remains to estimate $\mathcal{I}_0$ and $\mathcal{I}_{b,\theta}$.

Estimate of $\mathcal{I}_0$. Note that Assumption 2.5 with $\rho \ge 1$ implies

$$\|\xi-\xi^{Int}\| = \sup_{j\in\{-N,\dots,-1\}}\ \sup_{s\in[t_j,t_{j+1}]} |\xi(s) - \frac{t_{j+1}-s}{\Delta}\xi(t_j)$$

$$-\frac{s-t_j}{\Delta}\xi(t_{j+1})| \le K\Delta.$$

Using $\sup_{u\in\mathbb{R}^d}|f'(u)| \le K$ and Lemma 3.4, the term $\mathcal{I}_0$ is estimated as

$$|\mathcal{I}_0| = |\mathbb{E}[f(\varphi^{Int}(t_{N\Delta};0,\xi)] - \mathbb{E}[f(\varphi^{Int}(t_{N\Delta};0,\xi^{Int}))]|$$

$$= \Big|\int_0^1 \mathbb{E}\Big[f'(\varphi^{Int}(t_{N\Delta};0,\varsigma\xi+(1-\varsigma)\xi^{Int}))$$

$$
\mathcal{D}\varphi^{Int}(t_{N^\Delta};0,\varsigma\xi+(1-\varsigma)\xi^{Int})(\xi-\xi^{Int})\Big]\mathrm{d}\varsigma\Big|
$$

$$
\le K\int_0^1\mathbb{E}\Big[\big|\mathcal{D}\varphi^{Int}(t_{N^\Delta};0,\varsigma\xi+(1-\varsigma)\xi^{Int})(\xi-\xi^{Int})\big|\Big]\mathrm{d}\varsigma
$$

$$
\le K\Delta\int_0^1\mathbb{E}\Big[\big|\mathcal{D}\varphi^{Int}(t_{N^\Delta};0,\varsigma\xi+(1-\varsigma)\xi^{Int})\big|\Big]\mathrm{d}\varsigma
$$

$$
\le K\Delta.
$$

Estimate of $\mathcal{I}_{b,\theta}$. We rewrite $\mathcal{I}_{b,\theta}$ as

$$
\mathcal{I}_{b,\theta}=\sum_{i=1}^{N^\Delta}\int_0^1\mathbb{E}\Big[\Big\langle f'(\Phi(T;t_i,Y_i^\varsigma))\mathcal{D}\Phi(T;t_i,Y_i^\varsigma),\theta I^{[0]}\int_{t_{i-1}}^{t_i}\int_0^1\mathcal{D}b(Z_i^{\beta_2})
$$

$$
(\varphi_{t_{i-1}}^{Int}(0,\xi^{Int})-\Phi_{t_i}(t_{i-1},\varphi_{t_{i-1}}^{Int}(0,\xi^{Int})))\mathrm{d}\beta_2\mathrm{d}r\Big\rangle\Big]\mathrm{d}\varsigma,
$$

where $Z_{i,r}^{\beta_2}=\beta_2\varphi_{t_{i-1}}^{Int}(0,\xi^{Int})+(1-\beta_2)\Phi_{t_i}(t_{i-1},\varphi_{t_{i-1}}^{Int}(0,\xi^{Int}))$. Similar to (3.9), we have that for any $i\in\mathbb{N}_+$,

$$
\varphi_{t_{i-1}}^{Int}(0,\xi^{Int})-\Phi_{t_i}(t_{i-1},\varphi_{t_{i-1}}^{Int}(0,\xi^{Int}))
$$

$$
=\sum_{j=-N}^{-1}I^{[j]}[\varphi(t_{i+j-1};0,\xi^{Int})-\varphi(t_{i+j};0,\xi^{Int})]+I^{[0]}[\varphi(t_{i-1};0,\xi^{Int})
$$

$$
-\Phi(t_i;t_{i-1},\varphi_{t_{i-1}}^{Int}(0,\xi^{Int}))]
$$

$$
=-\sum_{j=-N}^{-1}\mathbf{1}_{\{i+j\ge 1\}}I^{[j]}
$$

$$
\Big(\int_{t_{i+j-1}}^{t_{i+j}}b(\varphi_u(0,\xi^{Int}))\mathrm{d}u+\int_{t_{i+j-1}}^{t_{i+j}}\sigma(\varphi_u(0,\xi^{Int}))\mathrm{d}W(u)\Big)
$$

$$
+\sum_{j=-N}^{-1}\mathbf{1}_{\{i+j\le 0\}}I^{[j]}\big(\xi(t_{i+j-1})-\xi(t_{i+j})\big)-I^{[0]}\Big((1-\theta)b(\varphi_{t_{i-1}}^{Int}(0,\xi^{Int}))\Delta
$$

$$
+\theta b(\Phi_{t_i}(t_{i-1},\varphi_{t_{i-1}}^{Int}(0,\xi^{Int})))\Delta+\sigma(\varphi_{t_{i-1}}^{Int}(0,\xi^{Int}))\delta W_{i-1}\Big).
$$

Hence, similar to the estimate of $\mathcal{I}_b$, we can derive that $|\mathcal{I}_{b,\theta}|\le K\Delta$.

Combining estimates of terms $\mathcal{I}_0$, $\mathcal{I}_b$, $\mathcal{I}_{b,\theta}$, and $\mathcal{I}_\sigma$, we complete the proof. □

3.3 Numerical Experiments

This section presents numerical algorithms together with numerical experiments aimed at verifying the weak convergence rate of the θ-EM method when applied to SFDEs. Two one-dimensional numerical examples are included to illustrate the corresponding theoretical results.

Algorithm 2 describes the numerical framework employed to verify the weak convergence rate of the θ-EM method for SFDEs. The exact solution is approximated by a reference solution computed with a sufficiently fine stepsize, while all numerical solutions share the same realization of the driving Brownian motion. The weak error is computed by evaluating the Monte Carlo approximation, where the test function is taken as $f(x) = \sin(|x|), x \in \mathbb{R}$.

Linear Stochastic Delay Differential Equation We first consider the one-dimensional linear stochastic delay differential equation:

$$\mathrm{d}x^{\xi}(t) = (-\alpha x^{\xi}(t) + \beta x^{\xi}(t-\tau))\mathrm{d}t + \sigma \mathrm{d}W(t), \quad t > 0, \tag{3.37}$$

where in the numerical experiment, constants $\alpha = 1.2, \beta = 0.8, \sigma = 0.5, \tau = 1$, and initial datum $\xi(t) = 1$ for $t \in [-\tau, 0]$. We employ $M = 2000$ independent trajectories to estimate expectations of errors. Figure 3.1a illustrates the weak convergence rate of the EM method for this linear model. The dashed line indicates the reference slope of rate 1. The weak errors exhibit a clear linear decay with respect to the stepsizes, which is consistent with the theoretical weak convergence rate one. Figure 3.1b depicts the longtime evolution of the weak error between the EM solution and the reference solution. It can be observed that the weak error remains uniformly bounded over long time intervals $[0, 500]$, confirming our longtime weak convergence result.

Nonlinear Stochastic Functional Differential Equation We then verify the weak convergence rate of the following one-dimensional nonlinear SFDE:

$$\begin{aligned}\mathrm{d}x^{\xi}(t) &= \Big(\frac{1}{\tau}\int_{-\tau}^{0} x^{\xi}(t+s)\mathrm{d}s - (x^{\xi}(t))^3 - 2x^{\xi}(t)\Big)\mathrm{d}t \\ &\quad + \Big(\frac{a}{\tau}\int_{-\tau}^{0} x^{\xi}(t+s)\mathrm{d}s + \frac{1}{2}x^{\xi}(t) + \frac{1}{2}\Big)\mathrm{d}W(t),\end{aligned} \tag{3.38}$$

where $\tau = 0.125, a = 0.1$, and the initial datum $\xi(t) = 1$ for $t \in [-\tau, 0]$. We employ $M = 1000$ independent trajectories to estimate expectations of errors. Figure 3.2a displays the weak convergence rate of the backward EM method applied to this nonlinear model (3.38). As shown in Fig. 3.2a, the numerical errors agree well with the reference slope 1. This confirms the theoretical weak convergence rate of the backward EM method for the nonlinear SFDE. Figure 3.2b illustrates the longtime evolution of the weak error between the solution of the backward EM

Algorithm 2 Weak convergence rate for θ-EM method applied to SFDEs

Require: SFDE parameters (b, σ, ξ, τ), parameter θ, final time T, stepsizes $\{\Delta_j\}_{j=1}^J$, reference stepsize Δ_{ref}, Monte Carlo size M, test function f.

Output: Weak errors $\{e_j\}_{j=1}^J$.

Generate M independent Brownian motion increments $\delta W_{\text{ref}}^{(\tilde{m})}(k) \sim \mathcal{N}(0, \Delta_{\text{ref}})$, $k = 0, \ldots, N_{\text{ref}} - 1$, with $N_{\text{ref}} = T/\Delta_{\text{ref}}$.

Set $\tilde{d}_{\text{ref}} = \tau/\Delta_{\text{ref}}$.

Initialize $x_{\text{ref}}^{(\tilde{m})}(t_k)$, $\quad k = -\tilde{d}_{\text{ref}}, \ldots, 0$, $\tilde{m} = 1, \ldots, M$.

for $m = 1, \ldots, M$ **do**

  for $k = 0, \ldots, N_{\text{ref}} - 1$ **do**

$$x_{\text{ref}}^{(\tilde{m})}(t_{k+1}) = x_{\text{ref}}^{(\tilde{m})}(t_k) + \big(\theta b(x_{\text{ref},t_{k+1}}^{(\tilde{m})}) + (1-\theta) b(x_{\text{ref},t_k}^{(\tilde{m})})\big)\Delta_{\text{ref}} + \sigma(x_{\text{ref},t_k}^{(\tilde{m})})\delta W_{\text{ref}}^{(\tilde{m})}(k).$$

  end for

end for

for $j = 1, \ldots, J$ **do**

  Set $\Delta = \Delta_j$, $N = T/\Delta$, $\tilde{d} = \tau/\Delta$, $r = \Delta/\Delta_{\text{ref}}$.

  Construct coarse Brownian motion increments

$$\delta W^{(\tilde{m})}(k) = \sum_{i=kr}^{(k+1)r-1} \delta W_{\text{ref}}^{(\tilde{m})}(i), \quad k = 0, \ldots, N-1.$$

  Initialize $y^{(\tilde{m})}(t_k)$, $\quad k = -\tilde{d}, \ldots, 0$.

  for $\tilde{m} = 1, \ldots, M$ **do**

    for $k = 0, \ldots, N-1$ **do**

$$y^{(\tilde{m})}(t_{k+1}) = y^{(\tilde{m})}(t_k) + \big(\theta b(y_{t_{k+1}}^{(\tilde{m})}) + (1-\theta) b(y_{t_k}^{(\tilde{m})})\big)\Delta + \sigma(y_{t_k}^{(\tilde{m})})\delta W^{(\tilde{m})}(k).$$

    end for

  end for

  Compute the weak error

$$e_j = \left| \frac{1}{M} \sum_{\tilde{m}=1}^{M} \big(f(y^{(\tilde{m})}(T)) - f(x_{\text{ref}}^{(\tilde{m})}(T)) \big) \right|.$$

end for

method and the reference solution. The weak error remains uniformly bounded over long time intervals [0, 500], validating the longtime weak convergence result for the nonlinear SFDE with multiplicative noise.

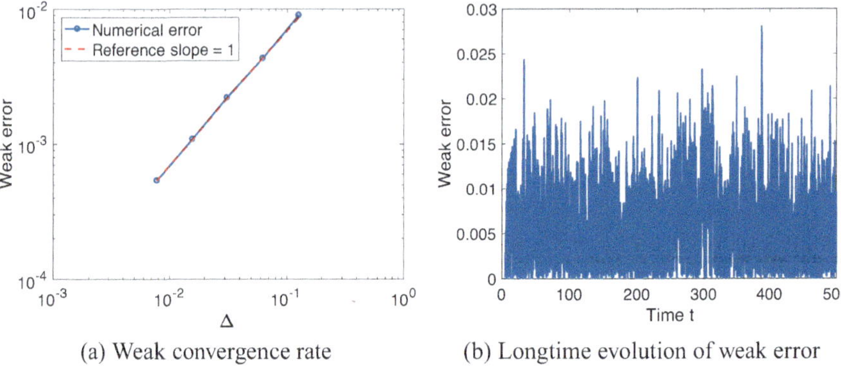

Fig. 3.1 Weak convergence for EM method of (3.37); $\alpha = 1.2, \beta = 0.8, \sigma = 0.5, \tau = 1, M = 2000, \xi \equiv 1, f(x) = \sin(|x|)$. (**a**) $\Delta = 2^{-3}, 2^{-4}, \dots, 2^{-7}, \Delta_{\text{ref}} = 2^{-12}, T = 1$, and (**b**) $\Delta = 2^{-4}, \Delta_{\text{ref}} = 2^{-8}, T = 500$. (**a**) Weak convergence rate. (**b**) Longtime evolution of weak error

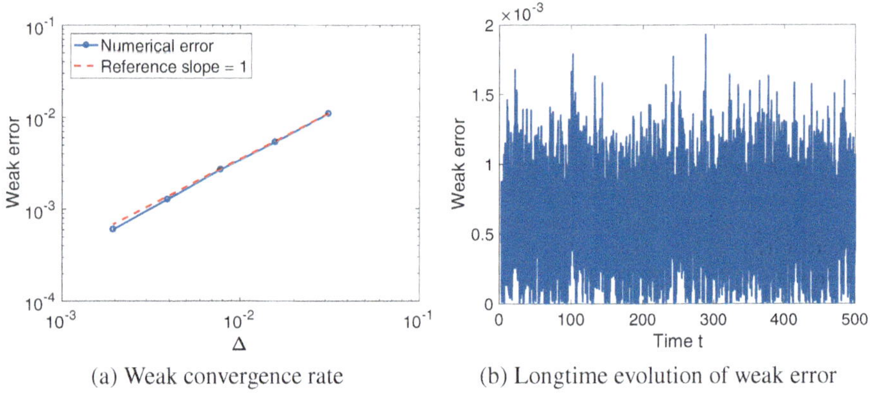

Fig. 3.2 Weak convergence for backward EM method of (3.38); $\tau = 0.125, a = 0.1, M = 1000, \xi \equiv 1, f(x) = \sin(|x|)$. (**a**) $\Delta = 2^{-5}, 2^{-6}, \dots, 2^{-9}, \Delta_{\text{ref}} = 2^{-13}, T = 1$, and (**b**) $\Delta = 2^{-7}, \Delta_{\text{ref}} = 2^{-10}, T = 500$. (**a**) Weak convergence rate. (**b**) Longtime evolution of weak error

3.4 Summary and Outlook

In this chapter, we have investigated the numerical approximation of the invariant measure for the SFDE (1.1) with the superlinearly growing drift coefficient. In particular, we have proved the existence and uniqueness of the invariant measure for the θ-EM functional solution and established the convergence rate of the numerical invariant measure via the weak convergence analysis.

Regarding the weak convergence analysis of numerical methods for SFDEs, several significant advances have been made. For example, authors in [3, 4] establish the weak convergence of the EM method in the finite time horizon for stochastic delay differential equations with Lipschitz continuous coefficients, and authors in [1] prove the result to SFDEs with Hölder continuous drift coefficient. In the context of numerical approximation of invariant measures for superlinear SFDEs, several results have been established. Authors in [2] investigate the convergence of the EM method in approximating the invariant measure of the functional solution for SFDEs. Authors in [12] obtain the half-order convergence for the numerical invariant measure of the backward EM method applied to SFDEs with distributed argument, based on the longtime mean-square convergence analysis. Authors in [7] prove the first-order convergence for the backward EM method applied to stochastic differential equations with piecewise continuous arguments, using longtime weak convergence analysis.

It is worth noting that the weak convergence analysis methodology developed in this chapter is also applicable to exploring the convergence properties of the numerical density function. A more detailed presentation of the theoretical derivations and explanations will be provided in Chap. 4. Moreover, we have observed that the ergodic measure precisely depicts the longtime stable state of the system and is intrinsically linked to the longtime probabilistic behavior of the functional solution for SFDEs. In light of this observation, our attention in the subsequent chapter will be directed towards this very aspect. Specifically, we will conduct a comprehensive examination of the relationship between the numerical time-average $\frac{1}{T}\int_0^T f(y_t^{\xi,\Delta})\mathrm{d}t$ and the ergodic limit $\mu(f) := \int_{\mathcal{C}^d} f(x)\mathrm{d}\mu(x)$ for the invariant measure μ of (1.1) and some class of suitable functionals f.

We remark that Assumption 3.4 is introduced mainly for technical reasons, primarily to facilitate the use of the dual property of the Malliavin derivative operator (i.e., (C.1), commonly referred to as the integration by parts formula) in the weak convergence analysis. In fact, there are still some problems associated with the weak convergence and the approximation of the invariant measure for the SFDE that are challenging and far from being well understood. For example, how to weaken or even eliminate Assumption 3.4 to include a more general class of SFDEs? Can we obtain the same weak convergence rate if the diffusion coefficient grows superlinearly? For the SFDE driven by noises rougher than Brownian motion, such as the fractional Brownian motion or rough path, does the invariant measure of the numerical method exist uniquely, and if so, does the numerical invariant measure converge to the exact one?

It is worth noting that the invariant measure of numerical methods, which is the main focus of this chapter, represents only one facet of numerical stochastic dynamics. Beyond invariant measures, there are several other fundamental dynamical objects that play a central role in understanding the longtime behavior of numerical methods, such as numerical random attractors, Sinai–Ruelle–Bowen (SRB) measures, and intermittency. In recent years, significant progress has been made in analyzing these numerical dynamical behaviors for various stochastic systems. For instance, [11] and [15] show that Euler-type discretizations of rough differential equations and stochastic sine-Gordon lattice equations generate a random dynamical system, and establish the existence and convergence of corresponding numerical attractors. Regarding SRB measures, [8] provides the first work on the existence of SRB measures for numerical discretizations of stochastic Hopf bifurcation. In terms of the intermittency, [5] establishes weak intermittency for numerical discretizations of stochastic heat equations for the first time. However, for SFDEs, such numerical stochastic dynamics remain largely unexplored to our knowledge, which presents a challenging and promising direction for future research.

References

1. J. Bao, J. Shao, Weak convergence of path-dependent SDEs with irregular coefficients. Numer. Algebra Control Optim. **15**, 108–129 (2025)
2. J. Bao, J. Shao, C. Yuan, Invariant probability measures for path-dependent random diffusions. Nonlinear Anal. **228**, Paper No. 113201 (2023)
3. E. Buckwar, T. Shardlow, Weak approximation of stochastic differential delay equations. IMA J. Numer. Anal. **25**, 57–86 (2005)
4. E. Buckwar, R. Kuske, S. Mohammed, T. Shardlow, Weak convergence of the Euler scheme for stochastic differential delay equations. LMS J. Comput. Math. **11**, 60–99 (2008)
5. C. Chen, T. Dang, J. Hong, Weak intermittency of stochastic heat equation under discretizations. J. Differ. Equ. **333**, 268–301 (2022)
6. C. Chen, T. Dang, J. Hong, T. Zhou, CLT for approximating ergodic limit of SPDEs via a full discretization. Stoch. Process. Appl. **157**, 1–41 (2023)
7. C. Chen, J. Hong, Y. Lu, Stochastic differential equation with piecewise continuous arguments: Markov property, invariant measure and numerical approximation. Discrete Contin. Dyn. Syst. Ser. B **28**, 765–807 (2023)
8. C. Chen, J. Hong, Y. Wang, Random attractor and SRB measure for stochastic Hopf bifurcation under discretization (2025). https://arxiv.org/abs/2509.12865
9. E. Clément, A. Kohatsu-Higa, D. Lamberton, A duality approach for the weak approximation of stochastic differential equations. Ann. Appl. Probab. **16**, 1124–1154 (2006)
10. G. Da Prato, J. Zabczyk, *Ergodicity for Infinite-dimensional Systems*, vol. 229 of London Mathematical Society Lecture Note Series (Cambridge University Press, Cambridge, 1996)
11. L.H. Duc, P. Kloeden, Numerical attractors for rough differential equations. SIAM J. Numer. Anal. **61**, 2381–2407 (2023)
12. B. Shi, Y. Wang, X. Mao, F. Wu, Approximation of invariant measures of a class of backward Euler–Maruyama scheme for stochastic functional differential equations. J. Differ. Equ. **389**, 415–456 (2024)
13. A.N. Shiryaev, *Probability*, vol. 95 of Graduate Texts in Mathematics, 2nd edn. (Springer, New York, 1996). Translated from the first (1980) Russian edition by R. P. Boas
14. C. Villani, *Optimal Transport: Old and New*, vol. 338 of Grundlehren der mathematischen Wissenschaften [Fundamental Principles of Mathematical Sciences] (Springer, Berlin, 2009)
15. S. Yang, Y. Li, Numerical attractors and approximations for stochastic or deterministic sine-Gordon lattice equations. Appl. Math. Comput. **413**, Paper No. 126640, 19 (2022)

Chapter 4
Numerical Central Limit Theorem

In Sect. 1.3.4, we have investigated the strong law of large numbers (SLLN) and the central limit theorem (CLT) for the functional solution of SFDEs, thereby clarifying the connection between these probabilistic limit theorems and the corresponding invariant measure. It is well-known that the functional solution of an autonomous SFDE, viewed in its natural infinite-dimensional state space, constitutes a time-homogeneous Markov process. However, due to its inherent dependence on the path history, the ergodicity and longtime behavior of such systems are routinely analyzed within this infinite-dimensional framework. For equations of this type, often characterized by highly degenerate coefficients, a natural question arises: can numerical discretizations preserve the probabilistic limit theorems of the original equation, particularly the SLLN and CLT?

In this chapter, we address this question by shifting our focus to general time-homogeneous Markov processes and studying the probabilistic limit behavior of their numerical discretizations. We provide a positive answer by developing a unified and transparent analytical approach for establishing corresponding limit theorems for numerical methods applied to such processes. In Sect. 4.1, we introduce some notation, and present the probabilistic limit theorems for time-homogeneous Markov processes, including the SLLN and the CLT. Section 4.2 presents our main results, including the SLLN and the CLT for numerical discretizations of time-homogeneous Markov processes, as well as the applications to the θ-EM method of the SFDE. Sections 4.3.1 and 4.3.2 are devoted to presenting proofs of the SLLN and the CLT of numerical discretizations, respectively. These proofs are based on the convergence of the numerical invariant measure and the decomposition of the normalized time-average of numerical discretizations.

C. Chen et al., *Numerical Analysis of Stochastic Functional Differential Equations*, Lecture Notes in Mathematics 2399, https://doi.org/10.1007/978-981-92-1592-8_4

4.1 Limit Theorems of Time-Homogeneous Markov Processes

In this section, we investigate the probabilistic limit theorems, including the SLLN and the CLT, for the Banach space-valued time-homogeneous Markov process.

We begin by introducing some preliminary notation. Let $(E, \|\cdot\|)$ be a real-valued separable Banach space and $\mathfrak{B}(E)$ be the Borel σ-algebra. Denote by $\mathcal{P}(E)$ the family of probability measures on $(E, \mathfrak{B}(E))$ and by $\nu(f) := \int_E f(u)\mathrm{d}\nu(u)$ the integral of the measurable functional f with respect to $\nu \in \mathcal{P}(E)$. Recall that the definition of $d_{p,\gamma}$ with $p \geq 1, \gamma \in (0, 1]$ is given in (1.28). Define the Wasserstein quasi-distance induced by the quasi-metric as

$$\mathbb{W}_{p,\gamma}(\nu_1, \nu_2) := \inf_{\pi \in \Pi(\nu_1,\nu_2)} \int_{E\times E} d_{p,\gamma}(u_1, u_2)\pi(\mathrm{d}u_1, \mathrm{d}u_2), \tag{4.1}$$

where $\nu_1, \nu_2 \in \mathcal{P}_{p,\gamma}(E) := \{\nu \in \mathcal{P}(E) : \int_{E\times E} d_{p,\gamma}(u_1, u_2)\mathrm{d}\nu(u_1)\mathrm{d}\nu(u_2) < \infty\}$ and $\Pi(\nu_1, \nu_2)$ denotes the collection of all probability measures on $E \times E$ with marginal measures ν_1 and ν_2.

Note that when $\nu_1(\|\cdot\|^p) < \infty$ and $\nu_2(\|\cdot\|^p) < \infty$,

$$\begin{aligned}
&\mathbb{W}_{p,\gamma}(\nu_1, \nu_2)\\
&\quad \leq \big(1+\nu_1(\|\cdot\|^p)+\nu_2(\|\cdot\|^p)\big)^{\frac{1}{2}}\Big(\inf_{\pi \in \Pi(\nu_1,\nu_2)} \int_{E\times E} 1 \wedge \|u_1-u_2\|^2\pi(\mathrm{d}u_1, \mathrm{d}u_2)\Big)^{\frac{\gamma}{2}}\\
&\quad = \big(1+\nu_1(\|\cdot\|^p)+\nu_2(\|\cdot\|^p)\big)^{\frac{1}{2}}(\mathbb{W}_2(\nu_1, \nu_2))^{\gamma},
\end{aligned}$$

where (1.28) and the Hölder inequality are used, and $\mathbb{W}_p$ is the bounded-Wasserstein distance defined by

$$\mathbb{W}_p(\nu_1, \nu_2) := \Big(\inf_{\pi \in \Pi(\nu_1,\nu_2)} \int_{E\times E} 1 \wedge \|u_1-u_2\|^p\pi(\mathrm{d}u_1, \mathrm{d}u_2)\Big)^{\frac{1}{p}}.$$

This leads to that for any $f \in C_{p,\gamma}$,

$$\begin{aligned}
|\nu_1(f) - \nu_2(f)| &= \inf_{\pi \in \Pi(\nu_1,\nu_2)} \Big|\int_{E\times E} (f(u_1) - f(u_2))\pi(\mathrm{d}u_1, \mathrm{d}u_2)\Big|\\
&\leq \|f\|_{p,\gamma}\mathbb{W}_{p,\gamma}(\nu_1, \nu_2)\\
&\leq \|f\|_{p,\gamma}\big(1+\nu_1(\|\cdot\|^p)+\nu_2(\|\cdot\|^p)\big)^{\frac{1}{2}}(\mathbb{W}_2(\nu_1, \nu_2))^{\gamma}.
\end{aligned} \tag{4.2}$$

Let $\{X_t^x\}_{t\geq 0}$ be the E-valued time-homogeneous Markov process with the deterministic initial datum $X_0^x = x \in E$ on a filtered probability space $(\Omega, \mathcal{F}, \{\mathcal{F}_t\}_{t\geq 0}, \mathbb{P})$, in which the corresponding expectation is denoted by $\mathbb{E}$.

Denote by μ_t^x the probability measure induced by X_t^x, i.e., for any $A \in \mathfrak{B}(E)$, $\mu_t^x(A) = \mathbb{P}\{\omega \in \Omega : X_t^x \in A\}$. Denote by $\mathcal{B}_b := \mathcal{B}_b(E;\mathbb{R})$ (resp. $C_b := C_b(E;\mathbb{R})$) the family of bounded Borel measurable (resp. bounded and continuous) functions on E.

Define the linear operators P_t induced by X_t as

$$P_t : \mathcal{B}_b \to \mathcal{B}_b, \quad P_t f(x) := \mathbb{E}\big[f(X_t^x)\big] = \int_E f(u)\mathrm{d}\mu_t^x(u) \quad \forall\, x \in E.$$

According to the Markov property of X_t^x, one deduces that P_t is a Markov semigroup. If X_t^x admits a unique invariant measure μ satisfying $\mu(\|\cdot\|^l) < \infty$ for some constant $l > 0$, it follows from [12, Theorem 5.8] that P_t is uniquely extendible to a bounded linear operator on $L^l(E,\mu)$ which is still denoted by P_t. Here, denote by $L^l(E,\mu)$ the family of Borel measurable functions $f : E \to \mathbb{R}$ such that $\int_E |f(u)|^l \mathrm{d}\mu(u) < \infty$. In addition, for the space $C_{p,\gamma}$ with some $p \geq 1$ and $\gamma \in (0,1]$, if there are functions $\Phi : E \to [0,+\infty)$ and $\bar{\rho} : [0,+\infty) \to [0,+\infty)$ satisfying $\int_0^\infty \bar{\rho}(t)\mathrm{d}t < \infty$ such that for any $f \in C_{p,\gamma}$,

$$|P_t f(x) - \mu(f)| \leq K\|f\|_{p,\gamma}\Phi(x)\bar{\rho}(t) \quad \forall\, x \in E,\ t \geq 0,$$

then the Markov process $\{X_t^x\}_{t\geq 0}$ is said to be *uniformly mixing* on the space $C_{p,\gamma}$; see [29, Definition 2.5]. This property characterizes that the process "forgets" its initial states as time goes on.

We first introduce the assumption on the time-homogeneous Markov process. This assumption guarantees the existence of a unique invariant measure and ensures that probabilistic limit theorems including the SLLN and the CLT hold.

Assumption 4.1 *Assume that the time-homogeneous Markov process* $\{X_t^x\}_{t\geq 0}$ *satisfies the following conditions.*

(i) *There exist constants* $r \geq 2$, $\tilde{r} \geq 1$, *and* $L_1 > 0$ *such that for any* $x \in E$,

$$\sup_{t\geq 0} \mathbb{E}[\|X_t^x\|^r] \leq L_1(1 + \|x\|^{\tilde{r}r}).$$

(ii) *There exist constants* $\gamma_1 \in (0,1]$, $\beta \in [0, r-1]$, $L_2 > 0$, *and a continuous function* $\rho : [0,+\infty) \to [0,+\infty)$ *with* $\int_0^\infty \rho^{\gamma_1}(t)\mathrm{d}t < \infty$ *such that for any* $x, y \in E$,

$$\big(\mathbb{E}[\|X_t^x - X_t^y\|^2]\big)^{\frac{1}{2}} \leq L_2\|x-y\|(1 + \|x\|^\beta + \|y\|^\beta)\rho(t).$$

The probabilistic limit theorems including the SLLN and the CLT of Markov processes have been well-studied in [19, 20]. For our convenience, we present a modified version of [20, Proposition 5.3.5] in the following proposition.

Proposition 4.1 *Let Assumption* 4.1 *hold. Then* $\{X_t^x\}_{t\geq 0}$ *admits a unique invariant measure* $\mu \in \mathcal{P}(E)$. *Moreover, if* $\gamma \in [\gamma_1, 1]$ *and* p *satisfy* $2\tilde{r}\big(p\tilde{r}+(2+3\beta)\gamma\big) \leq r$, *then the SLLN and the CLT hold: for any* $f \in C_{p,\gamma}$,

$$\frac{1}{T}\int_0^T f(X_t^x)\mathrm{d}t \overset{\mathbb{P}\text{-a.s.}}{\longrightarrow} \mu(f) \quad \text{as } T \to \infty,$$

$$\frac{1}{\sqrt{T}}\int_0^T (f(X_t^x) - \mu(f))\mathrm{d}t \overset{d}{\longrightarrow} \mathcal{N}(0, v^2) \quad \text{as } T \to \infty,$$

where $v^2 := 2\mu\big((f - \mu(f))\int_0^\infty (P_t f - \mu(f))\mathrm{d}t\big)$.

Remark 4.1 We remark that by Assumption 4.1, the Markov process $\{X_t^x\}_{t\geq 0}$ is uniformly mixing for the space $C_{p,\gamma}$ with $\gamma \in [\gamma_1, 1]$ and $p\tilde{r} + 2(1+\beta)\gamma \leq r$, which follows from

$$\begin{aligned} |P_t f(x) - \mu(f)| &= |\mu_t^x(f) - \mu(f)| \\ &\leq K\|f\|_{p,\gamma}(1 + \mathbb{E}[\|X_t^x\|^p] + \mu(\|\cdot\|^p))^{\frac{1}{2}}\big(\mathbb{W}_2(\mu_t^x, \mu)\big)^\gamma \\ &\leq K\|f\|_{p,\gamma}(1 + \|x\|^{\frac{p\tilde{r}}{2}+(1+\beta)\gamma})\rho^\gamma(t), \end{aligned} \tag{4.3}$$

due to (4.2). The function ρ characterizes the decay speed in time, ensuring that the influence on the initial state diminishes rapidly over longtime. Moreover, using the same technique as (4.3), we obtain that for any $x, y \in E$,

$$|P_t f(x) - P_t f(y)| \leq K\|f\|_{p,\gamma}\|x - y\|^\gamma(1 + \|x\|^{\frac{p\tilde{r}}{2}+\gamma\beta} + \|y\|^{\frac{p\tilde{r}}{2}+\gamma\beta})\rho^\gamma(t). \tag{4.4}$$

The following lemma plays an important role in investigating the SLLN and the CLT of the time-homogeneous Markov process, whose proof is based on [29, Proposition 2.6] and the uniformly mixing (4.3).

Lemma 4.2 *Let Assumption* 4.1 *hold,* $\gamma \in [\gamma_1, 1]$, $\bar{\varrho} \in \mathbb{N}_+$, *and* $p \geq 1$ *satisfy* $\bar{\varrho}\big(\frac{p}{2}(1+\tilde{r}) + (1+\beta)\gamma\big) \leq r$. *Then for any* $f \in C_{p,\gamma}$ *and* $x \in E$, *we have*

$$\begin{aligned} &\mathbb{E}\Big[\Big|\frac{1}{T}\int_0^T f(X_s^x)\mathrm{d}s - \mu(f)\Big|^{2\bar{\varrho}}\Big] \\ &\quad \leq K\|f\|_{p,\gamma}^{2\bar{\varrho}}(1 + \|x\|^{\bar{\varrho}\tilde{r}(\frac{p}{2}(1+\tilde{r})+(1+\beta)\gamma)})T^{-\bar{\varrho}} \quad \forall\, T > 0. \end{aligned}$$

Next, with this lemma in hand, we give the proof of the SLLN and the CLT of the time-homogeneous Markov process stated in Proposition 4.1.

Proof of Proposition 4.1 **Proof of the SLLN.** For any $T \geq 1$, one has

$$\Big|\frac{1}{T}\int_0^T f(X_s^x)\mathrm{d}s - \mu(f)\Big| \leq \Big|\frac{1}{T}\int_{\lfloor T\rfloor}^T f(X_s^x)\mathrm{d}s\Big| + \Big|\frac{1}{\lfloor T\rfloor}\int_0^{\lfloor T\rfloor} f(X_s^x)\mathrm{d}s - \mu(f)\Big| + \Big|\Big(\frac{1}{T}-\frac{1}{\lfloor T\rfloor}\Big)\int_0^{\lfloor T\rfloor} f(X_s^x)\mathrm{d}s\Big|. \tag{4.5}$$

We first estimate the term $\big|\frac{1}{T}\int_{\lfloor T\rfloor}^T f(X_s^x)\mathrm{d}s\big|$. For any $k \in \mathbb{N}_+$, denote

$$\mathscr{D}_k^x := \Big\{\omega \in \Omega : \sup_{T\in[k,k+1)}\Big|\int_k^T f(X_s^x)\mathrm{d}s\Big| > \|f\|_{p,\gamma}k^{\frac{3}{4}}\Big\}.$$

Using the Chebyshev inequality, the Hölder inequality, and Assumption 4.1 (i), we have that

$$\begin{aligned}\sum_{k=1}^{\infty}\mathbb{P}(\mathscr{D}_k^x) &\leq \sum_{k=1}^{\infty}\frac{\mathbb{E}\Big[\sup_{T\in[k,k+1)}\big|\int_k^T f(X_s^x)\mathrm{d}s\big|^2\Big]}{\|f\|_{p,\gamma}^2 k^{\frac{3}{2}}}\\ &\leq \sum_{k=1}^{\infty}\frac{\mathbb{E}\Big[\sup_{T\in[k,k+1)}\int_k^T |f(X_s^x)|^2\mathrm{d}s(T-k)\Big]}{\|f\|_{p,\gamma}^2 k^{\frac{3}{2}}}\\ &\leq K\sum_{k=1}^{\infty}k^{-\frac{3}{2}}\int_k^{k+1}(1+\mathbb{E}[\|X_s^x\|^p])\mathrm{d}s\\ &\leq K(1+\|x\|^{\tilde{r}p})\sum_{k=1}^{\infty}k^{-\frac{3}{2}} \leq K(1+\|x\|^{\tilde{r}p}).\end{aligned}$$

Define

$$\tilde{N}^x(\omega) := \inf\Big\{j \in \mathbb{N}_+ : \sup_{T\in[k,k+1)}\Big|\int_k^T f(X_s^x)\mathrm{d}s\Big| \leq \|f\|_{p,\gamma}k^{\frac{3}{4}} \text{ for all } k \geq j\Big\}.$$

Applying the Borel–Cantelli lemma leads to that for $\mathbb{P}$-a.s. $\omega \in \Omega$, $\tilde{N}^x(\omega) < \infty$, and when $k \geq \tilde{N}^x(\omega)$, $\sup_{T\in[k,k+1)}\Big|\int_k^T f(X_s^x)\mathrm{d}s\Big| \leq \|f\|_{p,\gamma}k^{\frac{3}{4}}$, which implies

$$\Big|\frac{1}{T}\int_{\lfloor T\rfloor}^T f(X_s^x)\mathrm{d}s\Big| \xrightarrow{\mathbb{P}\text{-a.s.}} 0 \quad \text{as } T \to \infty. \tag{4.6}$$

Next, we estimate the second term $\Big|\frac{1}{\lfloor T\rfloor}\int_0^{\lfloor T\rfloor} f(X_s^x)\mathrm{d}s - \mu(f)\Big|$. For any $N \in \mathbb{N}_+$ and $\delta \in (0, \frac{1}{4})$, denote $\mathscr{A}_N^x := \{\omega \in \Omega : |\frac{1}{N}\int_0^N f(X_s^x)\mathrm{d}s - \mu(f)| > \|f\|_{p,\gamma}N^{-\delta}\}$.

This, combining Lemma 4.2 with $\bar{\varrho} = 2$ gives

$$\sum_{N=1}^{\infty} \mathbb{P}(\mathcal{A}_N^x) \leq K(1 + \|x\|^{\tilde{r}(p(1+\tilde{r})+2(1+\beta)\gamma)}) \sum_{N=1}^{\infty} N^{4\delta-2}$$
$$\leq K(1 + \|x\|^{\tilde{r}(p(1+\tilde{r})+2(1+\beta)\gamma)}).$$

Define

$$\mathcal{K}^x(\omega) := \inf\left\{ j \in \mathbb{N}_+ : \left| \frac{1}{N} \int_0^N f(X_s^x) \mathrm{d}s - \mu(f) \right| (\omega) \leq \|f\|_{p,\gamma} N^{-\delta} \text{ for all } N \geq j \right\}.$$

Applying the Borel–Cantelli lemma yields that for $\mathbb{P}$-a.s. $\omega \in \Omega$, when $N \geq \mathcal{K}^x(\omega)$,

$$\left| \frac{1}{N} \int_0^N f(X_s^N) \mathrm{d}s - \mu(f) \right| \leq \|f\|_{p,\gamma} N^{-\delta}, \tag{4.7}$$

which implies

$$\left| \frac{1}{\lfloor T \rfloor} \int_0^{\lfloor T \rfloor} f(X_s^x) \mathrm{d}s - \mu(f) \right| \xrightarrow{\mathbb{P}\text{-a.s.}} 0 \text{ as } T \to \infty. \tag{4.8}$$

For the third term $\left| \left(\frac{1}{T} - \frac{1}{\lfloor T \rfloor} \right) \int_0^{\lfloor T \rfloor} f(X_s^x) \mathrm{d}s \right|$, by (4.7) we deduce that

$$\begin{aligned} \left| \left(\frac{1}{T} - \frac{1}{\lfloor T \rfloor} \right) \int_0^{\lfloor T \rfloor} f(X_s^x) \mathrm{d}s \right| &\leq \frac{1}{T \lfloor T \rfloor} \left| \int_0^{\lfloor T \rfloor} f(X_s^x) \mathrm{d}s \right| \\ &\leq \frac{1}{T} \left(|\mu(f)| + \|f\|_{p,\gamma} \lfloor T \rfloor^{-\delta} \right) \xrightarrow{\mathbb{P}\text{-a.s.}} 0 \text{ as } T \to \infty. \end{aligned} \tag{4.9}$$

Therefore, plugging (4.6), (4.8), and (4.9) into (4.5) we derive the SLLN of $\{X_t^x\}_{t \geq 0}$.

Proof of the CLT. For any $T \geq 1$, note that

$$\begin{aligned} &\frac{1}{\sqrt{T}} \int_0^T \left(f(X_s^x) - \mu(f) \right) \mathrm{d}s \\ &= \frac{1}{\sqrt{T}} \int_{\lfloor T \rfloor}^T \left(f(X_s^x) - \mu(f) \right) \mathrm{d}s + \left(\frac{1}{\sqrt{T}} - \frac{1}{\sqrt{\lfloor T \rfloor}} \right) \int_0^{\lfloor T \rfloor} \left(f(X_s^x) - \mu(f) \right) \mathrm{d}s \\ &\quad + \frac{1}{\sqrt{\lfloor T \rfloor}} \int_0^{\lfloor T \rfloor} \left(f(X_s^x) - \mu(f) \right) \mathrm{d}s. \end{aligned}$$

It follows from Assumption 4.1 (i) that

$$\mathbb{E}\Big[\Big|\frac{1}{\sqrt{T}}\int_{\lfloor T\rfloor}^{T}(f(X_s^x)-\mu(f))\mathrm{d}s\Big|\Big]\le\frac{1}{\sqrt{T}}\int_{\lfloor T\rfloor}^{T}K\|f\|_{p,\gamma}\big(1+\sup_{t\ge 0}\mathbb{E}[\|X_t^x\|^{\frac{p}{2}}]\big)\mathrm{d}s$$

$$\le K(1+\|x\|^{\frac{\tilde{r}p}{2}})\frac{1}{\sqrt{T}}\to 0\quad\text{as } T\to\infty.$$

Making use of Lemma 4.2 and $(\sqrt{T}-\sqrt{\lfloor T\rfloor})^2\le T-\lfloor T\rfloor\le 1$, we deduce

$$\begin{aligned}&\mathbb{E}\Big[\Big|\Big(\frac{1}{\sqrt{T}}-\frac{1}{\sqrt{\lfloor T\rfloor}}\Big)\int_0^{\lfloor T\rfloor}(f(X_s^x)-\mu(f))\mathrm{d}s\Big|^2\Big]\\&\le\frac{\lfloor T\rfloor}{T}\mathbb{E}\Big[\Big|\frac{1}{\lfloor T\rfloor}\int_0^{\lfloor T\rfloor}(f(X_s^x)-\mu(f))\mathrm{d}s\Big|^2\Big]\\&\le K\|f\|_{p,\gamma}^2(1+\|x\|^{\tilde{r}(\frac{p}{2}(1+\tilde{r})+(1+\beta)\gamma)})T^{-1}\to 0\quad\text{as } T\to\infty.\end{aligned}$$

Those imply that it suffices to prove the CLT for the case of $T=N\in\mathbb{N}_+$. We have the decomposition

$$\frac{1}{\sqrt{N}}\int_0^N(f(X_t^x)-\mu(f))\mathrm{d}t=\frac{1}{\sqrt{N}}\mathscr{M}_N^x+\frac{1}{\sqrt{N}}\mathscr{R}_N^x,$$

where

$$\mathscr{M}_N^x:=\int_0^\infty\Big(\mathbb{E}\big[f(X_t^x)|\mathcal{F}_N\big]-\mu(f)\Big)\mathrm{d}t-\int_0^\infty\Big(\mathbb{E}\big[f(X_t^x)|\mathcal{F}_0\big]-\mu(f)\Big)\mathrm{d}t,$$

$$\mathscr{R}_N^x:=-\int_N^\infty\Big(\mathbb{E}\big[f(X_t^x)|\mathcal{F}_N\big]-\mu(f)\Big)\mathrm{d}t+\int_0^\infty\Big(\mathbb{E}\big[f(X_t^x)|\mathcal{F}_0\big]-\mu(f)\Big)\mathrm{d}t.$$

It can be shown that $\{\mathscr{M}_N^x\}_{N\in\mathbb{N}}$ is an $\{\mathcal{F}_N\}_{N\in\mathbb{N}}$-adapted martingale with $\mathscr{M}_0^x=0$. Define the martingale difference sequence of $\{\mathscr{M}_N^x\}_{N\in\mathbb{N}}$ by

$$\mathcal{Z}_N^x:=\mathscr{M}_N^x-\mathscr{M}_{N-1}^x\quad\forall N\in\mathbb{N}_+,\quad\mathcal{Z}_0^x=0.$$

Since

$$\begin{aligned}\mathcal{Z}_{N+1}^x&=\int_N^{N+1}\Big(f(X_t^x)-\mu(f)\Big)\mathrm{d}t+\int_{N+1}^\infty\Big(\mathbb{E}[f(X_t^x)|\mathcal{F}_{N+1}]-\mu(f)\Big)\mathrm{d}t\\&\quad-\int_N^\infty\Big(\mathbb{E}[f(X_t^x)|\mathcal{F}_N]-\mu(f)\Big)\mathrm{d}t\\&=\int_N^{N+1}\Big(f(X_t^x)-\mu(f)\Big)\mathrm{d}t+\int_0^\infty\Big(P_tf(X_{N+1}^x)-\mu(f)\Big)\mathrm{d}t\end{aligned}$$

$$-\int_0^\infty \Big(P_t f(X_N^x) - \mu(f)\Big)\mathrm{d}t, \tag{4.10}$$

we deduce that $\sup_{N\in\mathbb{N}} \mathbb{E}[|\mathcal{Z}_N^x|^c] \le K$ when $c(\frac{p\tilde{r}}{2} + (1+\beta)\gamma) \le r$. Moreover,

$$\begin{aligned}\mathbb{E}\big[|\mathcal{Z}_{N+1}^x|^2|\mathcal{F}_N\big] = \mathbb{E}\Big[\Big|\int_0^1 (f(X_t^y) - \mu(f))\mathrm{d}t + \int_0^\infty (P_t f(X_1^y) - \mu(f))\mathrm{d}t \\ - \int_0^\infty (P_t f(y) - \mu(f))\mathrm{d}t\Big|^2\Big]\Big|_{y=X_N^x} =: H(X_N^x). \end{aligned}\tag{4.11}$$

Without loss of generality, below we assume that $\mu(f) = 0$. Otherwise, we let $\tilde{f} := f - \mu(f)$ and consider $\tilde{f}$ instead of f.

Claim 1: $H \in C_{2\bar{p}_\gamma,\gamma}$ and $\|H\|_{2\bar{p}_\gamma,\gamma} \le K\|f\|_{p,\gamma}^2$ with $\bar{p}_\gamma := \tilde{r}(p\tilde{r}+(2+3\beta)\gamma)$. It can be calculated that

$$\begin{aligned}H(u) = \mathbb{E}\Big[\Big|\int_0^1 f(X_t^u)\mathrm{d}t\Big|^2\Big] + \mathbb{E}\Big[\Big|\int_0^\infty P_t f(X_1^u)\mathrm{d}t\Big|^2\Big] + \Big|\int_0^\infty P_t f(u)\mathrm{d}t\Big|^2 \\ + 2\mathbb{E}\Big[\int_0^1 f(X_t^u)\mathrm{d}t \int_0^\infty P_t f(X_1^u)\mathrm{d}t\Big] - 2\mathbb{E}\Big[\int_0^1 f(X_t^u)\mathrm{d}t\Big]\int_0^\infty P_t f(u)\mathrm{d}t \\ - 2\mathbb{E}\Big[\int_0^\infty P_t f(X_1^u)\mathrm{d}t\Big]\int_0^\infty P_t f(u)\mathrm{d}t. \end{aligned}\tag{4.12}$$

By the property of the semigroup P_t, we have

$$\mathbb{E}\Big[\int_0^1 f(X_t^u)\mathrm{d}t\Big]\int_0^\infty P_t f(u)\mathrm{d}t = \int_0^1 P_t f(u)\mathrm{d}t \int_0^\infty P_t f(u)\mathrm{d}t \tag{4.13}$$

and

$$\begin{aligned}\mathbb{E}\Big[\int_0^\infty P_t f(X_1^u)\mathrm{d}t\Big]\int_0^\infty P_t f(u)\mathrm{d}t = \int_0^\infty P_{t+1} f(u)\mathrm{d}t \int_0^\infty P_t f(u)\mathrm{d}t \\ = \Big(\int_0^\infty P_t f(u)\mathrm{d}t\Big)^2 - \int_0^1 P_t f(u)\mathrm{d}t \int_0^\infty P_t f(u)\mathrm{d}t. \end{aligned}\tag{4.14}$$

Combining the property of the conditional expectation gives

$$\begin{aligned}\mathbb{E}\Big[\Big|\int_0^1 f(X_t^u)\mathrm{d}t\Big|^2\Big] = 2\mathbb{E}\Big[\int_0^1 f(X_t^u)\int_t^1 f(X_s^u)\mathrm{d}s\mathrm{d}t\Big] \\ = 2\int_0^1 \mathbb{E}\Big[f(X_t^u)\mathbb{E}\Big[\int_t^1 f(X_s^u)\mathrm{d}s\Big|\mathcal{F}_t\Big]\Big]\mathrm{d}t \end{aligned}$$

$$= 2\int_0^1 \int_0^{1-t} P_t\big(f P_s f\big)(u)\mathrm{d}s\mathrm{d}t \tag{4.15}$$

and

$$\begin{aligned}
&\mathbb{E}\Big[\int_0^1 f(X_t^u)\mathrm{d}t \int_0^\infty P_t f(X_1^u)\mathrm{d}t\Big] \\
&= \int_0^1 \mathbb{E}\Big[f(X_t^u)\mathbb{E}\big[\int_0^\infty P_s f(X_1^u)\mathrm{d}s|\mathcal{F}_t\big]\Big]\mathrm{d}t \\
&= \int_0^1 \mathbb{E}\Big[f(X_t^u)\int_0^\infty P_{1-t+s} f(X_t^u)\mathrm{d}s\Big]\mathrm{d}t \\
&= \int_0^1 \mathbb{E}\Big[f(X_t^u)\int_0^\infty P_s f(X_t^u)\mathrm{d}s\Big]\mathrm{d}t - \int_0^1 \mathbb{E}\Big[f(X_t^u)\int_0^{1-t} P_s f(X_t^u)\mathrm{d}s\Big]\mathrm{d}t \\
&= \int_0^1 \mathbb{E}\Big[f(X_t^u)\int_0^\infty P_s f(X_t^u)\mathrm{d}s\Big]\mathrm{d}t - \int_0^1 \int_0^{1-t} P_t\big(f P_s f\big)(u)\mathrm{d}s\mathrm{d}t. \qquad (4.16)
\end{aligned}$$

Plugging (4.13)–(4.16) into (4.12) yields

$$\begin{aligned}
H(u) &= \mathbb{E}\Big[\Big|\int_0^\infty P_t f(X_1^u)\mathrm{d}t\Big|^2\Big] - \Big|\int_0^\infty P_t f(u)\mathrm{d}t\Big|^2 \\
&\quad + 2\mathbb{E}\Big[\int_0^1 f(X_t^u)\int_0^\infty P_s f(X_t^u)\mathrm{d}s\mathrm{d}t\Big].
\end{aligned}$$

We derive from $\tilde{r} \geq 1$, $p\tilde{r} + 2(1+\beta)\gamma \leq r$, Assumption 4.1 (i), and (4.3) that

$$\begin{aligned}
|H(u)| &\leq K\mathbb{E}\Big[\Big|\int_0^\infty \|f\|_{p,\gamma}(1 + \|X_1^u\|^{\frac{p\tilde{r}}{2}+(1+\beta)\gamma})(\rho(t))^\gamma \mathrm{d}t\Big|^2\Big] \\
&\quad + K\Big|\int_0^\infty \|f\|_{p,\gamma}(1 + \|u\|^{\frac{p\tilde{r}}{2}+(1+\beta)\gamma})(\rho(t))^\gamma \mathrm{d}t\Big|^2 \\
&\quad + K\mathbb{E}\Big[\Big|\int_0^1 \int_0^\infty \|f\|_{p,\gamma}^2(1 + \|X_t^u\|^{\frac{p}{2}(1+\tilde{r})+(1+\beta)\gamma})(\rho(s))^\gamma \mathrm{d}s\mathrm{d}t\Big] \\
&\leq K\|f\|_{p,\gamma}^2(1 + \|u\|^{\tilde{r}(p\tilde{r}+2(1+\beta)\gamma)}), \qquad (4.17)
\end{aligned}$$

and hence $\sup_{u\in E} \frac{|H(u)|}{1+\|u\|^{\bar{p}\gamma}} \leq K\|f\|_{p,\gamma}^2$. For any $u_1, u_2 \in E$,

$$H(u_1) - H(u_2) = \sum_{i=1}^3 \tilde{I}_i(u_1, u_2),$$

where

$$\tilde{I}_1(u_1,u_2)=\mathbb{E}\Big[\big|\int_0^\infty P_t f(X_1^{u_1})\mathrm{d}t\big|^2\Big]-\mathbb{E}\Big[\big|\int_0^\infty P_t f(X_1^{u_2})\mathrm{d}t\big|^2\Big],$$

$$\tilde{I}_2(u_1,u_2)=-\big|\int_0^\infty P_t f(u_1)\mathrm{d}t\big|^2+\big|\int_0^\infty P_t f(u_2)\mathrm{d}t\big|^2,$$

$$\begin{aligned}\tilde{I}_3(u_1,u_2)=&\,2\mathbb{E}\Big[\int_0^1 f(X_t^{u_1})\int_0^\infty P_s f(X_t^{u_1})\mathrm{d}s\mathrm{d}t\Big]\\&-2\mathbb{E}\Big[\int_0^1 f(X_t^{u_2})\int_0^\infty P_s f(X_t^{u_2})\mathrm{d}s\mathrm{d}t\Big].\end{aligned}$$

Using (4.3) and (4.4) and the Hölder inequality leads to

$$|\tilde{I}_2(u_1,u_2)|\le K\|f\|_{p,\gamma}^2\|u_1-u_2\|^\gamma(1+\|u_1\|^{p\tilde{r}+(1+2\beta)\gamma}+\|u_2\|^{p\tilde{r}+(1+2\beta)\gamma}).$$

By the relation $\tilde{I}_1(u_1,u_2)=-\mathbb{E}[\tilde{I}_2(X_1^{u_1},X_1^{u_2})]$, $2p\tilde{r}+2(1+2\beta)\gamma\le r$, and Assumption 4.1, we obtain

$$\begin{aligned}&|\tilde{I}_1(u_1,u_2)|\\&\le K\|f\|_{p,\gamma}^2\mathbb{E}\Big[\|X_1^{u_1}-X_1^{u_2}\|^\gamma(1+\|X_1^{u_1}\|^{p\tilde{r}+(1+2\beta)\gamma}+\|X_1^{u_2}\|^{p\tilde{r}+(1+2\beta)\gamma})\Big]\\&\le K\|f\|_{p,\gamma}^2\|u_1-u_2\|^\gamma(1+\|u_1\|^{\tilde{r}(p\tilde{r}+(1+2\beta)\gamma)+\gamma\beta}\\&\quad+\|u_2\|^{\tilde{r}(p\tilde{r}+(1+2\beta)\gamma)+\gamma\beta}).\end{aligned}$$

Similarly, $\tilde{I}_3$ can be estimated as follows:

$$\begin{aligned}&|\tilde{I}_3(u_1,u_2)|\\&\le 2\mathbb{E}\Big[\int_0^1|f(X_t^{u_1})-f(X_t^{u_2})|\Big|\int_0^\infty P_s f(X_t^{u_1})\mathrm{d}s\Big|\mathrm{d}t\Big]\\&\quad+2\mathbb{E}\Big[\int_0^1|f(X_t^{u_2})|\int_0^\infty|P_s f(X_t^{u_1})-P_s f(X_t^{u_2})|\mathrm{d}s\mathrm{d}t\Big]\\&\le K\|f\|_{p,\gamma}^2\int_0^1\mathbb{E}\Big[\|X_t^{u_1}-X_t^{u_2}\|^\gamma(1+\|X_t^{u_1}\|^p+\|X_t^{u_2}\|^p)^{\frac{1}{2}}\\&\quad(1+\|X_t^{u_1}\|^{\frac{p\tilde{r}}{2}+\gamma(1+\beta)})+\|X_t^{u_1}-X_t^{u_2}\|^\gamma(1+\|X_t^{u_1}\|^{\frac{p\tilde{r}}{2}+\gamma\beta}+\|X_t^{u_2}\|^{\frac{p\tilde{r}}{2}+\gamma\beta})\\&\quad(1+\|X_t^{u_2}\|^{\frac{p}{2}})\Big]\mathrm{d}t\end{aligned}$$

$$\leq K\|f\|_{p,\gamma}^2\|u_1-u_2\|^\gamma(1+\|u_1\|^{\frac{\tilde{r}p}{2}(1+\tilde{r})+\tilde{r}(1+\beta)\gamma+\beta\gamma}$$
$$+\|u_2\|^{\frac{\tilde{r}p}{2}(1+\tilde{r})+\tilde{r}(1+\beta)\gamma+\beta\gamma}),$$

where we used (4.3) and (4.4), the Hölder inequality, $p(1+\tilde{r})+2(1+\beta)\gamma\leq r$, and Assumption 4.1. This, together with the definition of $\bar{p}_\gamma$ gives that $H\in C_{2\bar{p}_\gamma,\gamma}$ and $\|H\|_{2\bar{p}_\gamma,\gamma}\leq K\|f\|_{p,\gamma}^2$, which finishes the proof of *Claim 1*.

Claim 2: $\mu(H)=\upsilon^2$. In fact, H can be rewritten as

$$H=P_1\Big|\int_0^\infty P_tf\,\mathrm{d}t\Big|^2-\Big|\int_0^\infty P_tf\,\mathrm{d}t\Big|^2+2\int_0^1P_t\Big(\int_0^\infty fP_sf\,\mathrm{d}s\Big)\mathrm{d}t,$$

which together with the fact that $P_t^*\mu=\mu$ for $t>0$ finishes the proof of *Claim 2*, where P_t^* is the dual operator of P_t.

Claim 3: $\frac{1}{\sqrt{N}}\mathscr{R}_N^x\xrightarrow{\mathbb{P}}0$ *as* $N\to\infty$. Combining Assumption 4.1 and (4.3) leads to

$$\mathbb{E}\Big[\Big|\frac{1}{\sqrt{N}}\mathscr{R}_N^x\Big|\Big]\leq\frac{1}{\sqrt{N}}\mathbb{E}\Big[\Big|\int_0^\infty P_tf(X_N^x)\mathrm{d}t\Big|+\Big|\int_0^\infty P_tf(x)\mathrm{d}t\Big|\Big]\to0\quad\text{as }N\to\infty.$$

Claim 4: $\frac{1}{\sqrt{N}}\mathscr{M}_N^x\xrightarrow{d}\mathscr{N}(0,\upsilon^2)$ *as* $N\to\infty$. The aim is to show that

$$\lim_{N\to\infty}\mathbb{E}\big[\exp\big(\mathbf{i}\iota N^{-\frac{1}{2}}\mathscr{M}_N^x\big)\big]=e^{-\frac{\upsilon^2\iota^2}{2}}\quad\forall\,\iota\in\mathbb{R},$$

where $\mathbf{i}$ is the imaginary unit. It is decomposed as

$$e^{\frac{\upsilon^2\iota^2}{2}}\mathbb{E}\big[\exp\big(\mathbf{i}\iota N^{-\frac{1}{2}}\mathscr{M}_N^x\big)\big]-1=\tilde{I}_1(N)+\tilde{I}_2(N)+\tilde{I}_3(N),$$

where

$$\tilde{I}_1(N):=\sum_{j=0}^{N-1}e^{\frac{\upsilon^2\iota^2(j+1)}{2N}}(1-e^{-\frac{\upsilon^2\iota^2}{2N}}-\frac{\upsilon^2\iota^2}{2N})\mathbb{E}[\exp(\mathbf{i}\iota N^{-\frac{1}{2}}\mathscr{M}_j^x)],$$

$$\tilde{I}_2(N):=-\iota^2N^{-1}\sum_{j=0}^{N-1}e^{\frac{\upsilon^2\iota^2(j+1)}{2N}}\mathbb{E}\Big[\exp(\mathbf{i}\iota N^{-\frac{1}{2}}\mathscr{M}_j^x)(\mathcal{Z}_{j+1}^x)^2Q(\iota N^{-\frac{1}{2}}\mathcal{Z}_{j+1}^x)\Big],$$

$$\tilde{I}_3(N):=\frac{\iota^2}{2N}\sum_{j=0}^{N-1}e^{\frac{\upsilon^2\iota^2(j+1)}{2N}}\mathbb{E}\Big[\exp(\mathbf{i}\iota N^{-\frac{1}{2}}\mathscr{M}_j^x)\big(\upsilon^2-(\mathcal{Z}_{j+1}^x)^2\big)\Big].$$

Here, $Q(\zeta) = \zeta^{-2}\int_0^{\zeta}\int_0^{\zeta_2}(e^{\mathbf{i}\zeta_1} - 1)\mathrm{d}\zeta_1\mathrm{d}\zeta_2$ satisfies $\sup_{\zeta\in\mathbb{R}}|Q(\zeta)| \le 1, |Q(\zeta)| = O(|\zeta|)$ for $\zeta \ll 1$, where $O(\cdot)$ denotes a quantity of the same order as its argument. Applying the inequality $|1 - e^{-\zeta} - \zeta| \le \frac{1}{2}\zeta^2$ for $\zeta > 0$ gives

$$
\begin{aligned}
\lim_{N\to\infty}|\tilde{I}_1(N)| &\le \lim_{N\to\infty} e^{\frac{v^2\iota^2}{2}}\sum_{j=0}^{N-1}|1 - e^{-\frac{v^2\iota^2}{2N}} - \frac{v^2\iota^2}{2N}| \\
&\le \lim_{N\to\infty} e^{\frac{v^2\iota^2}{2}}\sum_{j=0}^{N-1}\frac{v^4\iota^4}{8N^2} \le \lim_{N\to\infty}\frac{K}{N} = 0.
\end{aligned}
$$

For the estimate of $\tilde{I}_2(N)$, we arrive at that for $\epsilon \ll 1$ and $c_1 > 2$,

$$
\begin{aligned}
|\tilde{I}_2(N)| \le & \iota^2 e^{\frac{v^2\iota^2}{2}} N^{-1}\Big[\sup_{|\zeta|\le\iota\epsilon}|Q(\zeta)|\sum_{j=0}^{N-1}\mathbb{E}[(\mathcal{Z}_{j+1}^x)^2] \\
&+ \sum_{j=0}^{N-1}(\mathbb{E}[(\mathcal{Z}_{j+1}^x)^{c_1}])^{\frac{2}{c_1}}\left(\mathbb{P}(|\mathcal{Z}_{j+1}^x| > \epsilon\sqrt{N})\right)^{\frac{c_1-2}{c_1}}\Big].
\end{aligned}
$$

Since the condition on p in Proposition 4.1 implies that there exists $c_1 > 2$ such that $c_1(\frac{p\tilde{r}}{2} + (1+\beta)\gamma) \le r$, we have $\sup_{j\in\mathbb{N}}\mathbb{E}[(\mathcal{Z}_j^x)^{c_1}] \le K$. Thus $|\tilde{I}_2(N)|$ tends to 0 as $N \to \infty$ and $\epsilon \to 0$.

For the estimate of $\tilde{I}_3(N)$, we define $\widetilde{N} := \lceil N^{1-\epsilon_0}\rceil$ with some $\epsilon_0 \ll 1$ and $\widetilde{M} := \lceil N/\widetilde{N}\rceil$, and denote

$$
\begin{aligned}
\widetilde{\mathbb{B}}_i &= \{(i-1)\widetilde{N}, (i-1)\widetilde{N}+1, \ldots, i\widetilde{N}-1\} \quad \text{for all } i \in \{1, \ldots, \widetilde{M}-1\}, \\
\widetilde{\mathbb{B}}_{\widetilde{M}} &= \{(\widetilde{M}-1)\widetilde{N}, \ldots, N-1\}.
\end{aligned}
$$

Then the term $\tilde{I}_3(N)$ can be estimated as

$$
\begin{aligned}
&|\tilde{I}_3(N)| \\
&\le \Big|\frac{\iota^2}{2N}\sum_{i=1}^{\widetilde{M}}\mathbb{E}\Big[\exp(\mathbf{i}\iota N^{-\frac{1}{2}}\mathcal{M}_{(i-1)\widetilde{N}}^x)\sum_{j\in\widetilde{\mathbb{B}}_i} e^{\frac{v^2\iota^2(j+1)}{2N}}\left(v^2 - (\mathcal{Z}_{j+1}^x)^2\right)\Big]\Big| \\
&\quad + \frac{\iota^2}{2N}e^{\frac{v^2\iota^2}{2}}\sum_{i=1}^{\widetilde{M}}\sum_{j\in\widetilde{\mathbb{B}}_i}\mathbb{E}\Big[\left|\exp\left(\mathbf{i}\iota N^{-\frac{1}{2}}(\mathcal{M}_j^x - \mathcal{M}_{(i-1)\widetilde{N}}^x)\right) - 1\right|\left(v^2 + (\mathcal{Z}_{j+1}^x)^2\right)\Big] \\
&=: \tilde{I}_{3,1}(N) + \tilde{I}_{3,2}(N).
\end{aligned}
$$

The condition of p in Proposition 4.1 implies $2\bar{p}_\gamma \le r$. Then, using the property of the conditional expectation, *Claims 1–2*, (4.3), and Assumption 4.1 yields that

$$\begin{aligned}
\tilde{I}_{3,1}(N) &\le \frac{\iota^2}{2N}\sum_{i=1}^{\widetilde{M}}\mathbb{E}\Big[\Big|\mathbb{E}\Big[\sum_{j\in\widetilde{\mathbb{B}}_i} e^{\frac{v^2\iota^2(j+1)}{2N}}\big(v^2-(\mathcal{Z}^x_{j+1})^2\big)\Big|\mathcal{F}_{(i-1)\widetilde{N}}\Big]\Big|\Big]\\
&\le \frac{\iota^2}{2N}e^{\frac{v^2\iota^2}{2}}\sum_{i=1}^{\widetilde{M}}\mathbb{E}\Big[\sum_{j\in\widetilde{\mathbb{B}}_1}|v^2-\mathbb{E}[(\mathcal{Z}^y_{j+1})^2]|\Big|_{y=X^x_{(i-1)\widetilde{N}}}\Big]\\
&\le \frac{\iota^2}{2N}e^{\frac{v^2\iota^2}{2}}\sum_{i=1}^{\widetilde{M}}\mathbb{E}\Big[\sum_{j\in\widetilde{\mathbb{B}}_1}|\mu(H)-P_jH(y)|\Big|_{y=X^x_{(i-1)\widetilde{N}}}\Big]\to 0 \quad \text{as } N\to\infty.
\end{aligned}$$

The term $\tilde{I}_{3,2}(N)$ can be estimated as

$$\begin{aligned}
\tilde{I}_{3,2}(N) &\le \frac{\iota^2}{2N}e^{\frac{v^2\iota^2}{2}}\sum_{i=1}^{\widetilde{M}}\sum_{j\in\widetilde{\mathbb{B}}_i}\Big[\iota\tilde{\epsilon}_0\mathbb{E}[v^2+(\mathcal{Z}^x_{j+1})^2]\\
&\quad + 2\mathbb{E}\Big[(v^2+(\mathcal{Z}^x_{j+1})^2)\mathbf{1}_{\{N^{-\frac{1}{2}}|\mathcal{M}^x_j-\mathcal{M}^x_{(i-1)\widetilde{N}}|>\tilde{\epsilon}_0\}}\Big]\Big]\\
&\le K\tilde{\epsilon}_0 + K\frac{\iota^2}{N}e^{\frac{v^2\iota^2}{2}}\sum_{i=1}^{\widetilde{M}}\sum_{j\in\widetilde{\mathbb{B}}_i}(\mathbb{E}[v^{c_1}+(\mathcal{Z}^x_{j+1})^{c_1}])^{\frac{2}{c_1}}\\
&\quad \times\big(\mathbb{E}[|\mathcal{M}^x_j-\mathcal{M}^x_{(i-1)\widetilde{N}}|^2](\tilde{\epsilon}_0^2N)^{-1}\big)^{\frac{c_1-2}{c_1}}\\
&\le K\tilde{\epsilon}_0 + K\big(\widetilde{N}/(\tilde{\epsilon}_0^2N)\big)^{\frac{c_1-2}{c_1}}\to 0 \quad \text{as } N\to\infty \text{ and } \tilde{\epsilon}_0\to 0,
\end{aligned}$$

where we used

$$\mathbb{E}\big[|\mathcal{M}^x_j-\mathcal{M}^x_{(i-1)\widetilde{N}}|^2\big]=\sum_{l=(l-1)\widetilde{N}}^{j}\mathbb{E}[(\mathcal{Z}^x_l)^2]\le\sum_{l=0}^{\widetilde{N}-1}\sup_{j\in\mathbb{N}}\mathbb{E}[(\mathcal{Z}^x_j)^2]\le K\widetilde{N}.$$

Combining *Claims 1–4* completes the proof of the CLT. □

4.2 Limit Theorems of Numerical Discretizations

In this section, we present the preservation of limit theorems, in particular the SLLN and the CLT for time-homogeneous Markov processes by numerical discretizations.

Table 4.1 Definitions of the uniform stepsize Δ and the state space E^h

	Temporal semi-discretization for $h = 0$	Spatial-temporal full discretization for $h \neq 0$
Stepsize Δ	$\Delta := (0, \tau)^\top$	$\Delta := (h, \tau)^\top$
State space E^h	$E^0 := E$	$E^h (\subset E)$ with norm $\|\cdot\|_h$ satisfying $\|u^h\|_h \le K\|u^h\|$ for $u^h \in E^h$

Then we apply the main results to numerical discretizations of the θ-EM method for SFDEs.

We begin with some notation for numerical discretizations. The temporal semi-discretization is considered for the finite or infinite dimensional case and the spatial-temporal full discretization is considered for the infinite dimensional case. Let $h \in [0, 1]$ and $\tau \in (0, 1]$ be stepsizes in spatial and temporal directions, respectively. To unify the notation, we define the uniform stepsize Δ and the state space E^h for the temporal semi-discretization or spatial-temporal full discretization as shown in Table 4.1.

When the initial datum $x \in E^h$, we always take $x^h = x$. Denote $t_k := k\tau$ for $k \in \mathbb{N}$. For each given Δ, let $\{Y_{t_k}^{x^h,\Delta}\}_{k\in\mathbb{N}}$ denote the E^h-valued discretization of $\{X_t^x\}_{t\ge 0}$ satisfying $Y_0^{x^h,\Delta} = x^h$, and define the time-homogeneous Markov process on the whole time horizon by

$$Y_t^{x^h,\Delta} = \sum_{k=0}^{\infty} Y_{t_k}^{x^h,\Delta} \mathbf{1}_{[t_k,t_{k+1})}(t), \quad t > 0.$$

Denote by $\mu_t^{x^h,\Delta}$ the probability measure induced by $Y_t^{x^h,\Delta}$, i.e., for any $A \in \mathfrak{B}(E)$, $\mu_t^{x^h,\Delta}(A) = \mathbb{P}\{\omega \in \Omega : Y_t^{x^h,\Delta} \in A\}$. Similar to the definition of P_t, we define the linear operator P_t^Δ induced by $Y_t^{\cdot,\Delta}$ as

$$P_t^\Delta : \mathcal{B}_b \to \mathcal{B}_b, \quad P_t^\Delta f(x^h) := \mathbb{E}\big[f(Y_t^{x^h,\Delta})\big] = \int_E f(u)\mathrm{d}\mu_t^{x^h,\Delta}(u) \quad \forall\, x^h \in E^h.$$

If $Y_t^{x^h,\Delta}$ admits a unique invariant measure μ^Δ satisfying $\mu^\Delta(\|\cdot\|^l) < \infty$ for some constant $l > 0$, P_t^Δ is uniquely extendible to a bounded linear operator on $L^l(E^h, \mu^\Delta)$ which is still denoted by P_t^Δ. When there is no confusion, we also denote $\mathcal{B}_b(E^h; \mathbb{R}), C_b(E^h; \mathbb{R}), C_{p,\gamma}(E^h; \mathbb{R})$ by $\mathcal{B}_b$, C_b, $C_{p,\gamma}$, respectively.

We propose sufficient conditions on numerical discretizations to ensure the preservation of the SLLN and the CLT of time-homogeneous Markov processes. These conditions are related to moment boundedness, attractiveness, and strong convergence of numerical discretizations.

Assumption 4.2 *Assume that there exists* $\tilde{\Delta} = (\tilde{h}, \tilde{\tau})^\top \in [0, 1] \times (0, 1]$ *such that for any* $h \in [0, \tilde{h}]$, $\tau \in (0, \tilde{\tau}]$, *and* $x^h \in E^h$, $\{Y_{t_k}^{x^h,\Delta}\}_{k\in\mathbb{N}}$ *is a time homogeneous Markov process and satisfies the following conditions.*

(i) *There exist constants* $q \geq 2, \tilde{q} \geq 1$, *and* $L_3 > 0$ *such that*

$$\sup_{k\geq 0} \mathbb{E}[\|Y_{t_k}^{x^h,\Delta}\|^q] \leq L_3(1 + \|x^h\|^{\tilde{q}q}).$$

(ii) *There exist constants* $\gamma_2 \in (0, 1], \kappa \in [0, q-1], L_4 > 0$, *and a function* $\rho^\Delta : [0, +\infty) \to [0, +\infty)$ *with*

$$\sup_{\{h\in[0,\tilde{h}],\tau\in(0,\tilde{\tau}]\}} \tau \sum_{k=0}^{\infty} \left(\rho^\Delta(t_k)\right)^{\gamma_2} < \infty$$

such that for any $x^h, y^h \in E^h$,

$$\left(\mathbb{E}[\|Y_{t_k}^{x^h,\Delta} - Y_{t_k}^{y^h,\Delta}\|^2]\right)^{\frac{1}{2}} \leq L_4\|x^h - y^h\|(1 + \|x^h\|^\kappa + \|y^h\|^\kappa)\rho^\Delta(t_k).$$

Below, when there is no confusion, we always assume that $h \in [0, \tilde{h}]$ and $\tau \in (0, \tilde{\tau}]$. And constants r and q in Assumptions 4.1 (i) and 4.2 (i), respectively, are always assumed to be large enough to meet the need.

Assumption 4.3 *Assume that there exist* $\alpha = (\alpha_1, \alpha_2)^\top \in \mathbb{R}_+^2$ *and* $L_5 > 0$ *such that for any* $x \in E$,

$$\sup_{t\geq 0} \left(\mathbb{E}[\|X_t^x - Y_t^{x^h,\Delta}\|^2]\right)^{\frac{1}{2}} \leq L_5(1 + \|x\|^{\tilde{r}\vee\tilde{q}})|\Delta^\alpha|,$$

where $\Delta^\alpha := (h^{\alpha_1}, \tau^{\alpha_2})^\top$ *and* $|\Delta^\alpha| := h^{\alpha_1} + \tau^{\alpha_2}$.

Denote the time-average of the numerical discretization by

$$\frac{1}{k} S_k^{x^h,\Delta} := \frac{1}{k} \sum_{i=0}^{k-1} f(Y_{t_i}^{x^h,\Delta})$$

for $f \in C_{p,\gamma}$ with suitable parameters $p \geq 1$ and $\gamma \in (0, 1]$. Then we obtain the SLLN of the time-average of numerical discretizations.

Theorem 4.3 *Let Assumption 4.1 hold for the time-homogeneous Markov process* $\{X_t^x\}_{t\geq 0}$, *Assumption 4.2 hold for the time-homogeneous Markov process*

$\{Y_{t_k}^{x^h,\Delta}\}_{k\in\mathbb{N}}$, *Assumption 4.3 hold for* $\{X_t^x\}_{t\geq 0}$ *and* $\{Y_{t_k}^{x^h,\Delta}\}_{k\in\mathbb{N}}$, $\gamma \in [\gamma_2, 1]$, *and* $p \geq 1$ *with* $p(1+\tilde{q}) + 2(1+\kappa)\gamma \leq q$. *Then for any* $f \in C_{p,\gamma}$, *we have*

$$\lim_{|\Delta|\to 0}\lim_{k\to\infty}\frac{1}{k}S_k^{x^h,\Delta} = \mu(f) \quad \mathbb{P}\text{-a.s.} \tag{4.18}$$

The following CLT characterizes the limit behavior of $\frac{1}{\sqrt{k\tau}}\sum_{i=0}^{k-1}\big(f(Y_{t_i}^{x^h,\Delta}) - \mu(f)\big)\tau$, which is called the normalized time-average of the numerical discretization. Here, $f \in C_{p,\gamma}$ with suitable parameters $p \geq 1$ and $\gamma \in (0,1]$. Recall that $v^2 = 2\mu\big((f-\mu(f))\int_0^\infty (P_t f - \mu(f))\mathrm{d}t\big)$ is given in Proposition 4.1.

Theorem 4.4 *Let Assumption 4.1 hold for the time-homogeneous Markov process* $\{X_t^x\}_{t\geq 0}$, *Assumption 4.2 hold for the time-homogeneous Markov process* $\{Y_{t_k}^{x^h,\Delta}\}_{k\in\mathbb{N}}$, *and Assumption 4.3 hold for* $\{X_t^x\}_{t\geq 0}$ *and* $\{Y_{t_k}^{x^h,\Delta}\}_{k\in\mathbb{N}}$. *For any fixed* $\lambda \in (0, \alpha_2\gamma)$ *with* $\gamma \in [\gamma_1 \vee \gamma_2, 1]$, *set* $k := \lceil \tau^{-1-2\lambda}\rceil$ *and* $h := \tau^{\alpha_2/\alpha_1}$ *(for* $h \neq 0$*). Let* $p \geq 1$ *satisfy*

$$\tilde{q}^2\Big(3 \vee (\frac{1}{\lambda}+1)\Big)\Big(p(\tilde{q}\vee\tilde{r}) + \big(3+4(\kappa\vee\beta)\big)\gamma\Big) \leq q \wedge r, \tag{4.19}$$

and $\varepsilon \ll \lambda$ *with* $\sqrt{\varepsilon} - \frac{\varepsilon}{\lambda} - \frac{\varepsilon\sqrt{\varepsilon}}{\lambda} > 0$ *satisfy*

$$\lim_{\tau\to 0}\tau^{\varepsilon}\sum_{i=\lceil \tau^{-1-\frac{\varepsilon}{4}}\rceil}^{k}(\rho^{\Delta}(t_i))^{\gamma} = 0. \tag{4.20}$$

Then for any $f \in C_{p,\gamma}$,

$$\frac{1}{\sqrt{k\tau}}\sum_{i=0}^{k-1}\big(f(Y_{t_i}^{x^h,\Delta}) - \mu(f)\big)\tau \xrightarrow{d} \mathcal{N}(0, v^2) \quad \text{as } \tau \to 0. \tag{4.21}$$

Remark 4.2

(i) If Assumption 4.1 (i) is replaced by: "for any $r \geq 2$, there exist $\tilde{r} \geq 1$ and $L_1 > 0$ such that $\sup_{t\geq 0}\mathbb{E}[\|X_t^x\|^r] \leq L_1(1+\|x\|^{\tilde{r}r})$", and Assumption 4.2 (i) is replaced by: "for any $q \geq 2$, there exist $\tilde{q} \geq 1$ and $L_3 > 0$ such that $\sup_{k\geq 0}\mathbb{E}[\|Y_{t_k}^{x^h,\Delta}\|^q] \leq L_3(1+\|x^h\|^{\tilde{q}q})$", then the restriction (4.19) on p is not necessary.

(ii) When $P_{t_k}^{\Delta}$ is exponential mixing, i.e., $\rho^{\Delta}(t_k) = e^{-ct_k}$ with some $c > 0$, then (4.20) is satisfied naturally.

Remark 4.3 Our main results in Theorems 4.3 and 4.4 are proved with a strong convergence condition but for test functionals with lower regularity, i.e., $f \in C_{p,\gamma}$. In general, strong convergence rates are no larger than the weak ones. The numerical SLLN and CLT still hold with certain trade-off between the convergence condition of numerical discretizations (i.e., Assumption 4.3) and the regularity of test functionals. Precisely, if Assumption 4.3 is replaced by:

$$\sup_{t\geq 0}|\mathbb{E}[f(X_t^x)-f(Y_t^{x^h,\Delta})]|\leq K|\Delta^\alpha| \tag{4.22}$$

for test functionals belonging to some space $\mathfrak{C}$, then Theorems 4.3 and 4.4 still hold for $f\in C_{p,\gamma}\cap\mathfrak{C}$.

The main difference between the proofs of this remark and Theorem 4.4 lies in the estimation of terms $|P_t f(u)-P_t^\Delta f(u)|$ and $|\mu(f)-\mu^\Delta(f)|$, where $u\in E^h$, and μ^Δ is the invariant measure of the numerical discretization (see Proposition 4.6). Under the assumption (4.22), we have that for $u\in E^h$,

$$|P_t f(u)-P_t^\Delta f(u)|\leq \sup_{t\geq 0}|\mathbb{E}[f(X_t^u)-f(Y_t^{u,\Delta})]|\leq K|\Delta^\alpha|.$$

Hence, by (4.3), we have that $|\mu(f)-\mu^\Delta(f)|$ can be estimated as

$$\begin{aligned}|\mu(f)-\mu^\Delta(f)|&\leq |P_{t_k}f(u)-\mu(f)|+|P_{t_k}^\Delta f(u)-\mu^\Delta(f)|+|P_{t_k}f(u)-P_{t_k}^\Delta f(u)|\\&\leq K|\Delta^\alpha|,\end{aligned}$$

where we used (4.24), and in the last step we let $k\to\infty$.

Remark 4.4 Based on the SLLN and CLT results of numerical discretizations (Theorems 4.3 and 4.4), we provide an analysis on the computational cost of the time-average of numerical discretizations for approximating the ergodic limit to achieve a given accuracy ϵ. For a significance level $\vartheta\in(0,1)$, let the Gaussian confidence interval $[-I_\vartheta, I_\vartheta]$ be such that $\mathbb{P}\{|Z|\leq I_\vartheta\}=1-\vartheta$, where Z is a standard Gaussian random variable. Then by the numerical CLT, an asymptotic confidence interval with level $1-\vartheta$, centered around $\mu(f)$, is $[\mu(f)-\frac{I_\vartheta v}{\sqrt{k\tau}}, \mu(f)+\frac{I_\vartheta v}{\sqrt{k\tau}}]$. This, along with the relation $k=\lceil\tau^{-1-2\lambda}\rceil$ implies that for a given accuracy ϵ, when τ is set to be the size of $(\frac{\epsilon}{I_\vartheta v})^{\frac{1}{\lambda}}$, one arrives at $\mathbb{P}\{|\frac{1}{k}S_k^{x^h,\Delta}-\mu(f)|\leq\epsilon\}=1-\vartheta$. Hence, for any $\lambda\in(0,\alpha_2\gamma)$, the computational cost of $\frac{1}{k}S_k^{x^h,\Delta}$ is estimated as $h^{-1}k=O(\epsilon^{-\frac{1}{\lambda}(\frac{\alpha_2}{\alpha_1}+1)-2})$ (resp. $k=O(\epsilon^{-\frac{1}{\lambda}-2})$) for $h\neq 0$ (resp. $h=0$), where $\alpha=(\alpha_1,\alpha_2)^\top$ is the weak convergence rate given in Remark 4.3 of the numerical discretization. Here, $O(|\zeta|)$ is a quantity of the same order as $|\zeta|$.

The main results in Theorems 4.3 and 4.4 can be applied to numerical discretizations of a large class of autonomous stochastic differential equations. Here, we focus on the application to the case of the θ-EM functional solution of the SFDE (1.1).

Theorem 4.5 *Under Assumptions 2.1 to 2.5, the θ-EM functional solution $\{y_{t_k}^{\xi,\Delta}\}_{k\geq 0}$ fulfills the SLLN and the CLT, namely, for any $p \geq 1, \gamma \in (0,1]$, and $f \in C_{p,\gamma}$,*

$$\lim_{\Delta\to 0}\lim_{m_1\to\infty}\frac{1}{m_1}\sum_{k=0}^{m_1-1} f(y_{t_k}^{\xi,\Delta}) = \mu(f) \quad \mathbb{P}\text{-a.s.},$$

$$\lim_{\Delta\to 0}\frac{1}{\sqrt{m_2\Delta}}\sum_{k=0}^{m_2-1}(f(y_{t_k}^{\xi,\Delta}) - \mu(f))\Delta = \mathcal{N}(0,v^2) \quad \text{in distribution},$$

where $m_2 = \lceil\Delta^{-1-2\lambda}\rceil$ with $\lambda \in (0,\frac{1}{2}\gamma)$, $v^2 = 2\mu\big((f-\mu(f))\int_0^\infty \mathbb{E}[f(x_t^{\cdot}) - \mu(f)]\mathrm{d}t\big)$.

Proof Assumption 4.1 is ensured by Propositions 1.9 and 1.12. By Proposition 2.8, we obtain that Assumption 4.2 (i) is satisfied. By Proposition 2.10, we have that Assumption 4.2 (ii) is satisfied with $\kappa = 0$, $\rho^\Delta(t) = e^{-ct}$ for some $c > 0$, and any $\gamma_2 \in (0,1]$. Theorem 2.14 ensures that Assumption 4.3 is satisfied with $\alpha_1 = 0, \alpha_2 = \frac{1}{2}$. Applying Theorems 4.3 and 4.4 finishes the proof. □

4.3 Proofs of Numerical Limit Theorems

In this section, we present the proofs of the numerical SLLN (Theorem 4.3) and the numerical CLT (Theorem 4.4) in Sects. 4.3.1 and 4.3.2, respectively. The proofs are respectively based on the convergence of the numerical invariant measure and the decomposition of the normalized time-average of numerical discretizations.

4.3.1 Proof of Theorem 4.3

This subsection is devoted to giving the proof of the preservation of the SLLN of numerical discretizations, based on the existence and uniqueness as well as the convergence of the numerical invariant measure.

Proposition 4.6 *Let Assumption 4.2 hold for the time-homogeneous Markov process $\{Y_{t_k}^{x^h,\Delta}\}_{k\in\mathbb{N}}$. Then for each fixed Δ, $\{Y_{t_k}^{x^h,\Delta}\}_{k\in\mathbb{N}}$ admits a unique numerical invariant measure $\mu^\Delta \in \mathcal{P}(E^h)$. Furthermore, if in addition Assumption 4.1 holds for the time-homogeneous Markov process $\{X_t^x\}_{t\geq 0}$ and Assumption 4.3 holds for $\{X_t^x\}_{t\geq 0}$ and $\{Y_{t_k}^{x^h,\Delta}\}_{k\in\mathbb{N}}$, then we have $\mathbb{W}_2(\mu,\mu^\Delta) \leq K|\Delta^\alpha|$, where $|\Delta^\alpha|$ is given in Assumption 4.3.*

Proof The proof of the existence and uniqueness of the numerical invariant measure is analogous to that of Theorem 3.2, and the argument for the convergence rate follows similarly to the proof of Theorem 3.3. Therefore, we omit the details. □

Note that Assumption 4.2 leads to

$$\mu^{\Delta}(\|\cdot\|^q) \le \lim_{t\to\infty} \int_{E^h} \|u\|^q \mathrm{d}\mu_t^{\mathbf{0},\Delta}(u) \le \sup_{t\ge 0} \mathbb{E}[\|Y_t^{\mathbf{0},\Delta}\|^q] \le L_3. \tag{4.23}$$

Similar to the proof of (4.3), we derive that for $f \in C_{p,\gamma}$ with $p \in [1, q]$ and $\gamma \in (0, 1]$,

$$|P_{t_k}^{\Delta} f(u) - \mu^{\Delta}(f)| \le K\|f\|_{p,\gamma}(1 + \|u\|^{\frac{p\tilde{q}}{2}+(1+\kappa)\gamma})(\rho^{\Delta}(t_k))^{\gamma}, \quad k \in \mathbb{N}. \tag{4.24}$$

This is the uniformly mixing property of numerical discretizations. Based on [29, Proposition 2.6 and Remark 2.7], we obtain the estimate between $\frac{1}{k}S_k^{x^h,\Delta}$ and $\mu^{\Delta}(f)$, which is stated in the following lemma.

Lemma 4.7 *Let Assumption 4.2 hold for the time-homogeneous Markov process* $\{Y_{t_k}^{x^h,\Delta}\}_{k\in\mathbb{N}}$, $\gamma \in [\gamma_2, 1]$, $\varrho \in \mathbb{N}_+$, *and* $p \ge 1$ *satisfy* $\varrho\big(\frac{p}{2}(1+\tilde{q}) + (1+\kappa)\gamma\big) \le q$. *Then for any* $f \in C_{p,\gamma}$ *and* $x^h \in E^h$, *we have*

$$\mathbb{E}\Big[\Big|\frac{1}{k}S_k^{x^h,\Delta} - \mu^{\Delta}(f)\Big|^{2\varrho}\Big] \le K\|f\|_{p,\gamma}^{2\varrho}(1 + \|x^h\|^{\varrho\tilde{q}(\frac{p}{2}(1+\tilde{q})+(1+\kappa)\gamma)})t_k^{-\varrho} \quad \forall\, k \in \mathbb{N}_+. \tag{4.25}$$

With these in hand, we present the proof of the SLLN, i.e., $\frac{1}{k}S_k^{x^h,\Delta} \overset{\mathbb{P}\text{-a.s.}}{\longrightarrow} \mu(f)$ as $k \to \infty$ and $|\Delta| \to 0$.

Proof of Theorem 4.3 Applying (4.2) and Proposition 4.6 leads to

$$|\mu(f) - \mu^{\Delta}(f)| \le K\|f\|_{p,\gamma}|\Delta^{\alpha}|^{\gamma}, \quad f \in C_{p,\gamma}. \tag{4.26}$$

Hence it suffices to prove that $\lim_{k\to\infty}\big|\frac{1}{k}S_k^{x^h,\Delta} - \mu^{\Delta}(f)\big| = 0$ $\mathbb{P}$-a.s. for each fixed stepsize Δ and $f \in C_{p,\gamma}$.

For any $\delta \in (0, \frac{1}{4})$ and $k \in \mathbb{N}$ with $k \ge \lceil 1/\tau \rceil$, define

$$\mathcal{A}_k^{x^h,\Delta} := \Big\{\omega \in \Omega : \Big|\frac{1}{k}S_k^{x^h,\Delta} - \mu^{\Delta}(f)\Big|(\omega) > \|f\|_{p,\gamma}t_k^{-\delta}\tau^{-\frac{1}{4}}\Big\}.$$

Note that the condition $p(1+\tilde{q})+2(1+\kappa)\gamma \le q$ coincides with the one in Lemma 4.7 with $\varrho=2$. By virtue of (4.25) with $\varrho=2$ one has

$$\mathbb{E}\big[\big|\frac{1}{k}S_k^{x^h,\Delta}-\mu^\Delta(f)\big|^4\big] \le K\|f\|_{p,\gamma}^4(1+\|x^h\|^{\tilde{q}(p(1+\tilde{q})+2(1+\kappa)\gamma)})t_k^{-2}.$$

Applying the Chebyshev inequality yields

$$\begin{aligned}\mathbb{P}(\mathcal{A}_k^{x^h,\Delta}) &\le \frac{t_k^{4\delta}\tau}{\|f\|_{p,\gamma}^4}\mathbb{E}\Big[\big|\frac{1}{k}S_k^{x^h,\Delta}-\mu^\Delta(f)\big|^4\Big]\\ &\le K(1+\|x^h\|^{\tilde{q}(p(1+\tilde{q})+2(1+\kappa)\gamma)})t_k^{2(2\delta-1)}\tau.\end{aligned} \tag{4.27}$$

Then it follows from $2(1-2\delta)>1$ that

$$\sum_{k=\lceil\frac{1}{\tau}\rceil}^{\infty}\mathbb{P}(\mathcal{A}_k^{x^h,\Delta}) \le K(1+\|x^h\|^{\tilde{q}(p(1+\tilde{q})+2(1+\kappa)\gamma)}). \tag{4.28}$$

Define the random variable $\mathcal{K}^{x^h,\Delta}$ by

$$\begin{aligned}\mathcal{K}^{x^h,\Delta}(\omega) := \inf\big\{j\in\mathbb{N} \text{ with } j\ge\lceil\frac{1}{\tau}\rceil : \big|\frac{1}{k}S_k^{x^h,\Delta}-\mu^\Delta(f)\big|(\omega)\\ \le \|f\|_{p,\gamma}t_k^{-\delta}\tau^{-\frac{1}{4}}\ \forall k\ge j+1\big\}.\end{aligned}$$

Inequality (4.28), along with the Borel–Cantelli lemma implies that $\mathcal{K}^{x^h,\Delta}<\infty$ $\mathbb{P}$-a.s. and

$$\mathbb{P}\Big\{\omega\in\Omega : \big|\frac{1}{k}S_k^{x^h,\Delta}-\mu^\Delta(f)\big|(\omega)\le\|f\|_{p,\gamma}t_k^{-\delta}\tau^{-\frac{1}{4}} \text{ for all } k\ge\mathcal{K}^{x^h,\Delta}(\omega)+1\Big\}=1. \tag{4.29}$$

Let $T^{x^h,\Delta}=\tau\mathcal{K}^{x^h,\Delta}$. Noting $2(2\delta-1)<-1$, there exists a constant $l>0$ (e.g., $l=-\frac{1+2(2\delta-1)}{2}>0$) such that $l+2(2\delta-1)<-1$. This, combining (4.27) implies

$$\begin{aligned}\mathbb{E}[(T^{x^h,\Delta})^l] &= \sum_{j=\lceil\frac{1}{\tau}\rceil}^{\infty}\mathbb{E}\big[(T^{x^h,\Delta})^l\mathbf{1}_{\{\mathcal{K}^{x^h,\Delta}=j\}}\big] = \sum_{j=\lceil\frac{1}{\tau}\rceil}^{\infty}(j\tau)^l\mathbb{P}\big(\mathcal{K}^{x^h,\Delta}=j\big)\\ &\le \sum_{j=\lceil\frac{1}{\tau}\rceil}^{\infty}(t_j)^l\mathbb{P}(\mathcal{A}_j^{x^h,\Delta}) \le K(1+\|x^h\|^{\tilde{q}(p(1+\tilde{q})+2(1+\kappa)\gamma)}).\end{aligned} \tag{4.30}$$

For $\tilde{\delta} > 0$ and $N \in \mathbb{N}_+$, define $\mathcal{D}_N^{x^h} := \{\omega \in \Omega : T^{x^h,\Delta} + 1 > N^{\tilde{\delta}}\}$. When $\tilde{\delta}$ is chosen so that $\tilde{\delta}l > 1$, it follows from (4.30) that

$$\sum_{N=1}^{\infty} \mathbb{P}(\mathcal{D}_N^{x^h}) \leq \sum_{N=1}^{\infty} (\frac{1}{N})^{\tilde{\delta}l} \mathbb{E}[(T^{x^h,\Delta} + 1)^l] \leq K(1 + \|x^h\|^{\tilde{q}(p(1+\tilde{q})+2(1+\kappa)\gamma)}).$$

Using the Borel–Cantelli lemma again yields that there exists a random variable $\mathfrak{N}^{x^h} < \infty$ $\mathbb{P}$-a.s. such that

$$\mathbb{P}\Big\{\omega \in \Omega : T^{x^h,\Delta}(\omega) + 1 \leq N^{\tilde{\delta}} \text{ for all } N \geq \mathfrak{N}^{x^h}(\omega)\Big\} = 1. \tag{4.31}$$

Combining (4.29) and (4.31) yields that for a.s. $\omega \in \Omega$, when $k \geq \frac{(\mathfrak{N}^{x^h}(\omega))^{\tilde{\delta}}}{\tau^{1+\frac{1}{4\delta}}} \geq \frac{T^{x^h,\Delta}(\omega)+1}{\tau} \geq \mathcal{K}^{x^h,\Delta}(\omega) + 1$, we have

$$\Big|\frac{1}{k} S_k^{x^h,\Delta} - \mu^{\Delta}(f)\Big|(\omega) \leq \|f\|_{p,\gamma} t_k^{-\delta} \tau^{-\frac{1}{4}} \leq \|f\|_{p,\gamma} (\mathfrak{N}^{x^h}(\omega))^{-\tilde{\delta}\delta},$$

which implies $\lim_{k\to\infty} \frac{1}{k} S_k^{x^h,\Delta} = \mu^{\Delta}(f)$ almost surely for fixed Δ. This finishes the proof. □

4.3.2 *Proof of Theorem 4.4*

In this section, we present the proof of the preservation of the CLT of numerical discretizations, based on the decomposition of the normalized time-average of numerical discretizations. Precisely, we introduce the following decomposition of the normalized time-average $\frac{1}{\sqrt{k\tau}} \sum_{i=0}^{k-1} \big(f(Y_{t_i}^{x^h,\Delta}) - \mu(f)\big)\tau$,

$$\frac{1}{\sqrt{k\tau}} \sum_{i=0}^{k-1} \big(f(Y_{t_i}^{x^h,\Delta}) - \mu(f)\big)\tau = \frac{1}{\sqrt{k\tau}} \mathcal{M}_k^{x^h,\Delta} + \frac{1}{\sqrt{k\tau}} \mathcal{R}_k^{x^h,\Delta}, \tag{4.32}$$

where

$$\mathcal{M}_k^{x^h,\Delta} := \tau \sum_{i=0}^{\infty} \Big(\mathbb{E}[f(Y_{t_i}^{x^h,\Delta})|\mathcal{F}_{t_k}] - \mu^{\Delta}(f)\Big) - \tau \sum_{i=0}^{\infty} \Big(\mathbb{E}[f(Y_{t_i}^{x^h,\Delta})|\mathcal{F}_0] - \mu^{\Delta}(f)\Big), \tag{4.33}$$

$$\begin{aligned}\mathscr{R}_k^{x^h,\Delta} := &-\tau\sum_{i=k}^{\infty}\Big(\mathbb{E}[f(Y_{t_i}^{x^h,\Delta})|\mathcal{F}_{t_k}]-\mu^\Delta(f)\Big)+\tau\sum_{i=0}^{\infty}\Big(\mathbb{E}[f(Y_{t_i}^{x^h,\Delta})|\mathcal{F}_0]-\mu^\Delta(f)\Big)\\ &+k\tau(\mu^\Delta(f)-\mu(f)).\end{aligned}$$

By virtue of the Slutsky theorem, the idea for proving Theorem 4.4 is to prove that $\frac{1}{\sqrt{k\tau}}\mathscr{M}_k^{x^h,\Delta}\xrightarrow{d}\mathcal{N}(0,v^2)$ and $\frac{1}{\sqrt{k\tau}}\mathscr{R}_k^{x^h,\Delta}\xrightarrow{\mathbb{P}}0$, where we recall that $\xrightarrow{d}$ and $\xrightarrow{\mathbb{P}}$ denote the convergence in distribution and in probability, respectively.

By fully utilizing the Markov property and the uniformly mixing property of numerical discretizations of the time-homogeneous Markov process, it is shown in the following proposition that $\mathscr{M}_k^{x^h,\Delta}$ is well-defined and is a martingale for $k\in\mathbb{N}$.

Proposition 4.8 *Let Assumption 4.2 hold for the time-homogeneous Markov process* $\{Y_{t_k}^{x^h,\Delta}\}_{k\in\mathbb{N}}$, $\gamma\in[\gamma_2,1]$, *and* $p\geq 1$ *satisfy* $\frac{p\tilde{q}}{2}+(1+\kappa)\gamma\leq q$. *Then for any* $x^h\in E^h$, $f\in C_{p,\gamma}$, *the sequence* $\{\mathscr{M}_k^{x^h,\Delta}\}_{k\in\mathbb{N}}$ *is an* $\{\mathcal{F}_{t_k}\}_{k\in\mathbb{N}}$*-adapted martingale with* $\mathscr{M}_0^{x^h,\Delta}=0$. *Moreover, we have*

$$\begin{aligned}|\mathscr{M}_k^{x^h,\Delta}|\leq{}&\tau\Big|\sum_{i=0}^{k-1}\big(f(Y_{t_i}^{x^h,\Delta})-\mu^\Delta(f)\big)\Big|\\ &+K\|f\|_{p,\gamma}\Big(1+\|Y_{t_k}^{x^h,\Delta}\|^{\frac{p\tilde{q}}{2}+(1+\kappa)\gamma}+\|x^h\|^{\frac{p\tilde{q}}{2}+(1+\kappa)\gamma}\Big).\end{aligned}\tag{4.34}$$

Proof From the definition of $\mathscr{M}_k^{x^h,\Delta}$, we have

$$\begin{aligned}\mathscr{M}_k^{x^h,\Delta}={}&\tau\sum_{i=0}^{k-1}\big(f(Y_{t_i}^{x^h,\Delta})-\mu^\Delta(f)\big)+\tau\sum_{i=0}^{\infty}\big(P_{t_i}^\Delta f(Y_{t_k}^{x^h,\Delta})-\mu^\Delta(f)\big)\\ &-\tau\sum_{i=0}^{\infty}\big(P_{t_i}^\Delta f(x^h)-\mu^\Delta(f)\big).\end{aligned}\tag{4.35}$$

It follows from (4.24) that

$$\begin{aligned}|\mathscr{M}_k^{x^h,\Delta}|\leq{}&\tau\Big|\sum_{i=0}^{k-1}\big(f(Y_{t_i}^{x^h,\Delta})-\mu^\Delta(f)\big)\Big|+K\|f\|_{p,\gamma}(1+\|Y_{t_k}^{x^h,\Delta}\|^{\frac{p\tilde{q}}{2}+(1+\kappa)\gamma})\\ &\times\tau\sum_{i=0}^{\infty}(\rho^\Delta(t_i))^\gamma+K\|f\|_{p,\gamma}(1+\|x^h\|^{\frac{p\tilde{q}}{2}+(1+\kappa)\gamma})\tau\sum_{i=0}^{\infty}(\rho^\Delta(t_i))^\gamma.\end{aligned}$$

Using the conditions $\gamma \in [\gamma_2, 1]$ and $\frac{p\tilde{q}}{2} + (1+\kappa)\gamma \le q$, and Assumption 4.2 leads to

$$\mathbb{E}[|\mathscr{M}_k^{x^h,\Delta}|] \le K\|f\|_{p,\gamma} k(1 + \|x^h\|^{\tilde{q}(\frac{p\tilde{q}}{2}+(1+\kappa)\gamma)}) < \infty.$$

In addition, by (4.35), we have

$$\begin{aligned}\mathscr{M}_k^{x^h,\Delta} - \mathscr{M}_{k-1}^{x^h,\Delta} &= \tau\big(f(Y_{t_{k-1}}^{x^h,\Delta}) - \mu^\Delta(f)\big) + \tau\sum_{i=0}^{\infty}\big(P_{t_i}^\Delta f(Y_{t_k}^{x^h,\Delta}) - \mu^\Delta(f)\big)\\ &\quad - \tau\sum_{i=0}^{\infty}\big(P_{t_i}^\Delta f(Y_{t_{k-1}}^{x^h,\Delta}) - \mu^\Delta(f)\big), \end{aligned} \tag{4.36}$$

which gives

$$\begin{aligned}\mathbb{E}\Big[(\mathscr{M}_{k+j}^{x^h,\Delta} - \mathscr{M}_k^{x^h,\Delta})|\mathcal{F}_{t_k}\Big] &= \tau\sum_{i=0}^{j-1}\big(P_{t_i}^\Delta f(Y_{t_k}^{x^h,\Delta}) - \mu^\Delta(f)\big)\\ &\quad + \tau P_{t_j}^\Delta\sum_{i=0}^{\infty}\big(P_{t_i}^\Delta f - \mu^\Delta(f)\big)(Y_{t_k}^{x^h,\Delta})\\ &\quad - \tau\sum_{i=0}^{\infty}\big(P_{t_i}^\Delta f(Y_{t_k}^{x^h,\Delta}) - \mu^\Delta(f)\big). \end{aligned} \tag{4.37}$$

Applying the dominated convergence theorem yields

$$\begin{aligned}\tau P_{t_j}^\Delta\sum_{i=0}^{\infty}\big(P_{t_i}^\Delta f - \mu^\Delta(f)\big)(Y_{t_k}^{x^h,\Delta}) &= \tau\sum_{i=0}^{\infty}\big(P_{t_{i+j}}^\Delta f - \mu^\Delta(f)\big)(Y_{t_k}^{x^h,\Delta})\\ &= \tau\sum_{i=j}^{\infty}\big(P_{t_i}^\Delta f(Y_{t_k}^{x^h,\Delta}) - \mu^\Delta(f)\big). \end{aligned}$$

Inserting the above equality into (4.37) leads to $\mathbb{E}\big[(\mathscr{M}_{k+j}^{x^h,\Delta} - \mathscr{M}_k^{x^h,\Delta})|\mathcal{F}_{t_k}\big] = 0$. This finishes the proof. □

Define the martingale difference sequence of $\{\mathscr{M}_k^{x^h,\Delta}\}_{k\in\mathbb{N}}$ by

$$\mathcal{Z}_k^{x^h,\Delta} := \mathscr{M}_k^{x^h,\Delta} - \mathscr{M}_{k-1}^{x^h,\Delta} \quad \forall\, k \in \mathbb{N}_+, \quad \mathcal{Z}_0^{x^h,\Delta} = 0.$$

Denote by $\mathbb{E}\big[|\mathcal{Z}_k^{x^h,\Delta}|^2|\mathcal{F}_{t_{k-1}}\big]$ the conditional variance of the martingale difference sequence, whose properties are presented in following propositions.

Proposition 4.9 *Let Assumption 4.2 hold for the time-homogeneous Markov process* $\{Y_{t_k}^{x^h,\Delta}\}_{k\in\mathbb{N}}$ *and* $\gamma \in [\gamma_2, 1]$.

(i) *For any constant c satisfying* $c\big(\frac{p\tilde{q}}{2} + (1+\kappa)\gamma\big) \le q$, *it holds that*

$$\sup_{k\ge 0} \mathbb{E}[|\mathcal{Z}_k^{x^h,\Delta}|^c] \le K\|f\|_{p,\gamma}^c(1 + \|x^h\|^{c\tilde{q}(\frac{p\tilde{q}}{2}+(1+\kappa)\gamma)}). \tag{4.38}$$

(ii) *Let* $p \ge 1$ *satisfy* $p\tilde{q} + 2(1+\kappa)\gamma \le q$. *Then* $\mathbb{E}[|\mathcal{Z}_{k+1}^{x^h,\Delta}|^2|\mathcal{F}_{t_k}] = H^\Delta(Y_{t_k}^{x^h,\Delta})$ *for all* $k \in \mathbb{N}$, *where* $H^\Delta : E^h \to \mathbb{R}$ *is defined as*

$$\begin{aligned} H^\Delta(u) = &-\tau^2|f(u) - \mu^\Delta(f)|^2 + \tau^2\mathbb{E}\Big[\Big|\sum_{i=0}^{\infty}\big(P_{t_i}^\Delta f(Y_{t_1}^{u,\Delta}) - \mu^\Delta(f)\big)\Big|^2\Big] \\ &- \tau^2\Big|\sum_{i=0}^{\infty}\big(P_{t_i}^\Delta f(u) - \mu^\Delta(f)\big)\Big|^2 \\ &+ 2\tau^2\big(f(u) - \mu^\Delta(f)\big)\sum_{i=0}^{\infty}\big(P_{t_i}^\Delta f(u) - \mu^\Delta(f)\big). \end{aligned} \tag{4.39}$$

(iii) *Let* $p \ge 1$ *satisfy* $2p\tilde{q}+2(1+2\kappa)\gamma \le q$. *Then* $H^\Delta \in C_{2\tilde{p}_\gamma,\gamma}$ *and* $\|H^\Delta\|_{2\tilde{p}_\gamma,\gamma} \le K\|f\|_{p,\gamma}^2$, *where* $\tilde{p}_\gamma := \tilde{q}\big(p\tilde{q} + (2+3\kappa)\gamma\big)$.

Proof

(i) It follows from (4.23), (4.24), and (4.36), Assumption 4.2, and the condition $\gamma \in [\gamma_2, 1]$ that

$$\begin{aligned} |\mathcal{Z}_k^{x^h,\Delta}| &\le \tau\big|f(Y_{t_{k-1}}^{x^h,\Delta}) - \mu^\Delta(f)\big| + \tau\sum_{i=0}^{\infty}\big|P_{t_i}^\Delta f(Y_{t_k}^{x^h,\Delta}) - \mu^\Delta(f)\big| \\ &\quad + \tau\sum_{i=0}^{\infty}\big|P_{t_i}^\Delta f(Y_{t_{k-1}}^{x^h,\Delta}) - \mu^\Delta(f)\big| \\ &\le K\|f\|_{p,\gamma}\tau\big(1 + \|Y_{t_{k-1}}^{x^h,\Delta}\|^{\frac{p}{2}}\big) + K\|f\|_{p,\gamma}\Big(1 + \|Y_{t_k}^{x^h,\Delta}\|^{\frac{p\tilde{q}}{2}+(1+\kappa)\gamma} \\ &\quad + \|Y_{t_{k-1}}^{x^h,\Delta}\|^{\frac{p\tilde{q}}{2}+(1+\kappa)\gamma}\Big). \end{aligned} \tag{4.40}$$

Taking any constant c satisfying $c\big(\frac{p\tilde{q}}{2} + (1+\kappa)\gamma\big) \le q$ and using Assumption 4.2 (i) finish the proof of (i).

(ii) The condition $p\tilde{q} + 2(1+\kappa)\gamma \le q$ coincides with the one in (i) with $c = 2$, thus $\mathbb{E}\big[|\mathcal{Z}_{l+1}^{x^h,\Delta}|^2|\mathcal{F}_{t_l}\big]$ is well-defined. It follows from (4.36) that

$$\mathbb{E}\big[|\mathcal{Z}_{l+1}^{x^h,\Delta}|^2|\mathcal{F}_{t_l}\big] = \mathbb{E}\Big[\Big|\tau\big(f(y^h) - \mu^\Delta(f)\big) + \tau\sum_{i=0}^{\infty}\big(P_{t_i}^\Delta f(Y_{t_1}^{y^h,\Delta}) - \mu^\Delta(f)\big) \\ - \tau\sum_{i=0}^{\infty}\big(P_{t_i}^\Delta f(y^h) - \mu^\Delta(f)\big)\Big|^2\Big]\Big|_{y^h = Y_{t_l}^{x^h,\Delta}} =: G^\Delta(Y_{t_l}^{x^h,\Delta}). \tag{4.41}$$

For any $u \in E^h$,

$$\begin{aligned} G^\Delta(u) &= \tau^2|f(u) - \mu^\Delta(f)|^2 + \tau^2\Big|\sum_{i=0}^{\infty}\big(P_{t_i}^\Delta f(u) - \mu^\Delta(f)\big)\Big|^2 \\ &+ \tau^2\mathbb{E}\Big[\Big|\sum_{i=0}^{\infty}\big(P_{t_i}^\Delta f(Y_{t_1}^{u,\Delta}) - \mu^\Delta(f)\big)\Big|^2\Big] \\ &+ 2\tau^2\big(f(u) - \mu^\Delta(f)\big)P_{t_1}^\Delta\Big(\sum_{i=0}^{\infty}\big(P_{t_i}^\Delta f - \mu^\Delta(f)\big)\Big)(u) \\ &- 2\tau^2\big(f(u) - \mu^\Delta(f)\big)\sum_{i=0}^{\infty}\big(P_{t_i}^\Delta f(u) - \mu^\Delta(f)\big) \\ &- 2\tau^2\sum_{i=0}^{\infty}\big(P_{t_i}^\Delta f(u) - \mu^\Delta(f)\big)P_{t_1}^\Delta\Big(\sum_{j=0}^{\infty}\big(P_{t_j}^\Delta f - \mu^\Delta(f)\big)\Big)(u). \end{aligned} \tag{4.42}$$

Applying the dominated convergence theorem yields

$$\begin{aligned} &\big(f(u) - \mu^\Delta(f)\big)P_{t_1}^\Delta\Big(\sum_{i=0}^{\infty}\big(P_{t_i}^\Delta f - \mu^\Delta(f)\big)\Big)(u) \\ &\quad = \big(f(u) - \mu^\Delta(f)\big)\sum_{i=0}^{\infty}\big(P_{t_i}^\Delta f(u) - \mu^\Delta(f)\big) - |f(u) - \mu^\Delta(f)|^2. \end{aligned} \tag{4.43}$$

Similarly,

$$\sum_{i=0}^{\infty}\big(P_{t_i}^{\Delta}f(u)-\mu^{\Delta}(f)\big)P_{t_1}^{\Delta}\Big(\sum_{i=0}^{\infty}\big(P_{t_i}^{\Delta}f-\mu^{\Delta}(f)\big)\Big)(u)$$
$$=\Big|\sum_{i=0}^{\infty}\big(P_{t_i}^{\Delta}f(u)-\mu^{\Delta}(f)\big)\Big|^2-\big(f(u)-\mu^{\Delta}(f)\big)\sum_{i=0}^{\infty}\big(P_{t_i}^{\Delta}f(u)-\mu^{\Delta}(f)\big). \tag{4.44}$$

Inserting (4.43) and (4.44) into (4.42) implies $G^{\Delta}=H^{\Delta}$, which finishes the proof of (ii).

(iii) For any $u\in E^h$, using (4.23), (4.24), the conditions $\gamma\in[\gamma_2,1]$ and $p\tilde{q}+2(1+\kappa)\gamma\le q$, and Assumption 4.2, we deduce

$$|H^{\Delta}(u)|\le K\|f\|_{p,\gamma}^2\tau^2(1+\|u\|^p)+K\|f\|_{p,\gamma}^2\big(1+\mathbb{E}[\|Y_{t_1}^{u,\Delta}\|^{p\tilde{q}+2(1+\kappa)\gamma}]\big)$$
$$\times\Big(\tau\sum_{i=0}^{\infty}\big(\rho^{\Delta}(t_i)\big)^{\gamma}\Big)^2$$
$$+K\|f\|_{p,\gamma}^2\big(1+\|u\|^{p\tilde{q}+2(1+\kappa)\gamma}\big)\Big(\tau\sum_{i=0}^{\infty}\big(\rho^{\Delta}(t_i)\big)^{\gamma}\Big)^2$$
$$+K\|f\|_{p,\gamma}^2\tau(1+\|u\|^{\frac{p}{2}})\big(1+\|u\|^{\frac{p}{2}(1+\tilde{q})+(1+\kappa)\gamma}\big)\Big(\tau\sum_{i=0}^{\infty}\big(\rho^{\Delta}(t_i)\big)^{\gamma}\Big)$$
$$\le K\|f\|_{p,\gamma}^2(1+\|u\|^{\tilde{q}(p\tilde{q}+2(1+\kappa)\gamma)})\le K\|f\|_{p,\gamma}^2(1+\|u\|^{\tilde{p}_{\gamma}}),$$

which implies

$$\sup_{u\in E^h}\frac{|H^{\Delta}(u)|}{1+\|u\|^{\tilde{p}_{\gamma}}}\le K\|f\|_{p,\gamma}^2. \tag{4.45}$$

By the definition of H^{Δ}, we deduce that for any $u_1,u_2\in E^h$,

$$H^{\Delta}(u_1)-H^{\Delta}(u_2)=\sum_{l=1}^{4}I_l(u_1,u_2), \tag{4.46}$$

where

$$I_1(u_1,u_2):=\ \tau^2\big(|f(u_2)-\mu^{\Delta}(f)|^2-|f(u_1)-\mu^{\Delta}(f)|^2\big),$$

$$
\begin{aligned}
I_2(u_1, u_2) :=\ & \tau^2\mathbb{E}\Big[\Big|\sum_{i=0}^{\infty}\big(P_{t_i}^{\Delta} f(Y_{t_1}^{u_1,\Delta}) - \mu^{\Delta}(f)\big)\Big|^2 \\
& - \Big|\sum_{i=0}^{\infty}\big(P_{t_i}^{\Delta} f(Y_{t_1}^{u_2,\Delta}) - \mu^{\Delta}(f)\big)\Big|^2\Big], \\
I_3(u_1, u_2) & \\
:=\ & -\tau^2\Big|\sum_{i=0}^{\infty}\big(P_{t_i}^{\Delta} f(u_1) - \mu^{\Delta}(f)\big)\Big|^2 + \tau^2\Big|\sum_{i=0}^{\infty}\big(P_{t_i}^{\Delta} f(u_2) - \mu^{\Delta}(f)\big)\Big|^2, \\
I_4(u_1, u_2) =\ & 2\tau^2\big(f(u_1) - \mu^{\Delta}(f)\big)\sum_{i=0}^{\infty}\big(P_{t_i}^{\Delta} f(u_1) - \mu^{\Delta}(f)\big) \\
& - 2\tau^2\big(f(u_2) - \mu^{\Delta}(f)\big)\sum_{i=0}^{\infty}\big(P_{t_i}^{\Delta} f(u_2) - \mu^{\Delta}(f)\big).
\end{aligned}
$$

By the definition of $C_{p,\gamma}$ and (4.23), we have

$$
\begin{aligned}
|I_1(u_1, u_2)| &= \tau^2|f(u_1) + f(u_2) - 2\mu^{\Delta}(f)||f(u_1) - f(u_2)| \\
&\leq K\|f\|_{p,\gamma}^2\tau^2(1 \wedge \|u_1 - u_2\|^{\gamma})(1 + \|u_1\|^{2p} + \|u_2\|^{2p})^{\frac{1}{2}}.
\end{aligned} \tag{4.47}
$$

It follows from (4.24) that

$$
\begin{aligned}
|I_3(u_1, u_2)| = \tau^2\Big|&\sum_{i=0}^{\infty}\big(P_{t_i}^{\Delta} f(u_1) - \mu^{\Delta}(f)\big) \\
&+ \sum_{i=0}^{\infty}\big(P_{t_i}^{\Delta} f(u_2) - \mu^{\Delta}(f)\big)\Big|\Big|\sum_{i=0}^{\infty}\big(P_{t_i}^{\Delta} f(u_2) - P_{t_i}^{\Delta} f(u_1)\big)\Big| \\
\leq\ & K\|f\|_{p,\gamma}^2(1 + \|u_1\|^{\frac{p\tilde{q}}{2}+(1+\kappa)\gamma} \\
&+ \|u_2\|^{\frac{p\tilde{q}}{2}+(1+\kappa)\gamma})(\tau\sum_{i=0}^{\infty}(\rho^{\Delta}(t_i))^{\gamma})^2 \\
&\times \|u_1 - u_2\|^{\gamma}(1 + \|u_1\|^{\frac{p\tilde{q}}{2}+\gamma\kappa} + \|u_2\|^{\frac{p\tilde{q}}{2}+\gamma\kappa}) \\
\leq\ & K\|f\|_{p,\gamma}^2\|u_1 - u_2\|^{\gamma}(1 + \|u_1\|^{p\tilde{q}+(1+2\kappa)\gamma} + \|u_2\|^{p\tilde{q}+(1+2\kappa)\gamma}),
\end{aligned} \tag{4.48}
$$

where we used

$$\begin{aligned}
&|P_{t_k}^{\Delta} f(u_1) - P_{t_k}^{\Delta} f(u_2)| \\
&= |\mu_{t_k}^{u_1,\Delta}(f) - \mu_{t_k}^{u_2,\Delta}(f)| \\
&\leq K\|f\|_{p,\gamma}\|u_1 - u_2\|^{\gamma}(1 + \|u_1\|^{\frac{p\tilde{q}}{2}+\gamma\kappa} \\
&\quad + \|u_2\|^{\frac{p\tilde{q}}{2}+\gamma\kappa})(\rho^{\Delta}(t_k))^{\gamma}, \ u_1, u_2 \in E^h.
\end{aligned}$$

Noting $I_2(u_1, u_2) = -\mathbb{E}\big[I_3(Y_{t_1}^{u_1,\Delta}, Y_{t_1}^{u_2,\Delta})\big]$, by (4.48), $2p\tilde{q}+2(1+2\kappa)\gamma \leq q$, the Hölder inequality, and Assumption 4.2, we obtain

$$\begin{aligned}
|I_2(u_1, u_2)| &\leq K\|f\|_{p,\gamma}^2 \\
&\quad \mathbb{E}\Big[\|Y_{t_1}^{u_1,\Delta} - Y_{t_1}^{u_2,\Delta}\|^{\gamma}(1 + \|Y_{t_1}^{u_1,\Delta}\|^{p\tilde{q}+(1+2\kappa)\gamma} + \|Y_{t_1}^{u_2,\Delta}\|^{p\tilde{q}+(1+2\kappa)\gamma})\Big] \\
&\leq K\|f\|_{p,\gamma}^2\|u_1 - u_2\|^{\gamma} \\
&\quad \big(1 + \|u_1\|^{\tilde{q}(p\tilde{q}+(1+2\kappa)\gamma)+\gamma\kappa} + \|u_2\|^{\tilde{q}(p\tilde{q}+(1+2\kappa)\gamma)+\gamma\kappa}\big). \qquad (4.49)
\end{aligned}$$

Arguments similar to those of I_1 and I_3 lead to

$$\begin{aligned}
|I_4(u_1, u_2)| &\leq 2\tau^2|f(u_1) - f(u_2)|\Big|\sum_{i=0}^{\infty}\big(P_{t_i}^{\Delta} f(u_1) - \mu^{\Delta}(f)\big)\Big| \\
&\quad + 2\tau^2|f(u_2) - \mu^{\Delta}(f)|\Big|\sum_{i=0}^{\infty}\big(P_{t_i}^{\Delta} f(u_1) - P_{t_i}^{\Delta} f(u_2)\big)\Big| \\
&\leq K\|f\|_{p,\gamma}^2\tau\|u_1 - u_2\|^{\gamma} \\
&\quad (1 + \|u_1\|^{\frac{p}{2}(1+\tilde{q})+(1+\kappa)\gamma} + \|u_2\|^{\frac{p}{2}(1+\tilde{q})+(1+\kappa)\gamma}). \qquad (4.50)
\end{aligned}$$

According to $\tilde{q} \geq 1$ and $p \geq 1$, one has $p\tilde{q}^2 \geq \frac{p}{2}(1+\tilde{q})$, which implies

$$\tilde{p}_{\gamma} := \tilde{q}\big(p\tilde{q} + (2 + 3\kappa)\gamma\big) \geq \frac{p}{2}(1 + \tilde{q}) + (1 + \kappa)\gamma.$$

Plugging (4.47)–(4.50) into (4.46) deduces

$$\sup_{\substack{u_1,u_2 \in E^h \\ u_1 \neq u_2}} \frac{|H^{\Delta}(u_1) - H^{\Delta}(u_2)|}{\|u_1 - u_2\|^{\gamma}(1 + \|u_1\|^{2\tilde{p}_{\gamma}} + \|u_2\|^{2\tilde{p}_{\gamma}})^{\frac{1}{2}}} \leq K\|f\|_{p,\gamma}^2.$$

This, along with (4.45) implies that

$$\begin{aligned}
&\sup_{\substack{u_1,u_2\in E^h\\ u_1\neq u_2}}\frac{|H^{\Delta}(u_1)-H^{\Delta}(u_2)|}{(1\wedge\|u_1-u_2\|^{\gamma})(1+\|u_1\|^{2\tilde{p}_\gamma}+\|u_2\|^{2\tilde{p}_\gamma})^{\frac{1}{2}}}\\
&\le K\|f\|_{p,\gamma}^2+\sup_{\substack{u_1,u_2\in E^h\\ u_1\neq u_2}}\frac{|H^{\Delta}(u_1)-H^{\Delta}(u_2)|\mathbf{1}_{\{\|u_1-u_2\|>1\}}}{(1+\|u_1\|^{2\tilde{p}_\gamma}+\|u_2\|^{2\tilde{p}_\gamma})^{\frac{1}{2}}}\\
&\le K\|f\|_{p,\gamma}^2+\sup_{\substack{u_1,u_2\in E^h\\ u_1\neq u_2}}\frac{|H^{\Delta}(u_1)|+|H^{\Delta}(u_2)|}{(1+\|u_1\|^{2\tilde{p}_\gamma}+\|u_2\|^{2\tilde{p}_\gamma})^{\frac{1}{2}}}\le K\|f\|_{p,\gamma}^2.
\end{aligned}$$

The proof is completed. □

Proposition 4.10 *Let Assumption 4.1 hold for the time-homogeneous Markov process* $\{X_t^x\}_{t\ge 0}$*, Assumption 4.2 hold for the time-homogeneous Markov process* $\{Y_{t_k}^{x^h,\Delta}\}_{k\in\mathbb{N}}$*, Assumption 4.3 hold for* $\{X_t^x\}_{t\ge 0}$ *and* $\{Y_{t_k}^{x^h,\Delta}\}_{k\in\mathbb{N}}$*,* $\gamma\in[\gamma_1\vee\gamma_2,1]$*, and* $p\ge 1$ *satisfy* $p(1+\tilde{q}\vee\tilde{r})+2(1+\kappa\vee\beta)\gamma\le q\wedge r$*. Then for any* $f\in C_{p,\gamma}$*,*

$$\mu^{\Delta}(H^{\Delta})=-\tau^2\mu^{\Delta}\big(|f-\mu^{\Delta}(f)|^2\big)+2\tau\mu^{\Delta}\Big(\big(f-\mu^{\Delta}(f)\big)\sum_{i=0}^{\infty}\big(P_{t_i}^{\Delta}f-\mu^{\Delta}(f)\big)\tau\Big). \tag{4.51}$$

Moreover, $|\mu^{\Delta}(H^{\Delta})|\le K\tau\|f\|_{p,\gamma}^2$ *and* $\lim_{|\Delta|\to 0}\frac{\mu^{\Delta}(H^{\Delta})}{\tau}=v^2$*.*

Proof *Step 1: Proof of* (4.51). The condition on p in Proposition 4.10 implies that the one in Proposition 4.9 (ii) holds, and then H^{Δ} is well-defined. Moreover, H^{Δ} can be rewritten as

$$\begin{aligned}
H^{\Delta}(u)&=-\tau^2|f(u)-\mu^{\Delta}(f)|^2+\tau^2P_{t_1}^{\Delta}\Big|\sum_{i=0}^{\infty}\big(P_{t_i}^{\Delta}f-\mu^{\Delta}(f)\big)\Big|^2(u)\\
&\quad-\tau^2\Big|\sum_{i=0}^{\infty}\big(P_{t_i}^{\Delta}f(u)-\mu^{\Delta}(f)\big)\Big|^2\\
&\quad+2\tau^2\big(f(u)-\mu^{\Delta}(f)\big)\sum_{i=0}^{\infty}\big(P_{t_i}^{\Delta}f(u)-\mu^{\Delta}(f)\big),
\end{aligned}$$

which together with $(P_{t_1}^\Delta)^*\mu^\Delta = \mu^\Delta$ gives

$$\mu^\Delta(H^\Delta) = -\tau^2\mu^\Delta(|f-\mu^\Delta(f)|^2) + 2\tau\mu^\Delta\Big((f-\mu^\Delta(f))\sum_{i=0}^{\infty}(P_{t_i}^\Delta f - \mu^\Delta(f))\tau\Big).$$

Here, $(P_t^\Delta)^*$ is the transpose operator of P_t^Δ; see [12, Section 5.2].

Step 2: Proof of $|\mu^\Delta(H^\Delta)| \le K\tau\|f\|_{p,\gamma}^2$. According to the Young inequality, we arrive at

$$|\mu^\Delta(H^\Delta)| \le 2\tau\mu^\Delta(|f-\mu^\Delta(f)|^2) + \tau\mu^\Delta\Big(\big(\sum_{i=0}^{\infty}(P_{t_i}^\Delta f - \mu^\Delta(f))\big)^2\tau^2\Big). \tag{4.52}$$

The fact $f \in C_{p,\gamma}$ implies that $f^2 \in C_{2p,\gamma}$ and $\|f^2\|_{2p,\gamma} \le K\|f\|_{p,\gamma}^2$. Together with (4.23), we derive

$$\begin{aligned}\mu^\Delta(|f-\mu^\Delta(f)|^2) &= \mu^\Delta(f^2) - |\mu^\Delta(f)|^2\\ &\le K\|f\|_{p,\gamma}^2\mu^\Delta(1+\|\cdot\|^p) + K\big(\|f\|_{p,\gamma}\mu^\Delta(1+\|\cdot\|^{\frac{p}{2}})\big)^2\\ &\le K\|f\|_{p,\gamma}^2. \end{aligned}\tag{4.53}$$

It follows from (4.24), $p\tilde{q} + 2(1+\kappa)\gamma \le q$, and (4.23) that

$$\mu^\Delta\Big(\big(\sum_{i=0}^{\infty}(P_{t_i}^\Delta f - \mu^\Delta(f))\big)^2\tau^2\Big) \le K\|f\|_{p,\gamma}^2\mu^\Delta(1+\|\cdot\|^{p\tilde{q}+2(1+\kappa)\gamma}) \le K\|f\|_{p,\gamma}^2. \tag{4.54}$$

Inserting (4.53) and (4.54) into (4.52) yields $|\mu^\Delta(H^\Delta)| \le K\tau\|f\|_{p,\gamma}^2$.

Step 3: Proof of $\lim_{|\Delta|\to 0}\frac{\mu^\Delta(H^\Delta)}{\tau} = v^2$. Without loss of generality, below we assume that $\mu(f) = 0$. Otherwise, we let $\tilde{f} := f - \mu(f)$ and consider $\tilde{f}$ instead of f. The condition on p in Proposition 4.10 leads to $\frac{p}{2}(1+\tilde{q}\vee\tilde{r}) + (1+\beta\vee\kappa)\gamma \le r$, which along with $f \in C_{p,\gamma}$, (4.3), and (4.24) implies that

$$\mu\Big(f\sum_{i=0}^{\infty}(P_{t_i}^\Delta f - \mu^\Delta(f))\tau\mathbf{1}_{\{E^h\}}\Big) < \infty \quad \text{and} \quad \mu\Big(f\int_0^\infty P_t f\,\mathrm{d}t\Big) < \infty.$$

Together with (4.51), we arrive at

$$\begin{aligned}&\Big|\frac{\mu^\Delta(H^\Delta)}{\tau} - v^2\Big|\\ &\le \tau\mu^\Delta(|f-\mu^\Delta(f)|^2) + 2|\mu^\Delta(f)|\mu^\Delta\Big(\sum_{i=0}^{\infty}|P_{t_i}^\Delta f - \mu^\Delta(f)|\tau\Big)\end{aligned}$$

$$
+2\Big|\mu^{\Delta}\Big(f\sum_{i=0}^{\infty}(P_{t_i}^{\Delta}f-\mu^{\Delta}(f))\tau\mathbf{1}_{\{E^h\}}\Big)-\mu\Big(f\sum_{i=0}^{\infty}(P_{t_i}^{\Delta}f-\mu^{\Delta}(f))\tau\mathbf{1}_{\{E^h\}}\Big)\Big|
$$

$$
+2\Big|\mu\Big(f\sum_{i=0}^{\infty}(P_{t_i}^{\Delta}f-\mu^{\Delta}(f))\tau\mathbf{1}_{\{E^h\}}\Big)-\mu\Big(f\int_0^{\infty}P_tf\mathrm{d}t\Big)\Big|=:\sum_{i=1}^{4}II_i.
$$

Due to (4.53), we have $II_1\to 0$ as $|\Delta|\to 0$. By (4.23), (4.24), and (4.26), we arrive at

$$
\begin{aligned}
II_2&\le K|\mu^{\Delta}(f)|\mu^{\Delta}\Big(\|f\|_{p,\gamma}(1+\|\cdot\|^{\frac{p\tilde{q}}{2}+(1+\kappa)\gamma})\sum_{i=0}^{\infty}(\rho^{\Delta}(t_i))^{\gamma}\tau\Big)\\
&\le K\|f\|_{p,\gamma}|\mu^{\Delta}(f)-\mu(f)|\le K\|f\|_{p,\gamma}^2|\Delta^{\alpha}|^{\gamma},
\end{aligned}
\tag{4.55}
$$

where we used $\mu(f)=0$. Denote

$$
F^{\Delta}:=f\sum_{i=0}^{\infty}(P_{t_i}^{\Delta}f-\mu^{\Delta}(f))\tau\mathbf{1}_{\{E^h\}}.
$$

Using similar techniques to the proof of Proposition 4.9, we obtain

$$
F^{\Delta}\in C_{p(1+\tilde{q})+2(1+\kappa)\gamma,\gamma}\quad\text{and}\quad\|F^{\Delta}\|_{p(1+\tilde{q})+2(1+\kappa)\gamma,\gamma}\le K\|f\|_{p,\gamma}^2.
$$

Then by $p(1+\tilde{q})+2(1+\kappa)\gamma\le r\wedge q$ and (4.26), we deduce

$$
II_3\le K\|F^{\Delta}\|_{p(1+\tilde{q})+2(1+\kappa)\gamma,\gamma}|\Delta^{\alpha}|^{\gamma}\le K\|f\|_{p,\gamma}^2|\Delta^{\alpha}|^{\gamma}.
\tag{4.56}
$$

For the term II_4, we have

$$
\begin{aligned}
II_4&\le 2\mu\Big(\mathbf{1}_{\{E^h\}}\int_0^{\infty}\big|f\big(P_t^{\Delta}f-P_tf\big)+f\big(\mu(f)-\mu^{\Delta}(f)\big)\big|\mathrm{d}t\Big)\\
&\quad+2\mu\Big(f\int_0^{\infty}P_tf\mathrm{d}t\mathbf{1}_{\{E\backslash E^h\}}\Big),
\end{aligned}
$$

where $E\backslash E^h$ represents the complementary set of E^h and we used $\mu(f)=0$. It follows from (4.2) and Assumptions 4.2 and 4.3 that for any $u\in E^h$,

$$
\begin{aligned}
|P_tf(u)-P_t^{\Delta}f(u)|&\le\|f\|_{p,\gamma}\big(1+\mathbb{E}[\|X_t^u\|^p]+\mathbb{E}[\|Y_t^u\|^p]\big)^{\frac{1}{2}}\big(\mathbb{W}_2(\mu_t^u,\mu_t^{u,\Delta})\big)^{\gamma}\\
&\le K\|f\|_{p,\gamma}(1+\|u\|^{(\frac{p}{2}+\gamma)(\tilde{r}\vee\tilde{q})})|\Delta^{\alpha}|^{\gamma}.
\end{aligned}
\tag{4.57}
$$

This, along with (4.26) implies that

$$\begin{aligned}
&\mathbf{1}_{\{u\in E^h\}}\big|f(u)\big(P_t^\Delta f(u)-P_tf(u)\big)+f(u)\big(\mu(f)-\mu^\Delta(f)\big)\big|\\
&\quad\le \mathbf{1}_{\{u\in E^h\}}|f(u)|\Big(\big|P_t^\Delta f(u)-P_tf(u)\big|+K\|f\|_{p,\gamma}|\Delta^\alpha|^\gamma\Big)\\
&\quad\le K\|f\|_{p,\gamma}^2(1+\|u\|^{(\frac{p}{2}+\gamma)(\tilde r\vee\tilde q)+\frac{p}{2}}\mathbf{1}_{\{u\in E^h\}})|\Delta^\alpha|^\gamma.
\end{aligned}$$

Applying the Fubini theorem and the Fatou lemma leads to

$$\begin{aligned}
&\overline{\lim_{|\Delta|\to0}}\,\mu\Big(\mathbf{1}_{\{E^h\}}\int_0^\infty\big|f\big(P_t^\Delta f-P_tf\big)-f\big(\mu^\Delta(f)-\mu(f)\big)\big|\mathrm{d}t\Big)\\
&\quad\le\int_0^\infty\mu\Big(\overline{\lim_{|\Delta|\to0}}\,\mathbf{1}_{\{E^h\}}\big|f\big(P_t^\Delta f-P_tf\big)+f\big(\mu(f)-\mu^\Delta(f)\big)\big|\Big)\mathrm{d}t=0,
\end{aligned}$$

which together with $\lim_{|\Delta|\to0}\mu\big(f\int_0^\infty P_tf\,\mathrm{d}t\mathbf{1}_{\{E\setminus E^h\}}\big)=0$ gives $II_4\to0$ as $|\Delta|\to0$. Combining estimates of terms II_i, $i=1,2,3,4$ completes the proof. □

With these results at hand, we are now in a position to prove Theorem 4.4, which follows directly from the combination of Propositions 4.11 and 4.12.

Proposition 4.11 *Under the conditions of Theorem 4.4, $\frac{1}{\sqrt{k\tau}}\mathscr{M}_k^{x^h,\Delta}$ converges in distribution to $\mathscr{N}(0,v^2)$ as $|\Delta|\to0$, where $k=\lceil\tau^{-2\lambda-1}\rceil$.*

Proof By the definition of k, it is equivalent to showing that $\tau^\lambda\mathscr{M}^{x^h,\Delta}_{\lceil\tau^{-2\lambda-1}\rceil}$ converges in distribution to $\mathscr{N}(0,v^2)$. To this end, we show that the characteristic function of $\tau^\lambda\mathscr{M}^{x^h,\Delta}_{\lceil\tau^{-2\lambda-1}\rceil}$ satisfies

$$\lim_{|\Delta|\to0}\mathbb{E}\big[\exp\big(\mathbf{i}\iota\tau^\lambda\mathscr{M}^{x^h,\Delta}_{\lceil\tau^{-2\lambda-1}\rceil}\big)\big]=e^{-\frac{v^2\iota^2}{2}}\quad\forall\,\iota\in\mathbb{R},\tag{4.58}$$

where $\mathbf{i}$ is the imaginary unit. Without loss of generality, we assume that $\mu(f)=0$. Otherwise, we let $\tilde f=f-\mu(f)$ and consider $\tilde f$ instead of f.

A direct calculation gives

$$\begin{aligned}
&e^{\frac{v^2\iota^2}{2}}\mathbb{E}\big[\exp\big(\mathbf{i}\iota\tau^\lambda\mathscr{M}_k^{x^h,\Delta}\big)\big]-1\\
&=\sum_{j=0}^{k-1}\Big(e^{\frac{v^2\iota^2(j+1)}{2k}}\mathbb{E}\big[\exp\big(\mathbf{i}\iota\tau^\lambda\mathscr{M}_{j+1}^{x^h,\Delta}\big)\big]-e^{\frac{v^2\iota^2j}{2k}}\mathbb{E}\big[\exp\big(\mathbf{i}\iota\tau^\lambda\mathscr{M}_j^{x^h,\Delta}\big)\big]\Big)\\
&=\sum_{j=0}^{k-1}e^{\frac{v^2\iota^2(j+1)}{2k}}\mathbb{E}\Big[\exp\big(\mathbf{i}\iota\tau^\lambda\mathscr{M}_j^{x^h,\Delta}\big)\Big(\exp\big(\mathbf{i}\iota\tau^\lambda\mathcal{Z}_{j+1}^{x^h,\Delta}\big)-1+\frac{v^2\iota^2}{2k}\Big)\Big]
\end{aligned}$$

$$+\sum_{j=0}^{k-1} e^{\frac{\upsilon^2\iota^2(j+1)}{2k}}(1-e^{-\frac{\upsilon^2\iota^2}{2k}}-\frac{\upsilon^2\iota^2}{2k})\mathbb{E}\big[\exp(\mathbf{i}\iota\tau^{\lambda}\mathscr{M}_j^{x^h,\Delta})\big]. \tag{4.59}$$

Note that for any $\zeta \in \mathbb{R}\setminus\{0\}$, $e^{\mathbf{i}\zeta} = 1+\mathbf{i}\zeta-\frac{\zeta^2}{2}-\zeta^2 Q(\zeta)$, where $Q(\zeta) = \zeta^{-2}\int_0^{\zeta}\int_0^{\zeta_2}(e^{\mathbf{i}\zeta_1}-1)\mathrm{d}\zeta_1\mathrm{d}\zeta_2$ satisfies $\sup_{\zeta\in\mathbb{R}}|Q(\zeta)|\le 1$, $|Q(\zeta)| = O(|\zeta|)$ for $\zeta \ll 1$. Along with $\mathbb{E}[\mathcal{Z}_{j+1}^{x^h,\Delta}|\mathcal{F}_{t_j}] = 0$, we obtain

$$\begin{aligned}
&\mathbb{E}\Big[\exp(\mathbf{i}\iota\tau^{\lambda}\mathscr{M}_j^{x^h,\Delta})\big(\exp(\mathbf{i}\iota\tau^{\lambda}\mathcal{Z}_{j+1}^{x^h,\Delta})-1+\frac{\upsilon^2\iota^2}{2k}\big)\Big]\\
&\quad=\mathbb{E}\Big[\exp(\mathbf{i}\iota\tau^{\lambda}\mathscr{M}_j^{x^h,\Delta})\mathbb{E}\Big[\Big(-\frac{\iota^2\tau^{2\lambda}(\mathcal{Z}_{j+1}^{x^h,\Delta})^2}{2}-\iota^2\tau^{2\lambda}(\mathcal{Z}_{j+1}^{x^h,\Delta})^2 Q_{j+1}+\frac{\upsilon^2\iota^2}{2k}\Big)\Big|\mathcal{F}_{t_j}\Big]\Big]\\
&\quad=\frac{\iota^2\tau^{2\lambda}}{2}\mathbb{E}\Big[\exp(\mathbf{i}\iota\tau^{\lambda}\mathscr{M}_j^{x^h,\Delta})\big(\frac{\upsilon^2}{k\tau^{2\lambda}}-(\mathcal{Z}_{j+1}^{x^h,\Delta})^2\big)\Big]\\
&\qquad-\iota^2\tau^{2\lambda}\mathbb{E}\Big[\exp(\mathbf{i}\iota\tau^{\lambda}\mathscr{M}_j^{x^h,\Delta})(\mathcal{Z}_{j+1}^{x^h,\Delta})^2 Q_{j+1}\Big],
\end{aligned} \tag{4.60}$$

where $Q_{j+1} := Q(\iota\tau^{\lambda}\mathcal{Z}_{j+1}^{x^h,\Delta})$. Plugging (4.60) into (4.59) yields

$$e^{\frac{\upsilon^2\iota^2}{2}}\mathbb{E}[\exp(\mathbf{i}\iota\tau^{\lambda}\mathscr{M}_k^{x^h,\Delta})]-1 = I_1(k)+I_2(k)+I_3(k), \tag{4.61}$$

where

$$\begin{aligned}
I_1(k) &:= \sum_{j=0}^{k-1} e^{\frac{\upsilon^2\iota^2(j+1)}{2k}}(1-e^{-\frac{\upsilon^2\iota^2}{2k}}-\frac{\upsilon^2\iota^2}{2k})\mathbb{E}[\exp(\mathbf{i}\iota\tau^{\lambda}\mathscr{M}_j^{x^h,\Delta})],\\
I_2(k) &:= -\iota^2\tau^{2\lambda}\sum_{j=0}^{k-1} e^{\frac{\upsilon^2\iota^2(j+1)}{2k}}\mathbb{E}\Big[\exp(\mathbf{i}\iota\tau^{\lambda}\mathscr{M}_j^{x^h,\Delta})(\mathcal{Z}_{j+1}^{x^h,\Delta})^2 Q_{j+1}\Big],\\
I_3(k) &:= \frac{\iota^2\tau^{2\lambda}}{2}\sum_{j=0}^{k-1} e^{\frac{\upsilon^2\iota^2(j+1)}{2k}}\mathbb{E}\Big[\exp(\mathbf{i}\iota\tau^{\lambda}\mathscr{M}_j^{x^h,\Delta})\big(\frac{\upsilon^2}{k\tau^{2\lambda}}-(\mathcal{Z}_{j+1}^{x^h,\Delta})^2\big)\Big].
\end{aligned}$$

Estimate of the Term $I_1(k)$ Applying the inequality $|1-e^{-\zeta}-\zeta|\le\frac{1}{2}\zeta^2$ for $\zeta>0$ gives

$$\begin{aligned}
\lim_{\tau\to 0}|I_1(k)| &\le \lim_{\tau\to 0} e^{\frac{\upsilon^2\iota^2}{2}}\sum_{j=0}^{k-1}|1-e^{-\frac{\upsilon^2\iota^2}{2k}}-\frac{\upsilon^2\iota^2}{2k}|\\
&\le \lim_{\tau\to 0} e^{\frac{\upsilon^2\iota^2}{2}}\sum_{j=0}^{k-1}\frac{\upsilon^4\iota^4}{8k^2}\le\lim_{\tau\to 0}\frac{K}{k}=0.
\end{aligned}$$

Estimate of the Term $I_2(k)$ It follows from $\sup_{\zeta\in\mathbb{R}}|Q(\zeta)|\le 1$ that for any $\epsilon>0$,

$$|I_2(k)|\le \iota^2 e^{\frac{\upsilon^2\iota^2}{2}}\tau^{2\lambda}\sum_{j=0}^{k-1}\mathbb{E}\big[(\mathcal{Z}_{j+1}^{x^h,\Delta})^2|Q_{j+1}|\big]\le I_{2,1}(k)+I_{2,2}(k), \tag{4.62}$$

where

$$I_{2,1}(k):=\iota^2 e^{\frac{\upsilon^2\iota^2}{2}}\tau^{2\lambda}\sum_{j=0}^{k-1}\mathbb{E}\big[(\mathcal{Z}_{j+1}^{x^h,\Delta})^2\mathbf{1}_{\{|\mathcal{Z}_{j+1}^{x^h,\Delta}|>\epsilon\tau^{-\lambda}\}}\big],$$

$$I_{2,2}(k):=\iota^2 e^{\frac{\upsilon^2\iota^2}{2}}\tau^{2\lambda}\sum_{j=0}^{k-1}\mathbb{E}\big[(\mathcal{Z}_{j+1}^{x^h,\Delta})^2|Q_{j+1}|\mathbf{1}_{\{|\mathcal{Z}_{j+1}^{x^h,\Delta}|\le\epsilon\tau^{-\lambda}\}}\big].$$

Using Propositions 4.9 and 4.10 leads to that for any $\varepsilon\in(0,\lambda)$,

$$\begin{aligned}I_{2,2}(k)&\le \iota^2 e^{\frac{\upsilon^2\iota^2}{2}}\tau^{2\lambda}\sup_{|\zeta|\le\iota\epsilon}|Q(\zeta)|\sum_{j=0}^{k-1}\mathbb{E}[(\mathcal{Z}_{j+1}^{x^h,\Delta})^2]\\&\le \iota^2 e^{\frac{\upsilon^2\iota^2}{2}}\sup_{|\zeta|\le\iota\epsilon}|Q(\zeta)|\Big[\tau^{2\lambda}\sum_{j=0}^{\lceil\tau^{-1-(\lambda-\varepsilon)}\rceil-1}\mathbb{E}[(\mathcal{Z}_{j+1}^{x^h,\Delta})^2]\\&\quad+\tau^{2\lambda}\sum_{j=\lceil\tau^{-1-(\lambda-\varepsilon)}\rceil}^{k-1}\big(P_{t_j}^{\Delta}H^{\Delta}(x^h)-\mu^{\Delta}(H^{\Delta})\big)+K\|f\|_{p,\gamma}^2\Big].\end{aligned} \tag{4.63}$$

Combining $\mathbb{E}\big[|\mathscr{M}^{x^h,\Delta}_{\lceil\tau^{-1-(\lambda-\varepsilon)}\rceil}|^2\big]=\sum_{j=0}^{\lceil\tau^{-1-(\lambda-\varepsilon)}\rceil-1}\mathbb{E}[(\mathcal{Z}_{j+1}^{x^h,\Delta})^2]$ and (4.34) yields that

$$\begin{aligned}\sum_{j=0}^{\lceil\tau^{-1-(\lambda-\varepsilon)}\rceil-1}\mathbb{E}[(\mathcal{Z}_{j+1}^{x^h,\Delta})^2]&\le K\tau^{1-(\lambda-\varepsilon)}\sum_{j=0}^{\lceil\tau^{-1-(\lambda-\varepsilon)}\rceil-1}\mathbb{E}\big[|f(Y_{t_j}^{x^h,\Delta})|^2+|\mu^{\Delta}(f)|^2\big]\\&\quad+K\|f\|_{p,\gamma}^2\mathbb{E}\Big[1+\|Y^{x^h,\Delta}_{t_{\lceil\tau^{-1-(\lambda-\varepsilon)}\rceil}}\|^{p\tilde{q}+2(1+\kappa)\gamma}+\|x^h\|^{p\tilde{q}+2(1+\kappa)\gamma}\Big].\end{aligned}$$

Since the condition (4.19) implies that $p\tilde{q}+2(1+\kappa)\gamma\le r$, by Assumption 4.2 (i) we obtain

$$\sum_{j=0}^{\lceil\tau^{-1-(\lambda-\varepsilon)}\rceil-1}\mathbb{E}[(\mathcal{Z}_{j+1}^{x^h,\Delta})^2]\le K\|f\|_{p,\gamma}^2(1+\|x^h\|^{\tilde{q}(p\tilde{q}+2(1+\kappa)\gamma)})\tau^{-2(\lambda-\varepsilon)}. \tag{4.64}$$

It follows from (4.19) and Proposition 4.9 (iii) that $H^\Delta \in C_{2\tilde{p}_\gamma,\gamma}$ with $2\tilde{p}_\gamma \le q$. Making use of (4.24) yields

$$\tau^{2\lambda} \sum_{j=\lceil \tau^{-1-(\lambda-\varepsilon)} \rceil}^{k-1} \left(P_{t_j}^\Delta H^\Delta(x^h) - \mu^\Delta(H^\Delta)\right)$$

$$\le K \|f\|_{p,\gamma}^2 (1 + \|x^h\|^{\tilde{q}\tilde{p}_\gamma + (1+\kappa)\gamma}) \tau^{2\lambda} \sum_{j=\lceil \tau^{-1-(\lambda-\varepsilon)} \rceil}^{k-1} \left(\rho^\Delta(t_j)\right)^\gamma. \tag{4.65}$$

Inserting (4.64) and (4.65) into (4.63) leads to

$$I_{2,2}(k) \le K\iota^2 e^{\frac{\upsilon^2\iota^2}{2}} \sup_{|\zeta| \le \iota\epsilon} |Q(\zeta)| \|f\|_{p,\gamma}^2 \left(1 + \|x^h\|^{\tilde{q}(p\tilde{q}+2(1+\kappa)\gamma)} + \|x^h\|^{\tilde{q}\tilde{p}_\gamma+(1+\kappa)\gamma}\right)$$

$$\times \left(\tau^{2\varepsilon} + \tau^{2\lambda} \sum_{j=\lceil \tau^{-1-(\lambda-\varepsilon)} \rceil}^{k-1} \left(\rho^\Delta(t_j)\right)^\gamma + 1\right). \tag{4.66}$$

Note that (4.20) implies $\lim_{\tau\to 0} \tau^{2\lambda} \sum_{j=\lceil \tau^{-1-(\lambda-\varepsilon)} \rceil}^{k-1} \left(\rho^\Delta(t_j)\right)^\gamma = 0$. This, along with (4.66) leads to

$$\lim_{\epsilon\to 0} \lim_{\tau\to 0} I_{2,2}(k) = 0. \tag{4.67}$$

By virtue of the condition (4.19) there exists a positive constant c such that $c > \frac{2}{\lambda} \vee 4$ and $c\left(\frac{p\tilde{q}}{2} + (1 + \frac{\kappa}{2})\gamma\right) \le q$. Applying the Hölder inequality, the Chebyshev inequality, and (4.38), we have

$$I_{2,1}(k) \le \iota^2 e^{\frac{\upsilon^2\iota^2}{2}} \tau^{2\lambda} \sum_{j=0}^{k-1} \left(\mathbb{E}[|\mathcal{Z}_{j+1}^{x^h,\Delta}|^4]\right)^{\frac{1}{2}} \left(\mathbb{P}\{|\mathcal{Z}_{j+1}^{x^h,\Delta}| > \epsilon\tau^{-\lambda}\}\right)^{\frac{1}{2}}$$

$$\le \iota^2 e^{\frac{\upsilon^2\iota^2}{2}} \frac{1}{\epsilon^{c/2}} \tau^{2\lambda+\frac{c\lambda}{2}} \sum_{j=0}^{k-1} \left(\mathbb{E}[|\mathcal{Z}_{j+1}^{x^h,\Delta}|^4]\right)^{\frac{1}{2}} \left(\mathbb{E}[|\mathcal{Z}_{j+1}^{x^h,\Delta}|^c]\right)^{\frac{1}{2}}$$

$$\le K\iota^2 e^{\frac{\upsilon^2\iota^2}{2}} \frac{1}{\epsilon^{c/2}} \tau^{\frac{c\lambda}{2}-1} \|f\|_{p,\gamma}^{\frac{4+c}{2}} (1 + \|x^h\|^{\frac{(4+c)\tilde{q}}{2}\left(\frac{p\tilde{q}}{2}+(1+\kappa)\gamma\right)}). \tag{4.68}$$

Plugging (4.67) and (4.68) into (4.62) yields

$$\lim_{\tau\to 0} |I_2(k)| \le \lim_{\epsilon\to 0} \lim_{\tau\to 0} K\iota^2 e^{\frac{\upsilon^2\iota^2}{2}} \frac{1}{\epsilon^{c/2}} \tau^{\frac{c\lambda}{2}-1} \|f\|_{p,\gamma}^{\frac{4+c}{2}} \left(1 + \|x^h\|^{\frac{(4+c)\tilde{q}}{2}\left(\frac{p\tilde{q}}{2}+(1+\kappa)\gamma\right)}\right)$$

$$+ \lim_{\epsilon\to 0} \lim_{\tau\to 0} |I_{2,2}(k)| = 0.$$

Estimate of the Term $I_3(k)$ To simplify the notation, we denote $\tilde{k} = \lceil \tau^{-1-\varepsilon} \rceil$ and $M = \lceil k/\tilde{k} \rceil$. Divide $\{0, 1, \ldots, k-1\}$ into M blocks, and denote

$$\mathbb{B}_i = \{(i-1)\tilde{k}, (i-1)\tilde{k}+1, \ldots, i\tilde{k}-1\} \text{ for } i \in \{1, \ldots, M-1\},$$
$$\mathbb{B}_M = \{(M-1)\tilde{k}, \ldots, k-1\}.$$

We estimate the term $I_3(k)$ as

$$|I_3(k)| \le I_{3,1}(k) + I_{3,2}(k), \tag{4.69}$$

where

$$I_{3,1}(k) := \Big|\frac{\iota^2\tau^{2\lambda}}{2}\sum_{i=1}^{M}\mathbb{E}\Big[\exp(\mathbf{i}\iota\tau^\lambda \mathscr{M}_{(i-1)\tilde{k}}^{x^h,\Delta})\sum_{j\in\mathbb{B}_i} e^{\frac{\upsilon^2\iota^2(j+1)}{2k}}\big(\frac{\upsilon^2}{k\tau^{2\lambda}} - (\mathcal{Z}_{j+1}^{x^h,\Delta})^2\big)\Big]\Big|,$$

$$I_{3,2}(k) := \Big|\frac{\iota^2\tau^{2\lambda}}{2}\sum_{i=1}^{M}\sum_{j\in\mathbb{B}_i}\mathbb{E}\Big[\big(\exp(\mathbf{i}\iota\tau^\lambda \mathscr{M}_j^{x^h,\Delta}) - \exp(\mathbf{i}\iota\tau^\lambda \mathscr{M}_{(i-1)\tilde{k}}^{x^h,\Delta})\big)e^{\frac{\upsilon^2\iota^2(j+1)}{2k}}$$
$$\times\big(\frac{\upsilon^2}{k\tau^{2\lambda}} - (\mathcal{Z}_{j+1}^{x^h,\Delta})^2\big)\Big]\Big|.$$

It follows from the property of the conditional expectation and the time-homogeneous Markov property of $\{\mathcal{Z}_k^{x^h,\Delta}\}_{k\in\mathbb{N}}$ that

$$\begin{aligned} I_{3,1}(k) &\le \frac{\iota^2\tau^{2\lambda}}{2}\sum_{i=1}^{M}\mathbb{E}\Big[\Big|\mathbb{E}\Big[\sum_{j\in\mathbb{B}_i} e^{\frac{\upsilon^2\iota^2(j+1)}{2k}}\big(\frac{\upsilon^2}{k\tau^{2\lambda}} - (\mathcal{Z}_{j+1}^{x^h,\Delta})^2\big)\Big|\mathcal{F}_{t_{(i-1)\tilde{k}}}\Big]\Big|\Big] \\ &\le \frac{\iota^2\tau^{2\lambda-\varepsilon}}{2}e^{\frac{\upsilon^2\iota^2}{2}}\sum_{i=1}^{M}\mathbb{E}\Big[\Big|\tau^\varepsilon\sum_{j\in\mathbb{B}_1}\Big|\frac{\upsilon^2}{k\tau^{2\lambda}} - \mathbb{E}[(\mathcal{Z}_{j+1}^{y^h,\Delta})^2]\Big|\Big|_{y^h = Y_{t_{(i-1)\tilde{k}}}^{x^h,\Delta}}\Big|\Big]. \end{aligned} \tag{4.70}$$

It is straightforward to see from Proposition 4.9 (ii) that

$$\begin{aligned} &\tau^\varepsilon\sum_{j\in\mathbb{B}_1}\Big|\frac{\upsilon^2}{k\tau^{2\lambda}} - \mathbb{E}[(\mathcal{Z}_{j+1}^{y^h,\Delta})^2]\Big| \\ &\le \tau^\varepsilon\sum_{j=0}^{k-1}\Big|\frac{\upsilon^2}{k\tau^{2\lambda}} - \mu^\Delta(H^\Delta)\Big| + \tau^\varepsilon\sum_{j=0}^{\tilde{k}-1}|\mu^\Delta(H^\Delta) - \mathbb{E}[(\mathcal{Z}_{j+1}^{y^h,\Delta})^2]| \end{aligned}$$

$$+\tau^{\varepsilon}\sum_{j=\lceil\tau^{-1-\frac{\varepsilon}{4}}\rceil}^{\tilde{k}-1}|\mu^{\Delta}(H^{\Delta})-\mathbb{E}[(\mathcal{Z}_{j+1}^{y^h,\Delta})^2]|$$

$$\leq \tau^{\varepsilon+1}\tilde{k}v^2|\frac{1}{k\tau^{2\lambda+1}}-1|+\tau^{\varepsilon+1}\tilde{k}|v^2-\frac{\mu^{\Delta}(H^{\Delta})}{\tau}|+\tau^{\varepsilon-1-\frac{\varepsilon}{4}}|\mu^{\Delta}(H^{\Delta})|$$

$$+\tau^{\varepsilon}\sum_{j=0}^{\lceil\tau^{-1-\frac{\varepsilon}{4}}\rceil-1}\mathbb{E}[(\mathcal{Z}_{j+1}^{y^h,\Delta})^2]+\tau^{\varepsilon}\sum_{j=\lceil\tau^{-1-\frac{\varepsilon}{4}}\rceil}^{\tilde{k}-1}|\mu^{\Delta}(H^{\Delta})-P_{t_j}^{\Delta}H^{\Delta}(y^h)|. \tag{4.71}$$

Using the similar arguments to (4.64) and (4.65), we derive

$$\sum_{j=0}^{\lceil\tau^{-1-\frac{\varepsilon}{4}}\rceil-1}\mathbb{E}[(\mathcal{Z}_{j+1}^{y^h,\Delta})^2]\leq K\|f\|_{p,\gamma}^2(1+\|y^h\|^{\tilde{q}(p\tilde{q}+2(1+\kappa)\gamma)})\tau^{-\frac{\varepsilon}{2}} \tag{4.72}$$

and

$$\tau^{\varepsilon}\sum_{j=\lceil\tau^{-1-\frac{\varepsilon}{4}}\rceil}^{\tilde{k}-1}|\mu^{\Delta}(H^{\Delta})-P_{t_j}^{\Delta}H^{\Delta}(y^h)|$$

$$\leq K\|f\|_{p,\gamma}^2(1+\|y^h\|^{\tilde{q}\tilde{p}_{\gamma}+(1+\kappa)\gamma})\tau^{\varepsilon}\sum_{j=\lceil\tau^{-1-\frac{\varepsilon}{4}}\rceil}^{\tilde{k}-1}\left(\rho^{\Delta}(t_j)\right)^{\gamma}. \tag{4.73}$$

Inserting (4.72) and (4.73) into (4.71) and using $|\mu^{\Delta}(H^{\Delta})|\leq K\tau\|f\|_{p,\gamma}^2$, we arrive at

$$\tau^{\varepsilon}\sum_{j\in\mathbb{B}_1}\left|\frac{v^2}{k\tau^{2\lambda}}-\mathbb{E}[(\mathcal{Z}_{j+1}^{y^h,\Delta})^2]\right|$$

$$\leq \tau^{\varepsilon+1}\tilde{k}v^2\left|\frac{1}{k\tau^{2\lambda+1}}-1\right|+\tau^{\varepsilon+1}\tilde{k}\left|v^2-\frac{\mu^{\Delta}(H^{\Delta})}{\tau}\right|+\tau^{\frac{3\varepsilon}{4}}K\|f\|_{p,\gamma}^2+K\|f\|_{p,\gamma}^2\tau^{\frac{\varepsilon}{2}}$$
$$\times(1+\|y^h\|^{\tilde{q}(p\tilde{q}+2(1+\kappa)\gamma)})+K\|f\|_{p,\gamma}^2(1+\|y^h\|^{\tilde{q}\tilde{p}_{\gamma}+(1+\kappa)\gamma})\tau^{\varepsilon}$$
$$\times\sum_{j=\lceil\tau^{-1-\frac{\varepsilon}{4}}\rceil}^{\tilde{k}-1}\left(\rho^{\Delta}(t_j)\right)^{\gamma}. \tag{4.74}$$

Since $\tilde{q}(p\tilde{q}+2(1+\kappa)\gamma) \le q$ and $\tilde{q}\tilde{p}_\gamma + (1+\kappa)\gamma \le q$, plugging (4.74) into (4.70) and applying Assumption 4.2 (i), we have

$$I_{3,1}(k) \le K\iota^2 e^{\frac{v^2\iota^2}{2}}\Big(v^2\Big|\frac{1}{k\tau^{2\lambda+1}} - 1\Big| + \Big|v^2 - \frac{\mu^\Delta(H^\Delta)}{\tau}\Big| + \tau^{\frac{3\varepsilon}{4}}\|f\|^2_{p,\gamma}$$
$$+ \|f\|^2_{p,\gamma}(1+\|x^h\|^{\tilde{q}^2(p\tilde{q}+2(1+\kappa)\gamma)})\tau^{\frac{\varepsilon}{2}}$$
$$+ \|f\|^2_{p,\gamma}(1+\|x^h\|^{\tilde{q}(q\tilde{p}_\gamma+(1+\kappa)\gamma)})\tau^{\varepsilon}\sum_{j=\lceil\tau^{-1-\frac{\varepsilon}{4}}\rceil}^{\tilde{k}-1}\big(\rho^\Delta(t_j)\big)^\gamma\Big),$$

where we used $k = \lceil\tau^{-2\lambda-1}\rceil$, $\tilde{k} = \lceil\tau^{-1-\varepsilon}\rceil$, and $M = \lceil k/\tilde{k}\rceil = O(\tau^{-2\lambda+\varepsilon})$. It follows from (4.20) that $\lim_{\tau\to 0}\tau^\varepsilon\sum_{j=\lceil\tau^{-1-\frac{\varepsilon}{4}}\rceil}^{\tilde{k}-1}\big(\rho^\Delta(t_j)\big)^\gamma = 0$, This, along with $p(1+\tilde{q}\vee\tilde{r}) + 2(1+\kappa\vee\beta)\gamma \le q\wedge r$ and Proposition 4.10 implies that

$$\lim_{|\Delta|\to 0} I_{3,1}(k) = 0. \tag{4.75}$$

Furthermore, by $|e^{\mathbf{i}u}-1| \le u$ and the Hölder inequality, one has that for any $\varepsilon_1 > 0$,

$$I_{3,2}(k) \le \frac{\iota^2\tau^{2\lambda}}{2}e^{\frac{v^2\iota^2}{2}}\sum_{i=1}^{M}\sum_{j\in\mathbb{B}_i}\mathbb{E}\Big[\Big|\exp\big(\mathbf{i}\iota\tau^\lambda(\mathscr{M}_j^{x^h,\Delta} - \mathscr{M}_{(i-1)\tilde{k}}^{x^h,\Delta})\big) - 1\Big|$$
$$\times\big(\frac{v^2}{k\tau^{2\lambda}} + (\mathcal{Z}_{j+1}^{x^h,\Delta})^2\big)\Big]$$
$$\le \varepsilon_1\iota\frac{\iota^2\tau^{2\lambda}}{2}e^{\frac{v^2\iota^2}{2}}\sum_{i=1}^{M}\sum_{j\in\mathbb{B}_i}\big(\frac{v^2}{k\tau^{2\lambda}} + \mathbb{E}[(\mathcal{Z}_{j+1}^{x^h,\Delta})^2]\big)$$
$$+ \iota^2\tau^{2\lambda}e^{\frac{v^2\iota^2}{2}}\sum_{i=1}^{M}\sum_{j\in\mathbb{B}_i}\mathbb{E}\Big[\big(\frac{v^2}{k\tau^{2\lambda}} + (\mathcal{Z}_{j+1}^{x^h,\Delta})^2\big)\mathbf{1}_{\{\tau^\lambda|\mathscr{M}_j^{x^h,\Delta} - \mathscr{M}_{(i-1)\tilde{k}}^{x^h,\Delta}| > \varepsilon_1\}}\Big]$$
$$\le K\varepsilon_1\iota^3 e^{\frac{v^2\iota^2}{2}} + \varepsilon_1\frac{\iota^3\tau^{2\lambda}}{2}e^{\frac{v^2\iota^2}{2}}\sum_{j=0}^{k-1}\mathbb{E}\big[H^\Delta(Y_{t_j}^{x^h,\Delta})\big]$$
$$+ K\iota^2\tau^{2\lambda}e^{\frac{v^2\iota^2}{2}}\sum_{i=1}^{M}\sum_{j\in\mathbb{B}_i}\big(v^4 + \mathbb{E}[(\mathcal{Z}_{j+1}^{x^h,\Delta})^4]\big)^{\frac{1}{2}}$$
$$\big(\mathbb{P}\{\tau^\lambda|\mathscr{M}_j^{x^h,\Delta} - \mathscr{M}_{(i-1)\tilde{k}}^{x^h,\Delta}| > \varepsilon_1\}\big)^{\frac{1}{2}}. \tag{4.76}$$

It follows from (4.24) and the definition of $\mathscr{M}_i^{x^h,\Delta}$ that for any $j \in \mathbb{B}_i \setminus \{(i-1)\tilde{k}\}$,

$$|\mathscr{M}_j^{x^h,\Delta} - \mathscr{M}_{(i-1)\tilde{k}}^{x^h,\Delta}| \leq \tau \Big| \sum_{l=(i-1)\tilde{k}}^{j-1} \big(f(Y_{t_l}^{x^h,\Delta}) - \mu^\Delta(f)\big)\Big| + K\|f\|_{p,\gamma}(1+\|Y_{t_j}^{x^h,\Delta}\|^{\frac{p\tilde{q}}{2}+(1+\kappa)\gamma} + \|Y_{t_{(i-1)\tilde{k}}}^{x^h,\Delta}\|^{\frac{p\tilde{q}}{2}+(1+\kappa)\gamma}).$$

Choose $\varepsilon \ll 1$ such that $\lceil \frac{1+\sqrt{\varepsilon}}{\lambda} \rceil \leq \frac{1}{\lambda} + 1$. This, along with the condition (4.19) implies that

$$2\lceil 3 \vee \frac{1+\sqrt{\varepsilon}}{\lambda} \rceil \left(\frac{p\tilde{q}}{2} + (1+\kappa)\gamma\right) \leq q.$$

Combining the Chebyshev inequality and Assumption 4.2 (i) yields

$$\begin{aligned}
&\mathbb{P}\{\tau^\lambda |\mathscr{M}_j^{x^h,\Delta} - \mathscr{M}_{(i-1)\tilde{k}}^{x^h,\Delta}| > \varepsilon_1\} \\
&\leq K\Big(\frac{\tau^{\lambda+1}}{\varepsilon_1}\Big)^{2\lceil 3\vee\frac{1+\sqrt{\varepsilon}}{\lambda}\rceil} \\
&\quad \times \mathbb{E}\Big[\mathbb{E}\Big[\Big|\sum_{l=0}^{j-(i-1)\tilde{k}-1} \big(f(Y_{t_l}^{y^h,\Delta}) - \mu^\Delta(f)\big)\Big|^{2\lceil 3\vee\frac{1+\sqrt{\varepsilon}}{\lambda}\rceil}\Big]\Big|_{y^h = Y_{t_{(i-1)\tilde{k}}}^{x^h,\Delta}}\Big] \\
&\quad + K\|f\|_{p,\gamma}^{2\lceil 3\vee\frac{1+\sqrt{\varepsilon}}{\lambda}\rceil}\Big(\frac{\tau^\lambda}{\varepsilon_1}\Big)^{2\lceil 3\vee\frac{1+\sqrt{\varepsilon}}{\lambda}\rceil}\Big(1 + \|x^h\|^{2\tilde{q}\lceil 3\vee\frac{1+\sqrt{\varepsilon}}{\lambda}\rceil\left(\frac{p\tilde{q}}{2}+(1+\kappa)\gamma\right)}\Big).
\end{aligned} \tag{4.77}$$

Note that condition (4.19) implies that the requirement on p in Lemma 4.7 is satisfied with $\varrho = \lceil 3 \vee \frac{1+\sqrt{\varepsilon}}{\lambda} \rceil$. By virtue of Lemma 4.7 with $\varrho = \lceil 3 \vee \frac{1+\sqrt{\varepsilon}}{\lambda} \rceil$, we obtain

$$\begin{aligned}
&\mathbb{E}\Big[\Big|\sum_{l=0}^{j-(i-1)\tilde{k}-1} \big(f(Y_{t_l}^{y^h,\Delta}) - \mu^\Delta(f)\big)\Big|^{2\lceil 3\vee\frac{1+\sqrt{\varepsilon}}{\lambda}\rceil}\Big] \\
&\leq K\Big(\|f\|_{p,\gamma}^2 (j-(i-1)\tilde{k})^2 t_{j-(i-1)\tilde{k}}^{-1}\Big)^{\lceil 3\vee\frac{1+\sqrt{\varepsilon}}{\lambda}\rceil}\Big(1 + \|y^h\|^{\lceil 3\vee\frac{1+\sqrt{\varepsilon}}{\lambda}\rceil\tilde{q}\left(\frac{p}{2}(1+\tilde{q})+(1+\kappa)\gamma\right)}\Big).
\end{aligned}$$

This, along with (4.77), $\tilde{q}\lceil 3 \vee \frac{1+\sqrt{\varepsilon}}{\lambda}\rceil \left(\frac{p}{2}(1+\tilde{q}) + (1+\kappa)\gamma\right) \le q$, and Assumption 4.2 (i) implies that

$$\sum_{j\in\mathbb{B}_i} \left(\mathbb{P}\{\tau^{\lambda}|\mathscr{M}_j^{x^h,\Delta} - \mathscr{M}_{(i-1)\tilde{k}}^{x^h,\Delta}| > \varepsilon_1\}\right)^{\frac{1}{2}}$$

$$\le K\|f\|_{p,\gamma}^{\lceil 3\vee\frac{1+\sqrt{\varepsilon}}{\lambda}\rceil} \sum_{j\in\mathbb{B}_i} \left(\frac{\tau^{\lambda+1}(j-(i-1)\tilde{k})}{\varepsilon_1}\right)^{\lceil 3\vee\frac{1+\sqrt{\varepsilon}}{\lambda}\rceil}$$

$$\times \left(\mathbb{E}[\|Y_{t_{(i-1)\tilde{k}}}^{x^h,\Delta}\|^{\tilde{q}\lceil 3\vee\frac{1+\sqrt{\varepsilon}}{\lambda}\rceil(\frac{p}{2}(1+\tilde{q})+(1+\kappa)\gamma)}] + 1\right)^{\frac{1}{2}} t_{j-(i-1)\tilde{k}}^{-\lceil 3\vee\frac{1+\sqrt{\varepsilon}}{\lambda}\rceil/2}$$

$$+ K\|f\|_{p,\gamma}^{\lceil 3\vee\frac{1+\sqrt{\varepsilon}}{\lambda}\rceil} \left(\frac{\tau^{\lambda}}{\varepsilon_1}\right)^{\lceil 3\vee\frac{1+\sqrt{\varepsilon}}{\lambda}\rceil} \left(1+\|x^h\|^{\tilde{q}\lceil 3\vee\frac{1+\sqrt{\varepsilon}}{\lambda}\rceil\left(\frac{p\tilde{q}}{2}+(1+\kappa)\gamma\right)}\right)$$

$$\le K\|f\|_{p,\gamma}^{\lceil 3\vee\frac{1+\sqrt{\varepsilon}}{\lambda}\rceil} \left(1+\|x^h\|^{\lceil 3\vee\frac{1+\sqrt{\varepsilon}}{\lambda}\rceil\frac{\tilde{q}^2}{2}(\frac{p}{2}(1+\tilde{q})+(1+\kappa)\gamma)}\right)$$

$$\times \sum_{j\in\mathbb{B}_1} \left(\frac{\tau^{\lambda+1}j}{\varepsilon_1}\right)^{\lceil 3\vee\frac{1+\sqrt{\varepsilon}}{\lambda}\rceil} t_j^{-\lceil 3\vee\frac{1+\sqrt{\varepsilon}}{\lambda}\rceil/2}$$

$$+ K\|f\|_{p,\gamma}^{\lceil 3\vee\frac{1+\sqrt{\varepsilon}}{\lambda}\rceil} \left(\frac{\tau^{\lambda}}{\varepsilon_1}\right)^{\lceil 3\vee\frac{1+\sqrt{\varepsilon}}{\lambda}\rceil} \left(1+\|x^h\|^{\tilde{q}\lceil 3\vee\frac{1+\sqrt{\varepsilon}}{\lambda}\rceil\left(\frac{p\tilde{q}}{2}+(1+\kappa)\gamma\right)}\right)$$

$$\le K\|f\|_{p,\gamma}^{\lceil 3\vee\frac{1+\sqrt{\varepsilon}}{\lambda}\rceil} \left(1+\|x^h\|^{\lceil 3\vee\frac{1+\sqrt{\varepsilon}}{\lambda}\rceil\tilde{q}^2\left(\frac{p\tilde{q}}{2}+(1+\kappa)\gamma\right)}\right)$$

$$\times \left(\left(\frac{\tau^{\lambda}}{\varepsilon_1}\right)^{\lceil 3\vee\frac{1+\sqrt{\varepsilon}}{\lambda}\rceil} \sum_{j=0}^{\lceil\tau^{-1}\rceil-1} t_j^{\lceil 3\vee\frac{1+\sqrt{\varepsilon}}{\lambda}\rceil/2}\right.$$

$$\left.+ \left(\frac{\tau^{\lambda-\varepsilon}}{\varepsilon_1}\right)^{\lceil 3\vee\frac{1+\sqrt{\varepsilon}}{\lambda}\rceil} \sum_{j=\lceil\tau^{-1}\rceil}^{\tilde{k}-1} \left(\frac{j}{\tau^{-1-\varepsilon}}\right)^{\lceil 3\vee\frac{1+\sqrt{\varepsilon}}{\lambda}\rceil} t_j^{-\lceil 3\vee\frac{1+\sqrt{\varepsilon}}{\lambda}\rceil/2} + \left(\frac{\tau^{\lambda}}{\varepsilon_1}\right)^{\lceil 3\vee\frac{1+\sqrt{\varepsilon}}{\lambda}\rceil}\right)$$

$$\le \frac{1}{\varepsilon_1^{\lceil 3\vee\frac{1+\sqrt{\varepsilon}}{\lambda}\rceil}} K\|f\|_{p,\gamma}^{\lceil 3\vee\frac{1+\sqrt{\varepsilon}}{\lambda}\rceil} \left(1+\|x^h\|^{\lceil 3\vee\frac{1+\sqrt{\varepsilon}}{\lambda}\rceil\tilde{q}^2\left(\frac{p\tilde{q}}{2}+(1+\kappa)\gamma\right)}\right)$$

$$(1+\tau^{\frac{(\lambda-\varepsilon)(1+\sqrt{\varepsilon})}{\lambda}-1}+\tau).$$

Inserting the above inequality into (4.76) leads to

$$I_{3,2}(k) \le K\varepsilon_1\iota^3 e^{\frac{\upsilon^2\iota^2}{2}} + \varepsilon_1\frac{\iota^3\tau^{2\lambda}}{2}e^{\frac{\upsilon^2\iota^2}{2}}\sum_{j=0}^{k-1}\mathbb{E}\big[H^{\Delta}(Y_{t_j}^{x^h,\Delta})\big]$$

$$+ K\iota^2 e^{\frac{\upsilon^2\iota^2}{2}}\big(\upsilon^4 + \sup_{n\geq 0}\mathbb{E}[(\mathcal{Z}_n^{x^h,\Delta})^4]\big)^{\frac{1}{2}}\frac{1}{\varepsilon_1^{\lceil 3\vee\frac{1+\sqrt{\varepsilon}}{\lambda}\rceil}}\|f\|_{p,\gamma}^{\lceil 3\vee\frac{1+\sqrt{\varepsilon}}{\lambda}\rceil}$$

$$\times (1 + \|x^h\|^{\lceil 3\vee\frac{1+\sqrt{\varepsilon}}{\lambda}\rceil\tilde{q}^2\left(\frac{p\tilde{q}}{2}+(1+\kappa)\gamma\right)})(1 + \tau^{\sqrt{\varepsilon}-\frac{\varepsilon}{\lambda}-\frac{\varepsilon\sqrt{\varepsilon}}{\lambda}} + \tau)M\tau^{2\lambda}.$$

Recalling $M = O(\tau^{-2\lambda+\varepsilon})$, letting $\varepsilon \in (0,1)$ satisfy $\sqrt{\varepsilon} - \frac{\varepsilon}{\lambda} - \frac{\varepsilon\sqrt{\varepsilon}}{\lambda} > 0$, and using (4.38) and (4.63)–(4.65) yield

$$\lim_{\varepsilon_1\to 0}\lim_{\tau\to 0} I_{3,2}(k) = 0. \tag{4.78}$$

We conclude from (4.69), (4.75), and (4.78) that $\lim_{|\Delta|\to 0}|I_3(k)| = 0$. The desired argument (4.58) follows from estimates of terms I_i, $i = 1, 2, 3$. □

Proposition 4.12 *Under the conditions of Theorem 4.4,* $\frac{1}{\sqrt{k\tau}}\mathscr{R}_k^{x^h,\Delta}$ *converges in probability to* 0 *as* $|\Delta| \to 0$, *where* $k = \lceil \tau^{-2\lambda-1}\rceil$.

Proof By the definition of k, it is equivalent to showing that $\tau^{\lambda}\mathscr{R}_{\lceil\tau^{-2\lambda-1}\rceil}^{x^h,\Delta}$ converges in probability to 0. Let $\mathscr{R}_k^{x^h,\Delta} = \mathscr{R}_{k,1}^{x^h,\Delta} + \mathscr{R}_{k,2}^{x^h,\Delta}$, where

$$\mathscr{R}_{k,1}^{x^h,\Delta} := -\tau\sum_{i=k}^{\infty}\Big(\mathbb{E}[f(Y_{t_i}^{x^h,\Delta})|\mathcal{F}_{t_k}] - \mu^{\Delta}(f)\Big) + \tau\sum_{i=0}^{\infty}\Big(\mathbb{E}[f(Y_{t_i}^{x^h,\Delta})|\mathcal{F}_0] - \mu^{\Delta}(f)\Big)$$

and $\mathscr{R}_{k,2}^{x^h,\Delta} := k\tau(\mu^{\Delta}(f) - \mu(f))$. By (4.24) we deduce

$$\begin{aligned}|\mathscr{R}_{k,1}^{x^h,\Delta}| &\leq \tau\sum_{i=0}^{\infty}\big|P_{t_i}^{\Delta}f(Y_{t_k}^{x^h,\Delta}) - \mu^{\Delta}(f)\big| + \tau\sum_{i=0}^{\infty}\big|P_{t_i}^{\Delta}f(x^h) - \mu^{\Delta}(f)\big| \\ &\leq K\|f\|_{p,\gamma}(1 + \|Y_{t_k}^{x^h,\Delta}\|^{\frac{p\tilde{q}}{2}+(1+\kappa)\gamma} + \|x^h\|^{\frac{p\tilde{q}}{2}+(1+\kappa)\gamma}).\end{aligned}$$

Taking expectations on both sides of $|\mathscr{R}_{k,1}^{x^h,\Delta}|$ and using Assumption 4.2 (i), we derive

$$\mathbb{E}[|\mathscr{R}_{k,1}^{x^h,\Delta}|] \leq K\|f\|_{p,\gamma}(1 + \|x^h\|^{\tilde{q}(\frac{p\tilde{q}}{2}+(1+\kappa)\gamma)}).$$

This, along with $\frac{p\tilde{q}}{2} + (1+\kappa)\gamma \leq q$ and Assumption 4.2 (i) leads to that

$$\lim_{|\Delta|\to 0}\frac{1}{\sqrt{k\tau}}\mathbb{E}[|\mathscr{R}_{k,1}^{x^h,\Delta}|] \leq \lim_{|\Delta|\to 0}K\|f\|_{p,\gamma}(1 + \|x^h\|^{\tilde{q}(\frac{p\tilde{q}}{2}+(1+\kappa)\gamma)})\tau^{\lambda} = 0. \tag{4.79}$$

It follows from (4.26) that

$$\frac{1}{\sqrt{k\tau}}|\mathscr{R}_{k,2}^{x^h,\Delta}| = \sqrt{k\tau}\left|\mu^{\Delta}(f) - \mu(f)\right| \le K\|f\|_{p,\gamma}\sqrt{k\tau}(h^{\alpha_1\gamma} + \tau^{\alpha_2\gamma}).$$

Since $h = \tau^{\alpha_2/\alpha_1}$ for $h \neq 0$, one has

$$K\|f\|_{p,\gamma}\sqrt{k\tau}(h^{\alpha_1\gamma} + \tau^{\alpha_2\gamma}) = K\|f\|_{p,\gamma}\tau^{\alpha_2\gamma-\lambda},$$

which tends to 0 as $\tau \to 0$ by letting $\lambda \in (0, \alpha_2\gamma)$. This finishes the proof. □

4.4 Numerical Experiments

From Theorem 1.16 we know that when the uniformly moment boundedness and the attractiveness of the functional solution hold, the functional solution of the SFDE (1.1) has the SLLN and the CLT. This section is devoted to the numerical verification of the SLLN and the CLT for numerical solutions of SFDEs. We propose algorithms designed to capture the longtime statistical properties for time-averages of numerical solutions. Then we apply the algorithms to representative SFDE models, with θ-EM methods applied to discretize the equations, to verify the limit behaviors of the functional solution.

We begin by providing an algorithm (Algorithm 3) to verify the numerical SLLN for SFDEs using the θ-EM method. The main objective is to demonstrate that, regardless of the initial condition, the time-average of a test function f along the numerical functional solution converges to the same deterministic value as time increases. The algorithm starts with initialization, including model parameters and simulation parameters. To ensure a fair comparison across different initial conditions, we generate a single sequence of Brownian increments and use it for all trajectories associated with these initial conditions. This guarantees that any differences in the time-averages are solely due to the initial conditions and not to stochastic variability. Then, for each initial condition ξ_j, the SFDE (1.1) is numerically approximated by using the θ-EM method. And at each time step t_k, the observable $f(y_{t_k}^{\Delta,\xi_j})$ is computed and the running time-average $A_t^{(j)} := \frac{1}{n}\sum_{k=0}^{n-1} f(y_{t_k}^{\Delta,\xi_j})$ is updated, where $t = n\Delta$. The time-averages $A_t^{(j)}$ for all initial conditions are finally plotted as functions of time.

Then we present a Monte Carlo algorithm for the verification of numerical CLT for SFDEs (1.1). The objective is to generate an empirical distribution of the normalized time-average of numerical solutions and compare it with the theoretical Gaussian limit. The overall simulation framework, summarized in Algorithm 4, begins by estimating the stationary mean $\mu(f)$, which is generally unknown. Invoking the ergodicity and the SLLN of the process, we approximate $\mu(f)$ by computing the time-average of the reference trajectory over a longtime T_{long}, where

Algorithm 3 Verification of numerical SLLN for SFDEs

Require: SFDE parameters (b, σ, τ), a set of initial conditions $\{\xi_1, \xi_2, \dots, \xi_M\}$, parameter θ, sufficiently large final time T, stepsizes Δ, $N = T/\Delta$, test function f.

Output: Numerical time-average $A_t^{(j)} = \frac{1}{n}\sum_{k=0}^{n-1} f(y_{t_k}^{\Delta,\xi_j})$ for each ξ_j with $j = 1, \dots, M$ and $t = n\Delta$ with $n = 1, 2, \dots, N$.

Generate a common Brownian motion increment sequence $\{\delta W_k\}_{k=0}^{N-1}$ to ensure identical noise for all initial conditions.

for $j = 1, \dots, M$ **do**

  Initialize $y_0^{\Delta,\xi_j} = \xi_j^{Int}$ with ξ_j^{Int} denoting the linear interpolation of ξ_j.

  for $k = 0, \dots, N-1$ **do**

$$y^{\Delta,\xi_j}(t_{k+1}) = y^{\Delta,\xi_j}(t_k) + \big(\theta b(y_{t_{k+1}}^{\Delta,\xi_j}) + (1-\theta) b(y_{t_k}^{\Delta,\xi_j})\big)\Delta + \sigma(y_{t_k}^{\Delta,\xi_j})\delta W_k.$$

  end for

  Compute the running time-average $A_t^{(j)} = \frac{1}{n}\sum_{k=0}^{n-1} f(y_{t_k}^{\Delta,\xi_j})$ for $t = n\Delta$ with $n = 1, 2, \dots, N$.

end for

Comparison:

Plot $A_t^{(j)}$ versus t for all j.

Examine whether all numerical time-averages $A_t^{(j)}$ converge to the same limit as t increases, indicating the validity of the numerical SLLN.

the evolution of the system at each step is performed using the θ-EM method. We collect the values of the observation function $f(x_t)$ on the time interval $[0, T_{\text{long}}]$. The time-average of these values yields a global estimate $\hat{\mu}_f^{\text{global}}$, which serves as a reference for the true mean $\mu(f)$ in the subsequent steps.

The second step constitutes the main Monte Carlo simulation. We execute M_{trials} independent trials, which can be performed in parallel to improve computational efficiency. For each trial $n = 1, \dots, M_{\text{trials}}$, we first initialize the system with the same initial datum. Then, we simulate the process by using the θ-EM method over time $T_{\max}$ (we suppose $T_{\text{long}} > T_{\max}$) and compute the sample mean $\hat{\mu}_f^{(n)}$ for this path. Finally, we compute the normalized CLT statistic $S_n = \sqrt{T_{\max}}(\hat{\mu}_f^{(n)} - \hat{\mu}_f^{\text{global}})$, and store it. The resulting collection of samples $\mathcal{S} = \{S_n\}_{n=1}^{M_{\text{trials}}}$ forms the empirical distribution.

In the third step, we analyze the statistical properties of the collected samples in $\mathcal{S}$. We construct a histogram to visualize the empirical probability density function and compute key statistical moments (the skewness and kurtosis), to assess the closeness to a centered Gaussian distribution. These provide a quantitative verification of the CLT for numerical solutions. In particular, the empirical distribution should exhibit near-zero skewness and near-three kurtosis, confirming the asymptotic normality of the time-average for numerical solutions.

To demonstrate the implementation of Algorithms 3 and 4, we conduct numerical experiments on two examples: a linear stochastic delay differential equation and a superlinear SFDE.

Algorithm 4 Verification of numerical CLT for SFDEs

Input: SFDE parameters: (b, σ, τ, ξ); simulation parameters: Δ, T_{long}, $T_{\max}$, M_{trials}; parameter θ, test function f.

Output: Empirical distribution of the numerical normalized sum

$$S_n = \sqrt{T_{\max}}\Big(\frac{1}{T_{\max}}\int_0^{T_{\max}} f(y_t^{(n)})dt - \mu(f)\Big), \quad n = 1, \ldots, M_{\text{trials}}.$$

Step 1: Estimation of the stationary mean

1: Generate a long reference trajectory $\{x(t)\}_{t=0}^{T_{\text{long}}}$ of the SFDE using the θ-EM method.

2: Estimate the stationary mean as $\hat{\mu}_f^{\text{global}} \approx \frac{1}{T_{\text{long}}}\int_0^{T_{\text{long}}} f(x_t)dt$.

Step 2: Generation of CLT samples

3: Initialize an empty set of samples $\mathcal{S}$.

4: **for** $n = 1, \ldots, M_{\text{trials}}$ **do** ▷ In parallel

5: Generate an independent sample path $\{y_t^{(n)}\}_{t=0}^{T_{\max}}$ of the SFDE using the θ-EM method.

6: Compute the sample mean for this path: $\hat{\mu}_f^{(n)} = \frac{1}{T_{\max}}\int_0^{T_{\max}} f(y_t^{(n)})dt$.

7: Compute the normalized statistic: $S_n = \sqrt{T_{\max}}(\hat{\mu}_f^{(n)} - \hat{\mu}_f^{\text{global}})$.

8: Add S_n to the set $\mathcal{S}$.

9: **end for**

Step 3: Statistical analysis

10: Analyze the empirical distribution of the samples in $\mathcal{S}$ (e.g., plot histogram, compute skewness and kurtosis) and compare it with a centered Gaussian distribution $\mathcal{N}(0, v^2)$ to assess CLT validity.

Linear Stochastic Delay Differential Equation We consider the following one-dimensional linear stochastic delay differential equation

$$\mathrm{d}x(t) = (-\alpha x(t) + \beta x(t-\tau))\mathrm{d}t + \sigma \mathrm{d}W(t), \quad t > 0, \tag{4.80}$$

with various initial data $\xi_j \in C^1$, $j = 1, \ldots, M$. From Theorem 1.15, we know that when $\alpha > \beta$, the exact solution has ergodicity, the SLLN and the CLT. Namely, the time-average of the exact solution converges almost surely to the ergodic limit $\mu(f)$ for all j. Then we apply Algorithm 3 to numerically verify this behavior based on the EM method. We take five different initial data: for $t \in [-\tau, 0]$,

$$\begin{aligned}
\xi_1(t) &= 0.5\sin(2\pi t) + 0.3t^2, \\
\xi_2(t) &= 0.8\cos(\pi t) - 0.2t, \\
\xi_3(t) &= 0.6\sin(3\pi t) + 0.1t, \\
\xi_4(t) &= e^t - 1.2, \\
\xi_5(t) &= 0.4\sin(4\pi t) + 0.2\cos(2\pi t),
\end{aligned} \tag{4.81}$$

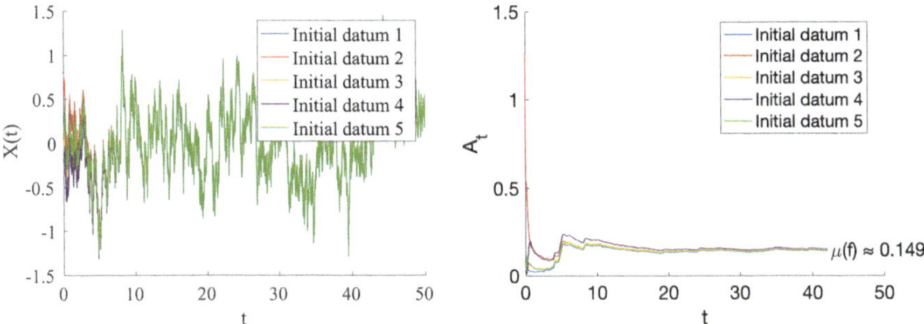

Fig. 4.1 Numerical SLLN of EM functional solutions for linear SFDE (4.80) with five different initial values; $\alpha = 2, \beta = 1, \sigma = 0.8, \tau = 1, T = 50, \Delta = 0.01$. (**a**) Five sample paths of numerical solutions starting from different initial values; (**b**) Corresponding time-averages $A_t^{(j)}$ of these solutions for $j = 1, \ldots, 5$ and $t \in [0, 50]$, where the test function $f(\phi) = \frac{1}{\tau}\int_{-\tau}^{0} \phi^2(s)\mathrm{d}s$. (**a**) Sample path of numerical solutions. (**b**) Time-average of numerical solutions

and the test function is taken as $f(\phi) = \frac{1}{\tau}\int_{-\tau}^{0} \phi^2(s)\mathrm{d}s$ for $\phi \in C^1$. The results are shown in Fig. 4.1. Subfigure (a) displays sample paths of numerical solutions starting from different initial values. After a period of time, the trajectories overlap and thus show that the system loses sensitivity to initial conditions. Subfigure (b) depicts the corresponding time-averages $A_t^{(j)}$ of these solutions for $j = 1, \ldots, 5$. The convergence of these time-averages towards a common value as time increases signifies the SLLN.

Then we verify the CLT of the numerical functional solution based on the proposed Algorithm 4. Figure 4.2 presents the empirical distribution of the normalized time-average of the numerical solution for $f(\phi) = \frac{1}{\tau}\int_{-\tau}^{0} \phi(s)\mathrm{d}s, \phi \in C^1$ at $T_{\max} = 100$, which approximates well to a Gaussian distribution with mean 0.0292 and variance 1.5021. Figures 4.3 and 4.4 further illustrate the evolution of empirical distributions of normalized time-averages as the stepsize decreases, for different test functions. Each subplot corresponds to a distinct stepsize, with the red curve representing the fitted normal density. From these two figures, we observe that the empirical distributions of the normalized time-average exhibit convergence toward a centered Gaussian distribution as the stepsize decays to zero. In comparison with the linear test function $f(\phi) = \frac{1}{\tau}\int_{-\tau}^{0} \phi(s)\mathrm{d}s, \phi \in C^1$ in Fig. 4.3, the nonlinear case $f(\phi) = \frac{1}{\tau}\int_{-\tau}^{0} \phi^2(s)\mathrm{d}s, \phi \in C^1$ in Fig. 4.4 requires longer simulation times for the empirical distributions to reach convergence toward a centered Gaussian distribution as the stepsize decreases.

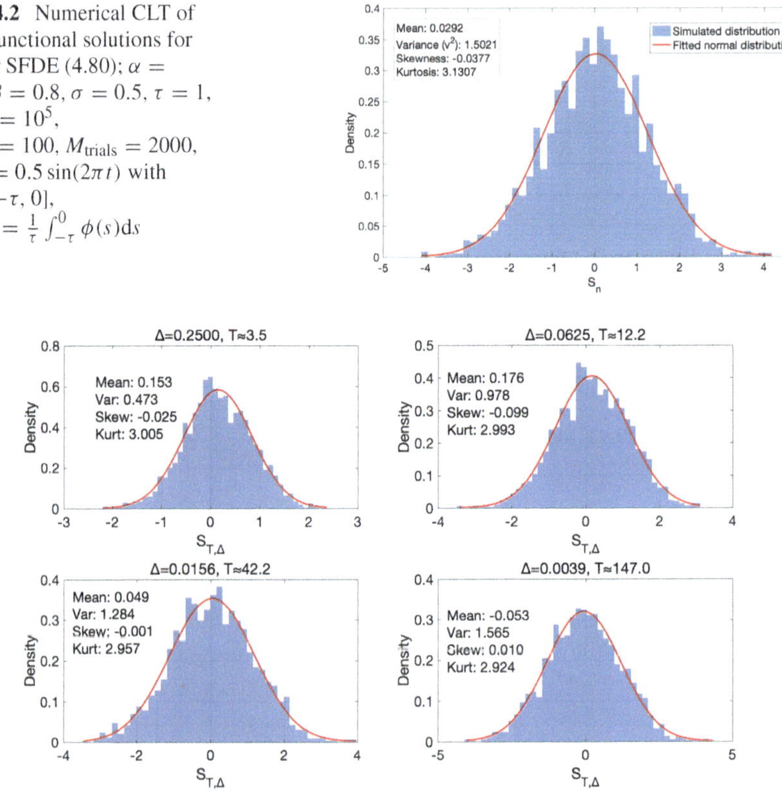

Fig. 4.2 Numerical CLT of EM functional solutions for linear SFDE (4.80); $\alpha = 1.2$, $\beta = 0.8$, $\sigma = 0.5$, $\tau = 1$, $T_{\text{long}} = 10^5$, $T_{\max} = 100$, $M_{\text{trials}} = 2000$, $\xi(t) = 0.5\sin(2\pi t)$ with $t \in [-\tau, 0]$, $f(\phi) = \frac{1}{\tau}\int_{-\tau}^{0}\phi(s)\mathrm{d}s$

Fig. 4.3 Numerical CLT of EM functional solutions for linear SFDE (4.80) under different stepsizes and running times; $\alpha = 1.2$, $\beta = 0.8$, $\tau = 1$, $\sigma = 0.5$, $\xi \equiv 1$, $\lambda = 0.45$, $T_{\text{long}} = 10^5$, $M_{\text{trials}} = 2000$, $f(\phi) = \frac{1}{\tau}\int_{-\tau}^{0}\phi(s)\mathrm{d}s$. (**a**) Top left: $\Delta = 2^{-2}$, $T_{\max} = \Delta^{-2\lambda}$; (**b**) Top right: $\Delta = 2^{-4}$, $T_{\max} = \Delta^{-2\lambda}$; (**c**) Bottom left: $\Delta = 2^{-6}$, $T_{\max} = \Delta^{-2\lambda}$; (**d**) Bottom right: $\Delta = 2^{-8}$, $T_{\max} = \Delta^{-2\lambda}$

Nonlinear Stochastic Functional Differential Equation We consider the following one-dimensional superlinear SFDE

$$\mathrm{d}x(t) = \Big(\frac{1}{\tau}\int_{-\tau}^{0} x(t+s)\mathrm{d}s - x^3(t) - 2x(t)\Big)\mathrm{d}t + \frac{a}{\tau}\int_{-\tau}^{0} x(t+s)\mathrm{d}s\mathrm{d}W(t), \tag{4.82}$$

with five different initial conditions ξ_j, $j = 1, \ldots, 5$ given by (4.81). Here, $a > 0$ represents the noise intensity. From Proposition 4.1, we derive that the exact functional solution of (4.82) admits a unique invariant measure and satisfies both the SLLN and the CLT. We now apply the backward EM method to generate numerical solutions. The results are shown in Fig. 4.5. To be specific, Subfigure (a) shows the simulated trajectories of the numerical solution for five different initial

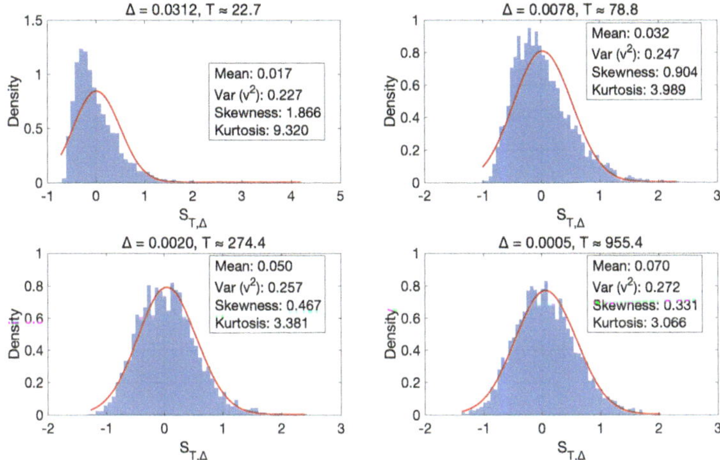

Fig. 4.4 Numerical CLT of EM functional solutions for linear SFDE (4.80) under different stepsizes and running times; $\alpha = 1.2$, $\beta = 0.8$, $\tau = 1$, $\sigma = 0.5$, $\xi \equiv 1$, $\lambda = 0.45$, $T_{\text{long}} = 10^5$, $M_{\text{trials}} = 2000$, $f(\phi) = \frac{1}{\tau}\int_{-\tau}^{0} \phi^2(s)\mathrm{d}s$. (**a**) Top left: $\Delta = 2^{-5}$, $T_{\max} = \Delta^{-2\lambda}$; (**b**) Top right: $\Delta = 2^{-7}$, $T_{\max} = \Delta^{-2\lambda}$; (c) Bottom left: $\Delta = 2^{-9}$, $T_{\max} = \Delta^{-2\lambda}$; (**d**) Bottom right: $\Delta = 2^{-11}$, $T_{\max} = \Delta^{-2\lambda}$

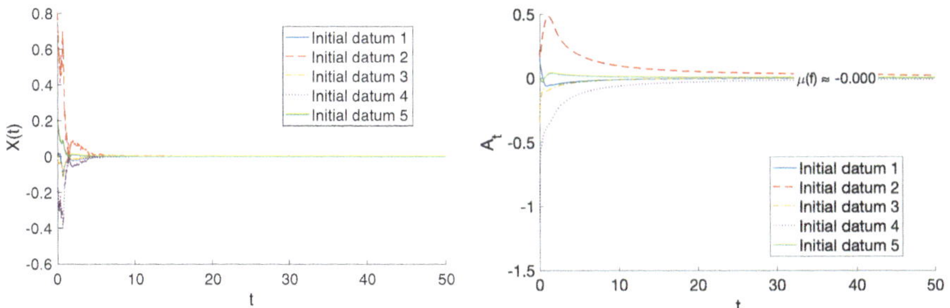

Fig. 4.5 Numerical SLLN of backward EM functional solutions for the nonlinear SFDE (4.82) with different initial values; $\tau = 1$, $a = 0.5$, $T = 50$, $\Delta = 0.01$. (**a**) Five sample paths of numerical solutions starting from different initial values; (**b**) Corresponding time-averages $A_t^{(j)}$ of these solutions for $j = 1, \ldots, 5$ and $t \in [0, 50]$, where the test function $f(\phi) = \frac{1}{\tau}\int_{-\tau}^{0} \phi(s)\mathrm{d}s$. (**a**) Sample path of numerical solutions. (**b**) Time-average of numerical solutions

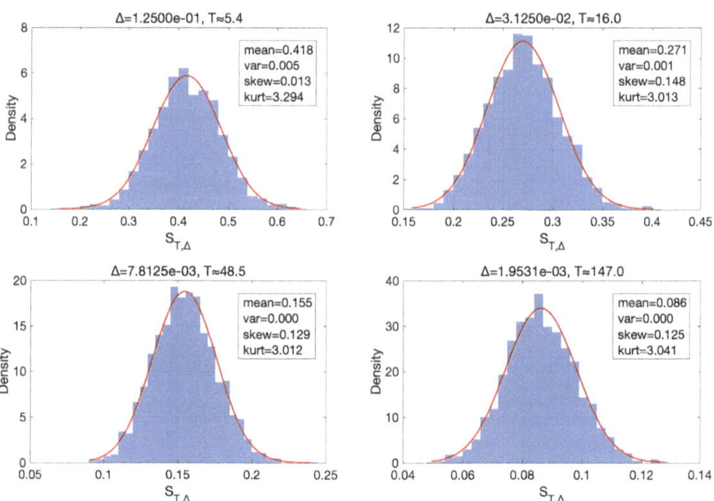

Fig. 4.6 Numerical CLT of backward EM functional solutions for nonlinear SFDE (4.82); $a = 0.1$, $\tau = 0.5$, $\xi \equiv 2$, $\lambda = 0.4$, $T_{\text{long}} = 1000$, $M_{\text{trials}} = 2000$, $f(\phi) = \frac{1}{\tau}\int_{-\tau}^{0}\phi(s)\mathrm{d}s$. (**a**) Top left: $\Delta = 2^{-3}$, $T_{\max} = \Delta^{-2\lambda}$; (**b**) Top right: $\Delta = 2^{-5}$, $T_{\max} = \Delta^{-2\lambda}$; (**c**) Bottom left: $\Delta = 2^{-7}$, $T_{\max} = \Delta^{-2\lambda}$; (**d**) Bottom right: $\Delta = 2^{-9}$, $T_{\max} = \Delta^{-2\lambda}$

values. The paths are distinct and noisy at the beginning, reflecting the influence of the multiplicative noise and different initial values. After a period of time, they are overlapped, which indicates that the system may attain an equilibrium state (constant zero). Subfigure (b) displays the evolution of the time-averages for each of the five paths. It is observed that despite starting from different initial values, all five time-averages converge towards the same value as time increases. These indicate that the system is ergodic, and the longtime behavior of the numerical functional solution is independent of the initial condition.

We next examine the CLT for the functional solution of the backward EM method with the linear test function $f(\phi) = \frac{1}{\tau}\int_{-\tau}^{0}\phi(s)\mathrm{d}s$, $\phi \in C^1$. The numerical solutions are computed using the backward EM method. Figure 4.6 displays the empirical distributions of the normalized time-averages of the numerical solutions for different stepsizes. As the stepsize decreases, the histograms become increasingly symmetric and centered around the mean 0, indicating that the normalized time-averages converge in distribution to a centered Gaussian distribution. This observation provides numerical verification supporting the CLT for the numerical functional solution of the nonlinear SFDE (4.82).

4.5 Summary and Outlook

This chapter has focused on general time-homogeneous Markov processes and has investigated probabilistic limit theorems, including the SLLN and the CLT, for numerical discretizations, with the aim of applying these results to the θ-EM method for the SFDE (1.1). The analysis is based on uniformly mixing properties and on strong or weak convergence of numerical discretizations, which are fundamental topics in numerical analysis; see e.g. [11, 14] and references therein. The main results in this chapter are adapted from [8].

There have been some works on studying numerical SLLN and numerical CLT of numerical methods for specific autonomous stochastic ordinary differential equations and stochastic partial differential equations. For example, for stochastic ordinary differential equations with globally Lipschitz continuous coefficients, the SLLN for Euler-type methods is derived in [4, 25]; [28] obtains the CLT for the EM method with decreasing stepsize of a wide class of Brownian ergodic diffusions; [23] proves the CLT and the self-normalized Cramér-type moderate deviation of the time-average for the EM method. For stochastic ordinary differential equations with non-globally Lipschitz continuous coefficients, [18] establishes the CLT of the BEM method. For the case of stochastic partial differential equations, a first attempt is [5], where the SLLN and the CLT for approximating the ergodic limit via a full discretization are established for the parabolic stochastic partial differential equation with the linearly growing drift coefficient. More recently, for the stochastic reaction-diffusion equation near sharp interface limit, [10] establishes a CLT for a temporal semi-discretization.

In addition to the considered SLLN and CLT, there are also other types of probabilistic limit theorems, such as the law of iterated logarithm [1, 2, 13, 21, 30, 31], the moderate deviations [3, 9, 15, 16], and the large deviations [17, 24, 26, 27], for the time-average of stochastic processes. These theorems can be viewed as important complements to the SLLN and the CLT. As for the numerical discretizations, [6] investigates the law of iterated logarithm for linear stochastic Hamiltonian systems and their numerical methods, providing a novel perspective for revealing the superiority of stochastic symplectic methods over their non-symplectic counterparts. The recent work [7] has investigated the law of iterated logarithm for numerical discretizations of time-homogeneous Markov processes, and presents verifiable criteria for preserving the pathwise asymptotic behavior. It is of interest to explore how the moderate and large deviations for the time-average of the Markov process are inherited by numerical discretizations in future studies.

The above discussions are mainly focused on autonomous equations and time-homogeneous Markov processes. In the non-autonomous case, however, the analysis is more intricate because the transition probability functions depend on both the initial and current times, which complicates the corresponding ergodic theory. We mention that for the non-autonomous two-dimensional stochastic Navier–Stokes equation on the torus, [22] establishes the SLLN, the CLT, and the law of iterated logarithm by using the martingale approximation procedure. This naturally raises the question of how these limit theorems can be established for non-autonomous SFDEs, and more generally for non-homogeneous Markov processes, as well as how they can be inherited by numerical discretizations.

References

1. N.T. Andersen, E. Giné, M. Ossiander, J. Zinn, The central limit theorem and the law of iterated logarithm for empirical processes under local conditions. Probab. Theory Related Fields **77**, 271–305 (1988)
2. J. Bao, J. Hao, Limit theorems for SDEs with irregular drifts. Bernoulli **31**, 1709–1734 (2025)
3. S.R. Behera, A.K. Majee, Stochastic fractional conservation laws: large deviation principle, central limit theorem and moderate deviation principle. Stoch. Partial Differ. Equ. Anal. Comput. **13**, 1361–1406 (2025)
4. C.-E. Bréhier, G. Vilmart, High order integrator for sampling the invariant distribution of a class of parabolic stochastic PDEs with additive space-time noise. SIAM J. Sci. Comput. **38**, A2283–A2306 (2016)
5. C. Chen, T. Dang, J. Hong, T. Zhou, CLT for approximating ergodic limit of SPDEs via a full discretization. Stoch. Process. Appl. **157**, 1–41 (2023)
6. C. Chen, X. Chen, T. Dang, J. Hong, Superiority of stochastic symplectic methods via the law of iterated logarithm (2024). https://arxiv.org/abs/2404.14842
7. C. Chen, X. Chen, J. Hong, The law of iterated logarithm for numerical approximation of time-homogeneous Markov process (2025). https://arxiv.org/abs/2511.05217
8. C. Chen, T. Dang, J. Hong, G. Song, On numerical discretizations that preserve probabilistic limit behaviors for time-homogeneous Markov processes. Bernoulli **31**, 3139–3164 (2025)
9. J. Cui, D. Sheng, Large and moderate deviation principles for stochastic partial differential equation on graph (2025). https://arxiv.org/abs/2509.05622
10. J. Cui, L. Sun, Weak approximation for stochastic reaction-diffusion equation near sharp interface limit (2023). https://arxiv.org/abs/2307.08241
11. J. Cui, J. Hong, L. Sun, Weak convergence and invariant measure of a full discretization for parabolic SPDEs with non-globally Lipschitz coefficients. Stoch. Process. Appl. **134**, 55–93 (2021)
12. G. Da Prato, *An Introduction to Infinite-dimensional Analysis*, Universitext (Springer, Berlin, 2006)
13. P. Deheuvels, D.M. Mason, Functional laws of the iterated logarithm for local empirical processes indexed by sets. Ann. Probab. **22**, 1619–1661 (1994)
14. W. Fang, M.B. Giles, Adaptive Euler–Maruyama method for SDEs with nonglobally Lipschitz drift. Ann. Appl. Probab. **30**, 526–560 (2020)
15. E. Gussetti, Pathwise central limit theorem and moderate deviations via rough paths for SPDEs with multiplicative noise. Electron. J. Probab. **30**, Paper No. 79 (2025)
16. Y. Inahama, Y. Xu, X. Yang, Moderate deviations for rough differential equations. Bull. Lond. Math. Soc. **56**, 2738–2748 (2024)
17. V. Jakšić, V. Nersesyan, C.-A. Pillet, A. Shirikyan, Large deviations from a stationary measure for a class of dissipative PDEs with random kicks. Comm. Pure Appl. Math. **68**, 2108–2143 (2015)
18. D. Jin, Central limit theorem for temporal average of backward Euler–Maruyama method. J. Comput. Math. **43**, 588–614 (2025)
19. T. Komorowski, A. Walczuk, Central limit theorem for Markov processes with spectral gap in the Wasserstein metric. Stoch. Process. Appl. **122**, 2155–2184 (2012)
20. A. Kulik, *Ergodic Behavior of Markov Processes: with Applications to Limit Theorems*, vol. 67 of De Gruyter Studies in Mathematics (De Gruyter, Berlin, 2018)

21. M. Ledoux, J. Zinn, Probabilistic limit theorems in the setting of Banach spaces, in *Handbook of the Geometry of Banach Spaces*, vol. 2 (North-Holland, Amsterdam, 2003), pp. 1177–1200
22. R. Liu, K. Lu, Limit theorems of the 2D stochastic Navier–Stokes equation driven by a time-periodic force and a degenerate noise. Sci. China Math. **68**, 2003–2018 (2025)
23. J. Lu, Y. Tan, L. Xu, Central limit theorem and self-normalized Cramér-type moderate deviation for Euler–Maruyama scheme. Bernoulli **28**, 937–964 (2022)
24. D. Martirosyan, V. Nersesyan, Local large deviations principle for occupation measures of the stochastic damped nonlinear wave equation. Ann. Inst. Henri Poincaré Probab. Stat. **54**, 2002–2041 (2018)
25. J.C. Mattingly, A.M. Stuart, M.V. Tretyakov, Convergence of numerical time-averaging and stationary measures via Poisson equations. SIAM J. Numer. Anal. **48**, 552–577 (2010)
26. V. Nersesyan, Large deviations for the Navier–Stokes equations driven by a white-in-time noise. Ann. H. Lebesgue **2**, 481–513 (2019)
27. V. Nersesyan, X. Peng, L. Xu, Large deviations principle via Malliavin calculus for the Navier–Stokes system driven by a degenerate white-in-time noise. J. Differ. Equ. **362**, 230–249 (2023)
28. G. Pagès, F. Panloup, Ergodic approximation of the distribution of a stationary diffusion: rate of convergence. Ann. Appl. Probab. **22**, 1059–1100 (2012)
29. A. Shirikyan, Law of large numbers and central limit theorem for randomly forced PDE's. Probab. Theory Related Fields **134**, 215–247 (2006)
30. T.-H. Tsai, Empirical law of the iterated logarithm for Markov chains with a countable state space. Stoch. Process. Appl. **89**, 175–191 (2000)
31. J.E. Yukich, The law of the iterated logarithm for empirical processes, in *Probability in Banach Spaces 6 (Sandbjerg, 1986)*, vol. 20 of Progr. Probab. (Birkhäuser Boston, Boston, 1990), pp. 265–282

Chapter 5
Numerical Density Function and Convergence Analysis

The density function for the solution of a stochastic differential equation, characterizing all the relevant probabilistic information, is one of the essential characteristics that reveals the probabilistic behavior of the underlying solution. It is of interest to investigate: (1) whether the numerical method can admit the density function when the exact solution is known to possess one, and (2) the error between density functions when both the exact and numerical solutions admit a density function.

This chapter focuses on studying the existence and convergence of the density function for the solution of the θ-EM method for the SFDE (1.1). In Sect. 5.1, we present the nondegeneracy of the numerical solution and prove the existence of the corresponding numerical density function by means of the Malliavin calculus. In Sect. 5.2, we investigate the convergence of the numerical density function. For the SFDE (1.1) with the multiplicative noise, we show that the numerical density function converges to the exact one in $L^1(\mathbb{R}^d)$ using a localization argument. Section 5.3 is devoted to the convergence rate of the numerical density function for the case of the additive noise. And we prove that the convergence rate of the numerical density function in the pointwise sense is 1, based on the test-functional-independent weak convergence analysis.

5.1 Existence of Numerical Density Function

In this section, we investigate the existence of density functions of solutions for (1.1) and the θ-EM method based on the technique of the Malliavin calculus. The deterministic initial datum ξ is fixed in this section. Consider the following assumptions in this section.

Assumption 5.1 *Assume that there exists a constant $a_5 > 0$ such that for any $\phi_1, \phi_2 \in C^d$,*

C. Chen et al., *Numerical Analysis of Stochastic Functional Differential Equations*, Lecture Notes in Mathematics 2399, https://doi.org/10.1007/978-981-92-1592-8_5

$$\langle b(\phi_1) - b(\phi_2), \phi_1(0) - \phi_2(0)\rangle$$
$$\leq a_5\Big(|\phi_1(0) - \phi_2(0)|^2 + \int_{-\tau}^{0} |\phi_1(r) - \phi_2(r)|^2 \mathrm{d}\nu_2(r)\Big).$$

Assumption 5.2 *Assume that the coefficient b has continuous Gâteaux derivative, and that there exists a constant $\tilde{\beta} \geq 0$ such that*

$$|\mathcal{D}b(\phi_1)\phi_2| \leq K(1 + \|\phi_1\|^{\tilde{\beta}})\|\phi_2\|,$$

where $\phi_1, \phi_2 \in C^d$ and $K > 0$.

Assumption 5.3 *Assume that the coefficient σ has continuous Gâteaux derivative, and that there exists some $\sigma_0 > 0$ such that*

$$\inf_{\phi \in C^d} \min_{u \in \mathbb{R}^d, |u|=1} u^\top \sigma(\phi)\sigma(\phi)^\top u \geq \sigma_0.$$

We present the existence of the density function of the θ-EM solution based on the criterion given by Theorem C.8. We first show the moment estimates on both the θ-EM functional solution and its Malliavin derivative.

Lemma 5.1 *Let Assumptions 2.1, 5.1, and 5.2 hold. Assume that the coefficient σ has continuous Gâteaux derivative. Then for $\Delta \in (0, \frac{1}{4\theta a_5})$, $p \geq 2$, and $T > 0$,*

$$\mathbb{E}\Big[\sup_{0 \leq k\Delta \leq T} \|y_{t_k}^{\xi,\Delta}\|^p\Big] \leq K_T, \qquad \sup_{0 \leq u \leq T} \mathbb{E}\Big[\sup_{u \leq k\Delta \leq T} \|D_u y_{t_k}^{\xi,\Delta}\|^p\Big] \leq K_T.$$

Proof The proof of $\mathbb{E}[\sup_{0 \leq k\Delta \leq T} \|y_{t_k}^{\xi,\Delta}\|^p] \leq K_T$ is similar to that of Proposition 2.2 and thus is omitted. Below we prove $\sup_{0 \leq u \leq T} \mathbb{E}[\sup_{u \leq k\Delta \leq T} \|D_u y_{t_k}^{\xi,\Delta}\|^p] \leq K_T$. The proof is split into three steps.

Step 1: We show that for $u \in [0, T]$,

$$\mathbb{E}[|D_u z^{\xi,\Delta}(t_{k+1})|^{2p}]$$
$$\leq K \sum_{i=0}^{k} (\Delta + \mathbf{1}_{[t_i, t_{i+1}]}(u))\mathbb{E}\big[|D_u z^{\xi,\Delta}(t_i)|^{2p} + |D_u y^{\xi,\Delta}(t_i)|^{2p}\big] + K_T. \tag{5.1}$$

For any $k \in \mathbb{N}$ with $0 \leq k\Delta \leq T$ and $u \in [0, T]$, from the definition of the auxiliary process $z^{\xi,\Delta}$ and the chain rule (see Lemma C.4), we have

$$|D_u z^{\xi,\Delta}(t_{k+1})|^2$$
$$= |D_u z^{\xi,\Delta}(t_k)|^2 + |\mathcal{D}b(y_{t_k}^{\xi,\Delta})D_u y_{t_k}^{\xi,\Delta}|^2\Delta^2$$

$$
\begin{aligned}
&+ |\mathcal{D}\sigma(y_{t_k}^{\xi,\Delta})D_u y_{t_k}^{\xi,\Delta}\delta W_k|^2 + |\sigma(y_{t_k}^{\xi,\Delta})\mathbf{1}_{[t_k,t_{k+1}]}(u)|^2 \\
&+ 2\big\langle D_u y^{\xi,\Delta}(t_k) - \theta\Delta\mathcal{D}b(y_{t_k}^{\xi,\Delta})D_u y_{t_k}^{\xi,\Delta}, \mathcal{D}b(y_{t_k}^{\xi,\Delta})D_u y_{t_k}^{\xi,\Delta}\Delta\big\rangle \\
&+ 2\big\langle D_u z^{\xi,\Delta}(t_k) + \mathcal{D}b(y_{t_k}^{\xi,\Delta})D_u y_{t_k}^{\xi,\Delta}\Delta, \sigma(y_{t_k}^{\xi,\Delta})\mathbf{1}_{[t_k,t_{k+1}]}(u)\big\rangle + \tilde{\mathcal{M}}_k \\
\le\ & |D_u z^{\xi,\Delta}(t_k)|^2 + |\mathcal{D}\sigma(y_{t_k}^{\xi,\Delta})D_u y_{t_k}^{\xi,\Delta}\delta W_k|^2 + |\sigma(y_{t_k}^{\xi,\Delta})\mathbf{1}_{[t_k,t_{k+1}]}(u)|^2 \\
&+ 2\big\langle D_u y^{\xi,\Delta}(t_k), \mathcal{D}b(y_{t_k}^{\xi,\Delta})D_u y_{t_k}^{\xi,\Delta}\Delta\big\rangle \\
&+ \Big\langle \frac{2(\theta-1)}{\theta}D_u z^{\xi,\Delta}(t_k) + \frac{2}{\theta}D_u y^{\xi,\Delta}(t_k), \sigma(y_{t_k}^{\xi,\Delta})\mathbf{1}_{[t_k,t_{k+1}]}(u)\Big\rangle + \tilde{\mathcal{M}}_k,
\end{aligned}
$$

where we used $1-2\theta \le 0$ for $\theta \in (\frac{1}{2}, 1]$, the notation

$$
\begin{aligned}
\tilde{\mathcal{M}}_k := 2\Big\langle & D_u z^{\xi,\Delta}(t_k) + \mathcal{D}b(y_{t_k}^{\xi,\Delta})D_u y_{t_k}^{\xi,\Delta}\Delta \\
&+ \sigma(y_{t_k}^{\xi,\Delta})\mathbf{1}_{[t_k,t_{k+1}]}(u), \mathcal{D}\sigma(y_{t_k}^{\xi,\Delta})D_u y_{t_k}^{\xi,\Delta}\delta W_k\Big\rangle,
\end{aligned}
$$

and relation

$$
D_u z^{\xi,\Delta}(t_k) = D_u y^{\xi,\Delta}(t_k) - \theta\Delta\mathcal{D}b(y_{t_k}^{\xi,\Delta})D_u y_{t_k}^{\xi,\Delta}, \quad k \ge 0. \tag{5.2}
$$

By Assumption 5.1, one has that for any $\phi, \phi_1 \in C^d$,

$$
\langle \phi(0), \mathcal{D}b(\phi_1)\phi\rangle \le a_5\Big(|\phi(0)|^2 + \int_{-\tau}^0 |\phi(r)|^2 \mathrm{d}\nu_2(r)\Big). \tag{5.3}
$$

This implies that for any $p \in \mathbb{N}_+$,

$$
|D_u z^{\xi,\Delta}(t_{k+1})|^{2p} \le |D_u z^{\xi,\Delta}(t_k)|^{2p} + \sum_{l=1}^{p} C_p^l I_{k,l} \le \sum_{i=0}^{k}\sum_{l=1}^{p} C_p^l I_{i,l}, \tag{5.4}
$$

where

$$
\begin{aligned}
I_{i,l} := |D_u z^{\xi,\Delta}(t_i)|^{2(p-l)}\Big(&|\mathcal{D}\sigma(y_{t_i}^{\xi,\Delta})D_u y_{t_i}^{\xi,\Delta}\delta W_i|^2 + |\sigma(y_{t_i}^{\xi,\Delta})\mathbf{1}_{[t_i,t_{i+1}]}(u)|^2 \\
&+ K\Delta\Big(|D_u y^{\xi,\Delta}(t_i)|^2 + \int_{-\tau}^0 |D_u y_{t_i}^{\xi,\Delta}(r)|^2 \mathrm{d}\nu_2(r)\Big) + \Big\langle \frac{2(\theta-1)}{\theta} D_u z^{\xi,\Delta}(t_i) \\
&+ \frac{2}{\theta}D_u y^{\xi,\Delta}(t_i), \sigma(y_{t_i}^{\xi,\Delta})\mathbf{1}_{[t_i,t_{i+1}]}(u)\Big\rangle + \tilde{\mathcal{M}}_i\Big)^l.
\end{aligned}
$$

For the term $I_{i,1}$, by the Young inequality and the property of the conditional expectation, we obtain

$$
\begin{aligned}
\mathbb{E}[I_{i,1}] \leq \mathbb{E}\Big[&|D_u z^{\xi,\Delta}(t_i)|^{2(p-1)}\Big(|\mathcal{D}\sigma(y_{t_i}^{\xi,\Delta})D_u y_{t_i}^{\xi,\Delta}|^2\Delta + |\sigma(y_{t_i}^{\xi,\Delta})\mathbf{1}_{[t_i,t_{i+1}]}(u)|^2 \\
&+ K\Delta\big(|D_u y^{\xi,\Delta}(t_i)|^2 + \int_{-\tau}^0 |D_u y_{t_i}^{\xi,\Delta}(r)|^2 \mathrm{d}\nu_2(r)\big) \\
&+ \big\langle\frac{2(\theta-1)}{\theta} D_u z^{\xi,\Delta}(t_i) + \frac{2}{\theta} D_u y^{\xi,\Delta}(t_i), \sigma(y_{t_i}^{\xi,\Delta})\mathbf{1}_{[t_i,t_{i+1}]}(u)\big\rangle\Big)\Big] \\
\leq K\big(&\Delta + \mathbf{1}_{[t_i,t_{i+1}]}(u)\big)\mathbb{E}\big[|D_u z^{\xi,\Delta}(t_i)|^{2p} + |D_u y^{\xi,\Delta}(t_i)|^{2p}\big] \\
&+ K\Delta\mathbb{E}\Big[\int_{-\tau}^0 |D_u y_{t_i}^{\xi,\Delta}(r)|^{2p}\mathrm{d}\nu_1(r) + \int_{-\tau}^0 |D_u y_{t_i}^{\xi,\Delta}(r)|^{2p}\mathrm{d}\nu_2(r)\Big] \\
&+ K_T\mathbf{1}_{[t_i,t_{i+1}]}(u),
\end{aligned}
\tag{5.5}
$$

where in the last inequality we used Assumptions 2.1 and

$$
\mathbb{E}\Big[\sup_{0\leq t_i\leq T} |y^{\xi,\Delta}(t_i)|^{2p}\Big] \leq K_T.
$$

For the term $I_{i,2}$, we have

$$
\begin{aligned}
\mathbb{E}[I_{i,2}] = \mathbb{E}\Big[&|D_u z^{\xi,\Delta}(t_i)|^{2(p-2)}\Big(|\mathcal{D}\sigma(y_{t_i}^{\xi,\Delta})D_u y_{t_i}^{\xi,\Delta}\delta W_i|^2 + |\sigma(y_{t_i}^{\xi,\Delta})\mathbf{1}_{[t_i,t_{i+1}]}(u)|^2 \\
&+ K\Delta\big(|D_u y^{\xi,\Delta}(t_i)|^2 + \int_{-\tau}^0 |D_u y_{t_i}^{\xi,\Delta}(r)|^2\mathrm{d}\nu_2(r)\big) + \big\langle\frac{2(\theta-1)}{\theta} D_u z^{\xi,\Delta}(t_i) \\
&+ \frac{2}{\theta} D_u y^{\xi,\Delta}(t_i), \sigma(y_{t_i}^{\xi,\Delta})\mathbf{1}_{[t_i,t_{i+1}]}(u)\big\rangle\Big)^2 + |D_u z^{\xi,\Delta}(t_i)|^{2(p-2)}\tilde{\mathcal{M}}_i^2\Big] \\
\leq K\big(&\Delta + \mathbf{1}_{[t_i,t_{i+1}]}(u)\big)\mathbb{E}\big[|D_u z^{\xi,\Delta}(t_i)|^{2p} + |D_u y^{\xi,\Delta}(t_i)|^{2p}\big] \\
&+ K\Delta\mathbb{E}\Big[\int_{-\tau}^0 |D_u y_{t_i}^{\xi,\Delta}(r)|^{2p}\mathrm{d}\nu_1(r) + \int_{-\tau}^0 |D_u y_{t_i}^{\xi,\Delta}(r)|^{2p}\mathrm{d}\nu_2(r)\Big] \\
&+ K_T\mathbf{1}_{[t_i,t_{i+1}]}(u).
\end{aligned}
\tag{5.6}
$$

Similarly, we deduce that for any $l \in \{3,\ldots,p\}$,

$$
\begin{aligned}
\mathbb{E}[|I_{i,l}|] \leq K\big(&\Delta + \mathbf{1}_{[t_i,t_{i+1}]}(u)\big)\mathbb{E}\big[|D_u z^{\xi,\Delta}(t_i)|^{2p} + |D_u y^{\xi,\Delta}(t_i)|^{2p}\big] + K\Delta \\
&\times\mathbb{E}\Big[\int_{\tau}^0 |D_u y_{t_i}^{\xi,\Delta}(r)|^{2p}\mathrm{d}\nu_1(r) + \int_{-\tau}^0 |D_u y_{t_i}^{\xi,\Delta}(r)|^{2p}\mathrm{d}\nu_2(r)\Big] \\
&+ K_T\mathbf{1}_{[t_i,t_{i+1}]}(u).
\end{aligned}
\tag{5.7}
$$

Inserting (5.5)–(5.7) into (5.4), we arrive at

$$
\begin{aligned}
&\mathbb{E}[|D_u z^{\xi,\Delta}(t_{k+1})|^{2p}] \\
&\quad \le K\sum_{i=0}^{k}\big(\Delta+\mathbf{1}_{[t_i,t_{i+1}]}(u)\big)\mathbb{E}\big[|D_u z^{\xi,\Delta}(t_i)|^{2p}+|D_u y^{\xi,\Delta}(t_i)|^{2p}\big] \\
&\qquad + K_T\sum_{i=0}^{k}\mathbf{1}_{[t_i,t_{i+1}]}(u) \\
&\qquad + K\Delta\sum_{i=0}^{k}\mathbb{E}\Big[\int_{-\tau}^{0}|D_u y_{t_i}^{\xi,\Delta}(r)|^{2p}\mathrm{d}\nu_1(r)+\int_{-\tau}^{0}|D_u y_{t_i}^{\xi,\Delta}(r)|^{2p}\mathrm{d}\nu_2(r)\Big] \\
&\quad \le K\sum_{i=0}^{k}\big(\Delta+\mathbf{1}_{[t_i,t_{i+1}]}(u)\big)\mathbb{E}\big[|D_u z^{\xi,\Delta}(t_i)|^{2p}+|D_u y^{\xi,\Delta}(t_i)|^{2p}\big]+K_T.
\end{aligned}
$$

Here, we used

$$
\sum_{i=0}^{k}\int_{-\tau}^{0}|D_u y_{t_i}^{\xi,\Delta}(r)|^{2p}\mathrm{d}\nu_j(r)\le\sum_{i=0}^{k}|D_u y^{\xi,\Delta}(t_i)|^{2p},\quad j=1,2,
$$

whose proof is similar to (2.15).

Step 2: We show that for $u\in[0,T]$,

$$
\begin{aligned}
&\sum_{i=0}^{k}(\Delta+\mathbf{1}_{[t_i,t_{i+1}]}(u))\mathbb{E}[|D_u y^{\xi,\Delta}(t_i)|^{2p}] \\
&\qquad \le K\sum_{i=0}^{k}(\Delta+\mathbf{1}_{[t_i,t_{i+1}]}(u))\mathbb{E}[|D_u z^{\xi,\Delta}(t_i)|^{2p}].
\end{aligned}
\tag{5.8}
$$

It follows from (5.2) and (5.3) that

$$
\begin{aligned}
|D_u z^{\xi,\Delta}(t_i)|^2 &> |D_u y^{\xi,\Delta}(t_i)|^2-2\theta\Delta\big\langle D_u y^{\xi,\Delta}(t_i),\mathcal{D}b(y_{t_i}^{\xi,\Delta})D_u y_{t_i}^{\xi,\Delta}\big\rangle \\
&\ge |D_u y^{\xi,\Delta}(t_i)|^2-2\theta\Delta a_5\big(|D_u y^{\xi,\Delta}(t_i)|^2 \\
&\qquad +\int_{-\tau}^{0}|D_u y_{t_i}^{\xi,\Delta}(r)|^2\mathrm{d}\nu_2(r)\big).
\end{aligned}
$$

Let $\Delta_1 \in (0, \frac{1}{4\theta a_5})$. Then for $\epsilon \in (0, \frac{1-2\theta a_5\Delta_1}{2\theta a_5\Delta_1} - 1)$ and $\Delta \in (0, \Delta_1)$,

$$\begin{aligned}&(1-2\theta a_5\Delta)^p|D_u y^{\xi,\Delta}(t_i)|^{2p}\\&\quad\leq K(\epsilon)|D_u z^{\xi,\Delta}(t_i)|^{2p} + (1+\epsilon)(2\theta a_5\Delta)^p \int_{-\tau}^{0} |D_u y_{t_i}^{\xi,\Delta}(r)|^{2p}\mathrm{d}\nu_2(r).\end{aligned} \tag{5.9}$$

This implies that

$$\begin{aligned}\sum_{i=0}^{k}\mathbb{E}[|D_u y^{\xi,\Delta}(t_i)|^{2p}] &\leq \frac{K}{(1-2\theta a_5\Delta_1)^p - (1+\epsilon)(2\theta a_5\Delta_1)^p}\\&\quad\times\sum_{i=0}^{k}\mathbb{E}[|D_u z^{\xi,\Delta}(t_i)|^{2p}]\\&\leq K\sum_{i=0}^{k}\mathbb{E}[|D_u z^{\xi,\Delta}(t_i)|^{2p}],\end{aligned} \tag{5.10}$$

and

$$\begin{aligned}&\sum_{i=0}^{k}\mathbf{1}_{[t_i,t_{i+1}]}(u)\mathbb{E}[|D_u y^{\xi,\Delta}(t_i)|^{2p}]\\&\quad\leq K\sum_{i=0}^{k}\mathbf{1}_{[t_i,t_{i+1}]}(u)\mathbb{E}[|D_u z^{\xi,\Delta}(t_i)|^{2p}] + K\Delta\sum_{i=0}^{k}\mathbb{E}[|D_u y^{\xi,\Delta}(t_i)|^{2p}]\\&\quad\leq K\sum_{i=0}^{k}\big(\Delta + \mathbf{1}_{[t_i,t_{i+1}]}(u)\big)\mathbb{E}[|D_u z^{\xi,\Delta}(t_i)|^{2p}].\end{aligned} \tag{5.11}$$

Combining (5.10) and (5.11) yields (5.8).

Step 3: We show $\sup_{0\leq u\leq T}\mathbb{E}\big[\sup_{u\leq t_{k+1}\leq T}|D_u y^{\xi,\Delta}(t_{k+1})|^{2p}\big] \leq K_T$. Inserting (5.8) into (5.1) and applying the discrete Gronwall inequality, we derive

$$\sup_{u\leq t_k\leq T}\mathbb{E}[|D_u z^{\xi,\Delta}(t_k)|^{2p}] \leq \sup_{u\leq t_k\leq T} K_T e^{Kt_{k+1}+1} \leq K_T. \tag{5.12}$$

It follows from (5.9) that

$$\sup_{u\leq t_k\leq T}\mathbb{E}[|D_u y^{\xi,\Delta}(t_k)|^{2p}] \leq K_T. \tag{5.13}$$

In addition, by (5.4), we obtain

$$\begin{aligned}
&\mathbb{E}\Big[\sup_{u\le t_{k+1}\le T}|D_u z^{\xi,\Delta}(t_{k+1})|^{2p}\Big]\\
&\le \mathbb{E}\Big[\sup_{u\le t_k\le T}\sum_{i=0}^{k}\big(C_p^1 I_{i,1}+C_p^2 I_{i,2}\big)\Big]+\mathbb{E}\Big[\sum_{i=0}^{\lfloor T/\Delta\rfloor}\sum_{l=3}^{p}C_p^l|I_{i,l}|\Big]\\
&\le \mathbb{E}\Big[\sup_{u\le t_k\le T}\sum_{i=0}^{k}\big(C_p^1 I_{i,1}+C_p^2 I_{i,2}\big)\Big]+\sum_{i=0}^{\lfloor T/\Delta\rfloor}\big(K\Delta+\mathbf{1}_{[t_i,t_{i+1}]}(u)\big)\\
&\quad\times\mathbb{E}\big[|D_u z^{\xi,\Delta}(t_i)|^{2p}+|D_u y^{\xi,\Delta}(t_i)|^{2p}\big]+K_T\\
&\le \mathbb{E}\Big[\sup_{u\le t_k\le T}\sum_{i=0}^{k}\big(C_p^1 I_{i,1}+C_p^2 I_{i,2}\big)\Big]+K_T. \qquad (5.14)
\end{aligned}$$

Utilizing the Burkholder–Davis–Gundy inequality, (5.12), and (5.13), we have

$$\begin{aligned}
&\mathbb{E}\Big[\sup_{u\le t_k\le T}\sum_{i=0}^{k}I_{i,1}\Big]\\
&\le K\sum_{i=0}^{\lfloor T/\Delta\rfloor}\big(\Delta+\mathbf{1}_{[t_i,t_{i+1}]}(u)\big)\mathbb{E}\Big[|D_u z^{\xi,\Delta}(t_i)|^{2p}+|D_u y^{\xi,\Delta}(t_i)|^{2p}\Big]\\
&\quad+K_T\sum_{i=0}^{\lfloor T/\Delta\rfloor}\mathbf{1}_{[t_i,t_{i+1}]}(u)+\mathbb{E}\Big[\sup_{u\le t_k\le T}\sum_{i=0}^{k}|D_u z^{\xi,\Delta}(t_i)|^{2(p-1)}\tilde{\mathcal{M}}_i\Big]\\
&\le K_T+K\mathbb{E}\Big[\Big(\sum_{i=\lfloor u/\Delta\rfloor}^{\lfloor T/\Delta\rfloor}|D_u z^{\xi,\Delta}(t_i)|^{4(p-1)}\big(|D_u z^{\xi,\Delta}(t_i)|+|D_u y^{\xi,\Delta}(t_i)|\\
&\quad+|\sigma(y_{t_i}^{\xi,\Delta})|\mathbf{1}_{[t_i,t_{i+1}]}(u)\big)^2|\mathcal{D}\sigma(y_{t_i}^{\xi,\Delta})D_u y_{t_i}^{\xi,\Delta}|^2\Delta\Big)^{\frac{1}{2}}\Big]\\
&\le K_T+\frac{1}{4C_p^1}\mathbb{E}\Big[\sup_{u\le t_i\le T}|D_u z^{\xi,\Delta}(t_i)|^{2p}\Big]\\
&\quad+K\mathbb{E}\Big[\Big(\sum_{i=\lfloor u/\Delta\rfloor}^{\lfloor T/\Delta\rfloor}|\mathcal{D}\sigma(y_{t_i}^{\xi,\Delta})D_u y_{t_i}^{\xi,\Delta}|^2\Delta\Big)^{p}\Big]\\
&\quad+K\mathbb{E}\Big[\Big(\sum_{i=\lfloor u/\Delta\rfloor}^{\lfloor T/\Delta\rfloor}(|\sigma(y_{t_i}^{\xi,\Delta})|^2\mathbf{1}_{[t_i,t_{i+1}]}(u)
\end{aligned}$$

$$+|D_u y^{\xi,\Delta}(t_i)|^2)|\mathcal{D}\sigma(y_{t_i}^{\xi,\Delta})D_u y_{t_i}^{\xi,\Delta}|^2\Delta\Big)^{\frac{p}{2}}\Big]$$
$$\leq K_T+\frac{1}{4C_p^1}\mathbb{E}\Big[\sup_{u\leq t_i\leq T}|D_u z^{\xi,\Delta}(t_i)|^{2p}\Big]. \tag{5.15}$$

Similarly, the summation of $I_{i,2}$ can be estimated as

$$\mathbb{E}\Big[\sup_{u\leq t_k\leq T}\sum_{i=0}^{k}I_{i,2}\Big]\leq K_T+\frac{1}{4C_p^2}\mathbb{E}\Big[\sup_{u\leq t_k\leq T}|D_u z^{\xi,\Delta}(t_k)|^{2p}\Big].$$

This, along with (5.14) and (5.15) gives that

$$\sup_{0\leq u\leq T}\mathbb{E}\Big[\sup_{u\leq t_{k+1}\leq T}|D_u z^{\xi,\Delta}(t_{k+1})|^{2p}\Big]\leq K_T.$$

It follows from (5.9) that $\sup_{0\leq u\leq T}\mathbb{E}\Big[\sup_{u\leq t_{k+1}\leq T}|D_u y^{\xi,\Delta}(t_{k+1})|^{2p}\Big]\leq K_T$. The proof is completed. □

As a consequence, we obtain $y^{\xi,\Delta}(t_k)\in\mathbb{D}^{1,p}(\mathbb{R}^d)$. Below we show the existence of the numerical density function by further verifying that the Malliavin covariance matrix of the numerical solution is almost surely invertible.

Theorem 5.2 *Under Assumptions 2.1, 5.1, 5.2, and 5.3, for* $\Delta\in(0,\frac{1}{4\theta a_5})$ *and* $k\in\mathbb{N}_+$, *the law of* $y^{\xi,\Delta}(t_k)$ *admits a density function, where* a_5 *is given in Assumption 5.1.*

Proof According to Theorem C.8, it suffices to show that the Malliavin covariance matrix of $y^{\xi,\Delta}(t_k)$, defined by

$$\gamma_k:=\int_0^{t_k}D_r y^{\xi,\Delta}(t_k)(D_r y^{\xi,\Delta}(t_k))^\top\mathrm{d}r,$$

is $\mathbb{P}$-a.s. invertible.

Taking the Malliavin derivative on both sides of the θ-EM solution, we derive from $D_r\delta W_k=\mathbf{1}_{[t_k,t_{k+1}]}(r)$ and the chain rule (see Lemma C.4) that for $r\geq 0$,

$$D_r y^{\xi,\Delta}(t_{k+1})=D_r y^{\xi,\Delta}(t_k)+(1-\theta)\mathcal{D}b(y_{t_k}^{\xi,\Delta})D_r y_{t_k}^{\xi,\Delta}\Delta+\theta\mathcal{D}b(y_{t_{k+1}}^{\xi,\Delta})D_r y_{t_{k+1}}^{\xi,\Delta}\Delta$$
$$+\mathcal{D}\sigma(y_{t_k}^{\xi,\Delta})D_r y_{t_k}^{\xi,\Delta}\delta W_k+\sigma(y_{t_k}^{\xi,\Delta})\mathbf{1}_{[t_k,t_{k+1}]}(r).$$

Similar to (3.9), we have $y_{t_{k+1}}^{\xi,\Delta}=\sum_{j=-N}^{0}I^{[j]}y_{t_{k+1}}^{\xi,\Delta}(t_j)$, where $I^{[j]}(\cdot)\in C([-\tau,0];\mathbb{R})$, $j=-N,\dots,0$ are given in (3.5). This implies

$$D_r y^{\xi,\Delta}_{t_{k+1}} = D_r\Big(\sum_{j=-N}^{-1} I^{[j]} y^{\xi,\Delta}_{t_{k+1}}(t_j)\Big) + D_r y^{\xi,\Delta}(t_{k+1}) I^{[0]}.$$

Using this expansion, we derive the following expression

$$\begin{aligned}
&D_r y^{\xi,\Delta}(t_{k+1}) - \theta \mathcal{D} b(y^{\xi,\Delta}_{t_{k+1}}) D_r y^{\xi,\Delta}(t_{k+1}) I^{[0]}\Delta \\
&\quad = D_r y^{\xi,\Delta}(t_k) + (1-\theta)\mathcal{D} b(y^{\xi,\Delta}_{t_k}) D_r y^{\xi,\Delta}_{t_k}\Delta + \sigma(y^{\xi,\Delta}_{t_k}) \mathbf{1}_{[t_k,t_{k+1}]}(r) \\
&\qquad + \theta \mathcal{D} b(y^{\xi,\Delta}_{t_{k+1}})\Big(\sum_{j=-N}^{-1} I^{[j]} D_r y^{\xi,\Delta}_{t_{k+1}}(t_j)\Big)\Delta + \mathcal{D}\sigma(y^{\xi,\Delta}_{t_k}) D_r y^{\xi,\Delta}_{t_k}\delta W_k \\
&\quad = D_r y^{\xi,\Delta}(t_k) + (1-\theta)\mathcal{D} b(y^{\xi,\Delta}_{t_k}) D_r\Big(\sum_{j=-N}^{0} I^{[j]} y^{\xi,\Delta}_{t_k}(t_j)\Big)\Delta \\
&\qquad + \theta \mathcal{D} b(y^{\xi,\Delta}_{t_{k+1}}) I^{[-1]} D_r y^{\xi,\Delta}(t_k)\Delta + \theta \mathcal{D} b(y^{\xi,\Delta}_{t_{k+1}}) \sum_{j=-N}^{-2} I^{[j]} D_r y^{\xi,\Delta}_{t_{k+1}}(t_j)\Delta \\
&\qquad + \mathcal{D}\sigma(y^{\xi,\Delta}_{t_k}) D_r\Big(\sum_{j=-N}^{0} I^{[j]} y^{\xi,\Delta}_{t_k}(t_j)\Big)\delta W_k + \sigma(y^{\xi,\Delta}_{t_k}) \mathbf{1}_{[t_k,t_{k+1}]}(r).
\end{aligned}$$

It is straightforward to see that

$$\begin{aligned}
&\Big(\mathrm{Id}_{d\times d} - \theta \mathcal{D} b(y^{\xi,\Delta}_{t_{k+1}})(I^{[0]}\mathrm{Id}_{d\times d})\Delta\Big) D_r y^{\xi,\Delta}(t_{k+1}) \\
&\quad = \sum_{j=-N}^{0} A_{j,k} + \sigma(y^{\xi,\Delta}_{t_k}) \mathbf{1}_{[t_k,t_{k+1}]}(r),
\end{aligned} \tag{5.16}$$

where

$$\begin{aligned}
A_{0,k} :=& \Big(\mathrm{Id}_{d\times d} + (1-\theta)\mathcal{D} b(y^{\xi,\Delta}_{t_k})(I^{[0]}\mathrm{Id}_{d\times d})\Delta \\
&+ \theta \mathcal{D} b(y^{\xi,\Delta}_{t_{k+1}})(I^{[-1]}\mathrm{Id}_{d\times d})\Delta\Big) D_r y^{\xi,\Delta}(t_k) \\
&+ \mathcal{D}\sigma(y^{\xi,\Delta}_{t_k})(I^{[0]} D_r y^{\xi,\Delta}(t_k))\delta W_k, \\
A_{-N,k} :=& \Big((1-\theta)\mathcal{D} b(y^{\xi,\Delta}_{t_k})(I^{[-N]}\mathrm{Id}_{d\times d})\Delta\Big) D_r y^{\xi,\Delta}(t_{k-N}) \\
&+ \mathcal{D}\sigma(y^{\xi,\Delta}_{t_k})(I^{[-N]} D_r y^{\xi,\Delta}(t_{k-N}))\delta W_k,
\end{aligned}$$

and

$$
\begin{aligned}
A_{j,k} :=& \Big((1-\theta)\mathcal{D}b(y_{t_k}^{\xi,\Delta})(I^{[j]}\mathrm{Id}_{d\times d})\Delta + \theta\mathcal{D}b(y_{t_{k+1}}^{\xi,\Delta})(I^{[j-1]}\mathrm{Id}_{d\times d})\Delta\Big)\\
&\times D_r y^{\xi,\Delta}(t_{k+j})\\
&+ \mathcal{D}\sigma(y_{t_k}^{\xi,\Delta})(I^{[j]}D_r y^{\xi,\Delta}(t_{k+j}))\delta W_k
\end{aligned}
$$

for $j = -1, \ldots, -N+1$. Here, $I^{[j]}\mathrm{Id}_{d\times d}, j = -N, \ldots, 0$ are elements of $C([-\tau, 0]; \mathbb{R}^{d\times d}) \cong C^d \otimes \mathbb{R}^d$ and thus can be acted upon by the operators $\mathcal{D}b(y_{t_k}^{\xi,\Delta})$, $\mathcal{D}b(y_{t_{k+1}}^{\xi,\Delta})$. Similar to (3.12), it follows from Assumption 5.1 that for any $u \in \mathbb{R}^d$ with $u \neq 0$, one has

$$
\begin{aligned}
&u^\top \mathcal{D}b(y_{t_{k+1}}^{\xi,\Delta})(I^{[0]}\mathrm{Id}_{d\times d})u\\
&\le a_5\Big(|u|^2 + \int_{-\tau}^0 |I^{[0]}\mathrm{Id}_{d\times d}u|^2 \mathrm{d}\nu_2(r)\Big)\\
&\le a_5\Big(|u|^2 + \int_{-\tau}^0 \Big|\frac{s-t_{-1}}{\Delta}\mathbf{1}_{\Delta_{-1}}(s)\Big|^2 |u|^2 \mathrm{d}\nu_2(r)\Big) \le 2a_5|u|^2,
\end{aligned}
$$

where we used $I^{[0]}(0) = 1$. Then for $\Delta \in (0, \frac{1}{4\theta a_5})$,

$$
u^\top\Big(\mathrm{Id}_{d\times d} - \theta\mathcal{D}b(y_{t_{k+1}}^{\xi,\Delta})(I^{[0]}\mathrm{Id}_{d\times d})\Delta\Big)u \ge (1 - 2\theta a_5\Delta)|u|^2 > 0,
$$

which implies that $\mathrm{Id}_{d\times d} - \theta\mathcal{D}b(y_{t_{k+1}}^{\xi,\Delta})(I^{[0]}\mathrm{Id}_{d\times d})\Delta$ is invertible. Denoting

$$
A_{1,k} := \big(\mathrm{Id}_{d\times d} - \theta\mathcal{D}b(y_{t_{k+1}}^{\xi,\Delta})(I^{[0]}\mathrm{Id}_{d\times d})\Delta\big)^{-1},
$$

we derive $\|A_{1,k}\|_{\mathcal{L}(\mathbb{R}^d;\mathbb{R}^d)} \in (0, \frac{1}{1-2\theta a_5\Delta})$. According to (5.16), we arrive at

$$
\begin{aligned}
\gamma_{k+1} &= \int_0^{t_{k+1}} D_r y^{\xi,\Delta}(t_{k+1})(D_r y^{\xi,\Delta}(t_{k+1}))^\top \mathrm{d}r\\
&= \int_0^{t_k} A_{1,k}\sum_{j=-N}^0 A_{j,k}\Big(\sum_{j=-N}^0 A_{j,k}\Big)^\top (A_{1,k})^\top \mathrm{d}r\\
&\quad + \int_{t_k}^{t_{k+1}} A_{1,k}\sigma(y_{t_k}^{\xi,\Delta})\sigma(y_{t_k}^{\xi,\Delta})^\top (A_{1,k})^\top \mathrm{d}r,
\end{aligned}
\tag{5.17}
$$

where we used $D_r y^{\xi,\Delta}(t_k) = 0$ for $r > t_k$. Since $\sigma(x)\sigma(x)^\top$ is positive definite, we deduce

$$u^\top \gamma_{k+1} u \geq u^\top A_{1,k}\sigma(y_{t_k}^{\xi,\Delta})\sigma(y_{t_k}^{\xi,\Delta})^\top (A_{1,k})^\top u\, \Delta > 0 \quad \mathbb{P}\text{-}a.s.$$

Moreover, utilizing again the invertibility of $\sigma\sigma^\top$, we have that

$$\gamma_1 = \int_0^{t_1} A_{1,0}\sigma(\xi)\sigma(\xi)^\top (A_{1,0})^\top \mathrm{d}r$$

is also $\mathbb{P}$-a.s. invertible. This finishes the proof. □

Remark 5.1 Under conditions in Theorem 5.2, if we further assume that coefficients b and σ are smooth with bounded derivatives of arbitrary orders, then for each $k \in \mathbb{N}_+$, the law of $y^{\xi,\Delta}(t_k)$ admits a smooth density function.

5.2 Density Convergence for Multiplicative Noise Case

In this section, we investigate the convergence of the numerical density function of the θ-EM method. We show the convergence of the density function in $L^1(\mathbb{R}^d)$ when SFDE (1.1) has superlinearly growing drift coefficient and multiplicative noise. We present the following assumptions on the second-order derivatives of coefficients and on the Hölder continuity of the initial datum.

Assumption 5.4 *Assume that coefficients b and σ have continuous Gâteaux derivatives up to order 2 satisfying*

$$|\mathcal{D}^2 b(\phi_1)(\phi_2, \phi_3)| \leq K(1 + \|\phi_1\|^{(\tilde{\beta}-1)\vee 0})\|\phi_2\|\|\phi_3\|,$$
$$|\mathcal{D}^2 \sigma(\phi_1)(\phi_2, \phi_3)| \leq K\|\phi_2\|\|\phi_3\|,$$

where $\phi_1, \phi_2, \phi_3 \in C^d$, $K > 0$, and $\tilde{\beta}$ is given in Assumption 5.2.

We note that the example

$$b(\phi) = |\phi(0)|^2\phi(0) + \int_{-\tau}^0 \phi(r)\mathrm{d}\nu_2(r), \quad \phi \in C^d$$

satisfies Assumption 5.4 with $\tilde{\beta} = 2$. The main result on the convergence of the density function concerned in this subsection is stated as follows.

Theorem 5.3 *Let Assumptions 2.1, 2.5 and 5.1 to 5.4 hold. Then for any* $T > 0$,

$$\lim_{\Delta\to 0}\sup_{0<k\Delta\le T}\int_{\mathbb{R}^d}|\mathfrak{p}^{\Delta}(t_k,x)-\mathfrak{p}(t_k,x)|\mathrm{d}x=0,$$

where $\mathfrak{p}(t_k,\cdot)$ *and* $\mathfrak{p}^{\Delta}(t_k,\cdot)$ *are density functions of* $x^{\xi}(t_k)$ *and* $y^{\xi,\Delta}(t_k)$*, respectively.*

Proof The proof is based on a localization argument and is divided into three steps.

Step 1. In this step, we introduce the truncated version of the SFDE (1.1) together with its corresponding θ-EM solution. We then estimate the error between the solutions of the truncated SFDE and that of its numerical approximation. Introduce the smooth cut-off functional $\Theta_R : C^d \to [0,1]$ with continuous derivatives and compact support. For $R > \|\xi\|$, let $\Theta_R(x) = 1$ for $\|x\| \le R$ and $\Theta_R(x) = 0$ for $\|x\| > R+1$. Then we consider the truncated version of (1.1) on $[0,T]$,

$$\begin{aligned}\mathrm{d}x^{\xi,R}(t) &= b^R(x_t^{\xi,R})\mathrm{d}t+\sigma^R(x_t^{\xi,R})\mathrm{d}W(t), \quad t\in(0,T],\\ x^{\xi,R}(t) &= \xi, \quad t\in[-\tau,0],\end{aligned} \tag{5.18}$$

where $b^R(\cdot) = \Theta_R(\cdot)b(\cdot)$ and $\sigma^R(\cdot) = \Theta_R(\cdot)\sigma(\cdot)$ are globally Lipschitz continuous for each R. Moreover, the coefficients b^R and σ^R satisfy

$$|\mathcal{D}b^R(\phi_1)\phi_2|^2 \le K_R\Big(|\phi_2(0)|^2+\int_{-\tau}^0|\phi_2(r)|^2\mathrm{d}\nu_2(r)\Big), \tag{5.19}$$

$$|\mathcal{D}\sigma^R(\phi_1)\phi_2|^2 \le K_R\Big(|\phi_2(0)|^2+\int_{-\tau}^0|\phi_2(r)|^2\mathrm{d}\nu_1(r)\Big), \tag{5.20}$$

where $\phi_1,\phi_2\in C^d$. In addition, the numerical solution and the numerical functional solution of the θ-EM method for (5.18) are denoted by $\{y^{\xi,\Delta,R}(t_k)\}_{k\ge -N}$ and $\{y_{t_k}^{\xi,\Delta,R}\}_{k\ge 0}$, respectively. We need to estimate the error between $x^{\xi,R}(t)$ and $y^{\xi,\Delta,R}(t)$ in $\|\cdot\|_{1,2}$. Here we recall the norm $\|G\|_{1,2} = (\mathbb{E}[|G|^2+\|DG\|^2_{H\otimes\mathbb{R}^d}])^{\frac{1}{2}}$ for all $\mathbb{R}^d$-valued random variable G, where $H = L^2([0,T];\mathbb{R}^m)$. For the strong convergence errors of the θ-EM solution and its truncated version, similar to the proof of Theorem 2.14, using Lemmas 1.17 and 5.1, we deduce from Assumptions 2.1, 2.5, 5.1, and 5.2 that

$$\mathbb{E}\Big[\sup_{0\le t_k\le T}|x^{\xi}(t_k)-y^{\xi,\Delta}(t_k)|^4\Big]\le K_T\Delta^2 \tag{5.21}$$

and

$$\mathbb{E}\Big[\sup_{0\le t_k\le T}|x^{\xi,R}(t_k)-y^{\xi,\Delta,R}(t_k)|^4\Big]\le K_{T,R}\Delta^2. \tag{5.22}$$

Now we estimate the term

$$\mathbb{E}[\|Dx^{\xi,R}(t_k)-Dy^{\xi,\Delta,R}(t_k)\|^2_{H\otimes\mathbb{R}^d}]=\int_0^T\mathbb{E}[|D_u x^{\xi,R}(t_k)-D_u y^{\xi,\Delta,R}(t_k)|^2]\mathrm{d}u.$$

Define the auxiliary process of the θ-EM discretization as follows:

$$\begin{cases} z^{\xi,\Delta,R}(t_k)= \xi(t_k), \quad -N\le k\le -1,\\ z^{\xi,\Delta,R}(t_k)= y^{\xi,\Delta,R}(t_k)-\theta b^R(y^{\xi,\Delta,R}_{t_k})\Delta, \quad 0\le k\le N^\Delta. \end{cases}$$

Then for $1\le k\le N^\Delta$,

$$z^{\xi,\Delta,R}(t_k)= z^{\xi,\Delta,R}(t_{k-1})+b^R(y^{\xi,\Delta,R}_{t_{k-1}})\Delta+\sigma^R(y^{\xi,\Delta,R}_{t_{k-1}})\delta W_{k-1},\ 1\le k\le N^\Delta,$$

and the continuous version $\{z^{\xi,\Delta,R}(t)\}_{t\ge-\tau}$ satisfies

$$z^{\xi,\Delta,R}(t)=z^{\xi,\Delta,R}(t_k)+b(y^{\xi,\Delta,R}_{t_k})(t-t_k)+\sigma(y^{\xi,\Delta,R}_{t_k})(W(t)-W(t_k)),$$

for $t\in[t_k,t_{k+1})$, $0\le k\le N^\Delta$ with the initial datum

$$z^{\xi,\Delta,R}(t)=\frac{t_{k+1}-t}{\Delta}z^{\xi,\Delta,R}(t_k)+\frac{t-t_k}{\Delta}z^{\xi,\Delta,R}(t_{k+1}),$$

for $t\in[t_k,t_{k+1}]$, $-N\le k\le -1$. Then by the Hölder inequality and Burkholder–Davis–Gundy inequality, the second moment of the error between Malliavin derivative of $x^{\xi,R}$ and $z^{\xi,\Delta,R}$ can be estimated as

$$\begin{aligned} &\mathbb{E}[|D_u x^{\xi,R}(t)-D_u z^{\xi,\Delta,R}(t)|^2]\\ &\le K(t-u)\mathbb{E}\Big[\int_u^t|\mathcal{D}b^R(x^{\xi,R}_s)D_u x^{\xi,R}_s-\mathcal{D}b^R(y^{\xi,\Delta,R}_{\lfloor s\rfloor})D_u y^{\xi,\Delta,R}_{\lfloor s\rfloor}|^2\mathrm{d}s\Big]\\ &\quad+K\mathbb{E}\Big[\int_u^t|\mathcal{D}\sigma^R(x^{\xi,R}_s)D_u x^{\xi,R}_s-\mathcal{D}\sigma^R(y^{\xi,\Delta,R}_{\lfloor s\rfloor})D_u y^{\xi,\Delta,R}_{\lfloor s\rfloor}|^2\mathrm{d}s\Big]\\ &\quad+K\mathbb{E}\Big[|\sigma^R(x^{\xi,R}_u)\mathbf{1}_{[0,t]}(u)-\sigma^R(y^{\xi,\Delta,R}_{\lfloor u\rfloor})\mathbf{1}_{[0,t]}(u)|^2\Big]. \end{aligned}\tag{5.23}$$

It follows from the Taylor formula that

$$\begin{aligned} &\Big|\mathcal{D}b^R(x^{\xi,R}_s)D_u x^{\xi,R}_s-\mathcal{D}b^R(y^{\xi,\Delta,R}_{\lfloor s\rfloor})D_u y^{\xi,\Delta,R}_{\lfloor s\rfloor}\Big|\\ &\le\Big|\int_0^1\mathcal{D}^2b^R\big(\varsigma x^{\xi,R}_{\lfloor s\rfloor}+(1-\varsigma)y^{\xi,\Delta,R}_s\big)(x^{\xi,R}_s-y^{\xi,\Delta,R}_{\lfloor s\rfloor},D_u x^{\xi,R}_s)\mathrm{d}\varsigma\Big|\\ &\quad+\Big|\mathcal{D}b^R(y^{\xi,\Delta,R}_{\lfloor s\rfloor})(D_u x^{\xi,R}_s-D_u y^{\xi,\Delta,R}_{\lfloor s\rfloor})\Big| \end{aligned}$$

$$\leq K_R\|x_s^{\xi,R}-y_{\lfloor s\rfloor}^{\xi,\Delta,R}\|\|D_ux_s^{\xi,R}\|+K_R\Big(|D_ux^{\xi,R}(s)-D_uy^{\xi,\Delta,R}(\lfloor s\rfloor)|^2$$
$$+\int_{-\tau}^0|D_ux_s^{\xi,R}(r)-D_uy_{\lfloor s\rfloor}^{\xi,\Delta,R}(r)|^2\mathrm{d}\nu_2(r)\Big)^{\frac{1}{2}},$$

where in the last inequality we used the boundedness of operators $\mathcal{D}b^R(\cdot)$ (see (5.19)) and $\mathcal{D}^2b^R(\cdot)$ on their compact support $\{x\in C^d:\|x\|\leq R+1\}$. Similar to the proof of Lemma 2.9, we obtain

$$\sup_{u\leq t\leq T}\mathbb{E}[\|D_uy_{\lfloor t\rfloor}^{\xi,\Delta,R}-D_uz_t^{\xi,\Delta,R}\|^2]\leq K_{T,R}\Delta.$$

Then

$$\mathbb{E}\Big[\big|\mathcal{D}b^R(x_s^{\xi,R})D_ux_s^{\xi,R}-\mathcal{D}b^R(y_{\lfloor s\rfloor}^{\xi,\Delta,R})D_uy_{\lfloor s\rfloor}^{\xi,\Delta,R}\big|^2\Big]$$
$$\leq K_R\mathbb{E}\big[\|x_s^{\xi,R}-y_{\lfloor s\rfloor}^{\xi,\Delta,R}\|^2\|D_ux_s^{\xi,R}\|^2\big]+K_R\mathbb{E}\Big[|D_ux^{\xi,R}(s)-D_uz^{\xi,\Delta,R}(s)|^2$$
$$+\int_{-\tau}^0|D_ux_s^{\xi,R}(r)-D_uz_s^{\xi,\Delta,R}(r)|^2\mathrm{d}\nu_2(r)\Big]+K_{T,R}\Delta. \tag{5.24}$$

Similarly, by using (5.20), we deduce

$$\mathbb{E}\Big[\big|\mathcal{D}\sigma^R(x_s^{\xi,R})D_ux_s^{\xi,R}-\mathcal{D}\sigma^R(y_{\lfloor s\rfloor}^{\xi,\Delta,R})D_uy_{\lfloor s\rfloor}^{\xi,\Delta,R}\big|\Big]$$
$$\leq K_R\mathbb{E}\big[\|x_s^{\xi,R}-y_{\lfloor s\rfloor}^{\xi,\Delta,R}\|^2\|D_ux_s^{\xi,R}\|^2\big]+K_R\mathbb{E}\Big[|D_ux^{\xi,R}(s)-D_uz^{\xi,\Delta,R}(s)|^2$$
$$+\int_{-\tau}^0|D_ux_s^{\xi,R}(r)-D_uz_s^{\xi,\Delta,R}(r)|^2\mathrm{d}\nu_1(r)\Big]+K_{T,R}\Delta. \tag{5.25}$$

Inserting (5.24) and (5.25) into (5.23), and then using Assumption 2.1, Lemmas 1.17 and 5.1, and (5.22), we obtain that

$$\mathbb{E}\Big[|D_ux^{\xi,R}(t)-D_uz^{\xi,\Delta,R}(t)|^2\Big]$$
$$\leq K_{T,R}\Delta+K_{T,R}\int_u^T\big(\mathbb{E}[\|x_{\lfloor s\rfloor}^{\xi,R}-y_{\lfloor s\rfloor}^{\xi,\Delta,R}\|^4]\big)^{\frac{1}{2}}\mathrm{d}s$$
$$+K_{T,R}\int_u^T\big(\mathbb{E}[\|x_{\lfloor s\rfloor}^{\xi,R}-x_s^{\xi,R}\|^4]\big)^{\frac{1}{2}}\mathrm{d}s$$
$$+K_{T,R}\mathbb{E}\Big[\int_u^T|D_ux^{\xi,R}(s)-D_uz^{\xi,\Delta,R}(s)|^2\mathrm{d}s\Big]$$
$$+K\sup_{0\leq s\leq T}\mathbb{E}[\|x_{\lfloor s\rfloor}^{\xi,R}-x_s^{\xi,R}\|^2]$$

$$\leq K_{T,R}\Delta + K_{T,R}\int_u^T \mathbb{E}\big[|D_u x^{\xi,R}(s) - D_u z^{\xi,\Delta,R}(s)|^2\big]\mathrm{d}s.$$

Applying the Gronwall inequality, we arrive at

$$\mathbb{E}\Big[|D_u x^{\xi,R}(t) - D_u z^{\xi,\Delta,R}(t)|^2\Big] \leq K_{T,R}\Delta,$$

which implies

$$\int_0^T \mathbb{E}\Big[|D_u x^{\xi,R}(t_k) - D_u y^{\xi,\Delta,R}(t_k)|^2\Big]\mathrm{d}u \leq K_{T,R}\Delta. \tag{5.26}$$

It follows from (5.22) and (5.26) that

$$\sup_{0\leq k\Delta\leq T} \|y^{\xi,\Delta,R}(t_k) - x^{\xi,R}(t_k)\|_{1,2} \leq K_{T,R}\Delta^{\frac{1}{2}}.$$

Step 2. In this step, we aim to apply [21, Lemma A.1] to estimate the error in $L^1(\mathbb{R}^d)$ between density functions of $x^{\xi,R}(\cdot)$ and $y^{\xi,\Delta,R}(\cdot)$. To this end, we first claim that for $\varrho \in (0, 1)$,

$$\mathbb{E}\Big[\Big(\int_0^T |D_r x^{\xi,R}(t)|^2\mathrm{d}r\Big)^{-\varrho}\Big] \leq K_{T,R,\varrho}. \tag{5.27}$$

In fact, by Proposition 1.18 with $u = \frac{1}{d}(1, \ldots, 1) \in \mathbb{R}^d$, we obtain that

$$\mathbb{P}\Big(\int_0^T |D_r x^{\xi,R}(t)|^2\mathrm{d}r \leq \varepsilon\Big) \leq K_{T,R}\varepsilon$$

for any $\varepsilon \in (0, 1)$. Hence for any $\varrho \in (0, 1)$, we derive that

$$\begin{aligned}&\sum_{n=1}^{\infty} n^{\varrho-1}\mathbb{P}\Big(\Big(\int_0^T |D_r x^{\xi,R}(t)|^2\mathrm{d}r\Big)^{-1} \geq n\Big)\\ &\quad\leq 1 + K_{T,R}\sum_{n=2}^{\infty} n^{\varrho-1}n^{-1} \leq K_{T,R,\varrho}.\end{aligned} \tag{5.28}$$

Then for any $\varrho \in (0, 1)$,

$$\begin{aligned}&\mathbb{E}\Big[\Big(\int_0^T |D_r x^{\xi,R}(t)|^2\mathrm{d}r\Big)^{-\varrho}\Big]\\ &\leq 1 + \sum_{n=1}^{\infty}(n+1)^{\varrho}\mathbb{P}\Big(n \leq \Big(\int_0^T |D_r x^{\xi,R}(t)|^2\mathrm{d}r\Big)^{-1} \leq n+1\Big)\end{aligned}$$

$$
\begin{aligned}
&\leq 2+\sum_{n=1}^{\infty}((n+1)^{\varrho}-n^{\varrho})\mathbb{P}\Big(\Big(\int_0^T |D_r x^{\xi,R}(t)|^2 \mathrm{d}r\Big)^{-1} \geq n\Big) \\
&\leq 2+\varrho\sum_{n=1}^{\infty} n^{\varrho-1}\mathbb{P}\Big(\Big(\int_0^T |D_r x^{\xi,R}(t)|^2 \mathrm{d}r\Big)^{-1} \geq n\Big) \leq K_{T,R,\varrho},
\end{aligned}
$$

where in the last step we used (5.28). This proves the claim (5.27).

Furthermore, similar to the proof of Lemma 1.17, it follows from Assumptions 2.1, 5.1, and 5.4 that $x^{\xi}(t) \in \mathbb{D}^{2,4}(\mathbb{R}^d)$ for all $t \in [0,T]$. Then, by [21, Eq. (6.1) and Lemma A.1], we arrive at

$$
\begin{aligned}
&\sup_{0<t_k\leq T}\int_{\mathbb{R}^d} |\mathfrak{p}^{\Delta,R}(t_k,x)-\mathfrak{p}^R(t_k,x)|\mathrm{d}x \\
&= \sup_{0<t_k\leq T} \mathrm{d}_{TV}(y^{\xi,\Delta,R}(t_k), x^{\xi,R}(t_k)) \\
&\leq K_{T,R}\sup_{0\leq t_k\leq T}\|y^{\xi,\Delta,R}(t_k)-x^{\xi,R}(t_k)\|_{1,2}^{\frac{2\varrho}{2\varrho+2}} \leq K_{T,R}\Delta^{\frac{\varrho}{2\varrho+2}},
\end{aligned}
\tag{5.29}
$$

where $\mathrm{d}_{TV}(X,Y)$ denotes the total variation distance between two $\mathbb{R}^d$-valued random variables X and Y, and $\mathfrak{p}^R(t_k,\cdot)$, $\mathfrak{p}^{\Delta,R}(t_k,\cdot)$ are density functions of $x^{\xi,R}(t_k)$ and $y^{\xi,\Delta,R}(t_k)$, respectively.

Step 3. In this step, we demonstrate that the numerical density function of the θ-EM solution of (1.1) converges to that of the exact solution of (1.1), building upon the convergence of the density function for the truncated solution established in Step 2. Denoting two sequences of events by

$$
\begin{aligned}
\Omega_R &:= \{\omega\in\Omega: \sup_{t\in[0,T]}|x^{\xi}(t)|\leq R\}, \\
\Omega_{R,y} &:= \{\omega\in\Omega: \sup_{t_k\in[0,T]}|y^{\xi,\Delta}(t_k)|\leq R\},
\end{aligned}
$$

we have

$$
\lim_{R\to\infty}\mathbb{P}(\Omega_R)=\mathbb{P}(\Omega)=\lim_{R\to\infty}\mathbb{P}(\Omega_{R,y})=1.
$$

According to [21, Eq. (6.1)], we derive

$$
\begin{aligned}
&\sup_{0<k\Delta\leq T}\int_{\mathbb{R}^d} |\mathfrak{p}^{\Delta}(t_k,x)-\mathfrak{p}(t_k,x)|\mathrm{d}x \\
&= \sup_{0<k\Delta\leq T} \mathrm{d}_{TV}(y^{\Delta,\xi}(t_k), x^{\xi}(t_k))
\end{aligned}
$$

$$\le 4\mathbb{P}(\Omega_{R-1}^c) + 2\mathbb{P}\Big(\sup_{0\le t_k\le T}|x^\xi(t_k) - y^{\xi,\Delta}(t_k)| \ge 1\Big)$$

$$+ \sup_{0<k\Delta\le T} d_{TV}(y^{\Delta,\xi,R}(t_k), x^{\xi,R}(t_k))$$

$$\le 4\frac{\mathbb{E}[\sup_{0\le t\le T}|x^\xi(t)|^2]}{(R-1)^2} + 2\mathbb{E}\Big[\sup_{0\le t_k\le T}|x^\xi(t_k) - y^{\xi,\Delta}(t_k)|^2\Big] + K_{T,R}\Delta^{\frac{\varrho}{2\varrho+2}}$$

$$\le \frac{K_T}{(R-1)^2} + K_T\Delta + K_{T,R}\Delta^{\frac{\varrho}{2\varrho+2}},$$

where we used Lemma 1.17, (5.21), and (5.29). Letting $\Delta \to 0$, $R \to \infty$, we obtain the desired argument. □

Remark 5.2 It follows from the proof of Theorem 5.3 that when the coefficients b and σ are globally Lipschitz continuous, the convergence rate of the numerical density function is almost 1/4 in $L^1(\mathbb{R}^d)$. We will show that for the additive noise case, the convergence rate could attain 1 in the pointwise sense (see Theorem 5.4).

It is known that the estimate between density functions of random variables is closely related to the total variation distance of random variables. Let

$$d_{TV}(X, Y) := \sup_{\Phi\in\mathcal{B}_b(\mathcal{H}),\|\Phi\|_\infty\le 1} |\mathbb{E}[\Phi(X)] - \mathbb{E}[\Phi(Y)]|$$

denote the total variation distance of two $\mathcal{H}$-valued random variables X, Y, where $\mathcal{H} = \mathbb{R}^d$ or $\mathcal{H} = C^d$, $\mathcal{B}_b(\mathcal{H})$ is the set of bounded and measurable mappings from $\mathcal{H}$ to $\mathbb{R}$, and $\|\Phi\|_\infty := \sup_{x\in\mathcal{H}}|\Phi(x)|$. As a result of Theorem 5.3, we obtain the convergence in the total variation distance

$$\lim_{\Delta\to 0} d_{TV}(x^\xi(t_n), y^{\xi,\Delta}(t_n)) = 0. \tag{5.30}$$

While for the θ-EM functional solution, as a C^d-valued random variable, its law does not converge in total variation distance, namely,

$$\limsup_{\Delta\to 0} d_{TV}(x^\xi_{t_n}, y^{\xi,\Delta}_{t_n}) > 0. \tag{5.31}$$

In order to illustrate (5.31), we consider the test function $\mathbf{1}_{\{\|x\|_{C_{1/2}([-\frac{\Delta}{2},0];\mathbb{R}^d)}<\infty\}}$, where $C_{1/2}([-\frac{\Delta}{2}, 0]; \mathbb{R}^d)$ is the space of all continuous functions $f : [-\frac{\Delta}{2}, 0] \to \mathbb{R}^d$ that is $\frac{1}{2}$-Hölder continuous, equipped with the norm $\|x\|_{C_{1/2}([-\frac{\Delta}{2},0];\mathbb{R}^d)} := \sup_{t,s\in[-\frac{\Delta}{2},0],t\ne s}\frac{|x(t)-x(s)|}{|t-s|^{\frac{1}{2}}}$. On the one hand, the θ-EM functional solution satisfies

$$\|y^{\xi,\Delta}_{t_n}\|_{C_{1/2}([-\frac{\Delta}{2},0];\mathbb{R}^d)} \leq \sup_{t,s\in[-\frac{\Delta}{2},0],t\neq s} \frac{|t-s|^{\frac{1}{2}}}{\Delta}(|y^{\xi,\Delta}(t_{n-1})|+|y^{\xi,\Delta}(t_n)|)$$
$$\leq C(\omega)\Delta^{-\frac{1}{2}} < \infty, \quad \omega\in\Omega,$$

due to the definition (1.39). On the other hand, by the Kolmogorov continuous theorem, there is a modification of the exact solution such that the path is almost surely $(\frac{1}{2}-\epsilon)$-Hölder continuous with any small constant $\epsilon\in(0,\frac{1}{2})$. This leads to

$$\|x^{\xi}_{t_n}\|_{C_{1/2}([-\frac{\Delta}{2},0];\mathbb{R}^d)} = \sup_{t,s\in[-\frac{\Delta}{2},0],t\neq s} \frac{|x(t_n+t)-x(t_n+s)|}{|t-s|^{\frac{1}{2}}} = \infty, \ \mathbb{P}\text{-a.s.}$$

Hence, recalling the definition of the total variation distance, we arrive at

$$\begin{aligned} d_{TV}(x^{\xi}_{t_n}, y^{\xi,\Delta}_{t_n}) \geq |&\mathbb{P}(\|x^{\xi}_{t_n}\|_{C_{1/2}([-\frac{\Delta}{2},0];\mathbb{R}^d)} < \infty) \\ &- \mathbb{P}(\|y^{\xi,\Delta}_{t_n}\|_{C_{1/2}([-\frac{\Delta}{2},0];\mathbb{R}^d)} < \infty)| = 1, \end{aligned}$$

which shows (5.31) by taking the upper limit.

5.3 Convergence Rate for Additive Noise Case

In this section, we consider the additive noise case and study the convergence rate for the density function of the θ-EM method, based on the test-functional-independent weak convergence analysis of the numerical method. Assumptions on the coefficients considered in this subsection are given below.

Assumption 5.5 *Assume that the noise of* (1.1) *is of additive type, i.e., $d=m$, and there exists some constant $\tilde{\sigma}>0$ such that for any $\phi\in C^d$, $\sigma(\phi)\equiv\tilde{\sigma}\mathrm{Id}_{d\times d}$.*

Since $\{-W(t)\}_{t\geq 0}$ is also a Brownian motion whenever $\{W(t)\}_{t\geq 0}$ is a Brownian motion, we only consider the case $\tilde{\sigma}>0$. The next assumption imposes higher-order regularity on the coefficients, which is needed to obtain estimates for the higher-order Malliavin derivatives of the solutions.

Assumption 5.6 *Assume that the coefficient b has continuous and bounded derivatives up to order* 4 *satisfying*

$$\sup_{\phi\in C^d} \|\mathcal{D}^k b(\phi)\|_{\mathcal{L}\left((C^d)^{\otimes k};\mathbb{R}^d\right)} \leq K$$

for $k \in \mathbb{N}_+$ *with* $k \leq 4$. *In addition, assume that b satisfies*

$$|b(\phi)|^2 \leq K\big(1 + |\phi(0)|^2 + \int_{-\tau}^{0} |\phi(r)|^2 \mathrm{d}\nu_2(r)\big), \tag{5.32}$$

for $\phi \in C^d$.

The longtime convergence rate of the numerical density function is stated as follows.

Theorem 5.4 *Let Assumption 2.2 with* $2a_1 - \frac{2\theta-1}{\theta^2} - 2a_2 > 0$, *Assumption 2.5 with* $\rho \geq 1$, *Assumption 3.4 with* $L_b := \sup_{i\in\{1,\ldots,n_b\}} \sup_{\phi\in C^d} |k_b^i(\phi)| < \infty$, *Assumptions 5.5 and 5.6 hold. In addition, assume that b satisfies*

$$\sup_{i\in\{1,\ldots,n_b\}} \sup_{l\in\{1,2,3\}} \sup_{\phi\in C^d} \|\mathcal{D}^l k_b^i(\phi)\|_{\mathcal{L}((C^d)^{\otimes l};\mathbb{R}^{d\times d})} \leq K.$$

Then there exists a constant $\tilde{\Delta} \in (0, 1]$ *such that for any* $\Delta \in (0, \tilde{\Delta}]$,

$$\sup_{T\geq T_0} \sup_{z\in\mathbb{R}^d} |\mathfrak{p}(T, z) - \mathfrak{p}^{\Delta}(T, z)| \leq K\Delta,$$

where $T_0 := \frac{\ln(\frac{3}{2})}{2L_b n_b(\theta+2)}$.

It is worth noting that the threshold T_0, arising from the lower bound estimate of the Malliavin covariance matrix (see Lemma 5.8), is necessary only for establishing the uniform estimate of the density convergence on the infinite time horizon $T > T_0$. This restriction is not required for the fixed time case. Indeed, for each fixed $T > 0$, there exists $K_T > 0$ such that $\sup_{z\in\mathbb{R}^d} |\mathfrak{p}(T, z) - \mathfrak{p}^{\Delta}(T, z)| \leq K_T\Delta$.

Remark 5.3 Under conditions in Theorem 5.4, we can also obtain (5.30). In fact, it follows from the Scheffé lemma and [9, Section 3.1] that for $T \geq T_0$,

$$\int_{\mathbb{R}^d} |\mathfrak{p}(T, z) - \mathfrak{p}^{\Delta}(T, z)|\mathrm{d}z \to 0 \text{ as } \Delta \to 0.$$

The proof of Theorem 5.4 is based on a weak convergence analysis of the θ-EM method. We would like to mention that there have been some works devoted to the weak convergence of the EM method for SFDEs. For instance, weak error estimates have been obtained that depend on a given test functional; see e.g. [7, 12]. When analyzing the convergence rate of the density function of the numerical solution, an effective approach is to apply the Malliavin integration by parts formula to derive a test-functional-independent weak convergence analysis; see [9] for the relevant study of the stochastic heat equation. For SFDEs, the high degeneracy of coefficients makes the derivation of the Malliavin integration by parts formula challenging. In this subsection, we will fully utilize the dimensionality reduction argument presented in Sect. 3.2 to establish the negative moment estimates for the determinant of the Malliavin covariance matrix, which allows us to derive the integration by parts formula; see Lemmas 5.8 and 5.9 for details. Then combining a priori estimates of both the exact solution and the numerical solution, we obtain the test-functional-independent weak convergence analysis for the θ-EM method.

We begin by presenting the relation between the error of density functions and the weak error of the numerical method. According to [9, 23], we have the following approximation of density functions for both the exact solution and the numerical solution: for fixed $T > 0$ and $z \in \mathbb{R}^d$,

$$\mathfrak{p}(T, z) = \lim_{n\to\infty} \int_{\mathbb{R}^d} g_{n^{-1}}(y - z)\mathfrak{p}(T, y)\mathrm{d}y = \lim_{n\to\infty} \mathbb{E}[g_{n^{-1}}(x^{\xi}(T) - z)],$$

$$\mathfrak{p}^{\Delta}(T, z) = \lim_{n\to\infty} \int_{\mathbb{R}^d} g_{n^{-1}}(y - z)\mathfrak{p}^{\Delta}(T, y)\mathrm{d}y = \lim_{n\to\infty} \mathbb{E}[g_{n^{-1}}(y^{\xi,\Delta}(T) - z)],$$

where g_{ζ} denotes the Gaussian density function with mean 0 and covariance matrix $\zeta \mathrm{Id}_{d\times d}$. This gives that

$$|\mathfrak{p}(T, z) - \mathfrak{p}^{\Delta}(T, z)| = \lim_{n\to\infty} |\mathbb{E}[g_{n^{-1}}(x^{\xi}(T) - z)] - \mathbb{E}[g_{n^{-1}}(y^{\xi,\Delta}(T) - z)]|. \tag{5.33}$$

Noting that for any $n \geq 1$, the function $\{g_{n^{-1}}(\cdot - z)\}_{n\geq 1, z\in\mathbb{R}^d}$ belongs to

$$\mathscr{C} := \Big\{ f : \mathbb{R}^d \to \mathbb{R} \,\Big|\, f \in C^{\infty}_{pol}(\mathbb{R}^d; \mathbb{R}), \exists\, F : \mathbb{R}^d \to \mathbb{R} \text{ with } 0 \leq F \leq 1 \text{ such that}$$
$$F(x_1, \ldots, x_d) = \int_{-\infty}^{x_1} \cdots \int_{-\infty}^{x_d} f(y_1, \ldots, y_d)\mathrm{d}y_d \cdots \mathrm{d}y_1 \Big\}, \tag{5.34}$$

we have

$$|\mathfrak{p}(T, z) - \mathfrak{p}^{\Delta}(T, z)| \leq \sup_{f\in\mathscr{C}} |\mathbb{E}[f(x^{\xi}(T))] - \mathbb{E}[f(y^{\xi,\Delta}(T))]|. \tag{5.35}$$

Hence, to obtain the convergence rate of the numerical density function, it suffices to estimate the error $|\mathbb{E}[f(x^{\xi}(T))] - \mathbb{E}[f(y^{\xi,\Delta}(T))]|$ for any $f \in \mathscr{C}$.

Here we rewrite the decomposition of the error in (3.11) for $f \in \mathscr{C}$ in the additive noise case as follows

$$\begin{aligned}
&\mathbb{E}[f(x^{\xi}(T))] - \mathbb{E}[f(y^{\xi,\Delta}(T))] \\
&= \mathbb{E}[f(\Phi(T; t_{N^{\Delta}}, \varphi^{Int}_{t_{N^{\Delta}}}(0, \xi)))] - \mathbb{E}[f(\Phi(T; t_{N^{\Delta}}, \varphi^{Int}_{t_{N^{\Delta}}}(0, \xi^{Int})))] \\
&\quad + \sum_{i=1}^{N^{\Delta}} \int_0^1 \mathbb{E}\Big[\big\langle f'(\Phi(T; t_i, Y_i^{\varsigma}))\mathcal{D}\Phi(T; t_i, Y_i^{\varsigma}), I^{[0]}(\mathcal{I}_b^i + \mathcal{I}_{b,\theta}^i)\big\rangle\Big]\mathrm{d}\varsigma \\
&= \mathcal{I}_0 + \mathcal{I}_b + \mathcal{I}_{b,\theta},
\end{aligned} \tag{5.36}$$

where $\mathcal{I}_0, \mathcal{I}_b$ and $\mathcal{I}_{b,\theta}$ coincide with those of (3.11).

Based on this decomposition of the weak error, we will prove that the test-functional-independent weak convergence rate is 1 (see Theorem 5.10). Then the proof of Theorem 5.4 follows from (5.35). To proceed, we make some preparations in the following subsection.

5.3.1 A Priori Estimates

This subsection gives some moment estimates for derivatives of the exact solution and numerical solution. In addition, we also present the negative moment estimates of the determinant of the Malliavin covariance matrix for the numerical solution. We begin by showing that the moments of high-order derivatives of the exact solution and numerical solution are bounded, which are stated in Lemmas 5.5 to 5.7.

Lemma 5.5 *Let Assumptions 2.2 and 5.5, and* (5.32) *in Assumption 5.6 hold. Then for any $p \in \mathbb{N}_+$ with $p \geq 2$,*

$$\sup_{t\geq 0} \mathbb{E}[\|x_t^\xi\|^{2p}] \leq K(1+\|\xi\|^{2p}). \tag{5.37}$$

In addition, there exists a constant $\tilde{\Delta} := \tilde{\Delta}(p) \in (0,1]$ such that for any $\Delta \in (0, \tilde{\Delta}]$,

$$\sup_{k\in\mathbb{N}} \mathbb{E}[\|y_{t_k}^{\xi,\Delta}\|^{2p}] \leq K(1+\|\xi\|^{2p}). \tag{5.38}$$

Proof The proof of (5.37) is similar to that of [28, Theorem 3] and is omitted. Below we show the proof of (5.38). The proof is split into three steps.

Step 1: In this step, we give a one-step recursion for the $2p$th moment of $z^{\xi,\Delta}$ and show that

$$\begin{aligned}
\mathbb{E}[|z^{\xi,\Delta}(t_{k+1})|^{2p}] \leq\ & \mathbb{E}[|y^{\xi,\Delta}(t_k)|^{2p}] \\
& + K_{\varepsilon_1}\Delta - \Big(2p(1-\theta)\big(a_1 - \frac{a_2(p-1)}{p}\big) - K\Delta - K\varepsilon_1\Big)\Delta \\
& \times \mathbb{E}[|y^{\xi,\Delta}(t_k)|^{2p}] \\
& + \Big(2a_2(1-\theta) + K\Delta\Big)\Delta\mathbb{E}\Big[\int_{-\tau}^{0} |y_{t_k}^{\xi,\Delta}(r)|^{2p} \mathrm{d}\nu_2(r)\Big].
\end{aligned} \tag{5.39}$$

By virtue of (1.43), we have that for any $k \in \mathbb{N}$,

$$
\begin{aligned}
|z^{\xi,\Delta}(t_{k+1})|^2 &= |y^{\xi,\Delta}(t_k) + (1-\theta)b(y^{\xi,\Delta}_{t_k})\Delta + \tilde{\sigma}\delta W_k|^2 \\
&= |y^{\xi,\Delta}(t_k)|^2 + (1-\theta)^2|b(y^{\xi,\Delta}_{t_k})|^2\Delta^2 + |\tilde{\sigma}\delta W_k|^2 \\
&\quad + 2(1-\theta)\Delta\langle y^{\xi,\Delta}(t_k), b(y^{\xi,\Delta}_{t_k})\rangle \\
&\quad + 2\langle y^{\xi,\Delta}(t_k) + (1-\theta)b(y^{\xi,\Delta}_{t_k})\Delta, \tilde{\sigma}\delta W_k\rangle.
\end{aligned}
$$

Then for any $p \in \mathbb{N}_+$ with $p \geq 2$,

$$
\mathbb{E}[|z^{\xi,\Delta}(t_{k+1})|^{2p}] = \mathbb{E}[|y^{\xi,\Delta}(t_k)|^{2p}] + \sum_{i=1}^{p} C_p^i \mathbb{E}[I_i], \tag{5.40}
$$

where

$$
\begin{aligned}
I_i := |y^{\xi,\Delta}(t_k)|^{2(p-i)}\Big(&(1-\theta)^2|b(y^{\xi,\Delta}_{t_k})|^2\Delta^2 + |\tilde{\sigma}\delta W_k|^2 \\
&+ 2(1-\theta)\Delta\langle y^{\xi,\Delta}(t_k), b(y^{\xi,\Delta}_{t_k})\rangle \\
&+ 2\langle y^{\xi,\Delta}(t_k) + (1-\theta)b(y^{\xi,\Delta}_{t_k})\Delta, \tilde{\sigma}\delta W_k\rangle\Big)^i.
\end{aligned}
$$

For the term I_1, it follows from the property of the conditional expectation, Assumption 2.2, and (5.32) that for $\varepsilon_1 \in (0,1)$,

$$
\begin{aligned}
\mathbb{E}[I_1] &\leq \mathbb{E}\Big[|y^{\xi,\Delta}(t_k)|^{2(p-1)}\Big((1-\theta)^2|b(y^{\xi,\Delta}_{t_k})|^2\Delta^2 + |\tilde{\sigma}|^2\Delta + K\Delta \\
&\quad - 2a_1(1-\theta)\Delta|y^{\xi,\Delta}(t_k)|^2 + 2a_2(1-\theta)\Delta\int_{-\tau}^{0}|y^{\xi,\Delta}_{t_k}(r)|^2 \mathrm{d}\nu_2(r)\Big)\Big] \\
&\leq \mathbb{E}\Big[|y^{\xi,\Delta}(t_k)|^{2(p-1)}\big(K\Delta + (1-\theta)^2|b(y^{\xi,\Delta}_{t_k})|^2\Delta^2\big)\Big] - 2(a_1 - \frac{a_2(p-1)}{p}) \\
&\quad \times (1-\theta)\Delta\mathbb{E}[|y^{\xi,\Delta}(t_k)|^{2p}] + \frac{2a_2}{p}(1-\theta)\Delta\mathbb{E}\Big[\int_{-\tau}^{0}|y^{\xi,\Delta}_{t_k}(r)|^{2p}\mathrm{d}\nu_2(r)\Big] \\
&\leq K_{\varepsilon_1}\Delta + K\Delta^2\Big(\mathbb{E}[|y^{\xi,\Delta}(t_k)|^{2p}] + \mathbb{E}\Big[\int_{-\tau}^{0}|y^{\xi,\Delta}_{t_k}(r)|^{2p}\mathrm{d}\nu_2(r)\Big]\Big) \\
&\quad - 2(a_1 - \frac{a_2(p-1)}{p} - \varepsilon_1)(1-\theta)\Delta\mathbb{E}[|y^{\xi,\Delta}(t_k)|^{2p}] \\
&\quad + \frac{2a_2}{p}(1-\theta)\Delta\mathbb{E}\Big[\int_{-\tau}^{0}|y^{\xi,\Delta}_{t_k}(r)|^{2p}\mathrm{d}\nu_2(r)\Big],
\end{aligned} \tag{5.41}
$$

where we used the Young inequality in the last step. For the term I_2, using (5.32) and the Young inequality again, we have

$$\begin{aligned}
\mathbb{E}[I_2] &= \mathbb{E}\Big[|y^{\xi,\Delta}(t_k)|^{2(p-2)}\Big((1-\theta)^2|b(y^{\xi,\Delta}_{t_k})|^2\Delta^2 + |\tilde{\sigma}\delta W_k|^2 \\
&\quad + 2(1-\theta)\Delta\langle y^{\xi,\Delta}(t_k), b(y^{\xi,\Delta}_{t_k})\rangle\Big)^2 \\
&\quad + |y^{\xi,\Delta}(t_k)|^{2(p-2)}\Big(2\langle y^{\xi,\Delta}(t_k) + (1-\theta)b(y^{\xi,\Delta}_{t_k})\Delta, \tilde{\sigma}\delta W_k\rangle\Big)^2\Big] \\
&\leq K\Delta^2\Big(1 + \mathbb{E}[|y^{\xi,\Delta}(t_k)|^{2p}] + \mathbb{E}\Big[\int_{-\tau}^{0}|y^{\xi,\Delta}_{t_k}(r)|^{2p}\mathrm{d}\nu_2(r)\Big]\Big) \\
&\quad + K\Big(\Delta\mathbb{E}[|y^{\xi,\Delta}(t_k)|^{2(p-1)}] + \Delta^3\mathbb{E}[|y^{\xi,\Delta}(t_k)|^{2(p-2)}|b(y^{\xi,\Delta}_{t_k})|^2]\Big) \\
&\leq K_{\varepsilon_1}\Delta + K\Delta^2\Big(\mathbb{E}[|y^{\xi,\Delta}(t_k)|^{2p}] + \mathbb{E}\Big[\int_{-\tau}^{0}|y^{\xi,\Delta}_{t_k}(r)|^{2p}\mathrm{d}\nu_2(r)\Big]\Big) \\
&\quad + \varepsilon_1\Delta\mathbb{E}[|y^{\xi,\Delta}(t_k)|^{2p}]. \qquad (5.42)
\end{aligned}$$

Similarly, we derive that for $i = 3, \ldots, p$,

$$\begin{aligned}
\mathbb{E}[I_i] &\leq K_{\varepsilon_1}\Delta + K\Delta^2\Big(\mathbb{E}[|y^{\xi,\Delta}(t_k)|^{2p}] + \mathbb{E}\Big[\int_{-\tau}^{0}|y^{\xi,\Delta}_{t_k}(r)|^{2p}\mathrm{d}\nu_2(r)\Big]\Big) \\
&\quad + \varepsilon_1\Delta\mathbb{E}[|y^{\xi,\Delta}(t_k)|^{2p}]. \qquad (5.43)
\end{aligned}$$

Inserting (5.41)–(5.43) into (5.40), one has (5.39).

Step 2: In this step, we show that the $2p$th moment of $z^{\xi,\Delta}(t_k)$ is bounded uniformly with $k \geq -N$, i.e. $\sup_{k\geq -N}\mathbb{E}[|z^{\xi,\Delta}(t_k)|^{2p}] \leq K(1 + \|\xi\|^{2p})$.

Noting that for $\epsilon \in (0,1)$, we have the identity

$$\begin{aligned}
&e^{\epsilon t_{k+1}}\mathbb{E}[|z^{\xi,\Delta}(t_{k+1})|^{2p}] \\
&= \sum_{i=0}^{k}\Big(e^{\epsilon t_{i+1}}\mathbb{E}[|z^{\xi,\Delta}(t_{i+1})|^{2p}] - e^{\epsilon t_i}\mathbb{E}[|z^{\xi,\Delta}(t_i)|^{2p}]\Big) + \mathbb{E}[|z^{\xi,\Delta}(0)|^{2p}].
\end{aligned}$$

Plugging (5.39) into the above identity and using (2.27), we derive

$$\begin{aligned}
&e^{\epsilon t_{k+1}}\mathbb{E}[|z^{\xi,\Delta}(t_{k+1})|^{2p}] \\
&\leq \sum_{i=0}^{k} e^{\epsilon t_{i+1}}\mathbb{E}[|y^{\xi,\Delta}(t_i)|^{2p}] - \sum_{i=0}^{k} e^{\epsilon t_i}\mathbb{E}[|z^{\xi,\Delta}(t_i)|^{2p}] + K_{\varepsilon_1}\Delta\sum_{i=0}^{k} e^{\epsilon t_{i+1}}
\end{aligned}$$

$$
\begin{aligned}
&- \Big(2p(1-\theta)\big(a_1 - \frac{a_2(p-1)}{p}\big) - K\Delta - K\varepsilon_1\Big)\Delta \sum_{i=0}^{k} e^{\epsilon t_{i+1}} \mathbb{E}[|y^{\xi,\Delta}(t_i)|^{2p}] \\
&+ \Big(2a_2(1-\theta) + K\Delta\Big)\Delta \sum_{i=0}^{k} e^{\epsilon t_{i+1}} \mathbb{E}\Big[\int_{-\tau}^{0} |y^{\xi,\Delta}_{t_i}(r)|^{2p} \mathrm{d}\nu_2(r)\Big] \\
&+ \mathbb{E}[|z^{\xi,\Delta}(0)|^{2p}] \\
\le &\sum_{i=0}^{k} e^{\epsilon t_{i+1}} \mathbb{E}[|y^{\xi,\Delta}(t_i)|^{2p}] - \sum_{i=0}^{k} e^{\epsilon t_i} \mathbb{E}[|z^{\xi,\Delta}(t_i)|^{2p}] + K_{\varepsilon_1,\epsilon} e^{\epsilon t_{k+1}} + K\|\xi\|^{2p} \\
&- \Big(2p(1-\theta)\big(a_1 - a_2 e^{\epsilon\tau}\big) - K\Delta - K\varepsilon_1\Big)\Delta \sum_{i=0}^{k} e^{\epsilon t_{i+1}} \mathbb{E}[|y^{\xi,\Delta}(t_i)|^{2p}].
\end{aligned}
\tag{5.44}
$$

According to $z^{\xi,\Delta}(t_k) = y^{\xi,\Delta}(t_k) - \theta b(y^{\xi,\Delta}_{t_k})\Delta$ and Assumption 2.2, we obtain

$$
|z^{\xi,\Delta}(t_k)|^2 \ge -K\Delta + (1 + 2a_1\theta\Delta)|y^{\xi,\Delta}(t_k)|^2 - 2a_2\theta\Delta \int_{\tau}^{0} |y^{\xi,\Delta}_{t_k}(r)|^2 \mathrm{d}\nu_2(r).
$$

This implies

$$
\begin{aligned}
&(1 + 2a_1\theta\Delta)^p \sum_{i=0}^{k} e^{\epsilon t_i} |y^{\xi,\Delta}(t_i)|^{2p} \\
&\le (1+\varepsilon_1) \sum_{i=0}^{k} e^{\epsilon t_i} \Big(|z^{\xi,\Delta}(t_i)|^2 + 2a_2\theta\Delta \int_{-\tau}^{0} |y^{\xi,\Delta}_{t_i}(r)|^2 \mathrm{d}\nu_2(r)\Big)^p \\
&\quad + K_{\varepsilon_1}\Delta^p \sum_{i=0}^{k} e^{\epsilon t_i} \\
&\le (1+\varepsilon_1) \sum_{i=0}^{k} \sum_{l=0}^{p} \frac{C_p^l (2a_2\theta\Delta)^l (p-l)}{p} e^{\epsilon t_i} |z^{\xi,\Delta}(t_i)|^{2p} \\
&\quad + (1+\varepsilon_1) \sum_{i=0}^{k} \sum_{l=0}^{p} \frac{C_p^l (2a_2\theta\Delta)^l l}{p} e^{\epsilon t_i} \int_{-\tau}^{0} |y^{\xi,\Delta}_{t_i}(r)|^{2p} \mathrm{d}\nu_2(r) + K_{\varepsilon_1}\Delta e^{\epsilon t_k} \\
&\le (1+\varepsilon_1) \sum_{i=0}^{k} \sum_{l=0}^{p} \frac{C_p^l (2a_2\theta\Delta)^l (p-l)}{p} e^{\epsilon t_i} |z^{\xi,\Delta}(t_i)|^{2p}
\end{aligned}
$$

$$
+ (1+\varepsilon_1)e^{\epsilon\tau}\sum_{i=0}^{k}\sum_{l=0}^{p}\frac{C_p^l(2a_2\theta\Delta)^l l}{p}e^{\epsilon t_i}|y^{\xi,\Delta}(t_i)|^{2p} + K\|\xi\|^{2p} + K_{\varepsilon_1}\Delta e^{\epsilon t_k}. \tag{5.45}
$$

Owing to $a_1 > a_2$, choose sufficiently small numbers $\varepsilon_1, \epsilon > 0$ such that $a_1 > a_2(1+\varepsilon_1)e^{\epsilon\tau}$. Combining (5.45) and the relation

$$
\begin{aligned}
(1+\varepsilon_1)e^{\epsilon\tau}\sum_{l=0}^{p}\frac{C_p^l(2a_2\theta\Delta)^l l}{p} &\le \sum_{l=0}^{p}\frac{C_p^l(2a_2\theta(1+\varepsilon_1)e^{\epsilon\tau}\Delta)^l l}{p} \\
&\le \sum_{l=0}^{p}\frac{C_p^l(2a_1\theta\Delta)^l l}{p}
\end{aligned}
$$

and

$$
\begin{aligned}
&(1+2a_1\theta\Delta)^p - (1+\varepsilon_1)e^{\epsilon\tau}\sum_{l=0}^{p}\frac{C_p^l(2a_2\theta\Delta)^l l}{p} \\
&\ge \sum_{l=0}^{p}C_p^l(2a_1\theta\Delta)^l - \sum_{l=0}^{p}\frac{C_p^l(2a_1\theta\Delta)^l l}{p} \\
&= \sum_{l=0}^{p}\frac{C_p^l(2a_1\theta\Delta)^l(p-l)}{p},
\end{aligned}
$$

we arrive at

$$
\begin{aligned}
&\sum_{i=0}^{k}\sum_{l=0}^{p}\frac{C_p^l(2a_1\theta\Delta)^l(p-l)}{p}e^{\epsilon t_i}|y^{\xi,\Delta}(t_i)|^{2p} \\
&\quad\le (1+\varepsilon_1)\sum_{i=0}^{k}\sum_{l=0}^{p}\frac{C_p^l(2a_2\theta\Delta)^l(p-l)}{p}e^{\epsilon t_i}|z^{\xi,\Delta}(t_i)|^{2p} \\
&\quad\quad + K\|\xi\|^{2p} + K_{\varepsilon_1}\Delta e^{\epsilon t_k}.
\end{aligned}
$$

Plugging the above inequality into (5.44), we deduce

$$
\begin{aligned}
&e^{\epsilon t_{k+1}}\mathbb{E}[|z^{\xi,\Delta}(t_{k+1})|^{2p}] \\
&\le \left(e^{\epsilon\Delta} - \frac{\sum_{l=0}^{p}C_p^l(2a_1\theta\Delta)^l(p-l)}{\sum_{l=0}^{p}C_p^l(2a_2(1+\varepsilon_1)\theta\Delta)^l(p-l)}\right)
\end{aligned}
$$

$$\times \sum_{i=0}^{k} e^{\epsilon t_i} \mathbb{E}[|y^{\xi,\Delta}(t_i)|^{2p}] + K_{\varepsilon_1,\epsilon} e^{\epsilon t_{k+1}} + K\|\xi\|^{2p}$$

$$-\Big(2p(1-\theta)\big(a_1 - a_2 e^{\epsilon\tau}\big) - K\Delta - K\varepsilon_1\Big)\Delta \sum_{i=0}^{k} e^{\epsilon t_{i+1}} \mathbb{E}|y^{\xi,\Delta}(t_i)|^{2p}].$$

Since

$$\begin{aligned}
&\frac{\sum_{l=0}^{p} C_p^l (2a_1\theta\Delta)^l (p-l)}{\sum_{l=0}^{p} C_p^l (2a_2(1+\varepsilon_1)\theta\Delta)^l (p-l)} \\
&\geq 1 + \frac{\sum_{l=1}^{p} C_p^l (2(a_1 - a_2(1+\varepsilon_1))\theta\Delta)^l (p-l)}{\sum_{l=0}^{p} C_p^l (2a_2(1+\varepsilon_1)\theta\Delta)^l (p-l)} \\
&\geq 1 + \frac{C_p^1 (p-1) 2(a_1 - a_2(1+\varepsilon_1))\theta}{\sum_{l=0}^{p} C_p^l (2a_2(1+\varepsilon_1)\theta)^l p}\Delta \\
&= 1 + \frac{(p-1)2(a_1 - a_2(1+\varepsilon_1))\theta}{(2a_2(1+\varepsilon_1)\theta + 1)^p}\Delta,
\end{aligned}$$

it is straightforward to see that

$$\begin{aligned}
&e^{\epsilon t_{k+1}} \mathbb{E}[|z^{\xi,\Delta}(t_{k+1})|^{2p}] \\
&\leq \Big(e^{\epsilon\Delta} - 1 - \frac{(p-1)2(a_1 - a_2(1+\varepsilon_1))\theta}{(2a_2(1+\varepsilon_1)\theta + 1)^p}\Delta\Big) \sum_{i=0}^{k} e^{\epsilon t_i} \mathbb{E}[|y^{\xi,\Delta}(t_i)|^{2p}] \\
&\quad + K_{\varepsilon_1,\epsilon} e^{\epsilon t_{k+1}} + K\|\xi\|^{2p} - \Big(2p(1-\theta)\big(a_1 - a_2 e^{\epsilon\tau}\big) - K\Delta - K\varepsilon_1\Big)\Delta \\
&\quad \times \sum_{i=0}^{k} e^{\epsilon t_{i+1}} \mathbb{E}[|y^{\xi,\Delta}(t_i)|^{2p}].
\end{aligned} \tag{5.46}$$

It follows from $e^{\epsilon\Delta} \leq 1 + \epsilon\Delta + \frac{e^{\epsilon\Delta}}{2}(\epsilon\Delta)^2$ that

$$\begin{aligned}
&e^{\epsilon\Delta} - 1 - \frac{(p-1)2(a_1 - a_2(1+\varepsilon_1))\theta}{(2a_2(1+\varepsilon_1)\theta + 1)^p}\Delta \\
&\quad - \Big(2p(1-\theta)\big(a_1 - a_2 e^{\epsilon\tau}\big) - K\Delta - K\varepsilon_1\Big)\Delta e^{\epsilon\Delta} \\
&\leq \Big(\epsilon - \frac{(p-1)2(a_1 - a_2(1+\varepsilon_1))\theta}{(2a_2(1+\varepsilon_1)\theta + 1)^p}\Big)\Delta \\
&\quad - \Big(2p(1-\theta)\big(a_1 - a_2 e^{\epsilon\tau}\big) - K\Delta - K\varepsilon_1 - \frac{\epsilon^2}{2}\Delta\Big)\Delta e^{\epsilon\Delta}.
\end{aligned}$$

There exist sufficiently small constants $\varepsilon_1, \epsilon > 0$ such that

$$\epsilon - \frac{(p-1)2(a_1 - a_2(1+\varepsilon_1))\theta}{(2a_2(1+\varepsilon_1)\theta+1)^p} \le 0,$$
$$p(1-\theta)\big(a_1 - a_2 e^{\epsilon\tau}\big) \ge K\varepsilon_1.$$

Further choose sufficiently small $\tilde{\Delta} := \tilde{\Delta}(p) \in (0, 1]$ such that

$$p(1-\theta)\big(a_1 - a_2 e^{\epsilon\tau}\big) \ge K\tilde{\Delta} - \frac{\epsilon^2}{2}\tilde{\Delta}.$$

Thus for any $\Delta \in (0, \tilde{\Delta}]$,

$$\begin{aligned} &e^{\epsilon\Delta} - 1 - \frac{(p-1)2(a_1 - a_2(1+\varepsilon_1))\theta}{(2a_2(1+\varepsilon_1)\theta+1)^p}\Delta \\ &- \Big(2p(1-\theta)\big(a_1 - a_2 e^{\epsilon\tau}\big) - K\Delta - K\varepsilon_1\Big)\Delta e^{\epsilon\Delta} \le 0. \end{aligned}$$

Combining (5.46), we conclude that

$$e^{\epsilon t_{k+1}}\mathbb{E}[|z^{\xi,\Delta}(t_{k+1})|^{2p}] \le K_{\varepsilon_1,\epsilon}e^{\epsilon t_{k+1}} + K\|\xi\|^{2p},$$

which yields $\sup_{k\ge -N}\mathbb{E}[|z^{\xi,\Delta}(t_k)|^{2p}] \le K(1+\|\xi\|^{2p})$.

Step 3: By using the moment estimate of $z^{\xi,\Delta}$ obtained in Step 2, along with $z^{\xi,\Delta}(t_k) = y^{\xi,\Delta}(t_k) - \theta b(y^{\xi,\Delta}_{t_k})\Delta$, we have that $\sup_{k\ge -N}\mathbb{E}[|y^{\xi,\Delta}(t_k)|^{2p}] \le K(1+\|\xi\|^{2p})$. Furthermore, similar to the proof of Proposition 2.4, we conclude that $\sup_{k\in\mathbb{N}}\mathbb{E}[\|y^{\xi,\Delta}_{t_k}\|^{2p}] \le K(1+\|\xi\|^{2p})$. The proof is completed. □

Lemma 5.6 *Let $n \in \{1, 2\}$. Under Assumptions 2.2, 5.5, and 5.6, there exists a constant $\hat{\lambda}_4 > 0$ such that for any $p \in \mathbb{N}_+$ with $p \ge 2$, the following hold.*

(i) For $t > r_1 \vee \cdots \vee r_n$, we have

$$\begin{aligned} &\mathbb{E}[\|D_{(r_1,\dots,r_n)}x^{\tilde{\xi}}_t\|^{2p}] \\ &\le Ke^{-\hat{\lambda}_4 p(t-r_1\vee\cdots\vee r_n)}\Big(1+\sum_{k=0}^{n}\sum_{1\le i_1<\cdots<i_k\le n}\mathbb{E}[\|D_{(r_{i_1},\dots,r_{i_k})}\tilde{\xi}\|^{2\tilde{p}}]\Big); \end{aligned}$$

(ii) For $t > 0$, we have

$$\mathbb{E}[\|\mathcal{D}x^{\tilde{\xi}}_t\cdot\eta\|^{2p}] \le Ke^{-\hat{\lambda}_4 pt}\|\eta\|^{2p},$$

and

$$\mathbb{E}[\|D_{(r_1,\cdots,r_n)}\mathcal{D}x_t^{\tilde{\xi}}\cdot\eta\|^{2p}]$$
$$\leq Ke^{-\hat{\lambda}_4 pt}\|\eta\|^{2p}\Big(1+\sum_{k=0}^{n}\sum_{1\leq i_1<\cdots<i_k\leq n}\mathbb{E}[\|D_{(r_{i_1}\ldots,r_{i_k})}\tilde{\xi}\|^{2\tilde{p}}]\Big).$$

Here, $r_1,\ldots,r_n>0$, $\tilde{p}\geq p$ is some constant, $\tilde{\xi}\in\mathbb{D}^{n,2\tilde{p}}(C^d)$, and we adopt the convention that $D_{(r_{i_1},\ldots,r_{i_k})}\tilde{\xi}:=\tilde{\xi}$ for $k=0$.

Proof The proofs are similar to those of Lemmas 3.4, 3.5, 3.10, and 3.11, and hence are omitted. □

Lemma 5.7 *Let $n\in\{1,2,3,4\}$ and $\tilde{n}\in\{1,2,3\}$. Let Assumption 2.2 with $2a_1-\frac{2\theta-1}{\theta^2}-2a_2>0$, Assumptions 5.5 and 5.6 hold. Then there exists a constant $\hat{\lambda}_5>0$ such that for any $p\in\mathbb{N}_+$ and $\Delta\in(0,\tilde{\Delta}]$ with $\tilde{\Delta}:=\tilde{\Delta}(p)$, the following hold.*

(i) For $t_k>r_1\vee\cdots\vee r_n$, we have

$$\mathbb{E}[\|D_{(r_1,\ldots,r_n)}y_{t_k}^{\tilde{\xi},\Delta}\|^{2p}]$$
$$\leq Ke^{-\hat{\lambda}_5 p(t_k-r_1\vee\cdots\vee r_n)}\Big(1+\sum_{k=0}^{n}\sum_{1\leq i_1<\cdots<i_k\leq n}\mathbb{E}[\|D_{(r_{i_1},\ldots,r_{i_k})}\tilde{\xi}\|^{2\tilde{p}}]\Big);$$

(ii) For $t_k>0$, we have

$$\mathbb{E}[\|\mathcal{D}y_{t_k}^{\tilde{\xi},\Delta}\cdot\eta\|^{2p}]\leq Ke^{-\hat{\lambda}_5 pt_k}\|\eta\|^{2p},$$

and

$$\mathbb{E}[\|D_{(r_1,\ldots,r_{\tilde{n}})}\mathcal{D}y_{t_k}^{\tilde{\xi},\Delta}\cdot\eta\|^{2p}]$$
$$\leq Ke^{-\hat{\lambda}_5 pt_k}\|\eta\|^{2p}\Big(1+\sum_{k=0}^{\tilde{n}}\sum_{1\leq i_1<\cdots<i_k\leq\tilde{n}}\mathbb{E}[\|D_{(r_{i_1},\ldots,r_{i_k})}\tilde{\xi}\|^{2\tilde{p}}]\Big).$$

Here, $r_1,\ldots,r_n>0$, $\tilde{p}\geq p$ is some constant, $\tilde{\xi}\in\mathbb{D}^{4,2\tilde{p}}(C^d)$, and we adopt the convention that $D_{(r_{i_1},\ldots,r_{i_k})}\tilde{\xi}:=\tilde{\xi}$ for $k=0$.

Proof We only show the estimate of $\mathcal{D}y_{t_k}^{\tilde{\xi},\Delta}\cdot\eta$, since other terms can be estimated similarly. Similar to the proof of (3.17), by Assumptions 2.2 and 5.5, we derive

$$
\begin{aligned}
&|\mathcal{D}z^{\tilde{\xi},\Delta}(t_{k+1})\cdot\eta|^2\\
&=|\mathcal{D}z^{\tilde{\xi},\Delta}(t_k)\cdot\eta|^2+|\mathcal{D}b(y_{t_k}^{\tilde{\xi},\Delta})\mathcal{D}y_{t_k}^{\tilde{\xi},\Delta}\cdot\eta|^2\Delta^2\\
&\quad+2\langle\mathcal{D}y^{\tilde{\xi},\Delta}(t_k)\cdot\eta-\theta\Delta\mathcal{D}b(y_{t_k}^{\tilde{\xi},\Delta})\mathcal{D}y_{t_k}^{\tilde{\xi},\Delta}\cdot\eta,\mathcal{D}b(y_{t_k}^{\tilde{\xi},\Delta})\mathcal{D}y_{t_k}^{\tilde{\xi},\Delta}\cdot\eta\rangle\Delta\\
&\le A_{\theta,\Delta}|\mathcal{D}z^{\tilde{\xi},\Delta}(t_k)\cdot\eta|^2-\Big(2a_1-\frac{2\theta-1}{\theta^2}\Big)\Delta|\mathcal{D}y^{\tilde{\xi},\Delta}(t_k)\cdot\eta|^2\\
&\quad+2a_2\Delta\int_{-\tau}^{0}|\mathcal{D}y_{t_k}^{\tilde{\xi},\Delta}(r)\cdot\eta|^2\mathrm{d}\nu_2(r),
\end{aligned}
$$

where $A_{\theta,\Delta}=\frac{(1-\theta)^2}{\theta^2}+\frac{2\theta-1}{\theta^2(1+\Delta)}$. For $\hat{\lambda}_5>0$, we arrive at

$$
\begin{aligned}
&e^{\hat{\lambda}_5 t_{k+1}}|\mathcal{D}z^{\tilde{\xi},\Delta}(t_{k+1})\cdot\eta|^2\\
&\le|\mathcal{D}z^{\tilde{\xi},\Delta}(0)\cdot\eta|^2+(A_{\theta,\Delta}e^{\hat{\lambda}_5\Delta}-1)\sum_{i=0}^{k}e^{\hat{\lambda}_5 t_i}|\mathcal{D}z^{\tilde{\xi},\Delta}(t_k)\cdot\eta|^2\\
&\quad-\Big(2a_1-\frac{2\theta-1}{\theta^2}\Big)\Delta\sum_{i=0}^{k}e^{\hat{\lambda}_5 t_{i+1}}|\mathcal{D}y^{\tilde{\xi},\Delta}(t_i)\cdot\eta|^2\\
&\quad+2a_2\Delta\sum_{i=0}^{k}e^{\hat{\lambda}_5 t_{i+1}}\int_{-\tau}^{0}|\mathcal{D}y_{t_i}^{\tilde{\xi},\Delta}(r)\cdot\eta|^2\mathrm{d}\nu_2(r)\\
&\le K\|\eta\|^2+(A_{\theta,\Delta}e^{\hat{\lambda}_5\Delta}-1)\sum_{i=0}^{k}e^{\hat{\lambda}_5 t_i}|\mathcal{D}z^{\tilde{\xi},\Delta}(t_k)\cdot\eta|^2\\
&\quad-\Big(2a_1-\frac{2\theta-1}{\theta^2}-2a_2e^{\hat{\lambda}_5\tau}\Big)\Delta\sum_{i=0}^{k}e^{\hat{\lambda}_5 t_{i+1}}|\mathcal{D}y^{\tilde{\xi},\Delta}(t_i)\cdot\eta|^2.
\end{aligned}
$$

Note that $A_{\theta,\Delta}=1-\frac{(2\theta-1)\Delta}{\theta^2(1+\Delta)}\le e^{-\frac{(2\theta-1)\Delta}{2\theta^2}}$. Choose a sufficiently small number $\hat{\lambda}_5>0$ such that $A_{\theta,\Delta}e^{\hat{\lambda}_5\Delta}<1$ and $2a_1-\frac{2\theta-1}{\theta^2}-2a_2e^{\hat{\lambda}_5\tau}>0$. Then

$$
e^{\hat{\lambda}_5 t_{k+1}}|\mathcal{D}z^{\tilde{\xi},\Delta}(t_{k+1})\cdot\eta|^2\le K\|\eta\|^2,\quad\mathbb{P}\text{-a.s.}\tag{5.47}
$$

According to $\mathcal{D}z^{\tilde{\xi},\Delta}(t_k)\cdot\eta=\mathcal{D}y^{\tilde{\xi},\Delta}(t_k)\cdot\eta-\theta\Delta\mathcal{D}b(y_{t_k}^{\tilde{\xi},\Delta})\mathcal{D}y_{t_k}^{\tilde{\xi},\Delta}\cdot\eta$ and Assumption 2.2, we obtain

$$
\begin{aligned}
&(1+2a_1\theta\Delta)\sup_{0\le i\le k} e^{\hat{\lambda}_5 t_i}|\mathcal{D}y^{\tilde{\xi},\Delta}(t_i)\cdot\eta|^2\\
&\le \sup_{0\le i\le k} e^{\hat{\lambda}_5 t_i}|\mathcal{D}z^{\tilde{\xi},\Delta}(t_i)\cdot\eta|^2+2a_2\theta\Delta\sup_{0\le i\le k} e^{\hat{\lambda}_5 t_i}\int_{-\tau}^{0}|\mathcal{D}y^{\tilde{\xi},\Delta}_{t_i}(r)\cdot\eta|^2\mathrm{d}\nu_2(r)\\
&\le K\|\eta\|^2+\sup_{0\le i\le k} e^{\hat{\lambda}_5 t_i}|\mathcal{D}z^{\tilde{\xi},\Delta}(t_i)\cdot\eta|^2+2a_2\theta e^{\hat{\lambda}_5\tau}\Delta\sup_{0\le i\le k} e^{\hat{\lambda}_5 t_i}|\mathcal{D}y^{\tilde{\xi},\Delta}(t_i)\cdot\eta|^2.
\end{aligned}
$$

This, along with $a_2 e^{\hat{\lambda}_5\tau}<a_1$ and (5.47), implies that

$$
\sup_{0\le i\le k} e^{\hat{\lambda}_5 t_i}|\mathcal{D}y^{\tilde{\xi},\Delta}(t_i)\cdot\eta|^2\le K\|\eta\|^2,\quad \mathbb{P}\text{-a.s.}
$$

Furthermore, we conclude

$$
\begin{aligned}
\sup_{0\le i\le k} e^{\hat{\lambda}_5 t_i}\|\mathcal{D}y^{\tilde{\xi},\Delta}_{t_i}\cdot\eta\|^2 &\le e^{\hat{\lambda}_5\tau}\sup_{0\le i\le k}\ \sup_{i-N\le l\le i} e^{\hat{\lambda}_5 t_l}\|\mathcal{D}y^{\tilde{\xi},\Delta}(t_l)\cdot\eta\|^2\\
&\le K\|\eta\|^2,\quad \mathbb{P}\text{-a.s}
\end{aligned}
$$

and

$$
\sup_{0\le i\le k} e^{\hat{\lambda}_5 p t_i}\|\mathcal{D}y^{\tilde{\xi},\Delta}_{t_i}\cdot\eta\|^{2p}\le K\|\eta\|^{2p},\quad \mathbb{P}\text{-a.s.}
$$

The desired argument follows by taking expectations. □

The following lemma shows the negative moment estimates of the determinant of the Malliavin covariance matrix for the numerical solution.

Lemma 5.8 *Let Assumption 3.4 with $L_b=\sup_{i\in\{1,\dots,n_b\}}\sup_{\phi\in C^d}|k_b^i(\phi)|<\infty$ and Assumption 5.5 hold. Then for any $u\in\mathbb{R}^d$ with $|u|=1$ and $\Delta\in(0,\frac{1}{2L_b n_b\theta}]$,*

$$
|u^\top\gamma_{\Phi(T;t_j,Y_j^\varsigma)}u|\ge\frac{1}{4}(T\wedge T_0)\tilde{\sigma}^2,\quad j=1,\dots,N^\Delta,\tag{5.48}
$$

where $T_0=\frac{\ln(\frac{3}{2})}{2L_b n_b(\theta+2)}$ is given in Theorem 5.4, and Y_j^ς is given in (3.8). In particular, we have $\det(\gamma_{\Phi(T;t_i,Y_i^\varsigma)})^{-1}\in L^{\infty-}(\Omega)$.

Proof We first prove (5.48) for the case of $d=1$. For $r\in(t_k,t_{k+1}]$ with $j\le k\le i-1$, we have

$$
\begin{aligned}
D_r\Phi(t_i;t_j,Y_j^\varsigma)=\sum_{n=k}^{i-1}\Big[&(1-\theta)\mathcal{D}b(\Phi_{t_n}(t_j,Y_j^\varsigma))D_r\Phi_{t_n}(t_j,Y_j^\varsigma)\Delta\\
&+\theta\mathcal{D}b(\Phi_{t_{n+1}}(t_j,Y_j^\varsigma))D_r\Phi_{t_{n+1}}(t_j,Y_j^\varsigma)\Delta\Big]+\tilde{\sigma}.
\end{aligned}
$$

This, together with $D_r\Phi_{t_n}(t_j, Y_j^\varsigma) = \sum_{l=-N}^{0} I^{[l]} D_r\Phi(t_{n+l}; t_j, Y_j^\varsigma)$ and Assumption 3.4, implies that

$$
\begin{aligned}
&D_r\Phi(t_i; t_j, Y_j^\varsigma)\\
&= \sum_{n=k}^{i-1}\Big[(1-\theta)\sum_{\ell=1}^{n_b}\sum_{l=-N}^{0}\int_{-\tau}^{0} k_b^\ell(\Phi_{t_n}(t_j, Y_j^\varsigma)(s)) I^{[l]}(s)\mathrm{d}\nu_3^\ell(s) D_r\Phi(t_{n+l}; t_j, Y_j^\varsigma)\Delta\\
&\quad + \theta\sum_{\ell=1}^{n_b}\sum_{l=-N}^{0}\int_{-\tau}^{0} k_b^\ell(\Phi_{t_{n+1}}(t_j, Y_j^\varsigma)(s)) I^{[l]}(s)\mathrm{d}\nu_3^\ell(s) D_r\Phi(t_{n+l+1}; t_j, Y_j^\varsigma)\Delta\Big]\\
&\quad + \tilde{\sigma}.
\end{aligned} \tag{5.49}
$$

Below we split the proof into three steps.

Step 1. In this step, we present a discrete comparison principle, which will be used in Step 3 to derive a lower bound of $D_r\Phi(t_i; t_j, Y_j^\varsigma)$. Define a two-parameter nonnegative sequence $\{A_i^k\}_{0\le k,i\le N^\Delta}$ as follows: when $i \le k$, define $A_i^k = 0$; when $0 \le k \le i-1$, define

$$
A_i^k = L_b\Delta\sum_{n=k}^{i-1}\sum_{\ell=1}^{n_b}\sum_{l=-N}^{0}\nu_3^\ell(\Delta_{l-1}+\Delta_l)\big((1-\theta)A_{n+l}^k + \theta A_{n+l+1}^k\big) + \tilde{\sigma},
$$

where we let $\nu_3^\ell(\Delta_{-N-1}) = \nu_3^\ell(\Delta_0) = 0$ and $L_b = \sup_{\ell\in\{1,\ldots,n_b\}}\sup_{\phi\in C^d}|k_b^\ell(\phi)|$. It follows from the definition that when $i_1 - k_1 = i_2 - k_2 > 0$, $A_{i_1}^{k_1} = A_{i_2}^{k_2} =: \mathcal{A}_{i_1-k_1}$. Then

$$
\begin{aligned}
\mathcal{A}_{i-k} = A_i^k &= L_b\Delta\sum_{\ell=1}^{n_b}\sum_{l=-N}^{0}\nu_3^\ell(\Delta_{l-1}+\Delta_l)\big((1-\theta)A_{i-1+l}^k + \theta A_{i+l}^k\big)\\
&\quad + L_b\Delta\sum_{n=k}^{i-2}\sum_{\ell=1}^{n_b}\sum_{l=-N}^{0}\nu_3^\ell(\Delta_{l-1}+\Delta_l)\big((1-\theta)A_{n+l}^k + \theta A_{n+1+l}^k\big) + \tilde{\sigma}\\
&= L_b\Delta\sum_{\ell=1}^{n_b}\sum_{l=-N}^{0}\nu_3^\ell(\Delta_{l-1}+\Delta_l)\big((1-\theta)A_{i-1+l}^k + \theta A_{i+l}^k\big) + \mathcal{A}_{i-k-1}.
\end{aligned}
$$

This gives that for any $\Delta \in (0, \frac{1}{L_b\theta n_b})$,

$$
\mathcal{A}_{i-k} = \Big(1 - L_b\Delta\theta\sum_{\ell=1}^{n_b}\nu_3^\ell(\Delta_{-1})\Big)^{-1}\Big\{\mathcal{A}_{i-k-1}
$$

$$+ L_b\Delta\Big[(1-\theta)\sum_{\ell=1}^{n_b}\sum_{l=-N}^{0}\nu_3^\ell(\Delta_{l-1}+\Delta_l)\mathcal{A}_{i-1-k+l}$$

$$+\theta\sum_{\ell=1}^{n_b}\sum_{l=-N+1}^{0}\nu_3^\ell(\Delta_{l-2}+\Delta_{l-1})\mathcal{A}_{i-k+l-1}\Big]\Big\}.$$

By the relation $(1-L_b\Delta\theta n_b)^{-1}=1+\frac{L_b\theta\Delta n_b}{1-L_b\Delta\theta n_b}$ and by iteration, we obtain

$$\mathcal{A}_{i-k}\le(1+\frac{L_b\theta\Delta n_b}{1-L_b\Delta\theta n_b})\mathcal{A}_{i-k-1}+(1-L_b\Delta\theta n_b)^{-1}$$

$$\times L_b\Delta\Big[(1-\theta)\sum_{\ell=1}^{n_b}\sum_{l=-N}^{0}\nu_3^\ell(\Delta_{l-1}+\Delta_l)\mathcal{A}_{i-1-k+l}$$

$$+\theta\sum_{\ell=1}^{n_b}\sum_{l=-N+1}^{0}\nu_3^\ell(\Delta_{l-2}+\Delta_{l-1})\mathcal{A}_{i-k+l-1}\Big]$$

$$\le\mathcal{A}_1+\sum_{n=1}^{i-k-1}\frac{L_b\theta\Delta n_b}{1-L_b\Delta\theta n_b}\mathcal{A}_n+(1-L_b\Delta\theta n_b)^{-1}$$

$$\times L_b\Delta\sum_{\ell=1}^{n_b}\sum_{n=1}^{i-k-1}\Big[(1-\theta)\sum_{l=-N}^{0}\nu_3^\ell(\Delta_{l-1}+\Delta_l)\mathcal{A}_{n+l}$$

$$+\theta\sum_{l=-N+1}^{0}\nu_3^\ell(\Delta_{l-2}+\Delta_{l-1})\mathcal{A}_{n+l}\Big].$$

Noting that

$$\sum_{n=1}^{i-k-1}\sum_{l=-N}^{0}\nu_3^\ell(\Delta_{l-1}+\Delta_l)\mathcal{A}_{n+l}=\sum_{l=-N}^{0}\sum_{n=1}^{i-k-1}\nu_3^\ell(\Delta_{l-1}+\Delta_l)\mathcal{A}_{n+l}$$

$$\le\sum_{l=-N}^{0}\sum_{m=-N}^{i-k-1}\nu_3^\ell(\Delta_{l-1}+\Delta_l)\mathcal{A}_m\le 2\sum_{n=1}^{i-k-1}\mathcal{A}_n$$

and $\mathcal{A}_1=(1-L_b\sum_{\ell=1}^{n_b}\nu_3^\ell(\Delta_{-1})\Delta\theta)^{-1}\tilde{\sigma}\le(1-L_b\Delta\theta n_b)^{-1}\tilde{\sigma}$, we arrive at

$$\mathcal{A}_{i-k}\le(1-L_b\Delta\theta n_b)^{-1}\tilde{\sigma}+\frac{L_bn_b(\theta+2)}{1-L_b\Delta\theta n_b}\sum_{n=1}^{i-k-1}\mathcal{A}_n\Delta.$$

Combining the discrete Gronwall inequality, we deduce

$$\begin{aligned}\mathcal{A}_{i-k} &\le (1 - L_b \Delta\theta n_b)^{-1} \tilde{\sigma} \exp\left\{\frac{L_b n_b(\theta+2)\Delta(i-k-1)}{1 - L_b\Delta\theta n_b}\right\}\\ &=: K_0 \exp\left\{\frac{L_b n_b(\theta+2)\Delta(i-k-1)}{1 - L_b\Delta\theta n_b}\right\}. \end{aligned} \tag{5.50}$$

Step 2. We claim that

$$|D_r\Phi(t_i; t_j, Y_j^\varsigma)| \le A_i^k, \text{ for any } r \in (t_k, t_{k+1}], \ j \le k \le i-1. \tag{5.51}$$

We prove the claim by induction argument on $i-k$. In fact, when $j \le i_1 \le k_1$ and $r \in (t_{k_1}, t_{k_1+1}]$, we have $D_r\Phi(t_{i_1}; t_j, Y_j^\varsigma) = 0 = A_{i_1}^{k_1}$. This gives $|D_r\Phi(t_{i_1}; t_j, Y_j^\varsigma)| = A_{i_1}^{k_1}$ for $i_1 - k_1 \le 0$. Suppose that (5.51) holds for integers k, i satisfying $0 \le i-k \le i-k'-1$. Then we show (5.51) holds for $i-k = i-k'$. By (5.49) and the induction assumption $|D_r\Phi(t_n; t_j, Y_j^\varsigma)| \le A_n^{k'}$, $n \le i-1$, we have

$$\begin{aligned}&|D_r\Phi(t_i; t_j, Y_j^\varsigma)|\\ &\le \sum_{n=k'}^{i-2}\sum_{\ell=1}^{n_b}\sum_{l=-N}^{0} L_b \nu_3^\ell(\Delta_{l-1}+\Delta_l)\big((1-\theta)A_{n+l}^{k'}\Delta + \theta A_{n+l+1}^{k'}\Delta\big) + \tilde{\sigma}\\ &\quad + (1-\theta)\sum_{\ell=1}^{n_b}\sum_{l=-N}^{0} L_b\nu_3^\ell(\Delta_{l-1}+\Delta_l)A_{i-1+l}^{k'}\Delta\\ &\quad + \theta\sum_{\ell=1}^{n_b}\sum_{l=-N}^{-1} L_b\nu_3^\ell(\Delta_{l-1}+\Delta_l)A_{i+l}^{k'}\Delta\\ &\quad + L_b\theta\Delta\sum_{\ell=1}^{n_b}\nu_3^\ell(\Delta_{-1})|D_r\Phi(t_i; t_j, Y_j^\varsigma)|,\end{aligned}$$

which yields

$$\begin{aligned}&|D_r\Phi(t_i; t_j, Y_j^\varsigma)|\\ &\le \Big(1 - L_b\Delta\theta\sum_{\ell=1}^{n_b}\nu_3^\ell(\Delta_{-1})\Big)^{-1}\\ &\quad\times\Big[A_{i-1}^{k'} + L_b\Delta(1-\theta)\sum_{\ell=1}^{n_b}\sum_{l=-N}^{0}\nu_3^\ell(\Delta_{l-1}+\Delta_l)A_{i-1+l}^{k'}\end{aligned}$$

$$+ L_b\Delta\theta \sum_{\ell=1}^{n_b} \sum_{l=-N}^{-1} \nu_3^\ell(\Delta_{l-1}+\Delta_l)A_{i+l}^{k'}\Big] = A_i^{k'}, \quad r \in (t_{k'}, t_{k'+1}].$$

This finishes the proof of the claim (5.51).

Step 3. In this step, we prove the lower bound of $|D_r\Phi(t_i;t_j,Y_j^\varsigma)|$ and then show (5.48). From (5.49) and (5.50), we obtain

$$\begin{aligned}
&|D_r\Phi(t_i;t_j,Y_j^\varsigma)| \\
&\geq \tilde{\sigma} - L_b\Delta\Big[\sum_{n=k}^{i-1}\sum_{\ell=1}^{n_b}\sum_{l=-N}^{0}\Big((1-\theta)\nu_3^\ell(\Delta_{l-1}+\Delta_l)A_{n+l}^k \\
&\quad + \theta\nu_3^\ell(\Delta_{l-1}+\Delta_l)A_{n+l+1}^k\Big)\Big] \\
&\geq \tilde{\sigma} - L_b\Delta\Big[\sum_{n=k}^{i-1}\sum_{\ell=1}^{n_b}\sum_{l=-N}^{0}\Big((1-\theta)\nu_3^\ell(\Delta_{l-1}+\Delta_l)K_0 e^{-\frac{L_bn_b(\theta+2)\Delta}{1-L_b\Delta\theta n_b}} \\
&\quad + \theta\nu_3^\ell(\Delta_{l-1}+\Delta_l)K_0\Big)\exp\{\frac{L_bn_b(\theta+2)\Delta(n+l-k)}{1-L_b\Delta\theta n_b}\}\Big] \\
&\geq \tilde{\sigma} - 2L_bn_b\Delta\Big((1-\theta)K_0e^{-\frac{L_bn_b(\theta+2)\Delta}{1-L_b\Delta\theta n_b}} + \theta K_0\Big)\frac{\exp\{\frac{L_bn_b(\theta+2)\Delta(i-k)}{1-L_b\Delta\theta n_b}\}-1}{\exp\{\frac{L_bn_b(\theta+2)\Delta}{1-L_b\Delta\theta n_b}\}-1} \\
&\geq \tilde{\sigma} - \frac{2}{\theta+2}\Big(\exp\Big\{\frac{L_bn_b(\theta+2)\Delta(i-k)}{1-L_b\Delta\theta n_b}\Big\}-1\Big)\tilde{\sigma} \\
&\geq \tilde{\sigma} - \Big(\exp\Big\{\frac{L_bn_b(\theta+2)\Delta(i-k)}{1-L_b\Delta\theta n_b}\Big\}-1\Big)\tilde{\sigma},
\end{aligned}$$

where we used $e^x \geq x+1, x \geq 0$. Then,

$$|D_r\Phi(t_i;t_j,Y_j^\varsigma)| \geq \frac{1}{2}\tilde{\sigma},$$

when $\Delta \leq \frac{1}{2L_b\theta n_b}$ and $i-k \leq \frac{\ln(\frac{3}{2})}{2L_bn_b(\theta+2)\Delta}$. Thus,

$$\begin{aligned}
\gamma_{\Phi(T;t_j,Y_j^\varsigma)} &:= \int_0^T D_r\Phi(T;t_j,Y_j^\varsigma)(D_r\Phi(T;t_j,Y_j^\varsigma))^\top \mathrm{d}r \\
&\geq \int_{(T-T_0)\vee 0}^T |D_r\Phi(T;t_j,Y_j^\varsigma)|^2\mathrm{d}r \geq \frac{1}{4}(T\wedge T_0)\tilde{\sigma}^2,
\end{aligned}$$

where $T_0 = \frac{\ln(\frac{3}{2})}{2L_bn_b(\theta+2)}$. This finishes the proof of (5.48) for $d=1$.

For the case of $d \geq 2$, by replacing $D_r\Phi$ with $u^\top D_r\Phi$ in the above argument, one can obtain (5.48).

Moreover, it follows from (5.48) that

$$\lambda_{\min}(\gamma_{\Phi(T;t_j,Y_j^\varsigma)}) = \min_{u\in\mathbb{R}^d,|u|=1} u^\top\gamma_{\Phi(T;t_j,Y_j^\varsigma)}u \geq \frac{1}{4}(T\wedge T_0)\tilde{\sigma}^2,$$

which implies

$$|(\gamma_{\Phi(T;t_j,Y_j^\varsigma)})^{-1}| \leq K|\det(\gamma_{\Phi(T;t_j,Y_j^\varsigma)})^{-1}| \leq K[\frac{1}{4}(T\wedge T_0)\sigma^2]^{-d}.$$

Thus the proof is completed. □

With Lemmas 5.7 and 5.8 in hand, we present the following Malliavin integration by parts formula, which plays an important role in the test-functional-independent weak convergence analysis.

Lemma 5.9 *Let $\alpha = (\alpha_1, \ldots, \alpha_d)$ be the multi-index with $\alpha_j \in \mathbb{N}$, $j = 1, \ldots, d$ and $|\alpha| := \sum_{j=1}^d \alpha_j \leq 2$, $f \in \mathscr{C}$, and $G_1 \in \mathbb{D}^{|\alpha|+1,\infty}(\mathbb{R}^d)$. Then under conditions in Lemmas 5.7 and 5.8, there exist a constant $\Delta_2 \in (0, 1]$ and an element $H_{|\alpha|+1}$ such that for $\Delta \in (0, \Delta_2]$, $i \in \mathbb{N}$ with $i \leq N^\Delta$,*

$$\mathbb{E}[\partial_\alpha f(\Phi(T; t_i, Y_i^\varsigma))G_1] = \mathbb{E}[F(\Phi(T; t_i, Y_i^\varsigma))H_{|\alpha|+1}(\Phi(T; t_i, Y_i^\varsigma), G_1)], \tag{5.52}$$

where F is an antiderivative of f given in the definition of $\mathscr{C}$ (5.34). Moreover,

$$\sup_{T\geq T_0} |\mathbb{E}[\partial_\alpha f(\Phi(T; t_i, Y_i^\varsigma))G_1]| \leq K\|G_1\|_{|\alpha|+1,2}, \tag{5.53}$$

where T_0 is given in Theorem 5.4, and Y_i^ς is given in (3.8).

Proof According to the definition of $\mathscr{C}$, and applying [30, Proposition 2.1.4, (2.29)–(2.32)] give (5.52). Moreover, for $q > q_1 \geq 1$, there exist constants $\eta_1, \eta_2 > 0$ and integers $n_1, n_2 > 0$ such that

$$\begin{aligned}&\|H_{|\alpha|+1}(\Phi(T; t_i, Y_i^\varsigma), G_1)\|_{q_1}\\&\leq K(q_1, q)\|\gamma^{-1}_{\Phi(T;t_i,Y_i^\varsigma)}\|^{n_1}_{0,\eta_1}\|D\Phi(T; t_i, Y_i^\varsigma)\|^{n_2}_{|\alpha|+1,\eta_2}\|G_1\|_{|\alpha|+1,q}.\end{aligned}$$

It follows from Lemmas 5.7 and 5.8 that for any $\Delta \in (0, \Delta_2]$ with $\Delta_2 := \tilde{\Delta}\wedge\frac{1}{2L_b n_b\theta}$,

$$\|H_{|\alpha|+1}(\Phi(T; t_i, Y_i^\varsigma), G_1)\|_{q_1} \leq K\Big[\frac{1}{4}(T\wedge T_0)\tilde{\sigma}^2\Big]^{-dn_1}\|G_1\|_{|\alpha|+1,q}$$

where $\tilde{\Delta} := \tilde{\Delta}(\eta_2)$ is given in Lemma 5.7. Taking $q_1 = 1, q = 2$ and supremum for $T \geq T_0$, we finish the proof. □

5.3.2 Weak Convergence Analysis

In this subsection, we give the test-functional-independent weak convergence rate of the θ-EM method.

Theorem 5.10 *Let conditions in Theorem 5.4 hold. Then there exists $\tilde{\Delta} > 0$ such that for $\Delta \in (0, \tilde{\Delta}]$,*

$$\sup_{f\in\mathscr{C}}\sup_{T\geq T_0}\left|\mathbb{E}[f(x^{\xi}(T))] - \mathbb{E}[f(y^{\xi,\Delta}(T))]\right| \leq K\Delta,$$

where T_0 is given in Theorem 5.4.

Proof In order to obtain the test-functional-independent weak convergence rate, we need to estimate terms $\mathcal{I}_0, \mathcal{I}_b$, and $\mathcal{I}_{b,\theta}$ in (5.36) by means of the Malliavin integration by parts formula (5.52) and the inequality (5.53).

Estimate of Term $\mathcal{I}_0$ Recalling

$$\mathcal{I}_0 = \mathbb{E}[f(\Phi(T; t_{N\Delta}, \varphi^{Int}_{t_{N\Delta}}(0, \xi)))] - \mathbb{E}[f(\Phi(T; t_{N\Delta}, \varphi^{Int}_{t_{N\Delta}}(0, \xi^{Int})))],$$

it follows from (5.53), Lemma 5.6, and $\|\xi - \xi^{Int}\| \leq K\Delta$ that

$$\begin{aligned}
|\mathcal{I}_0| &= |\mathbb{E}[f(\varphi^{Int}(t_{N\Delta}; 0, \xi)] - \mathbb{E}[f(\varphi^{Int}(t_{N\Delta}; 0, \xi^{Int}))]| \\
&\leq \int_0^1 \left|\mathbb{E}\Big[f'(\varphi^{Int}(t_{N\Delta}; 0, \varsigma\xi + (1-\varsigma)\xi^{Int}))\mathcal{D}\varphi^{Int}(t_{N\Delta}; 0, \varsigma\xi \right. \\
&\qquad \left. + (1-\varsigma)\xi^{Int})(\xi - \xi^{Int})\Big]\right| \mathrm{d}\varsigma \\
&\leq K\int_0^1 \|\mathcal{D}\varphi^{Int}(t_{N\Delta}; 0, \varsigma\xi + (1-\varsigma)\xi^{Int})(\xi - \xi^{Int})\|_{2,2}\mathrm{d}\varsigma \\
&\leq K\Delta\int_0^1 \|\mathcal{D}\varphi^{Int}(t_{N\Delta}; 0, \varsigma\xi + (1-\varsigma)\xi^{Int})\|_{2,2}\mathrm{d}\varsigma \leq K\Delta.
\end{aligned}$$

Estimate of Term $\mathcal{I}_b$ Recalling the definition of $\mathcal{I}_b$, we have that

$$\begin{aligned}
\mathcal{I}_b = \sum_{i=1}^{N\Delta}\int_0^1 \mathbb{E}\Big[\Big\langle f'(\Phi(T; t_i, Y_i^{\varsigma}))\mathcal{D}\Phi(T; t_i, Y_i^{\varsigma}), I^{[0]}\int_{t_{i-1}}^{t_i}\int_0^1 \mathcal{D}b(Z^{\beta_1}_{i,r}) \\
(\varphi_r(t_{i-1}, \varphi_{t_{i-1}}(0, \xi^{Int})) - \varphi^{Int}_{t_{i-1}}(0, \xi^{Int}))\mathrm{d}\beta_1\mathrm{d}r\Big\rangle\Big]\mathrm{d}\varsigma,
\end{aligned}$$

where

$$Z_{i,r}^{\beta_1} = \beta_1 \varphi_r(t_{i-1}, \varphi_{t_{i-1}}(0, \xi^{Int})) + (1 - \beta_1)\varphi_{t_{i-1}}^{Int}(0, \xi^{Int}).$$

Inserting (3.35) (with $\sigma \equiv \tilde{\sigma}\,\mathrm{Id}_{d\times d}$) into $\mathcal{I}_b$, we are in the position to estimate $\mathcal{I}_b$. We only estimate the sub-term

$$\mathcal{I}_b^0 := \sum_{i=1}^{N^\Delta} \int_0^1 \mathbb{E}\Big[\Big\langle f'(\Phi(T; t_i, Y_i^\varsigma))\mathcal{D}\Phi(T; t_i, Y_i^\varsigma), I^{[0]} \int_{t_{i-1}}^{t_i} \int_0^1 \mathcal{D}b(Z_{i,r}^{\beta_1}) \sum_{j=-N}^{-1} \mathbf{1}_{[t_{i+j-1}-r, t_{i+j}-r)}(\cdot) \frac{t_j - \cdot}{\Delta} \int_{t_{i+j-1}}^{t_{i+j}} b(\varphi_u(0, \xi^{Int}))\mathrm{d}u\mathrm{d}\beta_1\mathrm{d}r\Big\rangle\Big]\mathrm{d}\varsigma,$$

the sub-term

$$\mathcal{I}_b^1 := \sum_{i=1}^{N^\Delta} \int_0^1 \mathbb{E}\Big[\Big\langle f'(\Phi(T; t_i, Y_i^\varsigma))\mathcal{D}\Phi(T; t_i, Y_i^\varsigma), I^{[0]} \int_{t_{i-1}}^{t_i} \int_0^1 \mathcal{D}b(Z_{i,r}^{\beta_1}) \sum_{j=-N}^{-1} \mathbf{1}_{[t_{i+j-1}-r, t_{i+j}-r)}(\cdot) \frac{t_j - \cdot}{\Delta} \int_{t_{i+j-1}}^{t_{i+j}} \tilde{\sigma}\mathrm{d}W(u)\mathrm{d}\beta_1\mathrm{d}r\Big\rangle\Big]\mathrm{d}\varsigma,$$

and the sub-term

$$\mathcal{I}_b^2 := \sum_{i=1}^{N^\Delta} \int_0^1 \mathbb{E}\Big[\Big\langle f'(\Phi(T; t_i, Y_i^\varsigma))\mathcal{D}\Phi(T; t_i, Y_i^\varsigma), I^{[0]} \int_{t_{i-1}}^{t_i} \int_0^1 \mathcal{D}b(Z_{i,r}^{\beta_1}) \sum_{j=-N}^{-1} \mathbf{1}_{[t_{i+j-1}-r, t_{i+j}-r)}(\cdot) \int_{t_{i+j-1}}^{r+\cdot} \tilde{\sigma}\mathrm{d}W(u)\mathrm{d}\beta_1\mathrm{d}r\Big\rangle\Big]\mathrm{d}\varsigma,$$

since other sub-terms in $\mathcal{I}_b$ can be estimated similarly.

For the term $\mathcal{I}_b^0$, it follows from (5.52) and (5.53) that

$$|\mathcal{I}_b^0| \le K \sum_{i=1}^{N^\Delta} \int_0^1 \Big\| \mathcal{D}\Phi(T; t_i, Y_i^\varsigma) I^{[0]} \int_{t_{i-1}}^{t_i} \int_0^1 \mathcal{D}b(Z_{i,r}^{\beta_1}) \sum_{j=-N}^{-1} \mathbf{1}_{[t_{i+j-1}-r, t_{i+j}-r)}(\cdot) \frac{t_j - \cdot}{\Delta} \int_{t_{i+j-1}}^{t_{i+j}} b(\varphi_u(0, \xi^{Int}))\mathrm{d}u\mathrm{d}\beta_1\mathrm{d}r \Big\|_{2,2} \mathrm{d}\varsigma. \tag{5.54}$$

According to Lemma 5.7, $\sup_{\phi_1 \in C^d} |\mathcal{D}b(\phi_1)\phi_2| \le K\|\phi_2\|$, (5.32), and (5.37), we have

$$\begin{aligned}
&\Big\|\mathcal{D}\Phi(T;t_i,Y_i^{\varsigma})I^{[0]}\int_{t_{i-1}}^{t_i}\int_0^1 \mathcal{D}b(Z_{i,r}^{\beta_1})\sum_{j=-N}^{-1}\mathbf{1}_{[t_{i+j-1}-r,t_{i+j}-r)}(\cdot)\\
&\quad\times\frac{t_j-\cdot}{\Delta}\int_{t_{i+j-1}}^{t_{i+j}} b(\varphi_u(0,\xi^{Int}))\mathrm{d}u\mathrm{d}\beta_1\mathrm{d}r\Big\|_{0,2}\\
&\le K\Big(e^{-\hat{\lambda}_5(T-t_i)}\mathbb{E}\Big[\Big\|I^{[0]}\int_{t_{i-1}}^{t_i}\int_0^1 \mathcal{D}b(Z_{i,r}^{\beta_1})\sum_{j=-N}^{-1}\mathbf{1}_{[t_{i+j-1}-r,t_{i+j}-r)}(\cdot)\\
&\quad\times\frac{t_j-\cdot}{\Delta}\int_{t_{i+j-1}}^{t_{i+j}} b(\varphi_u(0,\xi^{Int}))\mathrm{d}u\mathrm{d}\beta_1\mathrm{d}r\Big\|^2\Big]\Big)^{\frac{1}{2}}\le Ke^{-\frac{\hat{\lambda}_5(T-t_i)}{2}}\Delta^2. \qquad (5.55)
\end{aligned}$$

In addition, similar to the proof of Lemma 3.13, using Assumptions 5.7 and 5.6, we derive that for some $\tilde{p}\ge 2$,

$$\begin{aligned}
&\Big(\int_{t_i}^T \mathbb{E}\Big[\Big\|D_{r_1}\Big[\mathcal{D}\Phi(T;t_i,Y_i^{\varsigma})I^{[0]}\int_{t_{i-1}}^{t_i}\int_0^1 \mathcal{D}b(Z_{i,r}^{\beta_1})\sum_{j=-N}^{-1}\mathbf{1}_{[t_{i+j-1}-r,t_{i+j}-r)}(\cdot)\\
&\quad\times\frac{t_j-\cdot}{\Delta}\int_{t_{i+j-1}}^{t_{i+j}} b(\varphi_u(0,\xi^{Int}))\mathrm{d}u\mathrm{d}\beta_1\mathrm{d}r\Big]\Big\|^2\Big]\mathrm{d}r_1\Big)^{\frac{1}{2}}\\
&\le K\Big\{\int_{t_i}^T\Big(e^{-\hat{\lambda}_5(T-t_i)}\mathbb{E}\Big[\Big\|I^{[0]}\int_{t_{i-1}}^{t_i}\int_0^1 \mathcal{D}b(Z_{i,r}^{\beta_1})\sum_{j=-N}^{-1}\mathbf{1}_{[t_{i+j-1}-r,t_{i+j}-r)}(\cdot)\\
&\quad\times\frac{t_j-\cdot}{\Delta}\int_{t_{i+j-1}}^{t_{i+j}} b(\varphi_u(0,\xi^{Int}))\mathrm{d}u\mathrm{d}\beta_1\mathrm{d}r\Big\|^2\big(1+\|D_{r_1}Y_i^{\varsigma}\|^{\tilde{p}}\big)\Big]\\
&\quad+e^{-\hat{\lambda}_5(T-t_i)}\mathbb{E}\Big[\Big\|I^{[0]}\int_{t_{i-1}}^{t_i}\int_0^1 \mathcal{D}^2b(Z_{i,r}^{\beta_1})\Big(D_{r_1}Z_{i,r}^{\beta_1},\sum_{j=-N}^{-1}\mathbf{1}_{[t_{i+j-1}-r,t_{i+j}-r)}(\cdot)\\
&\quad\times\frac{t_j-\cdot}{\Delta}\int_{t_{i+j-1}}^{t_{i+j}} b(\varphi_u(0,\xi^{Int}))\mathrm{d}u\Big)\mathrm{d}\beta_1\mathrm{d}r\Big\|^2\Big]\\
&\quad+e^{-\hat{\lambda}_5(T-t_i)}\mathbb{E}\Big[\Big\|I^{[0]}\int_{t_{i-1}}^{t_i}\int_0^1 \mathcal{D}b(Z_{i,r}^{\beta_1})\sum_{j=-N}^{-1}\mathbf{1}_{[t_{i+j-1}-r,t_{i+j}-r)}(\cdot)\\
&\quad\times\frac{t_j-\cdot}{\Delta}\int_{t_{i+j-1}}^{t_{i+j}} \mathcal{D}b(\varphi_u(0,\xi^{Int}))D_{r_1}\varphi_u(0,\xi^{Int})\mathrm{d}u\mathrm{d}\beta_1\mathrm{d}r\Big\|^2\Big]\Big)\mathrm{d}r_1\Big\}^{\frac{1}{2}}
\end{aligned}$$

$$\leq K\Big\{\int_{t_i}^{T}\Big(\Delta^4 e^{-\hat{\lambda}_5(T-t_i)}(1+\sup_{u\geq 0}\mathbb{E}[\|\varphi_u(0,\xi^{Int})\|^4])^{\frac{1}{2}}\big(1+\mathbb{E}[\|D_{r_1}Y_i^{\varsigma}\|^{2\tilde{p}}]\big)^{\frac{1}{2}}$$

$$+\Delta^3 e^{-\hat{\lambda}_5(T-t_i)}\int_{t_{i-1}}^{t_i}\int_0^1\mathbb{E}[\|D_{r_1}Z_{i,r}^{\beta_1}\|^4]^{\frac{1}{2}}\mathrm{d}\beta_1\mathrm{d}r(1+\sup_{u\geq 0}\mathbb{E}[\|\varphi_u(0,\xi^{Int})\|^4])^{\frac{1}{2}}$$

$$+\Delta^3 e^{-\hat{\lambda}_5(T-t_i)}\int_{t_{i+j-1}}^{t_{i+j}}\mathbb{E}[\|D_{r_1}\varphi_u(0,\xi^{Int})\|^2]\mathrm{d}u\Big)\mathrm{d}r_1\Big\}^{\frac{1}{2}}$$

$$\leq Ke^{-\frac{\hat{\lambda}_5(T-t_i)}{2}}(T-t_i)^{\frac{1}{2}}\Delta^2. \tag{5.56}$$

Similarly,

$$\Big\{\int_{t_i}^{T}\int_{t_i}^{T}\mathbb{E}\Big[\Big\|D_{r_1,r_2}\Big(\mathcal{D}\Phi(T;t_i,Y_i^{\varsigma})I^{[0]}\int_{t_{i-1}}^{t_i}\int_0^1\mathcal{D}b(Z_{i,r}^{\beta_1})$$

$$\sum_{j=-N}^{-1}\mathbf{1}_{[t_{i+j-1}-r,t_{i+j}-r)}(\cdot)\frac{t_j-\cdot}{\Delta}\int_{t_{i+j-1}}^{t_{i+j}}b(\varphi_u(0,\xi^{Int}))\mathrm{d}u\mathrm{d}\beta_1\mathrm{d}r\Big)\Big\|^2\Big]\mathrm{d}r_1\mathrm{d}r_2\Big\}^{\frac{1}{2}}$$

$$\leq Ke^{-\frac{\hat{\lambda}_5(T-t_i)}{2}}(T-t_i)\Delta^2. \tag{5.57}$$

Inserting (5.55)–(5.57) into (5.54), one has

$$|\mathcal{I}_b^0|\leq K\sum_{i=1}^{N^{\Delta}}e^{-\frac{\hat{\lambda}_5(T-t_i)}{2}}\Delta^2\big(1+(T-t_i)^{\frac{1}{2}}+(T-t_i)\big)\leq K\Delta,$$

where we used the fact

$$\sup_{T>0}\Delta\sum_{i=1}^{N^{\Delta}}e^{-\frac{\hat{\lambda}_5}{2}(T-t_i)}(T-t_i)^l\leq\sup_{T\geq 0}\int_0^T e^{\frac{\hat{\lambda}_5\Delta}{2}}e^{-\frac{\hat{\lambda}_5}{2}(T-s)}(T-s)^l\mathrm{d}s<\infty\quad\forall\, l\in\mathbb{N}.$$

For the sub-term $\mathcal{I}_b^1$, we have

$$\mathcal{I}_b^1=\sum_{i=1}^{N^{\Delta}}\int_0^1\int_{t_{i-1}}^{t_i}\int_0^1\sum_{j=-N}^{-1}\mathbf{1}_{[t_{i+j-1}-r,t_{i+j}-r)}(\cdot)\frac{t_j-\cdot}{\Delta}\mathbb{E}\Big[\Big\langle(I^{[0]}\mathcal{D}b(Z_{i,r}^{\beta_1}))^*$$

$$f'(\Phi(T;t_i,Y_i^{\varsigma}))\mathcal{D}\Phi(T;t_i,Y_i^{\varsigma}),\int_{t_{i+j-1}}^{t_{i+j}}\tilde{\sigma}\mathrm{d}W(u)\Big\rangle\Big]\mathrm{d}\beta_1\mathrm{d}r\mathrm{d}\varsigma$$

$$=\sum_{i=1}^{N^{\Delta}}\sum_{j=-N}^{-1}\int_0^1\int_{t_{i-1}}^{t_i}\int_0^1\int_{t_{i+j-1}}^{t_{i+j}}\mathbf{1}_{[t_{i+j-1}-r,t_{i+j}-r)}(\cdot)\frac{t_j-\cdot}{\Delta}$$

$$\times\, \mathbb{E}\Big[\Big\langle D_u\big[(I^{[0]}\mathcal{D}b(Z_{i,r}^{\beta_1}))^* f'(\Phi(T;t_i,Y_i^{\varsigma}))\mathcal{D}\Phi(T;t_i,Y_i^{\varsigma})\big], \tilde{\sigma}\,\mathrm{Id}_{d\times d}\Big\rangle\Big]$$
$$\mathrm{d}u\mathrm{d}\beta_1\mathrm{d}r\mathrm{d}\varsigma$$
$$=\sum_{i=1}^{N^\Delta}\sum_{j=-N}^{-1}\int_0^1\int_{t_{i-1}}^{t_i}\int_0^1\int_{t_{i+j-1}}^{t_{i+j}}\Big\{\mathbb{E}\Big[\Big\langle f'(\Phi(T;t_i,Y_i^{\varsigma}))\mathcal{D}\Phi(T;t_i,Y_i^{\varsigma}),$$
$$I^{[0]}\mathcal{D}^2b(Z_{i,r}^{\beta_1})\Big(D_u Z_{i,r}^{\beta_1}, \mathbf{1}_{[t_{i+j-1}-r,t_{i+j}-r)}(\cdot)\frac{t_j-\cdot}{\Delta}\tilde{\sigma}\,\mathrm{Id}_{d\times d}\Big)\Big\rangle\Big]$$
$$+\,\mathbb{E}\Big[\Big\langle f'(\Phi(T;t_i,Y_i^{\varsigma}))D_u\mathcal{D}\Phi(T;t_i,Y_i^{\varsigma}), I^{[0]}\mathcal{D}b(Z_{i,r}^{\beta_1})$$
$$\mathbf{1}_{[t_{i+j-1}-r,t_{i+j}-r)}(\cdot)\frac{t_j-\cdot}{\Delta}\tilde{\sigma}\,\mathrm{Id}_{d\times d}\Big\rangle\Big]$$
$$+\,\mathbb{E}\Big[\Big\langle f''(\Phi(T;t_i,Y_i^{\varsigma}))D_u\Phi(T;t_i,Y_i^{\varsigma})$$
$$\mathcal{D}\Phi(T;t_i,Y_i^{\varsigma}), I^{[0]}\mathcal{D}b(Z_{i,r}^{\beta_1})\mathbf{1}_{[t_{i+j-1}-r,t_{i+j}-r)}(\cdot)\frac{t_j-\cdot}{\Delta}\tilde{\sigma}\,\mathrm{Id}_{d\times d}\Big\rangle\Big]\Big\}$$
$$\mathrm{d}u\mathrm{d}\beta_1\mathrm{d}r\mathrm{d}\varsigma.$$

This, together with (5.52) and (5.53), implies that

$$|\mathcal{I}_b^1|\le K\sum_{i=1}^{N^\Delta}\sum_{j=-N}^{-1}\int_0^1\int_{t_{i-1}}^{t_i}\int_0^1\int_{t_{i+j-1}}^{t_{i+j}}\Big\{\Big\|\mathcal{D}\Phi(T;t_i,Y_i^{\varsigma})I^{[0]}\mathcal{D}^2b(Z_{i,r}^{\beta_1})$$
$$\Big(D_u Z_{i,r}^{\beta_1}, \mathbf{1}_{[t_{i+j-1}-r,t_{i+j}-r)}(\cdot)\frac{t_j-\cdot}{\Delta}\tilde{\sigma}\,\mathrm{Id}_{d\times d}\Big)\Big\|_{2,2}$$
$$+\Big\|D_u\mathcal{D}\Phi(T;t_i,Y_i^{\varsigma})I^{[0]}\mathcal{D}b(Z_{i,r}^{\beta_1})\mathbf{1}_{[t_{i+j-1}-r,t_{i+j}-r)}(\cdot)\frac{t_j-\cdot}{\Delta}\tilde{\sigma}\,\mathrm{Id}_{d\times d}\Big\|_{2,2}$$
$$+\Big\|D_u\Phi(T;t_i,Y_i^{\varsigma})\mathcal{D}\Phi(T;t_i,Y_i^{\varsigma})I^{[0]}\mathcal{D}b(Z_{i,r}^{\beta_1})\mathbf{1}_{[t_{i+j-1}-r,t_{i+j}-r)}(\cdot)$$
$$\times\frac{t_j-\cdot}{\Delta}\tilde{\sigma}\,\mathrm{Id}_{d\times d}\Big\|_{3,2}\Big\}\mathrm{d}u\mathrm{d}\beta_1\mathrm{d}r\mathrm{d}\varsigma =: \mathcal{I}_{b,1}^1+\mathcal{I}_{b,2}^1+\mathcal{I}_{b,3}^1.$$

By Assumption 5.6, Lemmas 5.6 and 5.7, and (5.37), the term $\mathcal{I}_{b,1}^1$ is estimated as

$$\mathcal{I}_{b,1}^1< K\sum_{i=1}^{N^\Delta}\sum_{j=-N}^{-1}\int_0^1\int_{t_{i-1}}^{t_i}\int_0^1\int_{t_{i+j-1}}^{t_{i+j}}\Big\{\Big(e^{-\hat{\lambda}_5(T-t_i)}\mathbb{E}\Big[\Big\|I^{[0]}\mathcal{D}^2b(Z_{i,r}^{\beta_1})$$
$$\Big(D_u Z_{i,r}^{\beta_1}, \mathbf{1}_{[t_{i+j-1}-r,t_{i+j}-r)}(\cdot)\frac{t_j-\cdot}{\Delta}\tilde{\sigma}\,\mathrm{Id}_{d\times d}\Big)\Big\|^2\Big]\Big)^{\frac{1}{2}}$$

$$+\Big(\int_{t_i}^{T}\mathbb{E}\Big[\Big\|D_{r_1}\Big(\mathcal{D}\Phi(T;t_i,Y_i^{\varsigma})I^{[0]}\mathcal{D}^2b(Z_{i,r}^{\beta_1})$$

$$\Big(D_u Z_{i,r}^{\beta_1},\mathbf{1}_{[t_{i+j-1}-r,t_{i+j}-r)}(\cdot)\frac{t_j-\cdot}{\Delta}\tilde{\sigma}\,\mathrm{Id}_{d\times d}\Big)\Big\|^2\Big]\mathrm{d}r_1\Big)^{\frac{1}{2}}$$

$$+\Big(\int_{t_i}^{T}\int_{t_i}^{T}\mathbb{E}\Big[\Big\|D_{r_1,r_2}\Big(\mathcal{D}\Phi(T;t_i,Y_i^{\varsigma})I^{[0]}\mathcal{D}^2b(Z_{i,r}^{\beta_1})$$

$$\Big(D_u Z_{i,r}^{\beta_1},\mathbf{1}_{[t_{i+j-1}-r,t_{i+j}-r)}(\cdot)\frac{t_j-\cdot}{\Delta}\tilde{\sigma}\,\mathrm{Id}_{d\times d}\Big)\Big\|^2\Big]\mathrm{d}r_1\mathrm{d}r_2\Big)^{\frac{1}{2}}\Big\}$$

$$\mathrm{d}u\mathrm{d}\beta_1\mathrm{d}r\mathrm{d}\varsigma\le K\Delta.$$

For the term $\mathcal{I}_{b,2}^1$, we have that for some $\tilde{p}\ge 2$,

$$\mathcal{I}_{b,2}^1\le K\sum_{i=1}^{N^\Delta}\sum_{j=-N}^{-1}\int_0^1\int_{t_{i-1}}^{t_i}\int_0^1\int_{t_{i+j-1}}^{t_{i+j}}\Big\{\Big(e^{-\hat{\lambda}_5(T-t_i)}\mathbb{E}\Big[\Big\|I^{[0]}\mathcal{D}b(Z_{i,r}^{\beta_1})$$

$$\mathbf{1}_{[t_{i+j-1}-r,t_{i+j}-r)}(\cdot)\frac{t_j-\cdot}{\Delta}\tilde{\sigma}\,\mathrm{Id}_{d\times d}\Big\|^2(1+\|D_u Y_i^{\varsigma}\|^{\tilde{p}})\Big]\Big)^{\frac{1}{2}}$$

$$+\Big(\int_{t_i}^{T}\mathbb{E}\Big[\Big\|D_{r_1}\Big(D_u\mathcal{D}\Phi(T;t_i,Y_i^{\varsigma})I^{[0]}\mathcal{D}b(Z_{i,r}^{\beta_1})$$

$$\mathbf{1}_{[t_{i+j-1}-r,t_{i+j}-r)}(\cdot)\frac{t_j-\cdot}{\Delta}\tilde{\sigma}\,\mathrm{Id}_{d\times d}\Big)\Big\|^2\Big]\mathrm{d}r_1\Big)^{\frac{1}{2}}+\Big(\int_{t_i}^{T}\int_{t_i}^{T}\mathbb{E}\Big[\Big\|D_{r_1,r_2}$$

$$\Big(D_u\mathcal{D}\Phi(T;t_i,Y_i^{\varsigma})I^{[0]}\mathcal{D}b(Z_{i,r}^{\beta_1})\mathbf{1}_{[t_{i+j-1}-r,t_{i+j}-r)}(\cdot)$$

$$\frac{t_j-\cdot}{\Delta}\tilde{\sigma}\,\mathrm{Id}_{d\times d}\Big)\Big\|^2\Big]\mathrm{d}r_1\mathrm{d}r_2\Big)^{\frac{1}{2}}\Big\}\mathrm{d}u\mathrm{d}\beta_1\mathrm{d}r\mathrm{d}\varsigma\le K\Delta.$$

Similar to estimates of $\mathcal{I}_{b,1}^1$ and $\mathcal{I}_{b,2}^1$, for the term $\mathcal{I}_{b,3}^1$, we obtain that for some $\tilde{p}\ge 2$,

$$\mathcal{I}_{b,3}^1\le K\sum_{i=1}^{N^\Delta}\sum_{j=-N}^{-1}\int_0^1\int_{t_{i-1}}^{t_i}\int_0^1\int_{t_{i+j-1}}^{t_{i+j}}\Big\|D_u\Phi(T;t_i,Y_i^{\varsigma})\Big\|_{3,\tilde{p}}\Big\|\mathcal{D}\Phi(T;t_i,Y_i^{\varsigma})I^{[0]}$$

$$\mathcal{D}b(Z_{i,r}^{\beta_1})\mathbf{1}_{[t_{i+j-1}-r,t_{i+j}-r)}(\cdot)\frac{t_j-\cdot}{\Delta}\tilde{\sigma}\,\mathrm{Id}_{d\times d}\Big\|_{3,\tilde{p}}\mathrm{d}u\mathrm{d}\beta_1\mathrm{d}r\mathrm{d}\varsigma\le K\Delta.$$

Other terms can be estimated similarly as above. Therefore, we have $|\mathcal{I}_b^1|\le K\Delta$.

For the sub-term $\mathcal{I}_b^2$, based on Assumption 3.4, we have

$$\mathcal{I}_b^2 = \sum_{i=1}^{N^\Delta} \int_0^1 \mathbb{E}\Big[\Big\langle f'(\Phi(T; t_i, Y_i^\varsigma))\mathcal{D}\Phi(T; t_i, Y_i^\varsigma), I^{[0]} \int_{t_{i-1}}^{t_i} \int_0^1 \int_{-\tau}^0 k_b^1(Z_{i,r}^{\beta_1})$$

$$\sum_{j=-N}^{-1} \mathbf{1}_{[t_{i+j-1}-r, t_{i+j}-r)}(s) \int_{t_{i+j-1}}^{r+s} \tilde{\sigma} \mathrm{d}W(u) \mathrm{d}v_3^1(s) \mathrm{d}\beta_1 \mathrm{d}r \Big\rangle\Big] \mathrm{d}\varsigma$$

$$= \sum_{i=1}^{N^\Delta} \sum_{j=-N}^{-1} \int_0^1 \int_{t_{i-1}}^{t_i} \int_0^1 \int_{-\tau \vee (t_{i+j-1}-r)}^{(t_{i+j}-r)\wedge 0} \int_{t_{i+j-1}}^{r+s} \mathbb{E}\Big[\Big\langle D_u[(I^{[0]} k_b^1(Z_{i,r}^{\beta_1}))^*$$

$$f'(\Phi(T; t_i, Y_i^\varsigma))\mathcal{D}\Phi(T; t_i, Y_i^\varsigma)], \tilde{\sigma} \mathrm{Id}_{d\times d} \Big\rangle\Big] \mathrm{d}u \mathrm{d}v_3^1(s) \mathrm{d}\beta_1 \mathrm{d}r \mathrm{d}\varsigma$$

$$= \sum_{i=1}^{N^\Delta} \sum_{j=-N}^{-1} \int_0^1 \int_{t_{i-1}}^{t_i} \int_0^1 \int_{-\tau \vee (t_{i+j-1}-r)}^{(t_{i+j}-r)\wedge 0} \int_{t_{i+j-1}}^{r+s} \Big\{\mathbb{E}\Big[\Big\langle f'(\Phi(T; t_i, Y_i^\varsigma))$$

$$\mathcal{D}\Phi(T; t_i, Y_i^\varsigma), I^{[0]} D_u k_b^1(Z_{i,r}^{\beta_1})\tilde{\sigma} \mathrm{Id}_{d\times d}\Big\rangle\Big] + \mathbb{E}\Big[\Big\langle f'(\Phi(T; t_i, Y_i^\varsigma))$$

$$D_u \mathcal{D}\Phi(T; t_i, Y_i^\varsigma), I^{[0]} k_b^1(Z_{i,r}^{\beta_1})\tilde{\sigma} \mathrm{Id}_{d\times d}\Big\rangle\Big] + \mathbb{E}\Big[\Big\langle f''(\Phi(T; t_i, Y_i^\varsigma))$$

$$D_u \Phi(T; t_i, Y_i^\varsigma)\mathcal{D}\Phi(T; t_i, Y_i^\varsigma), I^{[0]} k_b^1(Z_{i,r}^{\beta_1})\tilde{\sigma} \mathrm{Id}_{d\times d}\Big\rangle\Big]\Big\} \mathrm{d}u \mathrm{d}v_3^1(s) \mathrm{d}\beta_1 \mathrm{d}r \mathrm{d}\varsigma,$$

where without loss of generality, we took the parameter n_b in Assumption 3.4 to be $n_b = 1$. Combining (5.52), (5.53), and Lemma 5.7, we deduce

$$|\mathcal{I}_b^2| \le K \sum_{i=1}^{N^\Delta} \sum_{j=-N}^{-1} \int_0^1 \int_{t_{i-1}}^{t_i} \int_0^1 \int_{-\tau \vee (t_{i+j-1}-r)}^{(t_{i+j}-r)\wedge 0} \int_{t_{i+j-1}}^{r+s} \Big\{\Big\| \mathcal{D}\Phi(T; t_i, Y_i^\varsigma)$$

$$I^{[0]} D_u k_b^1(Z_{i,r}^{\beta_1})\tilde{\sigma} \mathrm{Id}_{d\times d}\Big\|_{2,2} + \Big\| D_u \mathcal{D}\Phi(T; t_i, Y_i^\varsigma) I^{[0]} k_b^1(Z_{i,r}^{\beta_1})\tilde{\sigma} \mathrm{Id}_{d\times d}\Big\|_{2,2}$$

$$+ \Big\| D_u \Phi(T; t_i, Y_i^\varsigma), \mathcal{D}\Phi(T; t_i, Y_i^\varsigma) I^{[0]} k_b^1(Z_{i,r}^{\beta_1})\tilde{\sigma} \mathrm{Id}_{d\times d}\Big\|_{3,2}\Big\}$$

$$\mathrm{d}u \mathrm{d}v_3^1(s) \mathrm{d}\beta_1 \mathrm{d}r \mathrm{d}\varsigma$$

$$\le K\Delta,$$

where we used the assumption

$$\sup_{l\in\{1,2,3\}} \sup_{\phi \in C^d} \Big(\|\mathcal{D}^l k_b^1(\phi)\|_{\mathcal{L}((C^d)^{\otimes l}; \mathbb{R}^{d\times d})} + |k_b^1(\phi)| \Big) \le K.$$

Hence, we have $|\mathcal{I}_b| \le K\Delta$.

Estimate of $\mathcal{I}_{b,\theta}$ Similar to proof of $\mathcal{I}_b$, we derive that $|\mathcal{I}_{b,\theta}| \le K\Delta$.

Combining estimates of terms $\mathcal{I}_0$, $\mathcal{I}_b$ and $\mathcal{I}_{b,\theta}$, and taking supremum for $T \ge T_0$ and $f \in \mathscr{C}$, we complete the proof. □

5.4 Numerical Experiments

In this section, we present the numerical algorithm and corresponding numerical experiments to verify the convergence rate of the numerical density function for the θ-EM methods applied to SFDEs. Both the L^1-error and L^∞-error are examined to illustrate the convergence results.

Algorithm 5 outlines the procedure for investigating the weak convergence of the θ-EM method in approximating the density function of SFDEs. The reference solution, computed with a fine time step, and the numerical solutions with coarser time steps are generated using a common Brownian motion trajectory. The values of all sample paths at the final time T are collected. Then the terminal density function is estimated using the kernel density estimation. Specifically, the empirical density function is constructed by

$$\mathfrak{p}^\Delta(T,z) \approx \frac{1}{Mh}\sum_{i=1}^{M}\mathcal{K}\Big(\frac{z-y_i^\Delta(T)}{h}\Big),$$

where $\mathcal{K}$ is a smooth kernel function (e.g., Gaussian kernel), $h > 0$ is the bandwidth parameter controlling the smoothness of the estimate, M is the total number of simulated trajectories, and $y_i^\Delta(T)$ denotes the θ-EM solution at time T for the i-th trajectory. The selection of h represents a trade-off between bias and variance, and it is chosen as $h = 0.4$ through numerical calibration in this work. In the numerical implementation, the probability density function is approximated using the MATLAB built-in function `ksdensity`, which performs kernel density estimation with an automatically selected bandwidth.

To quantify the convergence rate, we evaluate the error between the numerical density and the reference density. The error is measured in either the L^∞ or L^1 sense as follows:

$$\|\mathfrak{p}^\Delta(T,\cdot)-\mathfrak{p}(T,\cdot)\|_{L^\infty(\mathbb{R}^d)} = \max_{z\in\mathbb{R}^d}|\mathfrak{p}^\Delta(T,z)-\mathfrak{p}(T,z)|,$$

$$\|\mathfrak{p}^\Delta(T,\cdot)-\mathfrak{p}(T,\cdot)\|_{L^1(\mathbb{R}^d)} = \int_{z\in\mathbb{R}^d}|\mathfrak{p}^\Delta(T,z)-\mathfrak{p}(T,z)|\mathrm{d}z.$$

By repeating the procedure for multiple stepsizes, a log-log plot of the density error versus the stepsize is generated, and the convergence rate is estimated.

Algorithm 5 Density convergence of numerical method for SFDE

Require: SFDE parameters (b, σ, ξ, τ), parameter θ, final time T, stepsizes $\{\Delta_j\}_{j=1}^J$, reference stepsize Δ_{ref}, Monte Carlo size M.

Output: Error of terminal density function for each Δ_j.

Generate M independent Brownian motion increments $\delta W_{\text{ref}}^{(\tilde{m})}(k) \sim \mathcal{N}(0, \Delta_{\text{ref}})$, $k = 0, \ldots, N_{\text{ref}} - 1$, with $N_{\text{ref}} = T/\Delta_{\text{ref}}$ and $\tilde{m} = 1, \ldots, M$.
Set $\tilde{d}_{\text{ref}} = \tau/\Delta_{\text{ref}}$.
Initialize $x_{\text{ref}}^{(\tilde{m})}(t_k)$, $k = -\tilde{d}_{\text{ref}}, \ldots, 0$, $\tilde{m} = 1, \ldots, M$.
for $\tilde{m} = 1, \ldots, M$ **do**
  for $k = 0, \ldots, N_{\text{ref}} - 1$ **do**

$$x_{\text{ref}}^{(\tilde{m})}(t_{k+1}) = x_{\text{ref}}^{(\tilde{m})}(t_k) + \big(\theta b(x_{\text{ref},t_{k+1}}^{(\tilde{m})}) + (1-\theta) b(x_{\text{ref},t_k}^{(\tilde{m})})\big)\Delta_{\text{ref}} + \sigma(x_{\text{ref},t_k}^{(\tilde{m})})\delta W_{\text{ref}}^{(\tilde{m})}(k).$$

  end for
end for
Compute reference terminal density function $\mathfrak{p}_{\text{ref}}(T, \cdot)$ from $\{x_{\text{ref}}^{\tilde{m}}(T)\}_{\tilde{m}=1}^M$ using kernel density estimation
for $j = 1, \ldots, J$ **do**
  Set $\Delta = \Delta_j$, $N = T/\Delta$, $\tilde{d} = \tau/\Delta$, $r = \Delta/\Delta_{\text{ref}}$.
  Construct coarse Brownian motion increments

$$\delta W^{(\tilde{m})}(k) = \sum_{i=kr}^{(k+1)r-1} \delta W_{\text{ref}}^{(\tilde{m})}(i), \quad k = 0, \ldots, N-1.$$

  Initialize $y^{(\tilde{m})}(t_k)$, $k = -\tilde{d}, \ldots, 0$.
  for $\tilde{m} = 1, \ldots, M$ **do**
    for $k = 0, \ldots, N-1$ **do**

$$y^{(\tilde{m})}(t_{k+1}) = y^{(\tilde{m})}(t_k) + \big(\theta b(y_{t_{k+1}}^{(\tilde{m})}) + (1-\theta) b(y_{t_k}^{(\tilde{m})})\big)\Delta + \sigma(y_{t_k}^{(\tilde{m})})\delta W^{(\tilde{m})}(k).$$

    end for
  end for
  Compute terminal density function $\mathfrak{p}^\Delta(T, \cdot)$ using kernel density estimation
  Compute error $e_j = \max_z |\mathfrak{p}^\Delta(T, z) - \mathfrak{p}_{\text{ref}}(T, z)|$ or $\int |\mathfrak{p}^\Delta(T, z) - \mathfrak{p}_{\text{ref}}(T, z)| dz$
end for
return $\{e_j\}_{j=1}^J$

Linear Stochastic Delay Differential Equation In this part, we take the linear stochastic delay differential equation (4.80) as an example to demonstrate the convergence rate of the numerical density function obtained via the EM method. Figure 5.1 illustrates the convergence rate of the numerical density function for the EM method, with parameters $\alpha = 2, \beta = 1, \sigma = 0.5$, and delay $\tau = 1$. The terminal time is set to $T = 8$, and the time stepsizes are chosen as $\Delta = 2^{-3}, 2^{-4}, \ldots, 2^{-7}$, with a fine reference stepsize $\Delta_{\text{ref}} = 2^{-9}$. Subfigure (a) presents the L^1-error, and Subfigure (b) displays the L^∞-error between the numerical and reference terminal densities. The dashed line indicates a reference slope of order

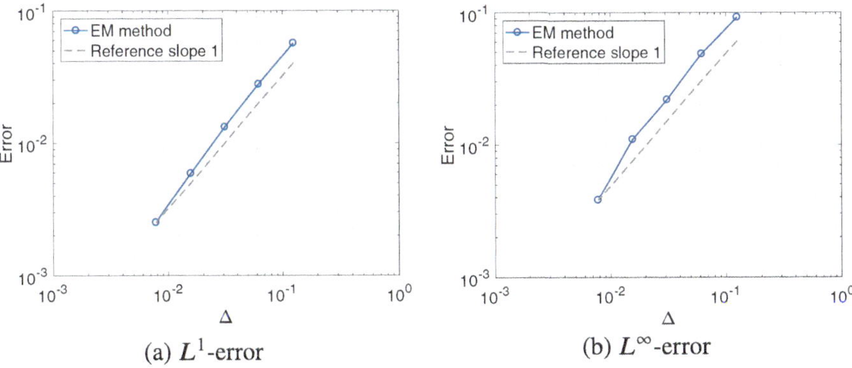

Fig. 5.1 Density convergence of EM method for (4.80): $\alpha = 2, \beta = 1, \sigma = 0.5, \tau = 1, \xi(t) = \sin(2\pi t)$ for $t \in [-\tau, 0]$, $M = 5000$, $T = 8$, $\Delta = 2^{-3}, 2^{-4}, \ldots, 2^{-7}$, $\Delta_{\text{ref}} = 2^{-9}$, $h = 0.4$. (**a**) L^1-error. (**b**) L^∞-error

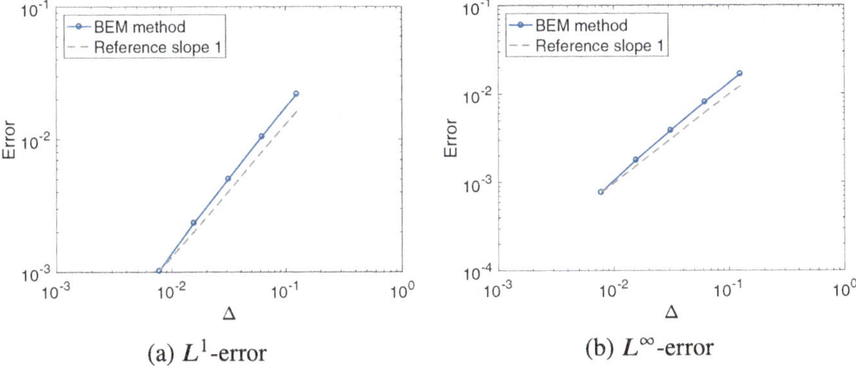

Fig. 5.2 Density convergence of backward EM method for (4.82): $a = 0.5, \tau = 1, \xi(t) = \sin(2\pi t)$ for $t \in [-\tau, 0]$, $T = 1$, $M = 5000$, $\Delta = 2^{-3}, 2^{-4}, \ldots, 2^{-7}$, $\Delta_{\text{ref}} = 2^{-9}$, $h = 0.4$. (**a**) L^1-error. (**b**) L^∞-error

one. Both plots show that the numerical density function achieves convergence with rate one.

Nonlinear Stochastic Functional Differential Equation In this part, we investigate the convergence rate of the numerical density function via the backward EM method for the nonlinear SFDE (4.82). Figure 5.2 demonstrates the convergence behavior of the numerical density function for the backward EM method, with parameters $a = 0.5$, $\tau = 1$, and terminal time $T = 1$. The stepsizes are chosen as $\Delta = 2^{-3}, 2^{-4}, \ldots, 2^{-7}$, while a reference solution is computed with $\Delta_{\text{ref}} = 2^{-9}$. Subfigure (a) shows the L^1-error and (b) the L^∞-error between the numerical and reference terminal densities. The dashed lines represent a reference slope of order one. Both error curves exhibit an approximately linear trend on the log-log scale, indicating that the numerical density function achieves convergence with rate one.

5.5 Summary and Outlook

This chapter has studied the existence and convergence of the density function for the θ-EM method of the SFDE. The content of this chapter is based on the results presented in [10]. For related works on density functions for solutions of SFDEs, we refer to [2, 11] for the study of the existence and smoothness of the density function of the exact solution of SFDEs with globally Lipschitz drift; and refer to [19] for the investigation of the asymptotic behavior for the perturbed densities for SFDEs with small noise under the globally Lipschitz drift condition. We also mention that for stochastic ordinary differential equations and stochastic partial differential equations, density approximations have been well studied in recent decades, and for interested readers, we refer to [1, 13, 18, 20, 22–25, 31–33] for the density approximations of stochastic ordinary differential equations and to [9, 21] for those of stochastic partial differential equations.

Our convergence results also indicate that the total variation distance between laws of solutions for the SFDE and its discretizations vanishes to zero as the time stepsize diminishes, while that between laws of functional solutions fails to vanish due to the high degeneracy of the equation. This finding highlights one of the main distinctions in asymptotic behaviors of the corresponding discretized systems when compared to stochastic ordinary (partial) differential equations. We mention some related results regarding the convergence of discretizations in total variation distance, building on existing studies concerning the convergence of densities for discretizations. For stochastic ordinary differential equations, the convergence in total variation distance for Itô–Taylor type discretizations can be obtained from the corresponding density convergence results in [1, 13, 20, 22–24], with the convergence rate detailed in [3, 4, 27]. For stochastic partial differential equations, utilizing the density convergence of discretizations, one can establish the convergence in total variation distance for random field solutions of discretizations, as shown in [9, 21]. The convergence behavior in total variation distance for the Hilbert-valued solution of temporal semi-discretizations depends on the choice of discretization methods, with some methods failing to converge; see e.g. [5, 6].

There remain several open problems concerning the study of density functions of SFDEs, both from theoretical and numerical perspectives. Building on the results established in this chapter, it is of interest to investigate whether the convergence rate 1 of the numerical density function can be achieved for SFDEs with multiplicative noise. Another important question is whether the convergence rate of density approximations can be established under weaker assumptions on the coefficients. Furthermore, when the driving noise is rougher than Brownian motion, such as SFDEs driven by fractional Brownian motion or rough paths, the density functions of solutions have been studied only to a very limited extent. In particular, for

SFDEs driven by fractional Brownian motion, the existence and smoothness of density functions have been obtained in [26]. However, for SFDEs in the rough path setting, such results appear to be unavailable to the best of our knowledge. Moreover, for both cases, the numerical approximations of density functions have not yet been addressed. We therefore propose these problems as open questions for future investigation.

The density functions of solution processes have a wide range of applications, and one of them is in the study of hitting probabilities, which have been deeply investigated for stochastic differential equations, see e.g. [14–17, 29]. In the numerical context of hitting probabilities, the first result is [8], which presents a noteworthy result: numerical discretizations can alter the hitting behaviors of the system. Specifically, for linear stochastic parabolic systems, the critical dimensions for the spatial semi-discretizations based on the finite difference method or the spectral Galerkin method, and temporal semi-discretizations by using the exponential EM method, are shown to be half of those for the original system. It is worth mentioning that for the finite dimensional case, the numerical solution derived from the continuous exponential EM method preserves the critical dimension of the exact solution of the original system. This motivates a compelling direction for future research to investigate the hitting probabilities for SFDEs. We note that in this chapter, based on the density functions, we have shown that the solution and the functional solution for the SFDE exhibit intrinsic distinctions. An interesting question is to explore how these differences manifest in hitting probabilities, and to understand the effect of numerical discretizations on hitting properties of the original SFDE.

References

1. V. Bally, D. Talay, The law of the Euler scheme for stochastic differential equations. II. Convergence rate of the density. Monte Carlo Methods Appl. **2**, 93–128 (1996)
2. D.R. Bell, S.-E.A. Mohammed, Smooth densities for degenerate stochastic delay equations with hereditary drift. Ann. Probab. **23**, 1875–1894 (1995)
3. O. Bencheikh, B. Jourdain, Convergence in total variation of the Euler–Maruyama scheme applied to diffusion processes with measurable drift coefficient and additive noise. SIAM J. Numer. Anal. **60**, 1701–1740 (2022)
4. P. Bras, G. Pagès, F. Panloup, Total variation distance between two diffusions in small time with unbounded drift: application to the Euler–Maruyama scheme. Electron. J. Probab. **27**, Paper No. 153 (2022)
5. C.-E. Bréhier, Influence of the regularity of the test functions for weak convergence in numerical discretization of SPDEs. J. Complexity **56**, 101424 (2020)
6. C.-E. Bréhier, Total variation error bounds for the accelerated exponential Euler scheme approximation of parabolic semilinear SPDEs. SIAM J. Numer. Anal. **62**, 1171–1190 (2024)
7. E. Buckwar, T. Shardlow, Weak approximation of stochastic differential delay equations. IMA J. Numer. Anal. **25**, 57–86 (2005)
8. C. Chen, J. Hong, D. Sheng, Influence of numerical discretizations on hitting probabilities for linear stochastic parabolic systems. J. Complexity **70**, Paper No. 101634 (2022)

9. C. Chen, J. Cui, J. Hong, D. Sheng, Accelerated exponential Euler scheme for stochastic heat equation: convergence rate of the density. IMA J. Numer. Anal. **43**, 1181–1220 (2023)
10. C. Chen, T. Dang, J. Hong, G. Song, Densities of stochastic functional differential equation and its discretizations. J. Differ. Equ. **459**, 1–38 (2026)
11. R. Chhaibi, I. Ekren, A Hörmander condition for delayed stochastic differential equations. Ann. H. Lebesgue **3**, 1023–1048 (2020)
12. E. Clément, A. Kohatsu-Higa, D. Lamberton, A duality approach for the weak approximation of stochastic differential equations. Ann. Appl. Probab. **16**, 1124–1154 (2006)
13. J. Cui, J. Hong, D. Sheng, Density function of numerical solution of splitting AVF scheme for stochastic Langevin equation. Math. Comput. **91**, 2283–2333 (2022)
14. R.C. Dalang, F. Pu, Optimal lower bounds on hitting probabilities for non-linear systems of stochastic fractional heat equations. Stoch. Process. Appl. **131**, 359–393 (2021)
15. R.C. Dalang, F. Pu, Hitting with probability one for stochastic heat equations with additive noise. J. Theoret. Probab. **37**, 3479–3495 (2024)
16. R.C. Dalang, D. Khoshnevisan, E. Nualart, Hitting probabilities for systems of non-linear stochastic heat equations with additive noise. ALEA Lat. Am. J. Probab. Math. Stat. **3**, 231–271 (2007)
17. R.C. Dalang, D. Khoshnevisan, E. Nualart, Hitting probabilities for systems for non-linear stochastic heat equations with multiplicative noise. Probab. Theory Related Fields **144**, 371–427 (2009)
18. M.A. El-Gebeily, H.E. Emara-Shabaik, Approximate solution of the Fokker–Planck–Kolmogorov equation by finite elements. Comm. Numer. Methods Eng. **10**, 763–771 (1994)
19. M. Ferrante, C. Rovira, M. Sanz-Solé, Stochastic delay equations with hereditary drift: estimates of the density. J. Funct. Anal. **177**, 138–177 (2000)
20. J. Guyon, Euler scheme and tempered distributions. Stoch. Process. Appl. **116**, 877–904 (2006)
21. J. Hong, D. Jin, D. Sheng, Density convergence of a fully discrete finite difference method for stochastic Cahn–Hilliard equation. Math. Comput. **93**, 2215–2264 (2024)
22. Y. Hu, S. Watanabe, Donsker's delta functions and approximation of heat kernels by the time discretization methods. J. Math. Kyoto Univ. **36**, 499–518 (1996)
23. A. Kebaier, A. Kohatsu-Higa, An optimal control variance reduction method for density estimation. Stoch. Process. Appl. **118**, 2143–2180 (2008)
24. A. Kohatsu-Higa, High order Itô–Taylor approximations to heat kernels. J. Math. Kyoto Univ. **37**, 129–150 (1997)
25. V. Konakov, E. Mammen, Edgeworth type expansions for Euler schemes for stochastic differential equations. Monte Carlo Methods Appl. **8**, 271–285 (2002)
26. J.A. León, S. Tindel, Malliavin calculus for fractional delay equations. J. Theoret. Probab. **25**, 854–889 (2012)
27. L. Li, M. Wang, Y. Wang, Error estimates of the Euler's method for stochastic differential equations with multiplicative noise via relative entropy (2024). https://arxiv.org/abs/2409.04991
28. Q. Luo, X. Mao, Y. Shen, Generalised theory on asymptotic stability and boundedness of stochastic functional differential equations. Autom. J. IFAC **47**, 2075–2081 (2011)
29. C. Mueller, R. Tribe, Hitting properties of a random string. Electron. J. Probab. **7**(10), 29 (2002)
30. D. Nualart, *The Malliavin Calculus and Related Topics*, Probability and its Applications, 2nd edn. (Springer, Berlin, 2006)
31. J. Reif, R. Barakat, Numerical solution of the Fokker–Planck equation via Chebyschev polynomial approximations with reference to first passage time probability density functions. J. Comput. Phys. **23**, 425–445 (1977)
32. L. Zeng, X. Wan, T. Zhou, Adaptive deep density approximation for fractional Fokker-Planck equations. J. Sci. Comput. **97**, Paper No. 68 (2023)
33. B. Zhou, H. Wang, T. Wang, D. Jiang, Stochastic generalized Kolmogorov systems with small diffusion: I. Explicit approximations for invariant probability density function. J. Differ. Equ. **382**, 141–210 (2024)

Chapter 6
Large Deviation Principle of Numerical Solution

Large deviations describe the asymptotic behavior of small probabilities of rare events on an exponential scale. In Theorem 1.20, we have established the Freidlin–Wentzell type large deviation principle (LDP) for the exact solution of the SFDE with small noise. In large deviation theory, the decay rates of probabilities associated with rare events are characterized by the large deviation rate function, which is typically defined through a constrained minimization problem and, in general, cannot be expressed explicitly. Therefore, it is of great interest to investigate whether numerical methods can asymptotically preserve the LDPs of the underlying systems, particularly when it comes to predicting rare-event probabilities that are of practical significance. In this chapter, based on the weak convergence approach, we study Freidlin–Wentzell type LDPs for θ-EM methods applied to SFDEs with small noise in the infinite time horizon.

In Sect. 6.1, we first establish the compactness of solutions for both the skeleton equation and the stochastic controlled equation of the θ-EM method in Banach space $C_{\xi^{Int}}([-\tau, +\infty); \mathbb{R}^d)$. Using the weak convergence approach, we then show that the θ-EM solution satisfies the LDP on $C_{\xi^{Int}}([-\tau, +\infty); \mathbb{R}^d)$ with the rate function given by the corresponding skeleton equation. In Sect. 6.2, we prove the existence and smoothness of the numerical density function for the θ-EM method applied to SFDEs with small noise. Combining techniques from the Malliavin calculus, we also present the logarithmic estimates of the numerical density function and reveal the relation between the logarithmic limit and the rate function of the LDP.

C. Chen et al., *Numerical Analysis of Stochastic Functional Differential Equations*, Lecture Notes in Mathematics 2399, https://doi.org/10.1007/978-981-92-1592-8_6

6.1 The Freidlin–Wentzell Large Deviation of Numerical Solution

In this section, we consider the SFDE with small noise (1.35) on the infinite time horizon. For convenience, we recall the equation here

$$\begin{cases} \mathrm{d}x^{\xi,\epsilon}(t) = b(x_t^{\xi,\epsilon})\mathrm{d}t + \sqrt{\epsilon}\sigma(x_t^{\xi,\epsilon})\mathrm{d}W(t), & t > 0, \\ x^{\xi,\epsilon}(0) = \xi \in C^d, \end{cases} \tag{6.1}$$

where $\epsilon \in (0, 1)$. The aim is to study the Freidlin–Wentzell large deviation of the θ-EM method of (6.1) by means of the weak convergence approach. The θ-EM method of (6.1) reads as

$$\begin{cases} y^{\Delta,\epsilon}(t_{k+1}) = y^{\Delta,\epsilon}(t_k) + (1-\theta)b(y_{t_k}^{\Delta,\epsilon})\Delta + \theta b(y_{t_{k+1}}^{\Delta,\epsilon})\Delta + \sqrt{\epsilon}\sigma(y_{t_k}^{\Delta,\epsilon})\delta W_k, \ k \in \mathbb{N}, \\ y^{\Delta,\epsilon}(t_k) = \xi(t_k), \quad k = -N, \ldots, 0, \end{cases} \tag{6.2}$$

where $y_{t_k}^{\Delta,\epsilon}(r) = \frac{t_{j+1}-r}{\Delta} y^{\Delta,\epsilon}(t_{k+j}) + \frac{r-t_j}{\Delta} y^{\Delta,\epsilon}(t_{k+j+1})$ for $r \in [t_j, t_{j+1}]$, $j \in \{-N, \ldots, -1\}$. The continuous version for the solution of (6.2) is defined by the linear interpolation of the solution. Here, for notational abbreviation, we omit the initial datum on the notation $y_{\cdot}^{\Delta,\epsilon}$ and $y^{\Delta,\epsilon}(\cdot)$ of the numerical solutions.

Fix the stepsize $\Delta \in (0, 1)$. We denote

$$\begin{aligned} &L_\Delta^2([0, +\infty); \mathbb{R}^m) \\ &:= \left\{ v : [0, +\infty) \to \mathbb{R}^m \text{ is measurable} \,\middle|\, \int_0^{+\infty} |v(\lfloor s \rfloor_\Delta)|^2 \mathrm{d}s < \infty \right\}, \end{aligned}$$

where $\lfloor s \rfloor_\Delta$ is the largest grid point that no larger than time s. For any $M \in (0, +\infty)$, set

$$\mathcal{S}_M := \left\{ v \in L_\Delta^2([0, +\infty); \mathbb{R}^m) \,\middle|\, \int_0^{+\infty} |v(\lfloor s \rfloor_\Delta)|^2 \mathrm{d}s \le M \right\}$$

and

$$\mathcal{A}_M := \left\{ v : \Omega \times [0, +\infty) \to \mathbb{R}^m \mid v(\cdot, t) \text{ is } \mathcal{F}_t\text{-measurable and } v \in \mathcal{S}_M \ \mathbb{P}\text{-a.s.} \right\}.$$

It follows from [22, Chapter III, Section 28, Theorem 1'] that $\mathcal{S}_M$ is a Polish space endowed with the weak topology. Introduce the space $C_{\xi^{Int}}([-\tau, +\infty); \mathbb{R}^d)$

consisting of continuous functions $u : [-\tau, \infty) \to \mathbb{R}^d$ with $u(r) = \xi^{Int}(r),\ r \in [-\tau, 0]$, which is a Polish space endowed with the norm

$$\|u\|_{\mathfrak{a}} := \sum_{k=1}^{\infty} e^{-\mathfrak{a}t_k}\Big(\sup_{t\in[-\tau,t_k]} |u(t)|\Big)\Delta$$

for some constant $\mathfrak{a} > 0$. Here, we recall that $\xi^{Int} \in C^{Int}$ is the linear interpolation of $\xi(t_i), i = -N, \ldots, 0$ (see Sect. 3.2).

In order to present the LDP for the solution of (6.2), we consider the following stochastic controlled equation

$$\begin{cases} y^{\Delta,v^\epsilon}(t_{k+1}) = y^{\Delta,v^\epsilon}(t_k) + (1-\theta)b(y_{t_k}^{\Delta,v^\epsilon})\Delta + \theta b(y_{t_{k+1}}^{\Delta,v^\epsilon})\Delta \\ \qquad\qquad +\sigma(y_{t_k}^{\Delta,v^\epsilon})\big(\sqrt{\epsilon}\delta W_k + v^\epsilon(t_k)\Delta\big), \quad k \in \mathbb{N}, \\ y^{\Delta,v^\epsilon}(t_k) = \xi(t_k), \quad k = -N, \ldots, 0, \end{cases} \tag{6.3}$$

and the skeleton equation

$$\begin{cases} w^{\Delta,v}(t_{k+1}) = w^{\Delta,v}(t_k) + (1-\theta)b(w_{t_k}^{\Delta,v})\Delta + \theta b(w_{t_{k+1}}^{\Delta,v})\Delta \\ \qquad\qquad +\sigma(w_{t_k}^{\Delta,v})v(t_k)\Delta, \quad k \in \mathbb{N}, \\ w^{\Delta,v}(t_k) = \xi(t_k), \quad k = -N, \ldots, 0, \end{cases} \tag{6.4}$$

where $v^\epsilon, v \in L^2_\Delta([0, +\infty); \mathbb{R}^m)$, and $y_{t_k}^{\Delta,v^\epsilon}$ and $w_{t_k}^{\Delta,v}$ are defined as follows: for $r \in [t_j, t_{j+1}], j \in \{-N, \ldots, -1\}$,

$$y_{t_k}^{\Delta,v^\epsilon}(r) := \frac{t_{j+1} - r}{\Delta} y^{\Delta,v^\epsilon}(t_{k+j}) + \frac{r - t_j}{\Delta} y^{\Delta,v^\epsilon}(t_{k+j+1}),$$
$$w_{t_k}^{\Delta,v}(r) := \frac{t_{j+1} - r}{\Delta} w^{\Delta,v}(t_{k+j}) + \frac{r - t_j}{\Delta} w^{\Delta,v}(t_{k+j+1}).$$

The continuous versions for solutions of (6.3) and (6.4) are defined by the linear interpolations of the solutions, respectively. Define measurable mappings $\mathcal{G}^{\Delta,\epsilon}, \mathcal{G}^{\Delta} : C([0, +\infty); \mathbb{R}^m) \to C_{\xi^{Int}}([-\tau, +\infty); \mathbb{R}^d)$ respectively, by

$$\mathcal{G}^{\Delta,\epsilon}\Big(\sqrt{\epsilon}W(\cdot) + \int_0^\cdot v^\epsilon(\lfloor s\rfloor_\Delta)ds\Big) = y^{\Delta,v^\epsilon}(\cdot), \quad \mathcal{G}^{\Delta}\Big(\int_0^\cdot v(\lfloor s\rfloor_\Delta)ds\Big) = w^{\Delta,v}(\cdot).$$

Introduce the auxiliary process of (6.4) as follows:

$$\begin{cases} z^{\Delta,v}(t_k) = \xi(t_k), \quad k = -N, \ldots, -1, \\ z^{\Delta,v}(t_k) = w^{\Delta,v}(t_k) - \theta b(w_{t_k}^{\Delta,v})\Delta, \quad k \in \mathbb{N}. \end{cases} \tag{6.5}$$

It follows from (6.5) that

$$z^{\Delta,v}(t_k) = z^{\Delta,v}(t_{k-1}) + b(w^{\Delta,v}_{t_{k-1}})\Delta + \sigma(w^{\Delta,v}_{t_{k-1}})v(t_{k-1})\Delta, \quad k \in \mathbb{N}_+.$$

The corresponding continuous version $\{z^{\Delta,v}(t)\}_{t\geq -\tau}$ is defined as: for $t \in [t_k, t_{k+1})$ with $k \in \mathbb{N}$,

$$z^{\Delta,v}(t) = z^{\Delta,v}(t_k) + b(w^{\Delta,v}_{t_k})(t - t_k) + \sigma(w^{\Delta,v}_{t_k})v(t_k)(t - t_k), \tag{6.6}$$

where for $t \in [-\tau, 0]$, the initial datum $z^{\Delta,v}(t)$ is defined by the linear interpolation of $z^{\Delta,v}(t_i)$, $i = -N, \ldots, 0$.

Theorem 6.1 *Let Assumptions 2.1 to 2.4 hold. Then for any* $\Delta \in (0, \frac{1}{4\theta a_2}]$ *with* a_2 *given in Assumption 2.2, the family of random variables* $\{y^{\Delta,\epsilon}(\cdot)\}_{\epsilon\in(0,1)}$ *satisfies the LDP on* $C_{\xi^{Int}}([-\tau, +\infty); \mathbb{R}^d)$*, i.e.,*

(i) for each closed subset F *of* $C_{\xi^{Int}}([-\tau, +\infty); \mathbb{R}^d)$,

$$\limsup_{\epsilon\to 0} \epsilon \log \mathbb{P}\big(y^{\Delta,\epsilon}(\cdot) \in F\big) \leq - \inf_{x\in F} I(x);$$

(ii) for each open subset O *of* $C_{\xi^{Int}}([-\tau, +\infty); \mathbb{R}^d)$,

$$\liminf_{\epsilon\to 0} \epsilon \log \mathbb{P}\big(y^{\Delta,\epsilon}(\cdot) \in O\big) \geq - \inf_{x\in O} I(x),$$

where the good rate function I *is given by*

$$I(f) = \inf_{\{v\in L^2_\Delta([0,+\infty);\mathbb{R}^m):\, f=\mathcal{G}^\Delta(\int_0^\cdot v(\lfloor s\rfloor_\Delta)\mathrm{d}s)\}} \frac{1}{2}\int_0^{+\infty} |v(\lfloor s\rfloor_\Delta)|^2 \mathrm{d}s.$$

According to the contraction principle (see e.g. [11, Theorem 4.2.1]), we obtain the following LDP for the family $\{y^{\Delta,\epsilon}(t)\}_{\epsilon\in(0,1)}$ for each $t > 0$.

Corollary 6.2 *Under conditions in Theorem 6.1, for any* $t > 0$*, the family of random variables* $\{y^{\Delta,\epsilon}(t)\}_{\epsilon\in(0,1)}$ *satisfies the LDP on* $\mathbb{R}^d$ *with the good rate function given by*

$$\tilde{I}(z) = \inf\{I(f) : f \in C_{\xi^{Int}}([-\tau, \infty); \mathbb{R}^d),\, z = f(t)\},$$

where $I(\cdot)$ *is given in Theorem 6.1.*

The proof of Theorem 6.1 is based on the equivalence of the LDP and the Laplace principle. According to [15, Condition 2.1, Theorem 2.10], it suffices to prove the compactness of solutions of the skeleton equation and the stochastic controlled equation (i.e., Propositions 6.4 and 6.6). We first give the a priori estimates of $w^{\Delta,v}$ and $z^{\Delta,v}$.

Lemma 6.3 *Let $M > 0$ and $v \in \mathcal{S}_M$. Under Assumptions 2.1, 2.2, and 2.4, it holds that for any $\Delta \in (0, \frac{1}{4\theta a_2}]$,*

$$\sup_{k\geq 0}|w^{\Delta,v}(t_k)| + \sup_{k\geq 0}|z^{\Delta,v}(t_k)| \leq (1+\|\xi\|^{\beta+1})K(M),$$

where a_2 is given in Assumption 2.2 and β is given in Assumption 2.4.

Proof By the definition of $z^{\Delta,v}$, we have

$$\begin{aligned}
\frac{1}{2}\mathrm{d}|z^{\Delta,v}(t)|^2 &= \langle b(w^{\Delta,v}_{\lfloor t\rfloor_\Delta}), z^{\Delta,v}(t)\rangle\mathrm{d}t + \langle \sigma(w^{\Delta,v}_{\lfloor t\rfloor_\Delta})v(\lfloor t\rfloor_\Delta), z^{\Delta,v}(t)\rangle\mathrm{d}t \\
&= \Big\langle b(w^{\Delta,v}_{\lfloor t\rfloor_\Delta}), w^{\Delta,v}(\lfloor t\rfloor_\Delta) - \theta b(w^{\Delta,v}_{\lfloor t\rfloor_\Delta})\Delta + b(w^{\Delta,v}_{\lfloor t\rfloor_\Delta})(t-\lfloor t\rfloor_\Delta) \\
&\quad + \sigma(w^{\Delta,v}_{\lfloor t\rfloor_\Delta})v(\lfloor t\rfloor_\Delta)(t-\lfloor t\rfloor_\Delta)\Big\rangle\mathrm{d}t \\
&\quad + \Big\langle \sigma(w^{\Delta,v}_{\lfloor t\rfloor_\Delta})v(\lfloor t\rfloor_\Delta), z^{\Delta,v}(\lfloor t\rfloor_\Delta) + b(w^{\Delta,v}_{\lfloor t\rfloor_\Delta})(t-\lfloor t\rfloor_\Delta) \\
&\quad + \sigma(w^{\Delta,v}_{\lfloor t\rfloor_\Delta})v(\lfloor t\rfloor_\Delta)(t-\lfloor t\rfloor_\Delta)\Big\rangle\mathrm{d}t \\
&= \Big[(t-\lfloor t\rfloor_\Delta - \theta\Delta)|b(w^{\Delta,v}_{\lfloor t\rfloor_\Delta})|^2 + \big\langle b(w^{\Delta,v}_{\lfloor t\rfloor_\Delta}), w^{\Delta,v}(\lfloor t\rfloor_\Delta)\big\rangle \\
&\quad + \big\langle 2b(w^{\Delta,v}_{\lfloor t\rfloor_\Delta})(t-\lfloor t\rfloor_\Delta) + z^{\Delta,v}(\lfloor t\rfloor_\Delta), \sigma(w^{\Delta,v}_{\lfloor t\rfloor_\Delta})v(\lfloor t\rfloor_\Delta)\big\rangle \\
&\quad + |\sigma(w^{\Delta,v}_{\lfloor t\rfloor_\Delta})v(\lfloor t\rfloor_\Delta)|^2(t-\lfloor t\rfloor_\Delta)\Big]\mathrm{d}t.
\end{aligned}$$

This implies

$$\begin{aligned}
|z^{\Delta,v}(t_{k+1})|^2 &= |z^{\Delta,v}(t_k)|^2 + 2\int_{t_k}^{t_{k+1}} |b(w^{\Delta,v}_{t_k})|^2(t-t_k-\theta\Delta)\mathrm{d}t \\
&\quad + 2\int_{t_k}^{t_{k+1}} \big\langle b(w^{\Delta,v}_{t_k}), w^{\Delta,v}(t_k)\big\rangle\mathrm{d}t + \int_{t_k}^{t_{k+1}} \big\langle 4b(w^{\Delta,v}_{t_k})(t-t_k) \\
&\quad + 2z^{\Delta,v}(t_k), \sigma(w^{\Delta,v}_{t_k})v(t_k)\big\rangle\mathrm{d}t \\
&\quad + 2\int_{t_k}^{t_{k+1}} |\sigma(w^{\Delta,v}_{t_k})v(t_k)|^2(t-t_k)\mathrm{d}t.
\end{aligned}$$

Using

$$b(w^{\Delta,v}_{\lfloor t\rfloor_\Delta}) = \frac{1}{\theta\Delta}\big(w^{\Delta,v}(\lfloor t\rfloor_\Delta) - z^{\Delta,v}(\lfloor t\rfloor_\Delta)\big),$$

$2\int_{t_k}^{t_{k+1}}(t-t_k-\theta\Delta)\mathrm{d}t=(1-2\theta)\Delta^2$, and $\int_{t_k}^{t_{k+1}}(t-t_k)\mathrm{d}t=\frac{1}{2}\Delta^2$, we derive

$$\begin{aligned}|z^{\Delta,v}(t_{k+1})|^2&=|z^{\Delta,v}(t_k)|^2+\frac{1-2\theta}{\theta^2}|w^{\Delta,v}(t_k)-z^{\Delta,v}(t_k)|^2\\&\quad+2\int_{t_k}^{t_{k+1}}\langle b(w_{t_k}^{\Delta,v}),w^{\Delta,v}(t_k)\rangle\mathrm{d}t\\&\quad+\int_{t_k}^{t_{k+1}}2\Big\langle\frac{1}{\theta}(w^{\Delta,v}(t_k)-z^{\Delta,v}(t_k))+z^{\Delta,v}(t_k),\sigma(w_{t_k}^{\Delta,v})v(t_k)\Big\rangle\mathrm{d}t\\&\quad+\Delta\int_{t_k}^{t_{k+1}}|\sigma(w_{t_k}^{\Delta,v})v(t_k)|^2\mathrm{d}t\\&\le\Big(\frac{(1-\theta)^2}{\theta^2}+\frac{2\theta-1}{\theta^2(1+\tilde{c}\Delta)}\Big)|z^{\Delta,v}(t_k)|^2\\&\quad+\Big(\frac{2\theta-1}{\theta^2}(1+\tilde{c}\Delta)+\frac{1-2\theta}{\theta^2}\Big)|w^{\Delta,v}(t_k)|^2\\&\quad+2\int_{t_k}^{t_{k+1}}\langle b(w_{t_k}^{\Delta,v}),w^{\Delta,v}(t_k)\rangle\mathrm{d}t+\Delta\int_{t_k}^{t_{k+1}}|\sigma(w_{t_k}^{\Delta,v})v(t_k)|^2\mathrm{d}t\\&\quad+\int_{t_k}^{t_{k+1}}2\Big\langle\frac{1}{\theta}(w^{\Delta,v}(t_k)-z^{\Delta,v}(t_k))+z^{\Delta,v}(t_k),\sigma(w_{t_k}^{\Delta,v})v(t_k)\Big\rangle\mathrm{d}t,\end{aligned}$$

where in the last step we used the inequality

$$2\langle a,b\rangle\le(1+\tilde{c}\Delta)|a|^2+\frac{1}{1+\tilde{c}\Delta}|b|^2$$

for $\tilde{c}>0$. Therefore, by iterating, we obtain that for $\iota\in(0,1)$,

$$\begin{aligned}&e^{\iota t_{k+1}}|z^{\Delta,v}(t_{k+1})|^2\\&=|z^{\Delta,v}(0)|^2+\sum_{i=0}^{k}\Big(e^{\iota t_{i+1}}|z^{\Delta,v}(t_{i+1})|^2-e^{\iota t_i}|z^{\Delta,v}(t_i)|^2\Big)\\&\le|z^{\Delta,v}(0)|^2+\Big(\frac{(1-\theta)^2}{\theta^2}+\frac{2\theta-1}{\theta^2(1+\tilde{c}\Delta)}-e^{-\iota\Delta}\Big)\sum_{i=0}^{k}e^{\iota t_{i+1}}|z^{\Delta,v}(t_i)|^2\\&\quad+\Delta\sum_{i=0}^{k}e^{\iota t_{i+1}}\Big[\frac{2\theta-1}{\theta^2}\tilde{c}|w^{\Delta,v}(t_i)|^2+2\langle b(w_{t_i}^{\Delta,v}),w^{\Delta,v}(t_i)\rangle\\&\quad+\Delta|\sigma(w_{t_i}^{\Delta,v})|^2|v(t_i)|^2\\&\quad+2\Big\langle\frac{1}{\theta}(w^{\Delta,v}(t_i)-z^{\Delta,v}(t_i))+z^{\Delta,v}(t_i),\sigma(w_{t_i}^{\Delta,v})v(t_i)\Big\rangle\Big].\end{aligned}$$

In virtue of (2.28), we can take sufficiently small numbers $\iota, \tilde{c} > 0$ independent of Δ such that $\frac{(1-\theta)^2}{\theta^2} + \frac{2\theta-1}{\theta^2(1+\tilde{c}\Delta)} - e^{-\iota\Delta} \le 0$. By Assumption 2.1 and the condition $a_1 - a_2 - L > 0$, we have

$$\begin{aligned}|\sigma(\phi) - \sigma(\mathbf{0}) + \sigma(\mathbf{0})|^2 &\le K + \frac{a_1 - a_2 + L}{2L}|\sigma(\phi) - \sigma(\mathbf{0})|^2 \\ &\le K + \frac{a_1 - a_2 + L}{2}\Big(|\phi(0)|^2 + \int_{-\tau}^{0}|\phi(r)|^2\mathrm{d}\nu_1(r)\Big),\end{aligned}$$

where we used the Young inequality in the first step. Recall that Assumption 2.2 implies (2.7). As a result, we obtain

$$\begin{aligned}2\langle\phi(0), b(\phi)\rangle + |\sigma(\phi)|^2 \le{}& K - (a_1 + a_2)|\phi(0)|^2 + \frac{a_1 - a_2 + L}{2}\int_{-\tau}^{0}|\phi(r)|^2\mathrm{d}\nu_1(r) \\ &+ 2a_2\int_{-\tau}^{0}|\phi(r)|^2\mathrm{d}\nu_2(r). \qquad (6.7)\end{aligned}$$

Then, utilizing the Young inequality, (6.7), and Assumption 2.1, we see that

$$\begin{aligned}&e^{\iota t_{k+1}}|z^{\Delta,v}(t_{k+1})|^2 \\ &\le |z^{\Delta,v}(0)|^2 + \int_0^{t_{k+1}} e^{\iota\lceil t\rceil_\Delta}\Big[\frac{2\theta-1}{\theta^2}\tilde{c}|w^{\Delta,v}(\lfloor t\rfloor_\Delta)|^2 + 2\langle b(w^{\Delta,v}_{\lfloor t\rfloor_\Delta}), w^{\Delta,v}(\lfloor t\rfloor_\Delta)\rangle \\ &\quad + \Delta|\sigma(w^{\Delta,v}_{\lfloor t\rfloor_\Delta})|^2|v(\lfloor t\rfloor_\Delta)|^2 + |\sigma(w^{\Delta,v}_{\lfloor t\rfloor_\Delta})|^2 \\ &\quad + \Big|\frac{1}{\theta}w^{\Delta,v}(\lfloor t\rfloor_\Delta) + \frac{\theta-1}{\theta}z^{\Delta,v}(\lfloor t\rfloor_\Delta)\Big|^2|v(\lfloor t\rfloor_\Delta)|^2\Big]\mathrm{d}t \\ &\le |\xi(0) - \theta b(\xi)\Delta|^2 + \int_0^{t_{k+1}} e^{\iota\lceil t\rceil_\Delta}\Big[K + \Big(\frac{2\theta-1}{\theta^2}\tilde{c} - a_1 - a_2\Big)|w^{\Delta,v}(\lfloor t\rfloor_\Delta)|^2 \\ &\quad + \frac{a_1 - a_2 + L}{2}\int_{-\tau}^{0}|w^{\Delta,v}_{\lfloor t\rfloor_\Delta}(r)|^2\mathrm{d}\nu_1(r) + 2a_2\int_{-\tau}^{0}|w^{\Delta,v}_{\lfloor t\rfloor_\Delta}(r)|^2\mathrm{d}\nu_2(r)\Big]\mathrm{d}t \\ &\quad + K\int_0^{t_{k+1}} e^{\iota\lceil t\rceil_\Delta}\Big(|w^{\Delta,v}(\lfloor t\rfloor_\Delta)|^2 + |z^{\Delta,v}(\lfloor t\rfloor_\Delta)|^2 + \int_{-\tau}^{0}|w^{\Delta,v}_{\lfloor t\rfloor_\Delta}(r)|^2\mathrm{d}\nu_1(r)\Big) \\ &\quad \times |v(\lfloor t\rfloor_\Delta)|^2\mathrm{d}t,\end{aligned}$$

where $\lceil t\rceil_\Delta := \min\{t_k > t : t_k = k\Delta, k = 0, 1, \ldots\}$ and we used

$$\sup_{t\ge 0}\Delta|v(\lfloor t\rfloor_\Delta)|^2 < \sum_{i=0}^{\infty}\Delta|v(t_i)|^2 = \int_0^{\infty}|v(\lfloor t\rfloor_\Delta)|^2\mathrm{d}t \le M.$$

Similar to the estimate of (2.15), we have that for $n = 1, 2$,

$$
\begin{aligned}
&\Delta \sum_{i=0}^{k} e^{\iota t_{i+1}} \int_{-\tau}^{0} |w_{t_i}^{\Delta,v}(r)|^2 \mathrm{d}\nu_n(r) \\
&= \Delta \sum_{j=-N}^{-1} \sum_{i=0}^{k} e^{\iota t_{i+1}} \int_{t_j}^{t_{j+1}} \Big|\frac{t_{j+1}-r}{\Delta} w^{\Delta,v}(t_{i+j}) + \frac{r-t_j}{\Delta} w^{\Delta,v}(t_{i+j+1})\Big|^2 \mathrm{d}\nu_n(r) \\
&\leq \Delta \sum_{j=-N}^{-1} \sum_{i=0}^{k} e^{\iota t_{i+1}} \int_{t_j}^{t_{j+1}} \frac{t_{j+1}-r}{\Delta} \mathrm{d}\nu_n(r) |w^{\Delta,v}(t_{i+j})|^2 \\
&\quad + \Delta \sum_{j=-N}^{-1} \sum_{i=0}^{k} e^{\iota t_{i+1}} \int_{t_j}^{t_{j+1}} \frac{r-t_j}{\Delta} \mathrm{d}\nu_n(r) |w^{\Delta,v}(t_{i+j+1})|^2 \\
&\leq \Delta \sum_{j=-N}^{-1} \sum_{l=-N}^{k} e^{\iota t_{l-j+1}} \int_{t_j}^{t_{j+1}} \frac{t_{j+1}-r}{\Delta} \mathrm{d}\nu_n(r) |w^{\Delta,v}(t_l)|^2 \\
&\quad + \Delta \sum_{j=-N}^{-1} \sum_{l=-N}^{k} e^{\iota t_{l-j}} \int_{t_j}^{t_{j+1}} \frac{r-t_j}{\Delta} \mathrm{d}\nu_n(r) |w^{\Delta,v}(t_l)|^2 \\
&\leq \Delta e^{\iota\tau} \sum_{j=-N}^{-1} \sum_{l=-N}^{k} e^{\iota t_{l+1}} \int_{t_j}^{t_{j+1}} \mathrm{d}\nu_n(r) |w^{\Delta,v}(t_l)|^2 \\
&\leq e^{\iota\tau} \tau \|\xi\|^2 + e^{\iota\tau} \Delta \sum_{l=0}^{k} e^{\iota t_{l+1}} |w^{\Delta,v}(t_l)|^2.
\end{aligned}
$$

Hence, utilizing this and the polynomial growth of b (see Assumption 2.4), we derive that

$$
\begin{aligned}
&e^{\iota t_{k+1}} |z^{\Delta,v}(t_{k+1})|^2 \\
&\quad \leq K(1 + \|\xi\|^{2(\beta+1)}) + \int_0^{t_{k+1}} e^{\iota \lceil t \rceil_\Delta} \Big[\Big(\frac{2\theta-1}{\theta^2}\tilde{c} - a_1 - a_2\Big) \\
&\quad + \Big(\frac{a_1 - a_2 + L}{2} + 2a_2\Big) e^{\iota\tau}\Big] \\
&\quad \times |w^{\Delta,v}(\lfloor t \rfloor_\Delta)|^2\Big] \mathrm{d}t + K e^{\iota t_{k+1}} + \Big(\frac{a_1 - a_2 + L}{2} + 2a_2\Big) e^{\iota\tau} \iota \|\xi\|^2 \\
&\quad + K \int_0^{t_{k+1}} \Big(\sup_{r \in [-\tau, t]} e^{\iota \lfloor r \rfloor_\Delta} |w^{\Delta,v}(\lfloor r \rfloor_\Delta)|^2 + \sup_{r \in [0,t]} e^{\iota \lfloor r \rfloor_\Delta} |z^{\Delta,v}(\lfloor r \rfloor_\Delta)|^2 \Big) \\
&\quad \times |v(\lfloor t \rfloor_\Delta)|^2 \mathrm{d}t.
\end{aligned}
$$

Analogously to the proof of Proposition 2.4, parameters $\tilde{c}$ and ι can further be taken such that

$$\frac{2\theta-1}{\theta^2}\tilde{c}-a_1-a_2+\Big(\frac{a_1-a_2+L}{2}+2a_2\Big)e^{\iota\tau}\le 0.$$

Consequently,

$$\begin{aligned}e^{\iota t_{k+1}}|z^{\Delta,v}(t_{k+1})|^2&\le K(1+\|\xi\|^{2(\beta+1)}+e^{\iota t_{k+1}})\\&\quad+K\int_0^{t_{k+1}}\Big(\sup_{r\in[-\tau,t]}e^{\iota\lfloor r\rfloor_\Delta}|w^{\Delta,v}(\lfloor r\rfloor_\Delta)|^2\\&\quad+\sup_{r\in[0,t]}e^{\iota\lfloor r\rfloor_\Delta}|z^{\Delta,v}(\lfloor r\rfloor_\Delta)|^2\Big)|v(\lfloor t\rfloor_\Delta)|^2\mathrm{d}t.\end{aligned}\tag{6.8}$$

By Assumption 2.2, one has that for any $\phi\in C^d$,

$$\langle\phi(0),b(\phi)\rangle\le|\phi(0)||b(0)|-a_1|\phi(0)|^2+a_2\int_{-\tau}^0|\phi(r)|^2\mathrm{d}\nu_2(r).$$

This, along with (6.5), implies that

$$\begin{aligned}&|z^{\Delta,v}(t_{k+1})|^2\\&\quad\ge|w^{\Delta,v}(t_{k+1})|^2-2\theta\Delta\langle w^{\Delta,v}(t_{k+1}),b(w^{\Delta,v}_{t_{k+1}})\rangle\\&\quad\ge|w^{\Delta,v}(t_{k+1})|^2-\theta\Delta\Big(K-2a_1|w^{\Delta,v}(t_{k+1})|^2+2a_2\int_{-\tau}^0|w^{\Delta,v}_{t_{k+1}}(r)|^2\mathrm{d}\nu_2(r)\Big)\\&\quad\ge|w^{\Delta,v}(t_{k+1})|^2-K\theta\Delta-2\theta a_2\Delta\int_{-\tau}^0|w^{\Delta,v}_{t_{k+1}}(r)|^2\mathrm{d}\nu_2(r),\end{aligned}$$

which yields

$$\begin{aligned}&e^{\iota t_{k+1}}|z^{\Delta,v}(t_{k+1})|^2\\&\quad\ge e^{\iota t_{k+1}}|w^{\Delta,v}(t_{k+1})|^2-K\theta\Delta e^{\iota t_{k+1}}-2\theta a_2e^{\iota\tau}\Delta\sup_{t\in[-\tau,t_{k+1}]}e^{\iota\lfloor t\rfloor_\Delta}|w^{\Delta,v}(\lfloor t\rfloor_\Delta)|^2.\end{aligned}$$

Plugging (6.8) into the above inequality leads to

$$\begin{aligned}&e^{\iota t_{k+1}}|w^{\Delta,v}(t_{k+1})|^2\\&\quad\le K\theta\Delta e^{\iota t_{k+1}}+2\theta a_2e^{\iota\tau}\Delta\sup_{t\in[-\tau,t_{k+1}]}e^{\iota\lfloor t\rfloor_\Delta}|w^{\Delta,v}(\lfloor t\rfloor_\Delta)|^2\\&\qquad+K(1+\|\xi\|^{2(\beta+1)}+e^{\iota t_{k+1}})+K\int_0^{t_{k+1}}\Big(\sup_{r\in[-\tau,t]}e^{\iota\lfloor r\rfloor_\Delta}|w^{\Delta,v}(\lfloor r\rfloor_\Delta)|^2\end{aligned}$$

$$+ \sup_{r\in[0,t]} e^{\iota\lfloor r\rfloor_\Delta}|z^{\Delta,v}(\lfloor r\rfloor_\Delta)|^2\Big)|v(\lfloor t\rfloor_\Delta)|^2\mathrm{d}t.$$

Noting that the above estimate holds for any $k \in \mathbb{N}$, we have

$$\begin{aligned}
&\sup_{t\in[-\tau,t_{k+1}]} e^{\iota\lfloor t\rfloor_\Delta}|w^{\Delta,v}(\lfloor t\rfloor_\Delta)|^2 \\
&\le K\theta\Delta e^{\iota t_{k+1}} + 2\theta a_2 e^{\iota\tau}\Delta \sup_{t\in[-\tau,t_{k+1}]} e^{\iota\lfloor t\rfloor_\Delta}|w^{\Delta,v}(\lfloor t\rfloor_\Delta)|^2 \\
&\quad + K(1+\|\xi\|^{2(\beta+1)} + e^{\iota t_{k+1}}) + K\int_0^{t_{k+1}}\Big(\sup_{r\in[-\tau,t]} e^{\iota\lfloor r\rfloor_\Delta}|w^{\Delta,v}(\lfloor r\rfloor_\Delta)|^2 \\
&\quad + \sup_{r\in[0,t]} e^{\iota\lfloor r\rfloor_\Delta}|z^{\Delta,v}(\lfloor r\rfloor_\Delta)|^2\Big)|v(\lfloor t\rfloor_\Delta)|^2\mathrm{d}t.
\end{aligned}$$

Now we choose a number ι such that $e^{\iota\tau} \le \frac{3}{2}$. Then for $\Delta \in (0, \frac{1}{4\theta a_2}]$,

$$\begin{aligned}
&\frac{1}{4}\sup_{t\in[-\tau,t_{k+1}]} e^{\iota\lfloor t\rfloor_\Delta}|w^{\Delta,v}(\lfloor t\rfloor_\Delta)|^2 \\
&\le K(1+\|\xi\|^{2(\beta+1)} + e^{\iota t_{k+1}}) + K\int_0^{t_{k+1}}\Big(\sup_{r\in[-\tau,t]} e^{\iota\lfloor r\rfloor_\Delta}|w^{\Delta,v}(\lfloor r\rfloor_\Delta)|^2 \\
&\quad + \sup_{r\in[0,t]} e^{\iota\lfloor r\rfloor_\Delta}|z^{\Delta,v}(\lfloor r\rfloor_\Delta)|^2\Big)|v(\lfloor t\rfloor_\Delta)|^2\mathrm{d}t,
\end{aligned}$$

which, together with (6.8) gives that

$$\begin{aligned}
&\sup_{t\in[-\tau,t_{k+1}]} e^{\iota\lfloor t\rfloor_\Delta}|w^{\Delta,v}(\lfloor t\rfloor_\Delta)|^2 + \sup_{t\in[0,t_{k+1}]} e^{\iota\lfloor t\rfloor_\Delta}|z^{\Delta,v}(\lfloor t\rfloor_\Delta)|^2 \\
&\le K(1+\|\xi\|^{2(\beta+1)} + e^{\iota t_{k+1}}) + K\int_0^{t_{k+1}}\Big(\sup_{r\in[-\tau,t]} e^{\iota\lfloor r\rfloor_\Delta}|w^{\Delta,v}(\lfloor r\rfloor_\Delta)|^2 \\
&\quad + \sup_{r\in[0,t]} e^{\iota\lfloor r\rfloor_\Delta}|z^{\Delta,v}(\lfloor r\rfloor_\Delta)|^2\Big)|v(\lfloor t\rfloor_\Delta)|^2\mathrm{d}t.
\end{aligned}$$

Applying the Gronwall inequality (see Proposition A.3) and using $v \in \mathcal{S}_M$, we deduce

$$\begin{aligned}
&\sup_{t\in[-\tau,t_{k+1}]} e^{\iota\lfloor t\rfloor_\Delta}|w^{\Delta,v}(\lfloor t\rfloor_\Delta)|^2 + \sup_{t\in[0,t_{k+1}]} e^{\iota\lfloor t\rfloor_\Delta}|z^{\Delta,v}(\lfloor t\rfloor_\Delta)|^2 \\
&\le K(1+\|\xi\|^{2(\beta+1)} + e^{\iota t_{k+1}})e^{K\int_0^{t_{k+1}}|v(\lfloor t\rfloor_\Delta)|^2\mathrm{d}t} \\
&\le K(1+\|\xi\|^{2(\beta+1)} + e^{\iota t_{k+1}})e^{KM}.
\end{aligned}$$

This implies that

$$\sup_{k\geq 0}|w^{\Delta,v}(t_k)|+\sup_{k\geq 0}|z^{\Delta,v}(t_k)|\leq (1+\|\xi\|^{\beta+1})K(M).$$

The proof is finished. □

Proposition 6.4 *Let* $M > 0$. *Under Assumptions 2.1, 2.2, and 2.4, for any* $\Delta \in (0, \frac{1}{4\theta a_2}]$, $\mathcal{K}_M := \{\mathcal{G}^{\Delta}(\int_0^{\cdot} v(\lfloor s\rfloor_{\Delta})\mathrm{d}s)|v \in \mathcal{S}_M\}$ *is compact in* $C_{\xi^{Int}}([-\tau,+\infty);\mathbb{R}^d)$.

Proof Since $\mathcal{S}_M$ is compact with respect to the weak topology, it suffices to show that the mapping $\mathcal{G}^{\Delta}\circ F$ is continuous, where $F : L^2_{\Delta}([0,+\infty);\mathbb{R}^m) \to C([0,\infty);\mathbb{R}^m)$ is defined by $F(v)(\cdot) := \int_0^{\cdot} v(\lfloor s\rfloor_{\Delta})\mathrm{d}s$. Namely, letting $v^{\epsilon} \to v$ in $\mathcal{S}_M$, we shall prove $w^{\Delta,v^{\epsilon}} \to w^{\Delta,v}$ in $C_{\xi^{Int}}([-\tau,+\infty);\mathbb{R}^d)$. First, we show that for any $k\in\mathbb{N}_+$, $\sup_{t\in[0,t_k]}|w^{\Delta,v^{\epsilon}}(\lfloor t\rfloor_{\Delta})-w^{\Delta,v}(\lfloor t\rfloor_{\Delta})| \to 0$ as $v^{\epsilon}\to v$ in $\mathcal{S}_M$. By (6.5) and the continuous version of $z^{\Delta,v}(\cdot)$, we have

$$\begin{aligned}
&\frac{1}{2}\mathrm{d}|z^{\Delta,v^{\epsilon}}(t)-z^{\Delta,v}(t)|^2\\
&=\langle b(w^{\Delta,v^{\epsilon}}_{\lfloor t\rfloor_{\Delta}})-b(w^{\Delta,v}_{\lfloor t\rfloor_{\Delta}}), z^{\Delta,v^{\epsilon}}(t)-z^{\Delta,v}(t)\rangle\mathrm{d}t\\
&\quad+\langle \sigma(w^{\Delta,v^{\epsilon}}_{\lfloor t\rfloor_{\Delta}})v^{\epsilon}(\lfloor t\rfloor_{\Delta})-\sigma(w^{\Delta,v}_{\lfloor t\rfloor_{\Delta}})v(\lfloor t\rfloor_{\Delta}), z^{\Delta,v^{\epsilon}}(t)-z^{\Delta,v}(t)\rangle\mathrm{d}t\\
&=\langle b(w^{\Delta,v^{\epsilon}}_{\lfloor t\rfloor_{\Delta}})-b(w^{\Delta,v}_{\lfloor t\rfloor_{\Delta}}), b(w^{\Delta,v^{\epsilon}}_{\lfloor t\rfloor_{\Delta}})-b(w^{\Delta,v}_{\lfloor t\rfloor_{\Delta}})\rangle(t-\lfloor t\rfloor_{\Delta})\mathrm{d}t\\
&\quad+\langle b(w^{\Delta,v^{\epsilon}}_{\lfloor t\rfloor_{\Delta}})-b(w^{\Delta,v}_{\lfloor t\rfloor_{\Delta}}), \sigma(w^{\Delta,v^{\epsilon}}_{\lfloor t\rfloor_{\Delta}})v^{\epsilon}(\lfloor t\rfloor_{\Delta})\\
&\quad-\sigma(w^{\Delta,v}_{\lfloor t\rfloor_{\Delta}})v(\lfloor t\rfloor_{\Delta})\rangle(t-\lfloor t\rfloor_{\Delta})\mathrm{d}t\\
&\quad+\langle b(w^{\Delta,v^{\epsilon}}_{\lfloor t\rfloor_{\Delta}})-b(w^{\Delta,v}_{\lfloor t\rfloor_{\Delta}}), w^{\Delta,v^{\epsilon}}(\lfloor t\rfloor_{\Delta})-w^{\Delta,v}(\lfloor t\rfloor_{\Delta})\rangle\mathrm{d}t\\
&\quad-\theta\Delta\langle b(w^{\Delta,v^{\epsilon}}_{\lfloor t\rfloor_{\Delta}})-b(w^{\Delta,v}_{\lfloor t\rfloor_{\Delta}}), b(w^{\Delta,v^{\epsilon}}_{\lfloor t\rfloor_{\Delta}})-b(w^{\Delta,v}_{\lfloor t\rfloor_{\Delta}})\rangle\mathrm{d}t\\
&\quad+\langle \sigma(w^{\Delta,v^{\epsilon}}_{\lfloor t\rfloor_{\Delta}})v^{\epsilon}(\lfloor t\rfloor_{\Delta})-\sigma(w^{\Delta,v}_{\lfloor t\rfloor_{\Delta}})v(\lfloor t\rfloor_{\Delta}), z^{\Delta,v^{\epsilon}}(t)-z^{\Delta,v}(t)\rangle\mathrm{d}t.
\end{aligned}$$

This implies

$$\begin{aligned}
&|z^{\Delta,v^{\epsilon}}(t_{k+1})-z^{\Delta,v}(t_{k+1})|^2\\
&=|z^{\Delta,v^{\epsilon}}(t_k)-z^{\Delta,v}(t_k)|^2+(1-2\theta)\Delta^2|b(w^{\Delta,v^{\epsilon}}_{t_k})-b(w^{\Delta,v}_{t_k})|^2\\
&\quad+\int_{t_k}^{t_{k+1}}2\langle b(w^{\Delta,v^{\epsilon}}_{\lfloor t\rfloor_{\Delta}})-b(w^{\Delta,v}_{\lfloor t\rfloor_{\Delta}}), w^{\Delta,v^{\epsilon}}(\lfloor t\rfloor_{\Delta})-w^{\Delta,v}(\lfloor t\rfloor_{\Delta})\rangle\mathrm{d}t\\
&\quad+\int_{t_k}^{t_{k+1}}2\Big\langle(\sigma(w^{\Delta,v^{\epsilon}}_{\lfloor t\rfloor_{\Delta}})-\sigma(w^{\Delta,v}_{\lfloor t\rfloor_{\Delta}}))v^{\epsilon}(\lfloor t\rfloor_{\Delta}), z^{\Delta,v^{\epsilon}}(t)-z^{\Delta,v}(t)
\end{aligned}$$

$$+\frac{1}{2}\Delta\big(b(w^{\Delta,v^\epsilon}_{\lfloor t\rfloor_\Delta})-b(w^{\Delta,v}_{\lfloor t\rfloor_\Delta})\big)\Big\rangle \mathrm{d}t$$
$$+\int_{t_k}^{t_{k+1}}2\Big\langle\sigma(w^{\Delta,v}_{\lfloor t\rfloor_\Delta})(v^\epsilon(\lfloor t\rfloor_\Delta)-v(\lfloor t\rfloor_\Delta)),z^{\Delta,v^\epsilon}(t)-z^{\Delta,v}(t)$$
$$+\frac{1}{2}\Delta\big(b(w^{\Delta,v^\epsilon}_{\lfloor t\rfloor_\Delta})-b(w^{\Delta,v}_{\lfloor t\rfloor_\Delta})\big)\Big\rangle \mathrm{d}t.$$

By $2\int_{t_k}^{t_{k+1}}(t-t_k)\mathrm{d}t=\Delta^2$,

$$\begin{aligned}&b(w^{\Delta,v^\epsilon}_{\lfloor t\rfloor_\Delta})-b(w^{\Delta,v}_{\lfloor t\rfloor_\Delta})\\&=\frac{1}{\theta\Delta}\Big(w^{\Delta,v^\epsilon}(\lfloor t\rfloor_\Delta)-z^{\Delta,v^\epsilon}(\lfloor t\rfloor_\Delta)-\big(w^{\Delta,v}(\lfloor t\rfloor_\Delta)-z^{\Delta,v}(\lfloor t\rfloor_\Delta)\big)\Big),\end{aligned}\tag{6.9}$$

and taking into account $1-2\theta\le 0$, one has

$$\begin{aligned}&(1-2\theta)\Delta^2|b(w^{\Delta,v^\epsilon}_{t_k})-b(w^{\Delta,v}_{t_k})|^2\\&\le\frac{(1-2\theta)}{\theta^2}(1-\frac{1}{(1+c_\lambda\Delta)})|z^{\Delta,v^\epsilon}(t_k)-z^{\Delta,v}(t_k)|^2\\&\quad+\frac{(1-2\theta)}{\theta^2}(1-(1+c_\lambda\Delta))|w^{\Delta,v^\epsilon}(t_k)-w^{\Delta,v}(t_k)|^2,\end{aligned}$$

where the constant c_λ is defined by (2.2). Then we obtain

$$\begin{aligned}&|z^{\Delta,v^\epsilon}(t_{k+1})-z^{\Delta,v}(t_{k+1})|^2\\&\le(\frac{(1-\theta)^2}{\theta^2}+\frac{2\theta-1}{\theta^2(1+c_\lambda\Delta)})|z^{\Delta,v^\epsilon}(t_k)-z^{\Delta,v}(t_k)|^2\\&\quad+c_\lambda\Delta|w^{\Delta,v^\epsilon}(t_k)-w^{\Delta,v}(t_k)|^2\\&\quad+\int_{t_k}^{t_{k+1}}2\langle b(w^{\Delta,v^\epsilon}_{\lfloor t\rfloor_\Delta})-b(w^{\Delta,v}_{\lfloor t\rfloor_\Delta}),w^{\Delta,v^\epsilon}(\lfloor t\rfloor_\Delta)-w^{\Delta,v}(\lfloor t\rfloor_\Delta)\rangle\mathrm{d}t\\&\quad+\int_{t_k}^{t_{k+1}}2\Big\langle(\sigma(w^{\Delta,v^\epsilon}_{\lfloor t\rfloor_\Delta})-\sigma(w^{\Delta,v}_{\lfloor t\rfloor_\Delta}))v^\epsilon(\lfloor t\rfloor_\Delta),z^{\Delta,v^\epsilon}(t)-z^{\Delta,v}(t)\\&\quad+\frac{1}{2}\Delta\big(b(w^{\Delta,v^\epsilon}_{\lfloor t\rfloor_\Delta})-b(w^{\Delta,v}_{\lfloor t\rfloor_\Delta})\big)\Big\rangle\mathrm{d}t\\&\quad+\int_{t_k}^{t_{k+1}}2\Big\langle\sigma(w^{\Delta,v}_{\lfloor t\rfloor_\Delta})(v^\epsilon(\lfloor t\rfloor_\Delta)-v(\lfloor t\rfloor_\Delta)),z^{\Delta,v^\epsilon}(t)-z^{\Delta,v}(t)\\&\quad+\frac{1}{2}\Delta\big(b(w^{\Delta,v^\epsilon}_{\lfloor t\rfloor_\Delta})-b(w^{\Delta,v}_{\lfloor t\rfloor_\Delta})\big)\Big\rangle\mathrm{d}t,\end{aligned}\tag{6.10}$$

Applying the Young inequality, one derives

$$\begin{aligned}
&\int_{t_k}^{t_{k+1}} 2\Big\langle\big(\sigma(w^{\Delta,v^\epsilon}_{\lfloor t\rfloor_\Delta})-\sigma(w^{\Delta,v}_{\lfloor t\rfloor_\Delta})\big)v^\epsilon(\lfloor t\rfloor_\Delta), z^{\Delta,v^\epsilon}(t)-z^{\Delta,v}(t)\\
&\quad+\frac{1}{2}\Delta\big(b(w^{\Delta,v^\epsilon}_{\lfloor t\rfloor_\Delta})-b(w^{\Delta,v}_{\lfloor t\rfloor_\Delta})\big)\Big\rangle\mathrm{d}t\\
&=\int_{t_k}^{t_{k+1}} 2\Big\langle\big(\sigma(w^{\Delta,v^\epsilon}_{\lfloor t\rfloor_\Delta})-\sigma(w^{\Delta,v}_{\lfloor t\rfloor_\Delta})\big)v^\epsilon(\lfloor t\rfloor_\Delta), z^{\Delta,v^\epsilon}(\lfloor t\rfloor_\Delta)-z^{\Delta,v}(\lfloor t\rfloor_\Delta)\\
&\quad+\Delta\big(b(w^{\Delta,v^\epsilon}_{\lfloor t\rfloor_\Delta})-b(w^{\Delta,v}_{\lfloor t\rfloor_\Delta})\big)\Big\rangle\mathrm{d}t+\int_{t_k}^{t_{k+1}}|\big(\sigma(w^{\Delta,v^\epsilon}_{\lfloor t\rfloor_\Delta})\\
&\quad-\sigma(w^{\Delta,v}_{\lfloor t\rfloor_\Delta})\big)v^\epsilon(\lfloor t\rfloor_\Delta)|^2\Delta\mathrm{d}t\\
&\quad+\int_{t_k}^{t_{k+1}}\big\langle\big(\sigma(w^{\Delta,v^\epsilon}_{\lfloor t\rfloor_\Delta})-\sigma(w^{\Delta,v}_{\lfloor t\rfloor_\Delta})\big)v^\epsilon(\lfloor t\rfloor_\Delta),\sigma(w^{\Delta,v}_{\lfloor t\rfloor_\Delta})\\
&\quad(v^\epsilon(\lfloor t\rfloor_\Delta)-v(\lfloor t\rfloor_\Delta))\big\rangle\Delta\mathrm{d}t\\
&\le\int_{t_k}^{t_{k+1}}|\sigma(w^{\Delta,v^\epsilon}_{\lfloor t\rfloor_\Delta})-\sigma(w^{\Delta,v}_{\lfloor t\rfloor_\Delta})|^2\mathrm{d}t\\
&\quad+K\int_{t_k}^{t_{k+1}}|v^\epsilon(\lfloor t\rfloor_\Delta)|^2\big(|z^{\Delta,v^\epsilon}(\lfloor t\rfloor_\Delta)-z^{\Delta,v}(\lfloor t\rfloor_\Delta)|^2\\
&\quad+|w^{\Delta,v^\epsilon}(\lfloor t\rfloor_\Delta)-w^{\Delta,v}(\lfloor t\rfloor_\Delta)|^2\big)\mathrm{d}t\\
&\quad+\int_{t_k}^{t_{k+1}}|\big(\sigma(w^{\Delta,v^\epsilon}_{\lfloor t\rfloor_\Delta})-\sigma(w^{\Delta,v}_{\lfloor t\rfloor_\Delta})\big)v^\epsilon(\lfloor t\rfloor_\Delta)|^2\Delta\mathrm{d}t\\
&\quad+\int_{t_k}^{t_{k+1}}\big\langle\big(\sigma(w^{\Delta,v^\epsilon}_{\lfloor t\rfloor_\Delta})-\sigma(w^{\Delta,v}_{\lfloor t\rfloor_\Delta})\big)v^\epsilon(\lfloor t\rfloor_\Delta),\sigma(w^{\Delta,v}_{\lfloor t\rfloor_\Delta})\\
&\quad(v^\epsilon(\lfloor t\rfloor_\Delta)-v(\lfloor t\rfloor_\Delta))\big\rangle\Delta\mathrm{d}t,
\end{aligned}\tag{6.11}$$

where we used (6.9) in the last step. Inserting (6.11) into (6.10), and then by iterating, we obtain that for $\gamma\in(0,1)$,

$$\begin{aligned}
&e^{\gamma t_{k+1}}|z^{\Delta,v^\epsilon}(t_{k+1})-z^{\Delta,v}(t_{k+1})|^2\\
&=\sum_{i=0}^{k}\Big(e^{\gamma t_{i+1}}|z^{\Delta,v^\epsilon}(t_{i+1})-z^{\Delta,v}(t_{i+1})|^2-e^{\gamma t_i}|z^{\Delta,v^\epsilon}(t_i)-z^{\Delta,v}(t_i)|^2\Big)\\
&\le\sum_{i=0}^{k}e^{\gamma t_{i+1}}\Big[\Big(\frac{(1-\theta)^2}{\theta^2}+\frac{2\theta-1}{\theta^2(1+c_\lambda\Delta)}-e^{-\gamma\Delta}\Big)|z^{\Delta,v^\epsilon}(t_i)-z^{\Delta,v}(t_i)|^2
\end{aligned}$$

$$
\begin{aligned}
&+ c_\lambda \Delta |w^{\Delta,v^\epsilon}(t_i) - w^{\Delta,v}(t_i)|^2\Big] \\
&+ \int_0^{t_{k+1}} e^{\gamma\lceil t\rceil_\Delta}\Big[2\langle b(w^{\Delta,v^\epsilon}_{\lfloor t\rfloor_\Delta}) - b(w^{\Delta,v}_{\lfloor t\rfloor_\Delta}), w^{\Delta,v^\epsilon}(\lfloor t\rfloor_\Delta) \\
&- w^{\Delta,v}(\lfloor t\rfloor_\Delta)\rangle + |\sigma(w^{\Delta,v^\epsilon}_{\lfloor t\rfloor_\Delta}) - \sigma(w^{\Delta,v}_{\lfloor t\rfloor_\Delta})|^2\Big]\mathrm{d}t + K\int_0^{t_{k+1}} e^{\gamma\lceil t\rceil_\Delta}|v^\epsilon(\lfloor t\rfloor_\Delta)|^2 \\
&\times \Big[|z^{\Delta,v^\epsilon}(\lfloor t\rfloor_\Delta) - z^{\Delta,v}(\lfloor t\rfloor_\Delta)|^2 + |w^{\Delta,v^\epsilon}(\lfloor t\rfloor_\Delta) - w^{\Delta,v}(\lfloor t\rfloor_\Delta)|^2\Big]\mathrm{d}t \\
&+ \int_0^{t_{k+1}} e^{\gamma\lceil t\rceil_\Delta}\Big|\sigma(w^{\Delta,v^\epsilon}_{\lfloor t\rfloor_\Delta}) \\
&- \sigma(w^{\Delta,v}_{\lfloor t\rfloor_\Delta})\Big|^2|v^\epsilon(\lfloor t\rfloor_\Delta)|^2\Delta\mathrm{d}t + J(t_{k+1}),
\end{aligned}
$$

where J is defined by

$$
\begin{aligned}
J(s) := \int_0^s 2e^{\gamma\lceil t\rceil_\Delta}\Big\langle&\sigma(w^{\Delta,v}_{\lfloor t\rfloor_\Delta})(v^\epsilon(\lfloor t\rfloor_\Delta) - v(\lfloor t\rfloor_\Delta)), \frac{1}{2}\Delta(\sigma(w^{\Delta,v^\epsilon}_{\lfloor t\rfloor_\Delta}) - \sigma(w^{\Delta,v}_{\lfloor t\rfloor_\Delta})) \\
&\times v^\epsilon(\lfloor t\rfloor_\Delta) + z^{\Delta,v^\epsilon}(t) - z^{\Delta,v}(t) + \frac{1}{2}\Delta\big(b(w^{\Delta,v^\epsilon}_{\lfloor t\rfloor_\Delta}) - b(w^{\Delta,v}_{\lfloor t\rfloor_\Delta})\big)\Big\rangle\mathrm{d}t.
\end{aligned}
$$

It follows from Assumptions 2.1 and 2.2 that

$$
\begin{aligned}
&e^{\gamma t_{k+1}}|z^{\Delta,v^\epsilon}(t_{k+1}) - z^{\Delta,v}(t_{k+1})|^2 \\
&\le \sum_{i=0}^{k} e^{\gamma t_{i+1}}\Big(\frac{(1-\theta)^2}{\theta^2} + \frac{2\theta-1}{\theta^2(1+c_\lambda\Delta)} - e^{-\gamma\Delta}\Big)|z^{\Delta,v^\epsilon}(t_i) - z^{\Delta,v}(t_i)|^2 \\
&- \sum_{i=0}^{k}\big(2a_1 - L - c_\lambda - (L+2a_2)e^{\gamma\tau}\big)\Delta e^{\gamma t_{i+1}}|w^{\Delta,v^\epsilon}(t_i) - w^{\Delta,v}(t_i)|^2 \\
&+ K\int_0^{t_{k+1}}\Big[|z^{\Delta,v^\epsilon}(\lfloor t\rfloor_\Delta) - z^{\Delta,v}(\lfloor t\rfloor_\Delta)|^2 + |w^{\Delta,v^\epsilon}(\lfloor t\rfloor_\Delta) - w^{\Delta,v}(\lfloor t\rfloor_\Delta)|^2\Big] \\
&\times e^{\gamma\lceil t\rceil_\Delta}|v^\epsilon(\lfloor t\rfloor_\Delta)|^2\mathrm{d}t + 2Le^{\gamma(\tau+\Delta)} \\
&\times \int_0^{t_{k+1}}\sup_{s\in[0,t]}\Big(e^{\gamma\lfloor s\rfloor_\Delta}|w^{\Delta,v^\epsilon}(\lfloor s\rfloor_\Delta) - w^{\Delta,v}(\lfloor s\rfloor_\Delta)|^2\Big) \\
&\times |v^\epsilon(\lfloor t\rfloor_\Delta)|^2\Delta\mathrm{d}t + J(t_{k+1}),
\end{aligned}
$$

Similar to the proof of Proposition 2.10, one can take a sufficiently small number γ independent of Δ such that

$$\frac{(1-\theta)^2}{\theta^2}+\frac{2\theta-1}{\theta^2(1+c_\lambda\Delta)}-e^{-\gamma\Delta}\le 0,\quad 2a_1-L-c_\lambda-(L+2a_2)e^{\gamma\tau}\ge 0.$$

Then, we see that

$$\begin{aligned}&e^{\gamma t_{k+1}}|z^{\Delta,v^\epsilon}(t_{k+1})-z^{\Delta,v}(t_{k+1})|^2\\&\le K\int_0^{t_{k+1}}|v^\epsilon(\lfloor t\rfloor_\Delta)|^2\sup_{s\in[0,t]}\Big[e^{\gamma\lfloor s\rfloor_\Delta}|z^{\Delta,v^\epsilon}(\lfloor s\rfloor_\Delta)-z^{\Delta,v}(\lfloor s\rfloor_\Delta)|^2\\&\quad+e^{\gamma\lfloor s\rfloor_\Delta}|w^{\Delta,v^\epsilon}(\lfloor s\rfloor_\Delta)-w^{\Delta,v}(\lfloor s\rfloor_\Delta)|^2\Big]dt+J(t_{k+1}).\end{aligned}\tag{6.12}$$

Denote $E^\epsilon(t)=\int_0^t\sigma(w^{\Delta,v}_{\lfloor s\rfloor_\Delta})(v^\epsilon(\lfloor s\rfloor_\Delta)-v(\lfloor s\rfloor_\Delta))ds$. Applying the integration by parts formula yields

$$\begin{aligned}J(t_{k+1})&=\sum_{i=0}^{k}\int_{t_i}^{t_{i+1}}2e^{\gamma t_{i+1}}\Big\langle\frac{1}{2}\Delta(\sigma(w^{\Delta,v^\epsilon}_{t_i})-\sigma(w^{\Delta,v}_{t_i}))v^\epsilon(t_i)+z^{\Delta,v^\epsilon}(t)-z^{\Delta,v}(t)\\&\quad+\frac{1}{2}\Delta\big(b(w^{\Delta,v^\epsilon}_{t_i})-b(w^{\Delta,v}_{t_i})\big),\mathrm{d}E^\epsilon(t)\Big\rangle\\&=\sum_{i=0}^{k}2e^{\gamma t_{i+1}}\Big[\Big\langle\frac{1}{2}\Delta(\sigma(w^{\Delta,v^\epsilon}_{t_i})-\sigma(w^{\Delta,v}_{t_i}))v^\epsilon(t_i)+z^{\Delta,v^\epsilon}(t)-z^{\Delta,v}(t)\\&\quad+\frac{1}{2}\Delta\big(b(w^{\Delta,v^\epsilon}_{t_i})-b(w^{\Delta,v}_{t_i})\big),E^\epsilon(t)\Big\rangle\Big|_{t=t_i}^{t=t_{i+1}}\\&\quad-\int_{t_i}^{t_{i+1}}\langle E^\epsilon(t),\mathrm{d}(z^{\Delta,v^\epsilon}(t)-z^{\Delta,v}(t))\rangle\Big]\\&=2e^{\gamma\lceil t\rceil_\Delta}\Big\langle E^\epsilon(t),\frac{1}{2}\Delta(\sigma(w^{\Delta,v^\epsilon}_{\lfloor t\rfloor_\Delta})-\sigma(w^{\Delta,v}_{\lfloor t\rfloor_\Delta}))v^\epsilon(\lfloor t\rfloor_\Delta)\\&\quad+z^{\Delta,v^\epsilon}(t)-z^{\Delta,v}(t)\\&\quad+\frac{1}{2}\Delta\big(b(w^{\Delta,v^\epsilon}_{\lfloor t\rfloor_\Delta})-b(w^{\Delta,v}_{\lfloor t\rfloor_\Delta})\big)\Big\rangle\Big|_{t=0}^{t=t_{k+1}}\\&\quad-2\int_0^{t_{k+1}}e^{\gamma\lceil t\rceil_\Delta}\Big\langle E^\epsilon(t),\frac{\mathrm{d}(z^{\Delta,v^\epsilon}(t)-z^{\Delta,v}(t))}{\mathrm{d}t}\Big\rangle\mathrm{d}t,\end{aligned}$$

which together with the differential form of $z^{\Delta,v^\epsilon}(t)-z^{\Delta,v}(t)$ and (6.9) gives that

$$
\begin{aligned}
J(t_{k+1}) \le\ & e^{\gamma t_{k+2}}|E^\epsilon(t_{k+1})||\sigma(w^{\Delta,v^\epsilon}_{t_{k+1}})-\sigma(w^{\Delta,v}_{t_{k+1}})|\int_{t_{k+1}}^{t_{k+2}}|v^\epsilon(\lfloor t\rfloor_\Delta)|\mathrm{d}t\\
&+Ke^{\gamma t_{k+1}}|E^\epsilon(t_{k+1})|^2\\
&+\frac{1}{8}\sup_{t\in[0,t_{k+1}]}e^{\gamma\lfloor t\rfloor_\Delta}\Big(|z^{\Delta,v^\epsilon}(\lfloor t\rfloor_\Delta)-z^{\Delta,v}(\lfloor t\rfloor_\Delta)|^2+|w^{\Delta,v^\epsilon}(\lfloor t\rfloor_\Delta)\\
&-w^{\Delta,v}(\lfloor t\rfloor_\Delta)|^2\Big)\\
&-2\int_0^{t_{k+1}}e^{\gamma\lceil t\rceil_\Delta}\Big\langle E^\epsilon(t),b(w^{\Delta,v^\epsilon}_{\lfloor t\rfloor_\Delta})-b(w^{\Delta,v}_{\lfloor t\rfloor_\Delta})+\sigma(w^{\Delta,v^\epsilon}_{\lfloor t\rfloor_\Delta})v^\epsilon(\lfloor t\rfloor_\Delta)\\
&-\sigma(w^{\Delta,v}_{\lfloor t\rfloor_\Delta})v(\lfloor t\rfloor_\Delta)\Big\rangle\mathrm{d}t.
\end{aligned}
$$

By $v^\epsilon(\cdot),v(\cdot)\in\mathcal{S}_M$, Assumptions 2.1 and 2.4, and Lemma 6.3, we arrive at

$$
\begin{aligned}
J(t_{k+1}) \le\ & K(M)e^{\gamma t_{k+1}}\sup_{j\ge -N}(1+|w^{\Delta,v^\epsilon}(t_j)|+|w^{\Delta,v}(t_j)|)|E^\epsilon(t_{k+1})|\\
&+Ke^{\gamma t_{k+1}}|E^\epsilon(t_{k+1})|^2+\frac{1}{8}\sup_{t\in[0,t_{k+1}]}e^{\gamma\lfloor t\rfloor_\Delta}\Big(|z^{\Delta,v^\epsilon}(\lfloor t\rfloor_\Delta)-z^{\Delta,v}(\lfloor t\rfloor_\Delta)|^2\\
&+|w^{\Delta,v^\epsilon}(\lfloor t\rfloor_\Delta)-w^{\Delta,v}(\lfloor t\rfloor_\Delta)|^2\Big)\\
&+K\int_0^{t_{k+1}}e^{\gamma\lceil t\rceil_\Delta}|E^\epsilon(t)|\Big[1+\sup_{k\ge -N}(|w^{\Delta,v^\epsilon}(t_k)|^{\beta+1}+|w^{\Delta,v}(t_k)|^{\beta+1})\\
&+(1+\sup_{k\ge -N}|w^{\Delta,v^\epsilon}(t_k)|)|v^\epsilon(\lfloor t\rfloor_\Delta)|\\
&+(1+\sup_{k\ge -N}|w^{\Delta,v}(t_k)|)|v(\lfloor t\rfloor_\Delta)|\Big]\mathrm{d}t\\
\le\ & K(M)e^{\gamma t_{k+1}}\sup_{t\in[0,t_{k+1}]}|E^\epsilon(t)|^2+K(M)e^{\gamma t_{k+1}}\sup_{t\in[0,t_{k+1}]}|E^\epsilon(t)|\\
&+\frac{1}{8}\sup_{t\in[0,t_{k+1}]}e^{\gamma\lfloor t\rfloor_\Delta}\\
&\times\Big(|z^{\Delta,v^\epsilon}(\lfloor t\rfloor_\Delta)-z^{\Delta,v}(\lfloor t\rfloor_\Delta)|^2+|w^{\Delta,v^\epsilon}(\lfloor t\rfloor_\Delta)-w^{\Delta,v}(\lfloor t\rfloor_\Delta)|^2\Big),
\end{aligned}
\tag{6.13}
$$

where we used the Hölder inequality in the last step. Noticing that

$$
\begin{aligned}
& e^{\gamma t_{k+1}}|z^{\Delta,v^\epsilon}(t_{k+1}) - z^{\Delta,v}(t_{k+1})|^2 \\
& \geq e^{\gamma t_{k+1}}|w^{\Delta,v^\epsilon}(t_{k+1}) - w^{\Delta,v}(t_{k+1})|^2 \\
& \quad - 2\theta e^{\gamma t_{k+1}}\Delta\langle w^{\Delta,v^\epsilon}(t_{k+1}) - w^{\Delta,v}(t_{k+1}), b(w^{\Delta,v^\epsilon}_{t_{k+1}}) - b(w^{\Delta,v}_{t_{k+1}})\rangle \\
& \geq e^{\gamma t_{k+1}}|w^{\Delta,v^\epsilon}(t_{k+1}) - w^{\Delta,v}(t_{k+1})|^2 \\
& \quad - 2\theta e^{\gamma t_{k+1}}\Delta\Big(- a_1|w^{\Delta,v^\epsilon}(t_{k+1}) - w^{\Delta,v}(t_{k+1})|^2 \\
& \quad + a_2\int_{-\tau}^{0}|w^{\Delta,v^\epsilon}_{t_{k+1}}(r) - w^{\Delta,v}_{t_{k+1}}(r)|^2 \mathrm{d}\nu_2(r)\Big) \\
& \geq e^{\gamma t_{k+1}}|w^{\Delta,v^\epsilon}(t_{k+1}) - w^{\Delta,v}(t_{k+1})|^2 \\
& \quad - 2\theta a_2 e^{\gamma\tau}\Delta \sup_{t\in[0,t_{k+1}]} e^{\gamma\lfloor t\rfloor_\Delta}|w^{\Delta,v^\epsilon}(\lfloor t\rfloor_\Delta) - w^{\Delta,v}(\lfloor t\rfloor_\Delta)|^2 \\
& \geq e^{\gamma t_{k+1}}|w^{\Delta,v^\epsilon}(t_{k+1}) - w^{\Delta,v}(t_{k+1})|^2 \\
& \quad - \frac{9}{16}\sup_{t\in[0,t_{k+1}]} e^{\gamma\lfloor t\rfloor_\Delta}|w^{\Delta,v^\epsilon}(\lfloor t\rfloor_\Delta) - w^{\Delta,v}(\lfloor t\rfloor_\Delta)|^2,
\end{aligned}
$$

where we used $\Delta \in (0, \frac{1}{4a_2\theta}]$ and we chose γ such that $e^{\gamma\tau} \leq \frac{9}{8}$. Thus inserting (6.13) into (6.12) and together with the above inequality, we obtain that for $\Delta \in (0, \frac{1}{4a_2\theta}]$,

$$
\begin{aligned}
& e^{\gamma t_{k+1}}|w^{\Delta,v^\epsilon}(t_{k+1}) - w^{\Delta,v}(t_{k+1})|^2 \\
& \leq K\int_0^{t_{k+1}}|v^\epsilon(\lfloor t\rfloor_\Delta)|^2 \sup_{s\in[0,t]}\Big[e^{\gamma\lfloor s\rfloor_\Delta}|z^{\Delta,v^\epsilon}(\lfloor s\rfloor_\Delta) - z^{\Delta,v}(\lfloor s\rfloor_\Delta)|^2 \\
& \quad + e^{\gamma\lfloor s\rfloor_\Delta}|w^{\Delta,v^\epsilon}(\lfloor s\rfloor_\Delta) \\
& \quad - w^{\Delta,v}(\lfloor s\rfloor_\Delta)|^2\Big]\mathrm{d}t + \frac{1}{8}\sup_{t\in[0,t_{k+1}]} e^{\gamma\lfloor t\rfloor_\Delta}|z^{\Delta,v^\epsilon}(\lfloor t\rfloor_\Delta) - z^{\Delta,v}(\lfloor t\rfloor_\Delta)|^2 \\
& \quad + (\frac{9}{16} + \frac{1}{8})\sup_{t\in[0,t_{k+1}]} e^{\gamma\lfloor t\rfloor_\Delta}|w^{\Delta,v^\epsilon}(\lfloor t\rfloor_\Delta) - w^{\Delta,v}(\lfloor t\rfloor_\Delta)|^2 \\
& \quad + K(M)e^{\gamma t_{k+1}}\sup_{t\in[0,t_{k+1}]}|E^\epsilon(t)| + K(M)e^{\gamma t_{k+1}}\sup_{t\in[0,t_{k+1}]}|E^\epsilon(t)|^2.
\end{aligned}
$$

Combining the estimates of $e^{\gamma t_{k+1}}|z^{\Delta,v^\epsilon}(t_{k+1})-z^{\Delta,v}(t_{k+1})|^2$ and $e^{\gamma t_{k+1}}|w^{\Delta,v^\epsilon}(t_{k+1}) - w^{\Delta,v}(t_{k+1})|^2$ yields that

$$\begin{aligned}
&\sup_{t\in[0,t_{k+1}]} e^{\gamma\lfloor t\rfloor_\Delta}|z^{\Delta,v^\epsilon}(\lfloor t\rfloor_\Delta) - z^{\Delta,v}(\lfloor t\rfloor_\Delta)|^2 \\
&\quad + \sup_{t\in[0,t_{k+1}]} e^{\gamma\lfloor t\rfloor_\Delta}|w^{\Delta,v^\epsilon}(\lfloor t\rfloor_\Delta) - w^{\Delta,v}(\lfloor t\rfloor_\Delta)|^2 \\
&\quad \le K\int_0^{t_{k+1}} |v^\epsilon(\lfloor t\rfloor_\Delta)|^2 \sup_{s\in[0,t]}\Big[e^{\gamma\lfloor s\rfloor_\Delta}|z^{\Delta,v^\epsilon}(\lfloor s\rfloor_\Delta) - z^{\Delta,v}(\lfloor s\rfloor_\Delta)|^2 \\
&\quad + e^{\gamma\lfloor s\rfloor_\Delta}|w^{\Delta,v^\epsilon}(\lfloor s\rfloor_\Delta) - w^{\Delta,v}(\lfloor s\rfloor_\Delta)|^2\Big]\mathrm{d}t \\
&\quad + \frac{1}{4}\sup_{t\in[0,t_{k+1}]} e^{\gamma\lfloor t\rfloor_\Delta}|z^{\Delta,v^\epsilon}(\lfloor t\rfloor_\Delta) - z^{\Delta,v}(\lfloor t\rfloor_\Delta)|^2\mathrm{d}t \\
&\quad + \frac{13}{16}\sup_{t\in[0,t_{k+1}]} e^{\gamma\lfloor t\rfloor_\Delta}|w^{\Delta,v^\epsilon}(\lfloor t\rfloor_\Delta) - w^{\Delta,v}(\lfloor t\rfloor_\Delta)|^2\mathrm{d}t \\
&\quad + K(M)e^{\gamma t_{k+1}}\sup_{t\in[0,t_{k+1}]}|E^\epsilon(t)| + K(M)e^{\gamma t_{k+1}}\sup_{t\in[0,t_{k+1}]}|E^\epsilon(t)|^2.
\end{aligned}$$

Simplifying the inequality by combining like terms on both sides, then applying the Gronwall inequality (see Proposition A.3) and the fact that $v^\epsilon(\cdot)\in\mathcal{S}_M$, we obtain

$$\begin{aligned}
&\sup_{t\in[0,t_{k+1}]} e^{\gamma\lfloor t\rfloor_\Delta}|z^{\Delta,v^\epsilon}(\lfloor t\rfloor_\Delta) - z^{\Delta,v}(\lfloor t\rfloor_\Delta)|^2 \\
&\quad + \sup_{t\in[0,t_{k+1}]} e^{\gamma\lfloor t\rfloor_\Delta}|w^{\Delta,v^\epsilon}(\lfloor t\rfloor_\Delta) - w^{\Delta,v}(\lfloor t\rfloor_\Delta)|^2 \\
&\le K(M)e^{\gamma t_{k+1}}\sup_{t\in[0,t_{k+1}]}|E^\epsilon(t)| + K(M)e^{\gamma t_{k+1}}\sup_{t\in[0,t_{k+1}]}|E^\epsilon(t)|^2. \qquad (6.14)
\end{aligned}$$

To proceed, we now prove that $\sup_{t\in[0,t_{k+1})}|E^\epsilon(t)| \to 0$ as $\epsilon\to 0$. We begin by showing the compactness of $\{E^\epsilon(\cdot)\}_{\epsilon\in(0,1)}$ in $C([0,t_{k+1}];\mathbb{R}^d)$, where $C([0,t_{k+1}];\mathbb{R}^d)$ denotes the space of continuous functions on $[0,t_{k+1}]$. According to Assumption 2.1, Lemma 6.3, $v^\epsilon(\cdot), v(\cdot)\in\mathcal{S}_M$, and the Hölder inequality, one has

$$\begin{aligned}
&\Big|\int_{r_1}^{r_2}\sigma(w^{\Delta,v}_{\lfloor s\rfloor_\Delta})(v^\epsilon(\lfloor s\rfloor_\Delta) - v(\lfloor s\rfloor_\Delta))\mathrm{d}s\Big| \\
&\quad \le K(M)(1+\|\xi\|^{\beta+1})\sqrt{|r_2-r_1|}, \\
&\sup_{\epsilon\in(0,1)}\sup_{t\in[0,+\infty)}\Big|\int_0^{+\infty}\sigma(w^{\Delta,v}_{\lfloor s\rfloor_\Delta})(v^\epsilon(\lfloor s\rfloor_\Delta) - v(\lfloor s\rfloor_\Delta))\mathrm{d}s\Big| \\
&\quad \le K(M)(1+\|\xi\|^{\beta+1}),
\end{aligned}$$

which, respectively, implies the equicontinuity of $\{E^{\epsilon}(t), t \in [0, +\infty)\}_{\epsilon\in(0,1)}$ and the uniform boundedness of $\{E^{\epsilon}(\cdot)\}_{\epsilon\in(0,1)}$. Hence, according to the Ascoli–Arzela theorem, we obtain that $\{E^{\epsilon}(\cdot)\}_{\epsilon\in(0,1)}$ is compact in $C([0, t_{k+1}]; \mathbb{R}^d)$.

We next show that $\sigma(w^{\Delta,v}_{\lfloor\cdot\rfloor_\Delta})(v^{\epsilon}(\lfloor\cdot\rfloor_\Delta) - v(\lfloor\cdot\rfloor_\Delta)) \to 0$ as $\epsilon \to 0$ with respect to the weak topology of $\mathcal{S}_{M'}$ with some $M' > 0$. For any $h \in L^2_\Delta([0, +\infty); \mathbb{R}^m)$, by Lemma 6.3, we have

$$\begin{aligned}&\int_0^{\infty} |\sigma(w^{\Delta,v}_{\lfloor t\rfloor_\Delta})^* h(t)|^2 \mathrm{d}t\\ &\quad \le \int_0^{\infty} K(1 + \sup_{t\ge 0} \|w^{\Delta,v}_{\lfloor t\rfloor_\Delta}\|^2)|h(t)|^2 \mathrm{d}t \le K(M)(1 + \|\xi\|^{2(\beta+1)}),\end{aligned}$$

which together with $v^{\epsilon} \to v$ as $\epsilon \to 0$ in $\mathcal{S}_M$ yields that for any $h \in L^2_\Delta([0, +\infty); \mathbb{R}^m)$,

$$\begin{aligned}&\int_0^{\infty} \langle h(t), \sigma(w^{\Delta,v}_{\lfloor t\rfloor_\Delta})(v^{\epsilon}(\lfloor t\rfloor_\Delta) - v(\lfloor t\rfloor_\Delta))\rangle \mathrm{d}t\\ &\quad = \int_0^{\infty} \langle \sigma(w^{\Delta,v}_{\lfloor t\rfloor_\Delta})^* h(t), v^{\epsilon}(\lfloor t\rfloor_\Delta) - v(\lfloor t\rfloor_\Delta)\rangle \mathrm{d}t \to 0 \quad \text{as} \quad \epsilon \to 0.\end{aligned}$$

This is to say that

$$\sigma(w^{\Delta,v}_{\lfloor\cdot\rfloor_\Delta})(v^{\epsilon}(\lfloor\cdot\rfloor_\Delta) - v(\lfloor\cdot\rfloor_\Delta)) \to 0 \quad \text{as} \quad \epsilon \to 0$$

with respect to the weak topology of $\mathcal{S}_{M'}$ for some $M' > 0$. By [9, Chapter VI, Proposition 3.3], we conclude that $\sup_{t\in[0,t_{k+1})} |E^{\epsilon}(t)| \to 0$.

Together with (6.14), we derive that for each $k \in \mathbb{N}_+$,

$$\sup_{t\in[0,t_{k+1}]} |w^{\Delta,v^{\epsilon}}(\lfloor t\rfloor_\Delta) - w^{\Delta,v}(\lfloor t\rfloor_\Delta)|^2 \to 0 \text{ as } \epsilon \to 0.$$

Thus, by the dominated convergence theorem, we arrive at

$$\begin{aligned}&\lim_{\epsilon\to 0} \|w^{\Delta,v^{\epsilon}} - w^{\Delta,v}\|_{\mathfrak{a}}\\ &\quad = \lim_{\epsilon\to 0} \sum_{k=1}^{\infty} e^{-\mathfrak{a}t_k}\Big(\sup_{t\in[0,t_k]} |w^{\Delta,v^{\epsilon}}(\lfloor t\rfloor_\Delta) - w^{\Delta,v}(\lfloor t\rfloor_\Delta)|\Big)\Delta\\ &\quad = \sum_{k=1}^{\infty} e^{-\mathfrak{a}t_k}\Big(\lim_{\epsilon\to 0}\sup_{t\in[0,t_k]} |w^{\Delta,v^{\epsilon}}(\lfloor t\rfloor_\Delta) - w^{\Delta,v}(\lfloor t\rfloor_\Delta)|\Big)\Delta = 0.\end{aligned}$$

This finishes the proof. □

In order to obtain the boundedness of the moments of y^{Δ,v^ϵ}, we introduce the auxiliary process

$$\begin{cases} Z^{\Delta,v^\epsilon}(t_k) = \xi(t_k), \quad k = -N, \dots, -1, \\ Z^{\Delta,v^\epsilon}(t_k) = y^{\Delta,v^\epsilon}(t_k) - \theta b(y^{\Delta,v^\epsilon}_{t_k})\Delta, \quad k \in \mathbb{N}, \end{cases} \tag{6.15}$$

which satisfies that for $k \in \mathbb{N}_+$,

$$\begin{aligned} Z^{\Delta,v^\epsilon}(t_k) = Z^{\Delta,v^\epsilon}(t_{k-1}) + b(y^{\Delta,v^\epsilon}_{t_{k-1}})\Delta + \sqrt{\epsilon}\sigma(y^{\Delta,v^\epsilon}_{t_{k-1}})\delta W_{k-1} \\ + \sigma(y^{\Delta,v^\epsilon}_{t_{k-1}})v^\epsilon(t_{k-1})\Delta. \end{aligned}$$

Define the continuous version $Z^{\Delta,v^\epsilon}(\cdot)$ as follows: for $t \in [t_k, t_{k+1})$ with $k \in \mathbb{N}$,

$$\begin{aligned} Z^{\Delta,v^\epsilon}(t) := Z^{\Delta,v^\epsilon}(t_k) + b(y^{\Delta,v^\epsilon}_{t_k})(t - t_k) + \sqrt{\epsilon}\sigma(y^{\Delta,v^\epsilon}_{t_k})\big(W(t) - W(t_k)\big) \\ + \sigma(y^{\Delta,v^\epsilon}_{t_k})v^\epsilon(t_k)(t - t_k), \end{aligned}$$

and for $t \in [-\tau, 0]$, $Z^{\Delta,v^\epsilon}(t)$ is defined as the linear interpolation of $Z^{\Delta,v^\epsilon}(t_i)$, $i = -N, \dots, 0$.

Lemma 6.5 *Let $M > 0$ and $\{v^\epsilon\}_{\epsilon\in(0,1)} \subset \mathcal{A}_M$. Under Assumptions 2.1 to 2.4, for any $\Delta \in (0, 1]$ and $p \in \mathbb{N}_+$, we have*

$$\begin{aligned} &\mathbb{E}\Big[\sup_{t\in[0,t_{k+1}]} |Z^{\Delta,v^\epsilon}(\lfloor t\rfloor_\Delta)|^{2p}\Big] + \mathbb{E}\Big[\sup_{t\in[0,t_{k+1}]} |y^{\Delta,v^\epsilon}(\lfloor t\rfloor_\Delta)|^{2p}\Big] \\ &\quad \le K(M)(1 + \|\xi\|^{2p(\beta+1)})e^{\epsilon K(M)t_{k+1}}. \end{aligned}$$

As a consequence, for sufficiently small $\epsilon > 0$, $Z^{\Delta,v^\epsilon}(\cdot)$ and $y^{\Delta,v^\epsilon}(\cdot)$ belong to $L^{2p}(\Omega; C_{\xi^{Int}}([-\tau, +\infty); \mathbb{R}^d))$.

Proof Step 1. We show that for $\Delta \in (0, \Delta^*]$ with some $\Delta^* \in (0, 1)$,

$$\begin{aligned} &\mathbb{E}\Big[\sup_{t\in[0,t_{k+1}]} |Z^{\Delta,v^\epsilon}(\lfloor t\rfloor_\Delta)|^{2p}\Big] + \mathbb{E}\Big[\sup_{t\in[0,t_{k+1}]} |y^{\Delta,v^\epsilon}(\lfloor t\rfloor_\Delta)|^{2p}\Big] \\ &\quad \le K(M)(1 + \|\xi\|^{2p(\beta+1)})e^{\epsilon K(M)t_{k+1}}. \end{aligned}$$

It follows from the definition of Z^{Δ,v^ϵ} and the Young inequality that

$$\begin{aligned} &|Z^{\Delta,v^\epsilon}(t_{k+1})|^2 \\ &\quad = |Z^{\Delta,v^\epsilon}(t_k)|^2 + |b(y^{\Delta,v^\epsilon}_{t_k})|^2\Delta^2 + \epsilon|\sigma(y^{\Delta,v^\epsilon}_{t_k})\delta W_k|^2 \\ &\qquad + 2\langle y^{\Delta,v^\epsilon}(t_k) - \theta\Delta b(y^{\Delta,v^\epsilon}_{t_k}), b(y^{\Delta,v^\epsilon}_{t_k})\Delta\rangle + 2\langle Z^{\Delta,v^\epsilon}(t_k), \sqrt{\epsilon}\sigma(y^{\Delta,v^\epsilon}_{t_k})\delta W_k\rangle \end{aligned}$$

$$
\begin{aligned}
&+2\sqrt{\epsilon}\langle b(y_{t_k}^{\Delta,v^\epsilon})\Delta,\sigma(y_{t_k}^{\Delta,v^\epsilon})\delta W_k\rangle+|\sigma(y_{t_k}^{\Delta,v^\epsilon})v^\epsilon(t_k)|^2\Delta^2\\
&+2\langle Z^{\Delta,v^\epsilon}(t_k)+b(y_{t_k}^{\Delta,v^\epsilon})\Delta+\sqrt{\epsilon}\sigma(y_{t_k}^{\Delta,v^\epsilon})\delta W_k,\sigma(y_{t_k}^{\Delta,v^\epsilon})v^\epsilon(t_k)\Delta\rangle\\
\le&\Big(\frac{(1-\theta)^2}{\theta^2}+\frac{2\theta-1}{\theta^2(1+\Delta)}\Big)|Z^{\Delta,v^\epsilon}(t_k)|^2\\
&+\Big(\frac{1-2\theta}{\theta^2}+\frac{2\theta-1}{\theta^2}(1+\Delta)\Big)|y^{\Delta,v^\epsilon}(t_k)|^2\\
&+2\Delta\langle y^{\Delta,v^\epsilon}(t_k),b(y_{t_k}^{\Delta,v^\epsilon})\rangle+\Delta|\sigma(y_{t_k}^{\Delta,v^\epsilon})|^2+\Delta|v^\epsilon(t_k)|^2\\
&\times\Big(|Z^{\Delta,v^\epsilon}(t_k)+b(y_{t_k}^{\Delta,v^\epsilon})\Delta|^2+|\sigma(y_{t_k}^{\Delta,v^\epsilon})|^2\Delta\Big)+\epsilon|\sigma(y_{t_k}^{\Delta,v^\epsilon})\delta W_k|^2+2\widetilde{\mathcal{M}}_k,
\end{aligned}
$$

where

$$
\widetilde{\mathcal{M}}_k:=\big\langle\sqrt{\epsilon}\sigma(y_{t_k}^{\Delta,v^\epsilon})\delta W_k,\sigma(y_{t_k}^{\Delta,v^\epsilon})v^\epsilon(t_k)\Delta+Z^{\Delta,v^\epsilon}(t_k)+b(y_{t_k}^{\Delta,v^\epsilon})\Delta\big\rangle.
$$

Recalling the notation $A_{\theta,\Delta}=\frac{(1-\theta)^2}{\theta^2}+\frac{2\theta-1}{\theta^2(1+\Delta)}$ and using Assumptions 2.1 to 2.3, we have

$$
\begin{aligned}
&|Z^{\Delta,v^\epsilon}(t_{k+1})|^2\\
&\le A_{\theta,\Delta}|Z^{\Delta,v^\epsilon}(t_k)|^2+\frac{2\theta-1}{\theta^2}\Delta|y^{\Delta,v^\epsilon}(t_k)|^2+K\Delta\\
&\quad-2a_3\Delta|y^{\Delta,v^\epsilon}(t_k)|^{2+\ell}+2L\Delta|y^{\Delta,v^\epsilon}(t_k)|^2+2L\Delta\int_{-\tau}^0|y_{t_k}^{\Delta,v^\epsilon}(r)|^2\mathrm{d}\nu_1(r)\\
&\quad+2a_4\Delta\int_{-\tau}^0|y_{t_k}^{\Delta,v^\epsilon}(r)|^2\mathrm{d}\nu_2(r)+\Delta|v^\epsilon(t_k)|^2\Big(|Z^{\Delta,v^\epsilon}(t_k)+b(y_{t_k}^{\Delta,v^\epsilon})\Delta|^2\\
&\quad+|\sigma(y_{t_k}^{\Delta,v^\epsilon})|^2\Delta\Big)+\epsilon|\sigma(y_{t_k}^{\Delta,v^\epsilon})\delta W_k|^2+2\widetilde{\mathcal{M}}_k\\
&\le A_{\theta,\Delta}|Z^{\Delta,v^\epsilon}(t_k)|^2+K\Delta-a_3\Delta|y^{\Delta,v^\epsilon}(t_k)|^{2+l}-R\Delta|y^{\Delta,v^\epsilon}(t_k)|^2\\
&\quad+2L\Delta\int_{-\tau}^0|y_{t_k}^{\Delta,v^\epsilon}(r)|^2\mathrm{d}\nu_1(r)+2a_4\Delta\\
&\quad\times\int_{-\tau}^0|y_{t_k}^{\Delta,v^\epsilon}(r)|^2\mathrm{d}\nu_2(r)+\epsilon|\sigma(y_{t_k}^{\Delta,v^\epsilon})\delta W_k|^2\\
&\quad+\Delta|v^\epsilon(t_k)|^2\Big(|Z^{\Delta,v^\epsilon}(t_k)+b(y_{t_k}^{\Delta,v^\epsilon})\Delta|^2+|\sigma(y_{t_k}^{\Lambda,v^\epsilon})|^2\Delta\Big)+2\widetilde{\mathcal{M}}_k,
\end{aligned}
\tag{6.16}
$$

where R is a positive constant with $R > 2L$ (which is similar to R_2 in (2.38)) and we used the Young inequality. From this, we obtain that for $p \in \mathbb{N}$ with $p > 1$,

$$(|Z^{\Delta,v^\epsilon}(t_{k+1})|^2 + a_3\Delta|y^{\Delta,v^\epsilon}(t_k)|^{2+\ell})^p \leq A_{\theta,\Delta}^p|Z^{\Delta,v^\epsilon}(t_k)|^{2p} + \sum_{i=1}^{p} C_p^i \mathcal{I}_i^\epsilon,$$

where for $i = 1, \ldots, p$,

$$\mathcal{I}_i^\epsilon := A_{\theta,\Delta}^{p-i}|Z^{\Delta,v^\epsilon}(t_k)|^{2(p-i)}\Big[K\Delta - R\Delta|y^{\Delta,v^\epsilon}(t_k)|^2 + 2L\Delta\int_{-\tau}^{0}|y_{t_k}^{\Delta,v^\epsilon}(r)|^2 \mathrm{d}\nu_1(r)$$
$$+ 2a_4\Delta\int_{-\tau}^{0}|y_{t_k}^{\Delta,v^\epsilon}(r)|^2\mathrm{d}\nu_2(r) + \epsilon|\sigma(y_{t_k}^{\Delta,v^\epsilon})\delta W_k|^2$$
$$+ \Delta|v^\epsilon(t_k)|^2\Big(|Z^{\Delta,v^\epsilon}(t_k) + b(y_{t_k}^{\Delta,v^\epsilon})\Delta|^2 + |\sigma(y_{t_k}^{\Delta,v^\epsilon})|^2\Delta\Big) + 2\widetilde{\mathcal{M}}_k\Big]^i.$$

For the term $\mathcal{I}_1^\epsilon$, applying the Young inequality yields that for $\gamma_1 \in (0, 1)$,

$$\mathcal{I}_1^\epsilon \leq \gamma_1\Delta A_{\theta,\Delta}^p|Z^{\Delta,v^\epsilon}(t_k)|^{2p} + K(\gamma_1)\Delta - R\Delta A_{\theta,\Delta}^{p-1}|Z^{\Delta,v^\epsilon}(t_k)|^{2(p-1)}|y^{\Delta,v^\epsilon}(t_k)|^2$$
$$+ K(\gamma_1)\Delta\int_{-\tau}^{0}|y_{t_k}^{\Delta,v^\epsilon}(r)|^{2p}\mathrm{d}\nu_1(r) + K(\gamma_1)\Delta\int_{-\tau}^{0}|y_{t_k}^{\Delta,v^\epsilon}(r)|^{2p}\mathrm{d}\nu_2(r)$$
$$+ K\Delta|v^\epsilon(t_k)|^2\Big(|Z^{\Delta,v^\epsilon}(t_k)|^{2p} + |y^{\Delta,v^\epsilon}(t_k)|^{2p} + \int_{-\tau}^{0}|y_{t_k}^{\Delta,v^\epsilon}(r)|^{2p}\mathrm{d}\nu_1(r)\Big)$$
$$+ A_{\theta,\Delta}^{p-1}|Z^{\Delta,v^\epsilon}(t_k)|^{2(p-1)}(\epsilon|\sigma(y_{t_k}^{\Delta,v^\epsilon})\delta W_k|^2 + 2\widetilde{\mathcal{M}}_k).$$

By (6.15) and Assumption 2.3 (see also (2.41)), we have

$$|y^{\Delta,v^\epsilon}(t_k)|^2 + 2a_3\theta\Delta|y^{\Delta,v^\epsilon}(t_k)|^{2+\ell}$$
$$\leq |Z^{\Delta,v^\epsilon}(t_k)|^2 + K\Delta + 2a_4\theta\Delta\int_{-\tau}^{0}|y_{t_k}^{\Delta,v^\epsilon}(r)|^2\mathrm{d}\nu_2(r), \tag{6.17}$$

which implies

$$|y^{\Delta,v^\epsilon}(t_k)|^{2p} \leq \Big(|Z^{\Delta,v^\epsilon}(t_k)|^2 + K\Delta + 2a_4\theta\Delta\int_{-\tau}^{0}|y_{t_k}^{\Delta,v^\epsilon}(r)|^2\mathrm{d}\nu_2(r)\Big)^{p-1}$$
$$\times |y^{\Delta,v^\epsilon}(t_k)|^2$$
$$\leq 2|Z^{\Delta,v^\epsilon}(t_k)|^{2(p-1)}|y^{\Delta,v^\epsilon}(t_k)|^2$$
$$+ K|y^{\Delta,v^\epsilon}(t_k)|^2\Big(K\Delta + 2a_4\theta\Delta\int_{-\tau}^{0}|y_{t_k}^{\Delta,v^\epsilon}(r)|^2\mathrm{d}\nu_2(r)\Big)^{p-1},$$

where we used $(|a|+|b|)^q \le (1+\varepsilon)|a|^q + C_{\varepsilon,q}|b|^q$ for $q \ge 1$ with $\varepsilon = 1$ in the last step. In consequence,

$$\begin{aligned}
&-|Z^{\Delta,v^\epsilon}(t_k)|^{2(p-1)}|y^{\Delta,v^\epsilon}(t_k)|^2 \\
&\quad\le -\frac{1}{2}|y^{\Delta,v^\epsilon}(t_k)|^{2p} + K|y^{\Delta,v^\epsilon}(t_k)|^2\Big(K\Delta + 2a_4\theta\Delta\int_{-\tau}^0 |y_{t_k}^{\Delta,v^\epsilon}(r)|^2 \mathrm{d}\nu_2(r)\Big)^{p-1}.
\end{aligned}$$

Inserting the above inequality into $\mathcal{I}_1^\epsilon$, we derive

$$\begin{aligned}
\mathcal{I}_1^\epsilon &\le \gamma_1\Delta A_{\theta,\Delta}^p|Z^{\Delta,v^\epsilon}(t_k)|^{2p} + K(\gamma_1)\Delta - \frac{1}{2}R\Delta A_{\theta,\Delta}^{p-1}|y^{\Delta,v^\epsilon}(t_k)|^{2p} + K(\gamma_1)\Delta \\
&\quad\times\Big(|y^{\Delta,v^\epsilon}(t_k)|^{2p} + \int_{-\tau}^0 |y_{t_k}^{\Delta,v^\epsilon}(r)|^{2p}\mathrm{d}\nu_1(r) + \int_{-\tau}^0 |y_{t_k}^{\Delta,v^\epsilon}(r)|^{2p}\mathrm{d}\nu_2(r)\Big) \\
&\quad + K\Delta|v^\epsilon(t_k)|^2\Big(|Z^{\Delta,v^\epsilon}(t_k)|^{2p} + |y^{\Delta,v^\epsilon}(t_k)|^{2p} + \int_{-\tau}^0 |y_{t_k}^{\Delta,v^\epsilon}(r)|^{2p}\mathrm{d}\nu_1(r)\Big) \\
&\quad + A_{\theta,\Delta}^{p-1}|Z^{\Delta,v^\epsilon}(t_k)|^{2(p-1)}(\epsilon|\sigma(y_{t_k}^{\Delta,v^\epsilon})\delta W_k|^2 + 2\widetilde{\mathcal{M}}_k).
\end{aligned}$$

For the term $\mathcal{I}_i^\epsilon$ with $i = 2, \dots, p$,

$$\begin{aligned}
\mathcal{I}_i^\epsilon &\le \gamma_1\Delta A_{\theta,\Delta}^p|Z^{\Delta,v^\epsilon}(t_k)|^{2p} + K(\gamma_1)\Delta^2 \\
&\quad\times\Big(1 + |y^{\Delta,v^\epsilon}(t_k)|^{2p} + \int_{-\tau}^0 |y_{t_k}^{\Delta,v^\epsilon}(r)|^{2p}\mathrm{d}\nu_1(r) \\
&\quad + \int_{-\tau}^0 |y_{t_k}^{\Delta,v^\epsilon}(r)|^{2p}\mathrm{d}\nu_2(r)\Big) + K(M)\Delta|v^\epsilon(t_k)|^2 \\
&\quad\times\Big(|Z^{\Delta,v^\epsilon}(t_k)|^{2p} + |y^{\Delta,v^\epsilon}(t_k)|^{2p} \\
&\quad + \int_{-\tau}^0 |y_{t_k}^{\Delta,v^\epsilon}|^{2p}\mathrm{d}\nu_1(r)\Big) + KA_{\theta,\Delta}^{p-i}|Z^{\Delta,v^\epsilon}(t_k)|^{2(p-i)} \\
&\quad\times(\epsilon|\sigma(y_{t_k}^{\Delta,v^\epsilon})\delta W_k|^{2i} + |\widetilde{\mathcal{M}}_k|^i),
\end{aligned}$$

where we used $\Delta|v^\epsilon(t_k)|^2 \le M$ $\mathbb{P}$-a.s. Combining the estimates of the terms $\mathcal{I}_i^\epsilon, i = 1, \dots, p$, we arrive at

$$\begin{aligned}
&|Z^{\Delta,v^\epsilon}(t_{k+1})|^{2p} \\
&\quad\le (1 + K\gamma_1\Delta)A_{\theta,\Delta}^p|Z^{\Delta,v^\epsilon}(t_k)|^{2p} + K(\gamma_1)\Delta + \mathcal{J}_{k1}(R) + \mathcal{J}_{k2} \\
&\qquad + K(M)\Delta|v^\epsilon(t_k)|^2\Big(|Z^{\Delta,v^\epsilon}(t_k)|^{2p} + |y^{\Delta,v^\epsilon}(t_k)|^{2p} + \int_{-\tau}^0 |y_{t_k}^{\Delta,v^\epsilon}(r)|^{2p}\mathrm{d}\nu_1(r)\Big)
\end{aligned}$$

$$
\begin{aligned}
&+ K|Z^{\Delta,v^\epsilon}(t_k)|^{2(p-1)}\Big(\epsilon|\sigma(y_{t_k}^{\Delta,v^\epsilon})\delta W_k|^2 + \widetilde{\mathcal{M}}_k\Big)\\
&+ K\sum_{i=2}^{p} C_p^i|Z^{\Delta,v^\epsilon}(t_k)|^{2(p-i)}\Big(\epsilon|\sigma(y_{t_k}^{\Delta,v^\epsilon})\delta W_k|^{2i} + |\widetilde{\mathcal{M}}_k|^i\Big),
\end{aligned}
$$

where

$$
\begin{aligned}
\mathcal{J}_{k1}(R) &:= -\Big(\frac{p}{2}RA_{\theta,\Delta}^{p-1} - K(\gamma_1)\Big)\Delta|y^{\Delta,v^\epsilon}(t_k)|^{2p}\\
&\quad + K(\gamma_1)\Delta\Big(\int_{-\tau}^{0}|y_{t_k}^{\Delta,v^\epsilon}(r)|^{2p}\mathrm{d}\nu_1(r) + \int_{-\tau}^{0}|y_{t_k}^{\Delta,v^\epsilon}(r)|^{2p}\mathrm{d}\nu_2(r)\Big),\\
\mathcal{J}_{k2} &:= K(\gamma_1)\Delta^2\Big(|y^{\Delta,v^\epsilon}(t_k)|^{2p} + \int_{-\tau}^{0}|y_{t_k}^{\Delta,v^\epsilon}(r)|^{2p}\mathrm{d}\nu_1(r)\\
&\quad + \int_{-\tau}^{0}|y_{t_k}^{\Delta,v^\epsilon}(r)|^{2p}\mathrm{d}\nu_2(r)\Big).
\end{aligned}
$$

Then for $\gamma_2 > 0$,

$$
\begin{aligned}
&e^{\gamma_2 t_{k+1}}|Z^{\Delta,v^\epsilon}(t_{k+1})|^{2p}\\
&\le |Z^{\Delta,v^\epsilon}(0)|^{2p} + \big((1+K\gamma_1\Delta)A_{\theta,\Delta}^{p}e^{\gamma_2\Delta} - 1\big)\sum_{l=0}^{k} e^{\gamma_2 t_l}|Z^{\Delta,v^\epsilon}(t_l)|^{2p}\\
&\quad + K(\gamma_1)e^{\gamma_2 t_{k+1}} + \sum_{l=0}^{k} e^{\gamma_2 t_{l+1}}\big(\mathcal{J}_{l1}(R) + \mathcal{J}_{l2}\big) + K(M)\sum_{l=0}^{k} e^{\gamma_2 t_{l+1}}\Delta|v^\epsilon(t_l)|^2\\
&\quad \times\Big(|Z^{\Delta,v^\epsilon}(t_l)|^{2p} + |y^{\Delta,v^\epsilon}(t_l)|^{2p} + \int_{-\tau}^{0}|y_{t_l}^{\Delta,v^\epsilon}(r)|^{2p}\mathrm{d}\nu_1(r)\Big)\\
&\quad + K\sum_{l=0}^{k} e^{\gamma_2 t_{l+1}}|Z^{\Delta,v^\epsilon}(t_l)|^{2(p-1)}\Big(\epsilon|\sigma(y_{t_l}^{\Delta,v^\epsilon})\delta W_l|^2 + \widetilde{\mathcal{M}}_l\Big)\\
&\quad + K\sum_{l=0}^{k} e^{\gamma_2 t_{l+1}}\sum_{i=2}^{p} C_p^i|Z^{\Delta,v^\epsilon}(t_l)|^{2(p-i)}\Big(\epsilon|\sigma(y_{t_l}^{\Delta,v^\epsilon})\delta W_l|^{2i} + |\widetilde{\mathcal{M}}_l|^i\Big).
\end{aligned}
$$

Similar to the estimate of (2.47) in the proof of Lemma 2.5, take γ_1 and γ_2 independent of Δ such that $(1+K\gamma_1\Delta)A_{\theta,\Delta}^{p}e^{\gamma_2\Delta} - 1 \le 0$. Similar to the proof of (2.50), by choosing R sufficiently large and further choosing Δ^* sufficiently small, we obtain that for $\Delta \le \Delta^*$,

$$
\sum_{l=0}^{k} e^{\gamma_2 t_{l+1}}\big(\mathcal{J}_{l1}(R) + \mathcal{J}_{l2}\big) \le K\|\xi\|^{2p}.
$$

Hence,

$$
\begin{aligned}
&e^{\gamma_2 t_{k+1}}|Z^{\Delta,v^\epsilon}(t_{k+1})|^{2p} \\
&\quad\le |Z^{\Delta,v^\epsilon}(0)|^{2p} + K\|\xi\|^{2p} + Ke^{\gamma_2 t_{k+1}} + K(M)\int_0^{t_{k+1}} |v^\epsilon(\lfloor s\rfloor_\Delta)|^2 \\
&\quad\times\Big(\sup_{r\in[0,s]} e^{\gamma_2\lfloor r\rfloor_\Delta}|Z^{\Delta,v^\epsilon}(\lfloor r\rfloor_\Delta)|^{2p} + \sup_{r\in[0,s]} e^{\gamma_2\lfloor r\rfloor_\Delta}|y^{\Delta,v^\epsilon}(\lfloor r\rfloor_\Delta)|^{2p}\Big)\mathrm{d}s \\
&\quad + K\sum_{l=0}^{k} e^{\gamma_2 t_l}|Z^{\Delta,v^\epsilon}(t_l)|^{2(p-1)}\Big(\epsilon|\sigma(y_{t_l}^{\Delta,v^\epsilon})\delta W_l|^2 + \widetilde{\mathcal{M}}_l\Big) \\
&\quad + K\sum_{l=0}^{k} e^{\gamma_2 t_l}\sum_{i=2}^{p} C_p^i|Z^{\Delta,v^\epsilon}(t_l)|^{2(p-i)}\Big(\epsilon|\sigma(y_{t_l}^{\Delta,v^\epsilon})\delta W_l|^{2i} + |\widetilde{\mathcal{M}}_l|^i\Big).
\end{aligned}
$$

Moreover, it follows from (6.17) and the $\tilde{\gamma}$-Young inequality that

$$
\begin{aligned}
&e^{\frac{\gamma_2}{p}t_k}|y^{\Delta,v^\epsilon}(t_k)|^2 + 2a_3\theta\Delta e^{\frac{\gamma_2}{p}t_k}|y^{\Delta,v^\epsilon}(t_k)|^{2+\ell} \\
&\quad\le e^{\frac{\gamma_2}{p}t_k}|Z^{\Delta,v^\epsilon}(t_k)|^2 + Ke^{\frac{\gamma_2}{p}t_k}\Delta + 2a_4\theta\Delta e^{\frac{\gamma_2}{p}t_k} \\
&\quad\times\int_{-\tau}^{0}\big(\tilde{\gamma}|y_{t_k}^{\Delta,v^\epsilon}(r)|^{2+\ell} + K(\tilde{\gamma})\big)\mathrm{d}\nu_2(r) \\
&\quad\le \sup_{t_l\in[0,t_k]} e^{\frac{\gamma_2}{p}t_l}|Z^{\Delta,v^\epsilon}(t_l)|^2 + \tilde{\gamma}2a_4\theta\Delta e^{\frac{\gamma_2}{p}\tau}\sup_{t_l\in[0,t_k]}\big(e^{\frac{\gamma_2}{p}t_l}|y^{\Delta,v^\epsilon}(t_l)|^{2+\ell}\big) \\
&\quad + K(\tilde{\gamma})\Delta e^{\frac{\gamma_2}{p}t_k},
\end{aligned}
$$

which implies

$$
\sup_{t_l\in[0,t_k]} e^{\frac{\gamma_2}{p}t_l}|y^{\Delta,v^\epsilon}(t_l)|^2 \le \sup_{t_l\in[0,t_k]} e^{\frac{\gamma_2}{p}t_l}|Z^{\Delta,v^\epsilon}(t_l)|^2 + K\Delta e^{\frac{\gamma_2}{p}t_k}
$$

by taking $\tilde{\gamma}$ such that $2\tilde{\gamma}a_4e^{\frac{\gamma_2}{p}\tau} \le 2a_3$. Then taking the pth power on both sides yields that

$$
\sup_{t\in[0,t_k]} e^{\gamma_2\lfloor t\rfloor_\Delta}|y^{\Delta,v^\epsilon}(\lfloor t\rfloor_\Delta)|^{2p} \le Ke^{\gamma_2 t_k} + K\sup_{t\in[0,t_k]} e^{\gamma_2\lfloor t\rfloor_\Delta}|Z^{\Delta,v^\epsilon}(\lfloor t\rfloor_\Delta)|^{2p}.
$$

We conclude

$$
\sup_{t\in[0,t_{k+1}]} e^{\gamma_2\lfloor t\rfloor_\Delta}|Z^{\Delta,v^\epsilon}(\lfloor t\rfloor_\Delta)|^{2p} + \sup_{t\in[0,t_{k+1}]} e^{\gamma_2\lfloor t\rfloor_\Delta}|y^{\Delta,v^\epsilon}(\lfloor t\rfloor_\Delta)|^{2p}
$$

$$
\begin{aligned}
&\leq Ke^{\gamma_2 t_{k+1}} + K(1+\|\xi\|^{2p(\beta+1)}) + K(M)\int_0^{t_{k+1}} |v^\epsilon(\lfloor s\rfloor_\Delta)|^2 \\
&\quad \times \Big(\sup_{r\in[0,s]} e^{\gamma_2\lfloor r\rfloor_\Delta}|Z^{\Delta,v^\epsilon}(\lfloor r\rfloor_\Delta)|^{2p} + \sup_{r\in[0,s]} e^{\gamma_2\lfloor r\rfloor_\Delta}|y^{\Delta,v^\epsilon}(\lfloor r\rfloor_\Delta)|^{2p}\Big)\mathrm{d}s \\
&\quad + K\sup_{j\leq k}\sum_{l=0}^{j} e^{\gamma_2 t_l}|Z^{\Delta,v^\epsilon}(t_l)|^{2(p-1)}\Big(\epsilon|\sigma(y_{t_l}^{\Delta,v^\epsilon})\delta W_l|^2 + \widetilde{\mathcal{M}}_l\Big) \\
&\quad + K\sum_{l=0}^{k} e^{\gamma_2 t_l}\sum_{i=2}^{p} C_p^i|Z^{\Delta,v^\epsilon}(t_l)|^{2(p-i)}\Big(\epsilon|\sigma(y_{t_l}^{\Delta,v^\epsilon})\delta W_l|^{2i} + |\widetilde{\mathcal{M}}_l|^i\Big),
\end{aligned}
\tag{6.18}
$$

where we used $|Z^{\Delta,v^\epsilon}(0)| \leq K(1+\|\xi\|^{\beta+1})$ due to Assumption 2.4. By virtue of the Gronwall inequality (see Proposition A.3), one has

$$
\begin{aligned}
&\sup_{t\in[0,t_{k+1}]} e^{\gamma_2\lfloor t\rfloor_\Delta}|Z^{\Delta,v^\epsilon}(\lfloor t\rfloor_\Delta)|^{2p} + \sup_{t\in[0,t_{k+1}]} e^{\gamma_2\lfloor t\rfloor_\Delta}|y^{\Delta,v^\epsilon}(\lfloor t\rfloor_\Delta)|^{2p} \\
&\leq e^{K(M)M}\Big[K(1+\|\xi\|^{2p(\beta+1)}) + Ke^{\gamma_2 t_{k+1}} + K\sup_{j\leq k}\sum_{l=0}^{j} e^{\gamma_2 t_l}|Z^{\Delta,v^\epsilon}(t_l)|^{2(p-1)} \\
&\quad \times\Big(\epsilon|\sigma(y_{t_l}^{\Delta,v^\epsilon})\delta W_l|^2 + \widetilde{\mathcal{M}}_l\Big) + K\sum_{l=0}^{k} e^{\gamma_2 t_l}\sum_{i=2}^{p} C_p^i|Z^{\Delta,v^\epsilon}(t_l)|^{2(p-i)} \\
&\quad \times\Big(\epsilon|\sigma(y_{t_l}^{\Delta,v^\epsilon})\delta W_l|^{2i} + |\widetilde{\mathcal{M}}_l|^i\Big)\Big].
\end{aligned}
\tag{6.19}
$$

Applying the Burkholder–Davis–Gundy inequality (see Proposition A.5) and the Young inequality, we arrive at that for $\gamma_3 \in (0,1)$,

$$
\begin{aligned}
&\mathbb{E}\Big[\sup_{j\leq k}\sum_{l=0}^{j} e^{\gamma_2 t_l}|Z^{\Delta,v^\epsilon}(t_l)|^{2(p-1)}\Big(\epsilon|\sigma(y_{t_l}^{\Delta,v^\epsilon})\delta W_l|^2 + \widetilde{\mathcal{M}}_l\Big)\Big] \\
&\leq \epsilon\Delta\mathbb{E}\Big[\sum_{l=0}^{k} e^{\gamma_2 t_l}|Z^{\Delta,v^\epsilon}(t_l)|^{2(p-1)}|\sigma(y_{t_l}^{\Delta,v^\epsilon})|^2\Big] \\
&\quad + K\mathbb{E}\Big[\Big(\sum_{l=0}^{k} e^{2\gamma_2 t_l}|Z^{\Delta,v^\epsilon}(t_l)|^{4(p-1)} \\
&\qquad \times \epsilon|\sigma(y_{t_l}^{\Delta,v^\epsilon})|^2\Big(|\sigma(y_{t_l}^{\Delta,v^\epsilon})|^2 M + |Z^{\Delta,v^\epsilon}(t_l)|^2 + |y^{\Delta,v^\epsilon}(t_l)|^2\Big)\Delta\Big)^{\frac{1}{2}}\Big]
\end{aligned}
$$

$$
\begin{aligned}
&\le \epsilon K(M,\gamma_3)\int_0^{t_{k+1}}\mathbb{E}\Big[\sup_{r\in[0,s]} e^{\gamma_2\lfloor r\rfloor_\Delta}|Z^{\Delta,v^\epsilon}(\lfloor r\rfloor_\Delta)|^{2p}\\
&\quad+\sup_{r\in[0,s]} e^{\gamma_2\lfloor r\rfloor_\Delta}|y^{\Delta,v^\epsilon}(\lfloor r\rfloor_\Delta)|^{2p}\Big]\mathrm{d}s+\gamma_3 K(M)\\
&\quad\times\Big(\sup_{t\in[0,t_{k+1}]} e^{\gamma_2\lfloor t\rfloor_\Delta}|Z^{\Delta,v^\epsilon}(\lfloor t\rfloor_\Delta)|^{2p}\\
&\quad+\sup_{t\in[0,t_{k+1}]} e^{\gamma_2\lfloor t\rfloor_\Delta}|y^{\Delta,v^\epsilon}(\lfloor t\rfloor_\Delta)|^{2p}\Big)+K\|\xi\|^{2p}.
\end{aligned}
$$

Similarly,

$$
\begin{aligned}
&\mathbb{E}\Big[\sum_{l=0}^{k} e^{\gamma_2 t_l}\sum_{i=2}^{p} C_p^i|Z^{\Delta,v^\epsilon}(t_l)|^{2(p-i)}\Big(\epsilon|\sigma(y_{t_l}^{\Delta,v^\epsilon})\delta W_l|^{2i}+|\widetilde{\mathcal{M}}_l|^i\Big)\Big]\\
&\quad\le \epsilon K(M)\int_0^{t_{k+1}}\mathbb{E}\Big[\sup_{r\in[0,s]} e^{\gamma_2\lfloor r\rfloor_\Delta}|Z^{\Delta,v^\epsilon}(\lfloor r\rfloor_\Delta)|^{2p}\\
&\quad\quad+\sup_{r\in[0,s]} e^{\gamma_2\lfloor r\rfloor_\Delta}|y^{\Delta,v^\epsilon}(\lfloor r\rfloor_\Delta)|^{2p}\Big]\mathrm{d}s\\
&\quad\quad+\gamma_3 K(M)\Big(\sup_{t\in[0,t_{k+1}]} e^{\gamma_2\lfloor t\rfloor_\Delta}|Z^{\Delta,v^\epsilon}(\lfloor t\rfloor_\Delta)|^{2p}\\
&\quad\quad+\sup_{t\in[0,t_{k+1}]} e^{\gamma_2\lfloor t\rfloor_\Delta}|y^{\Delta,v^\epsilon}(\lfloor t\rfloor_\Delta)|^{2p}\Big)\\
&\quad\quad+K\|\xi\|^{2p}.
\end{aligned}
$$

Taking expectation on both sides of (6.19), inserting the above two inequalities into (6.19), letting $\gamma_3>0$ be sufficiently small, and using the Gronwall inequality (see Proposition A.3) again, we deduce

$$
\begin{aligned}
&\mathbb{E}\Big[\sup_{t\in[0,t_{k+1}]} e^{\gamma_2\lfloor t\rfloor_\Delta}|Z^{\Delta,v^\epsilon}(\lfloor t\rfloor_\Delta)|^{2p}\Big]+\mathbb{E}\Big[\sup_{t\in[0,t_{k+1}]} e^{\gamma_2\lfloor t\rfloor_\Delta}|y^{\Delta,v^\epsilon}(\lfloor t\rfloor_\Delta)|^{2p}\Big]\\
&\quad\le Ke^{K(M)}(1+\|\xi\|^{2p(\beta+1)}+e^{\gamma_2 t_{k+1}})e^{\epsilon K(M)t_{k+1}},
\end{aligned}
$$

which implies

$$
\begin{aligned}
&\mathbb{E}\Big[\sup_{t\in[0,t_{k+1}]}|Z^{\Delta,v^\epsilon}(\lfloor t\rfloor_\Delta)|^{2p}\Big]+\mathbb{E}\Big[\sup_{t\in[0,t_{k+1}]}|y^{\Delta,v^\epsilon}(\lfloor t\rfloor_\Delta)|^{2p}\Big]\\
&\quad\le K(M)(1+\|\xi\|^{2p(\beta+1)})e^{\epsilon K(M)t_{k+1}}.
\end{aligned}
$$

We finish the proof of Step 1.

Step 2. We show that for $\Delta \in [\Delta^*, 1]$,

$$\mathbb{E}\Big[\sup_{t\in[0,t_{k+1}]}|Z^{\Delta,v^\epsilon}(\lfloor t\rfloor_\Delta)|^{2p}\Big]+\mathbb{E}\Big[\sup_{t\in[0,t_{k+1}]}|y^{\Delta,v^\epsilon}(\lfloor t\rfloor_\Delta)|^{2p}\Big]$$
$$\leq K(M)(1+\|\xi\|^{2p(\beta+1)})e^{\epsilon K(M)t_{k+1}}.$$

The proof is similar to that of Lemma 2.6, and thus we only give the sketch of the proof.

Similar to the estimate of $\mathcal{I}_i^\epsilon$ for $\Delta \in (0, \Delta^*]$, we have that for small constant $\bar{\gamma}_1 \in (0, 1)$,

$$\mathcal{I}_1^\epsilon \leq \bar{\gamma}_1 A_{\theta,\Delta}^p|Z^{\Delta,v^\epsilon}(t_k)|^{2p}+K(\bar{\gamma}_1)\Delta-\frac{1}{2}R\Delta A_{\theta,\Delta}^{p-1}|y^{\Delta,v^\epsilon}(t_k)|^{2p}$$
$$+K(\bar{\gamma}_1)\Delta^p\Big(|y^{\Delta,v^\epsilon}(t_k)|^{2p}+\int_{-\tau}^0|y_{t_k}^{\Delta,v^\epsilon}(r)|^{2p}\mathrm{d}\nu_2(r)$$
$$+\int_{-\tau}^0|y_{t_k}^{\Delta,v^\epsilon}(r)|^{2p}\mathrm{d}\nu_1(r)\Big)$$
$$+K(M)\Delta|v^\epsilon(t_k)|^2\Big(|Z^{\Delta,v^\epsilon}(t_k)|^{2p}+|y^{\Delta,v^\epsilon}(t_k)|^{2p}+\int_{-\tau}^0|y_{t_k}^{\Delta,v^\epsilon}(r)|^{2p}\mathrm{d}\nu_1(r)\Big)$$
$$+A_{\theta,\Delta}^{p-1}|Z^{\Delta,v^\epsilon}(t_k)|^{2(p-1)}\Big(\epsilon|\sigma(y_{t_k}^{\Delta,v^\epsilon})\delta W_k|^2+2\widetilde{\mathcal{M}}_k\Big),$$

and for $i \in \{2, \ldots, p\}$,

$$\mathcal{I}_i^\epsilon \leq \bar{\gamma}_1 A_{\theta,\Delta}^p|Z^{\Delta,v^\epsilon}(t_k)|^{2p}+K(\bar{\gamma}_1)\Delta^p\Big(|y^{\Delta,v^\epsilon}(t_k)|^{2p}+\int_{-\tau}^0|y_{t_k}^{\Delta,v^\epsilon}(r)|^{2p}\mathrm{d}\nu_1(r)$$
$$+\int_{-\tau}^0|y_{t_k}^{\Delta,v^\epsilon}(r)|^{2p}\mathrm{d}\nu_2(r)\Big)+K(M)\Delta|v^\epsilon(t_k)|^2\Big(|Z^{\Delta,v^\epsilon}(t_k)|^{2p}+|y^{\Delta,v^\epsilon}(t_k)|^{2p}$$
$$+\int_{-\tau}^0|y_{t_k}^{\Delta,v^\epsilon}|^{2p}\mathrm{d}\nu_1(r)\Big)+KA_{\theta,\Delta}^{p-i}|Z^{\Delta,v^\epsilon}(t_k)|^{2(p-i)}$$
$$\times\Big(\epsilon|\sigma(y_{t_k}^{\Delta,v^\epsilon})\delta W_k|^{2i}+|\widetilde{\mathcal{M}}_k|^i\Big).$$

It follows from $(|a|+|b|)^p \geq |a|^p+|b|^p$ for $p \in \mathbb{N}_+$ that

$$|Z^{\Delta,v^\epsilon}(t_{k+1})|^{2p}+a_3^p\Delta^p|y^{\Delta,v^\epsilon}(t_{k+1})|^{p(2+\ell)}$$
$$\leq(1+Kp\bar{\gamma}_1)A_{\theta,\Delta}^p|Z^{\Delta,v^\epsilon}(t_k)|^{2p}+K\Delta+\widetilde{\mathfrak{J}}_{k1}(R)+\widetilde{\mathfrak{J}}_{k2}$$
$$+K(M)\Delta|v^\epsilon(t_k)|^2\Big(|Z^{\Delta,v^\epsilon}(t_k)|^{2p}+|y^{\Delta,v^\epsilon}(t_k)|^{2p}+\int_{-\tau}^0|y_{t_k}^{\Delta,v^\epsilon}(r)|^{2p}\mathrm{d}\nu_1(r)\Big)$$

$$+ K|Z^{\Delta,v^\epsilon}(t_k)|^{2(p-1)}\Big(\epsilon|\sigma(y_{t_k}^{\Delta,v^\epsilon})\delta W_k|^2 + 2\widetilde{\mathcal{M}}_k\Big)$$

$$+ K\sum_{i=2}^{p} C_p^i|Z^{\Delta,v^\epsilon}(t_k)|^{2(p-i)}\Big(\epsilon|\sigma(y_{t_k}^{\Delta,v^\epsilon})\delta W_k|^{2i} + |\widetilde{\mathcal{M}}_k|^i\Big),$$

where

$$\widetilde{\mathcal{J}}_{k1}(R) = -\frac{p}{2}RA_{\theta,\Delta}^{p-1}\Delta|y^{\Delta,v^\epsilon}(t_k)|^{2p},$$

$$\widetilde{\mathcal{J}}_{k2} = K(\bar{\gamma}_1)\Delta^p\Big(|y^{\Delta,v^\epsilon}(t_k)|^{2p} + \int_{-\tau}^{0}|y_{t_k}^{\Delta,v^\epsilon}(r)|^{2p}\mathrm{d}v_1(r) + \int_{-\tau}^{0}|y_{t_k}^{\Delta,v^\epsilon}(r)|^{2p}\mathrm{d}v_2(r)\Big).$$

Similar to (2.54), by taking $\bar{\gamma}_1$ and $\bar{\gamma}_2$ sufficiently small such that $(1+Kp\bar{\gamma}_1)A_{\theta,\Delta^*}^p e^{\bar{\gamma}_2} - 1 \le 0$, we obtain that for $\Delta \in (\Delta^*, 1]$,

$$\begin{aligned}
&e^{\bar{\gamma}_2 t_{k+1}}|Z^{\Delta,v^\epsilon}(t_{k+1})|^{2p} \\
&\le |Z^{\Delta,v^\epsilon}0)|^{2p} + Ke^{\bar{\gamma}_2 t_{k+1}} + \sum_{l=0}^{k} e^{\bar{\gamma}_2 t_{l+1}}\widetilde{\mathcal{J}}_{l1}(R) \\
&\quad + \sum_{l=0}^{k} e^{\bar{\gamma}_2 t_{l+1}}(\widetilde{\mathcal{J}}_{l2} - a_3^p\Delta^p|y^{\Delta,v^\epsilon}(t_l)|^{p(2+\ell)}) + K(M)\sum_{l=0}^{k} e^{\bar{\gamma}_2 t_l}\Delta|v^\epsilon(t_l)|^2 \\
&\quad \times \Big(|Z^{\Delta,v^\epsilon}(t_l)|^{2p} + |y^{\Delta,v^\epsilon}(t_l)|^{2p} + \int_{-\tau}^{0}|y_{t_l}^{\Delta,v^\epsilon}(r)|^{2p}\mathrm{d}v_1(r)\Big) \\
&\quad + K\sum_{l=0}^{k} e^{\bar{\gamma}_2 t_l}|Z^{\Delta,v^\epsilon}(t_l)|^{2(p-1)}\Big(\epsilon|\sigma(y_{t_l}^{\Delta,v^\epsilon})\delta W_l|^2 + 2\widetilde{\mathcal{M}}_l\Big) \\
&\quad + K\sum_{l=0}^{k} e^{\bar{\gamma}_2 t_l}\sum_{i=2}^{p} C_p^i|Z^{\Delta,v^\epsilon}(t_l)|^{2(p-i)}\Big(\epsilon|\sigma(y_{t_l}^{\Delta,v^\epsilon})\delta W_l|^{2i} + |\widetilde{\mathcal{M}}_l|^i\Big).
\end{aligned}$$

Then by the similar procedure to the proofs of (2.56) and (2.57), and choosing a sufficiently large number R independent of Δ, we conclude that

$$\begin{aligned}
&\sum_{l=0}^{k} e^{\bar{\gamma}_2 t_{l+1}}\widetilde{\mathcal{J}}_{l1}(R) + \sum_{l=0}^{k} e^{\bar{\gamma}_2 t_{l+1}}(\widetilde{\mathcal{J}}_{l2} - a_3^p\Delta^p|y^{\Delta,v^\epsilon}(t_l)|^{p(2+\ell)}) \\
&\quad \le K(\|\xi\|^{2p} + 1 + e^{\bar{\gamma}_2 t_{k+1}}).
\end{aligned}$$

The remaining proof is similar to that of Step 1 and is omitted.

Combining Steps 1–2 finishes the proof. □

Proposition 6.6 *Let $M > 0$ and $\{v^\epsilon\}_{\epsilon\in(0,1)} \subset \mathcal{A}_M$ such that $v^\epsilon \xrightarrow[\epsilon\to 0]{d} v$ as $\mathbb{S}_M$-valued random variables. Under Assumptions 2.1 to 2.4, it holds that for any $\Delta \in (0, \frac{1}{4a_2\theta}]$,*

$$\mathcal{G}^{\Delta,\epsilon}\Big(\sqrt{\epsilon}W + \int_0^{\cdot} v^\epsilon(\lfloor s\rfloor_\Delta)\mathrm{d}s\Big) \xrightarrow[\epsilon\to 0]{d} \mathcal{G}^\Delta\Big(\int_0^{\cdot} v(\lfloor s\rfloor_\Delta)\mathrm{d}s\Big).$$

Proof To show $y^{\Delta,v^\epsilon} \xrightarrow[\epsilon\to 0]{d} w^{\Delta,v}$ as $v^\epsilon \xrightarrow[\epsilon\to 0]{d} v$, by the duality formula for the Kantorovich–Rubinstein distance (see e.g. [27, Remark 6.5]), it suffices to prove that $\mathbb{E}[1 \wedge \|y^{\Delta,v^\epsilon} - w^{\Delta,v}\|_{\mathfrak{a}}] \to 0$ as $\epsilon \to 0$. Note that

$$\begin{aligned}
&\lim_{\epsilon\to 0}\mathbb{E}\big[1 \wedge \|y^{\Delta,v^\epsilon} - w^{\Delta,v}\|_{\mathfrak{a}}\big] \\
&\le \sum_{k=1}^{\infty}\lim_{\epsilon\to 0}\mathbb{E}\Big[1 \wedge \Big(\Delta e^{-\mathfrak{a}t_k}\sup_{t\in[0,t_k]}|y^{\Delta,v^\epsilon}(t) - w^{\Delta,v}(t)|\Big)\Big] \\
&\le \sum_{k=1}^{\infty}\Delta e^{-\mathfrak{a}t_k}\lim_{\epsilon\to 0}\mathbb{E}\Big[(e^{\mathfrak{a}t_k}\Delta^{-1}) \wedge \sup_{t\in[0,t_k]}|y^{\Delta,v^\epsilon}(t) - w^{\Delta,v}(t)|\Big].
\end{aligned}$$

Hence, we only need to show that for any $k \in \mathbb{N}_+$,

$$\lim_{\epsilon\to 0}\mathbb{E}\Big[(e^{\mathfrak{a}t_k}\Delta^{-1}) \wedge \sup_{t\in[0,t_k]}|y^{\Delta,v^\epsilon}(t) - w^{\Delta,v}(t)|\Big] = 0.$$

By virtue of

$$\begin{aligned}
Z^{\Delta,v^\epsilon}(t) - z^{\Delta,v}(t) = {} & Z^{\Delta,v^\epsilon}(\lfloor t\rfloor_\Delta) - z^{\Delta,v}(\lfloor t\rfloor_\Delta) + \big(b(y^{\Delta,v^\epsilon}_{\lfloor t\rfloor_\Delta}) \\
& - b(w^{\Delta,v}_{\lfloor t\rfloor_\Delta})\big)(t - \lfloor t\rfloor_\Delta) \\
& + \sqrt{\epsilon}\sigma(y^{\Delta,v^\epsilon}_{\lfloor t\rfloor_\Delta})\big(W(t) - W(\lfloor t\rfloor_\Delta)\big) \\
& + \big(\sigma(y^{\Delta,v^\epsilon}_{\lfloor t\rfloor_\Delta})v^\epsilon(\lfloor t\rfloor_\Delta) - \sigma(w^{\Delta,v}_{\lfloor t\rfloor_\Delta})v(\lfloor t\rfloor_\Delta)\big)(t - \lfloor t\rfloor_\Delta),
\end{aligned}$$

and the Itô formula, we have that

$$\begin{aligned}
&\mathrm{d}|Z^{\Delta,v^\epsilon}(t) - z^{\Delta,v}(t)|^2 \\
&= \Big[2\big\langle Z^{\Delta,v^\epsilon}(t) - z^{\Delta,v}(t), b(y^{\Delta,v^\epsilon}_{\lfloor t\rfloor_\Delta}) - b(w^{\Delta,v}_{\lfloor t\rfloor_\Delta})\big\rangle \\
&\quad + \epsilon|\sigma(y^{\Delta,v^\epsilon}_{\lfloor t\rfloor_\Delta})|^2 + 2\big\langle Z^{\Delta,v^\epsilon}(t) - z^{\Delta,v}(t), \sigma(y^{\Delta,v^\epsilon}_{\lfloor t\rfloor_\Delta})v^\epsilon(\lfloor t\rfloor_\Delta)\big\rangle
\end{aligned}$$

$$- \sigma(w^{\Delta,v}_{\lfloor t \rfloor_\Delta})v(\lfloor t \rfloor_\Delta)\rangle\big]\mathrm{d}t$$
$$+ 2\sqrt{\epsilon}\langle Z^{\Delta,v^\epsilon}(t) - z^{\Delta,v}(t), \sigma(y^{\Delta,v^\epsilon}_{\lfloor t \rfloor_\Delta})\mathrm{d}W(t)\rangle.$$

Further utilizing

$$\begin{aligned} &Z^{\Delta,v^\epsilon}(\lfloor t \rfloor_\Delta) - z^{\Delta,v}(\lfloor t \rfloor_\Delta) \\ &= y^{\Delta,v^\epsilon}(\lfloor t \rfloor_\Delta) - w^{\Delta,v}(\lfloor t \rfloor_\Delta) - \theta\big(b(y^{\Delta,v^\epsilon}_{\lfloor t \rfloor_\Delta}) - b(w^{\Delta,v}_{\lfloor t \rfloor_\Delta})\big)\Delta, \end{aligned}$$

we deduce

$$\begin{aligned} &\mathrm{d}|Z^{\Delta,v^\epsilon}(t) - z^{\Delta,v}(t)|^2 \\ &= \Big[2\langle y^{\Delta,v^\epsilon}(\lfloor t \rfloor_\Delta) - w^{\Delta,v}(\lfloor t \rfloor_\Delta), b(y^{\Delta,v^\epsilon}_{\lfloor t \rfloor_\Delta}) - b(w^{\Delta,v}_{\lfloor t \rfloor_\Delta})\rangle \\ &\quad - 2\theta\Delta|b(y^{\Delta,v^\epsilon}_{\lfloor t \rfloor_\Delta}) - b(w^{\Delta,v}_{\lfloor t \rfloor_\Delta})|^2 + 2\langle(b(y^{\Delta,v^\epsilon}_{\lfloor t \rfloor_\Delta}) \\ &\quad - b(w^{\Delta,v}_{\lfloor t \rfloor_\Delta}))(t - \lfloor t \rfloor_\Delta), b(y^{\Delta,v^\epsilon}_{\lfloor t \rfloor_\Delta}) - b(w^{\Delta,v}_{\lfloor t \rfloor_\Delta})\rangle \\ &\quad + 2\big\langle\sigma(y^{\Delta,v^\epsilon}_{\lfloor t \rfloor_\Delta})v^\epsilon(\lfloor t \rfloor_\Delta) - \sigma(w^{\Delta,v}_{\lfloor t \rfloor_\Delta})v(\lfloor t \rfloor_\Delta), (b(y^{\Delta,v^\epsilon}_{\lfloor t \rfloor_\Delta}) - b(w^{\Delta,v}_{\lfloor t \rfloor_\Delta}))(t - \lfloor t \rfloor_\Delta) \\ &\quad + Z^{\Delta,v^\epsilon}(t) - z^{\Delta,v}(t)\big\rangle + \epsilon|\sigma(y^{\Delta,v^\epsilon}_{\lfloor t \rfloor_\Delta})|^2\Big]\mathrm{d}t \\ &\quad + 2\sqrt{\epsilon}\big\langle b(y^{\Delta,v^\epsilon}_{\lfloor t \rfloor_\Delta}) - b(w^{\Delta,v}_{\lfloor t \rfloor_\Delta}), \sigma(y^{\Delta,v^\epsilon}_{\lfloor t \rfloor_\Delta})(W(t) - W(\lfloor t \rfloor_\Delta))\big\rangle\mathrm{d}t \\ &\quad + 2\sqrt{\epsilon}\langle Z^{\Delta,v^\epsilon}(t) - z^{\Delta,v}(t), \sigma(y^{\Delta,v^\epsilon}_{\lfloor t \rfloor_\Delta})\mathrm{d}W(t)\rangle. \end{aligned}$$

By the Young inequality, we have

$$\begin{aligned} &|Z^{\Delta,v^\epsilon}(s) - z^{\Delta,v}(s)|^2 \\ &= \int_0^s \Big[2\big\langle y^{\Delta,v^\epsilon}(\lfloor t \rfloor_\Delta) - w^{\Delta,v}(\lfloor t \rfloor_\Delta), b(y^{\Delta,v^\epsilon}_{\lfloor t \rfloor_\Delta}) - b(w^{\Delta,v}_{\lfloor t \rfloor_\Delta})\big\rangle \\ &\quad + 2\big(t - \lfloor t \rfloor_\Delta - \theta\Delta\big)\big|b(y^{\Delta,v^\epsilon}_{\lfloor t \rfloor_\Delta}) - b(w^{\Delta,v}_{\lfloor t \rfloor_\Delta})\big|^2\Big]\mathrm{d}t + \int_0^s \Big[2\big\langle\sigma(y^{\Delta,v^\epsilon}_{\lfloor t \rfloor_\Delta})v^\epsilon(\lfloor t \rfloor_\Delta) \\ &\quad - \sigma(w^{\Delta,v}_{\lfloor t \rfloor_\Delta})v(\lfloor t \rfloor_\Delta), \big(b(y^{\Delta,v^\epsilon}_{\lfloor t \rfloor_\Delta}) - b(w^{\Delta,v}_{\lfloor t \rfloor_\Delta})\big)(t - \lfloor t \rfloor_\Delta) + Z^{\Delta,v^\epsilon}(t) - z^{\Delta,v}(t)\big\rangle \\ &\quad + \epsilon|\sigma(y^{\Delta,v^\epsilon}_{\lfloor t \rfloor_\Delta})|^2\Big]\mathrm{d}t + 2\sqrt{\epsilon}\int_0^s \big\langle b(y^{\Delta,v^\epsilon}_{\lfloor t \rfloor_\Delta}) - b(w^{\Delta,v}_{\lfloor t \rfloor_\Delta}), \sigma(y^{\Delta,v^\epsilon}_{\lfloor t \rfloor_\Delta})(W(t) \\ &\quad - W(\lfloor t \rfloor_\Delta))\big\rangle\mathrm{d}t \\ &\quad + 2\sqrt{\epsilon}\int_0^s \langle Z^{\Delta,v^\epsilon}(t) - z^{\Delta,v}(t), \sigma(y^{\Delta,v^\epsilon}_{\lfloor t \rfloor_\Delta})\mathrm{d}W(t)\rangle \end{aligned}$$

$$\leq \int_0^s \Big[2\big\langle y^{\Delta,v^\epsilon}(\lfloor t\rfloor_\Delta) - w^{\Delta,v}(\lfloor t\rfloor_\Delta), b(y^{\Delta,v^\epsilon}_{\lfloor t\rfloor_\Delta}) - b(w^{\Delta,v}_{\lfloor t\rfloor_\Delta})\big\rangle$$

$$+ |\sigma(y^{\Delta,v^\epsilon}_{\lfloor t\rfloor_\Delta}) - \sigma(w^{\Delta,v}_{\lfloor t\rfloor_\Delta})|^2$$

$$+ \epsilon|\sigma(y^{\Delta,v^\epsilon}_{\lfloor t\rfloor_\Delta})|^2\Big]\mathrm{d}t + K\int_0^s |v^\epsilon(\lfloor t\rfloor_\Delta)|^2\Big|\frac{1}{\theta}\big(y^{\Delta,v^\epsilon}(\lfloor t\rfloor_\Delta) - w^{\Delta,v}(\lfloor t\rfloor_\Delta)$$

$$+ Z^{\Delta,v^\epsilon}(\lfloor t\rfloor_\Delta) - z^{\Delta,v}(\lfloor t\rfloor_\Delta)\big) + Z^{\Delta,v^\epsilon}(t) - z^{\Delta,v}(t)\Big|^2\mathrm{d}t$$

$$+ \int_0^s 2\big\langle\sigma(w^{\Delta,v}_{\lfloor t\rfloor_\Delta})\big(v^\epsilon(\lfloor t\rfloor_\Delta) - v(\lfloor t\rfloor_\Delta)\big), \big(b(y^{\Delta,v^\epsilon}_{\lfloor t\rfloor_\Delta}) - b(w^{\Delta,v}_{\lfloor t\rfloor_\Delta})\big)(t - \lfloor t\rfloor_\Delta)$$

$$+ Z^{\Delta,v^\epsilon}(t)$$

$$- z^{\Delta,v}(t)\big\rangle\mathrm{d}t + 2\sqrt{\epsilon}\int_0^s \big\langle b(y^{\Delta,v^\epsilon}_{\lfloor t\rfloor_\Delta}) - b(w^{\Delta,v}_{\lfloor t\rfloor_\Delta}), \sigma(y^{\Delta,v^\epsilon}_{\lfloor t\rfloor_\Delta})(W(t) - W(\lfloor t\rfloor_\Delta))\big\rangle\mathrm{d}t$$

$$+ 2\sqrt{\epsilon}\int_0^s \langle Z^{\Delta,v^\epsilon}(t) - z^{\Delta,v}(t), \sigma(y^{\Delta,v^\epsilon}_{\lfloor t\rfloor_\Delta})\mathrm{d}W(t)\rangle,$$

where we used $\int_{t_k}^{t_{k+1}} 2(t - \lfloor t\rfloor_\Delta - \theta\Delta)\mathrm{d}t = (1-2\theta)\Delta^2 \leq 0$. It follows from Assumptions 2.1 and 2.2 and

$$\int_0^s \int_{-\tau}^0 |y^{\Delta,v^\epsilon}_{\lfloor t\rfloor_\Delta}(r) - w^{\Delta,v}_{\lfloor t\rfloor_\Delta}(r)|^2\mathrm{d}\nu_i(r)\mathrm{d}t$$

$$\leq \int_0^s |y^{\Delta,v^\epsilon}(\lfloor t\rfloor_\Delta) - w^{\Delta,v}(\lfloor t\rfloor_\Delta)|^2\mathrm{d}t \quad \forall i = 1, 2$$

that

$$|Z^{\Delta,v^\epsilon}(s) - z^{\Delta,v}(s)|^2$$

$$\leq \int_0^s \Big[-2(a_1 - a_2 - L)|y^{\Delta,v^\epsilon}(\lfloor t\rfloor_\Delta) - w^{\Delta,v}(\lfloor t\rfloor_\Delta)|^2 + \epsilon|\sigma(y^{\Delta,v^\epsilon}_{\lfloor t\rfloor_\Delta})|^2\Big]\mathrm{d}t$$

$$+ K\int_0^s |v^\epsilon(\lfloor t\rfloor_\Delta)|^2 \sup_{r\in[0,t]}\big(|y^{\Delta,v^\epsilon}(\lfloor r\rfloor_\Delta) - w^{\Delta,v}(\lfloor r\rfloor_\Delta)|^2$$

$$+ |Z^{\Delta,v^\epsilon}(r) - z^{\Delta,v}(r)|^2\big)\mathrm{d}t$$

$$+ E_1(s) + 2\sqrt{\epsilon}\int_0^s \big\langle b(y^{\Delta,v^\epsilon}_{\lfloor t\rfloor_\Delta}) - b(w^{\Delta,v}_{\lfloor t\rfloor_\Delta}), \sigma(y^{\Delta,v^\epsilon}_{\lfloor t\rfloor_\Delta})(W(t) - W(\lfloor t\rfloor_\Delta))\big\rangle\mathrm{d}t$$

$$+ 2\sqrt{\epsilon}\int_0^s \langle Z^{\Delta,v^\epsilon}(t) - z^{\Delta,v}(t), \sigma(y^{\Delta,v^\epsilon}_{\lfloor t\rfloor_\Delta})\mathrm{d}W(t)\rangle,$$

where

$$E_1(s) := \int_0^s 2\big\langle \sigma(w^{\Delta,v}_{\lfloor t\rfloor_\Delta})(v^\epsilon(\lfloor t\rfloor_\Delta) - v(\lfloor t\rfloor_\Delta)), \big(b(y^{\Delta,v^\epsilon}_{\lfloor t\rfloor_\Delta}) - b(w^{\Delta,v}_{\lfloor t\rfloor_\Delta})\big)(t - \lfloor t\rfloor_\Delta)$$
$$+ Z^{\Delta,v^\epsilon}(t) - z^{\Delta,v}(t)\big\rangle \mathrm{d}t.$$

By $a_1 > a_2 + L$, we arrive at

$$|Z^{\Delta,v^\epsilon}(s) - z^{\Delta,v}(s)|^2$$
$$\le \int_0^s K|v^\epsilon(\lfloor t\rfloor_\Delta)|^2 \sup_{r\in[0,t]} \Big(|y^{\Delta,v^\epsilon}(\lfloor r\rfloor_\Delta) - w^{\Delta,v}(\lfloor r\rfloor_\Delta)|^2 + |Z^{\Delta,v^\epsilon}(r)$$
$$- z^{\Delta,v}(r)|^2\Big)\mathrm{d}t + \int_0^s \epsilon|\sigma(y^{\Delta,v^\epsilon}_{\lfloor t\rfloor_\Delta})|^2 \mathrm{d}t + E_1(s)$$
$$+ 2\sqrt{\epsilon}\int_0^s \big\langle b(y^{\Delta,v^\epsilon}_{\lfloor t\rfloor_\Delta}) - b(w^{\Delta,v}_{\lfloor t\rfloor_\Delta}), \sigma(y^{\Delta,v^\epsilon}_{\lfloor t\rfloor_\Delta})(W(t) - W(\lfloor t\rfloor_\Delta))\big\rangle \mathrm{d}t$$
$$+ 2\sqrt{\epsilon}\int_0^s \langle Z^{\Delta,v^\epsilon}(t) - z^{\Delta,v}(t), \sigma(y^{\Delta,v^\epsilon}_{\lfloor t\rfloor_\Delta})\mathrm{d}W(t)\rangle.$$

Denote $E^\epsilon(t) := \int_0^t \sigma(w^{\Delta,v}_{\lfloor r\rfloor_\Delta})(v^\epsilon(\lfloor r\rfloor_\Delta) - v(\lfloor r\rfloor_\Delta))\mathrm{d}r$. By virtue of the Malliavin integration by parts formula, one has

$$E_1(s) = 2\langle E^\epsilon(s), (b(y^{\Delta,v^\epsilon}_{\lfloor s\rfloor_\Delta}) - b(w^{\Delta,v}_{\lfloor s\rfloor_\Delta}))(s - \lfloor s\rfloor_\Delta) + Z^{\Delta,v^\epsilon}(s) - z^{\Delta,v}(s)\rangle$$
$$-2\int_0^s \Big\langle E^\epsilon(t), 2\big(b(y^{\Delta,v^\epsilon}_{\lfloor t\rfloor_\Delta}) - b(w^{\Delta,v}_{\lfloor t\rfloor_\Delta})\big)\mathrm{d}t + \sqrt{\epsilon}\sigma(y^{\Delta,v^\epsilon}_{\lfloor t\rfloor_\Delta})\mathrm{d}W(t)$$
$$+\big(\sigma(y^{\Delta,v^\epsilon}_{\lfloor t\rfloor_\Delta})v^\epsilon(\lfloor t\rfloor_\Delta) - \sigma(w^{\Delta,v}_{\lfloor t\rfloor_\Delta})v(\lfloor t\rfloor_\Delta)\big)\mathrm{d}t\Big\rangle$$
$$\le K|E^\epsilon(s)|^2 + \frac{1}{4}|y^{\Delta,v^\epsilon}(\lfloor s\rfloor_\Delta) - w^{\Delta,v}(\lfloor s\rfloor_\Delta)|^2$$
$$+ \frac{1}{4}|Z^{\Delta,v^\epsilon}(\lfloor s\rfloor_\Delta) - z^{\Delta,v}(\lfloor s\rfloor_\Delta)|^2$$
$$+ \frac{1}{4}|Z^{\Delta,v^\epsilon}(s) - z^{\Delta,v}(s)|^2 - 4\int_0^s \langle E^\epsilon(t), b(y^{\Delta,v^\epsilon}_{\lfloor t\rfloor_\Delta}) - b(w^{\Delta,v}_{\lfloor t\rfloor_\Delta})\rangle \mathrm{d}t$$
$$- 2\int_0^s \Big\langle E^\epsilon(t), \big(\sigma(y^{\Delta,v^\epsilon}_{\lfloor t\rfloor_\Delta}) - \sigma(w^{\Delta,v}_{\lfloor t\rfloor_\Delta})\big)v^\epsilon(\lfloor t\rfloor_\Delta)\Big\rangle \mathrm{d}t$$
$$- 2\int_0^s \Big\langle E^\epsilon(t), \sigma(w^{\Delta,v}_{\lfloor t\rfloor_\Delta})$$
$$\times \big(v^\epsilon(\lfloor t\rfloor_\Delta) - v(\lfloor t\rfloor_\Delta)\big)\Big\rangle \mathrm{d}t - 2\sqrt{\epsilon}\int_0^s \langle E^\epsilon(t), \sigma(y^{\Delta,v^\epsilon}_{\lfloor t\rfloor_\Delta})\mathrm{d}W(t)\rangle$$

$$
\begin{aligned}
&\le K \sup_{r\in[0,s]} |E^\epsilon(r)|^2 + \frac{1}{4}|y^{\Delta,v^\epsilon}(\lfloor s\rfloor_\Delta) - w^{\Delta,v}(\lfloor s\rfloor_\Delta)|^2 \\
&\quad + \frac{1}{2}\sup_{r\in[0,s]} |Z^{\Delta,v^\epsilon}(r) - z^{\Delta,v}(r)|^2 \\
&\quad + K\int_0^s (|v^\epsilon(\lfloor t\rfloor_\Delta)|^2 + 1) \\
&\quad \times \sup_{r\in[0,t]} \Big(|y^{\Delta,v^\epsilon}(\lfloor r\rfloor_\Delta) - w^{\Delta,v}(\lfloor r\rfloor_\Delta)|^2 + |Z^{\Delta,v^\epsilon}(r) \\
&\quad - z^{\Delta,v}(r)|^2\Big)\mathrm{d}t - 2\sqrt{\epsilon}\int_0^s \langle E^\epsilon(t), \sigma(y^{\Delta,v^\epsilon}_{\lfloor t\rfloor_\Delta})\mathrm{d}W(t)\rangle,
\end{aligned}
$$

where we used

$$
\begin{aligned}
&2\int_0^s \langle E^\epsilon(t), \sigma(w^{\Delta,v}_{\lfloor t\rfloor_\Delta})(v^\epsilon(\lfloor t\rfloor_\Delta) - v(\lfloor t\rfloor_\Delta))\rangle \mathrm{d}t \\
&\quad = 2\int_0^s \langle E^\epsilon(t), \mathrm{d}E^\epsilon(t)\rangle = |E^\epsilon(s)|^2.
\end{aligned}
$$

Note that

$$
\begin{aligned}
&|Z^{\Delta,v^\epsilon}(t_k) - z^{\Delta,v}(t_k)|^2 \\
&\quad \ge |y^{\Delta,v^\epsilon}(t_k) - w^{\Delta,v}(t_k)|^2 - 2\theta\Delta\langle y^{\Delta,v^\epsilon}(t_k) - w^{\Delta,v}(t_k), b(y^{\Delta,v^\epsilon}_{t_k}) - b(w^{\Delta,v}_{t_k})\rangle \\
&\quad \ge |y^{\Delta,v^\epsilon}(t_k) - w^{\Delta,v}(t_k)|^2 - 2\theta\Delta\Big(-a_1|y^{\Delta,v^\epsilon}(t_k) - w^{\Delta,v}(t_k)|^2 \\
&\quad + a_2\int_{-\tau}^0 |y^{\Delta,v^\epsilon}_{t_k}(r) - w^{\Delta,v}_{t_k}(r)|^2 \mathrm{d}\nu_2(r)\Big),
\end{aligned}
$$

which implies

$$
\begin{aligned}
&|Z^{\Delta,v^\epsilon}(t_k) - z^{\Delta,v}(t_k)|^2 \\
&\quad \ge |y^{\Delta,v^\epsilon}(t_k) - w^{\Delta,v}(t_k)|^2 - 2a_2\theta\Delta \sup_{r\in[0,t_k]} |y^{\Delta,v^\epsilon}(\lfloor r\rfloor_\Delta) - w^{\Delta,v}(\lfloor r\rfloor_\Delta)|^2.
\end{aligned}
$$

Then,

$$
\begin{aligned}
&|y^{\Delta,v^\epsilon}(t_k) - w^{\Delta,v}(t_k)|^2 \\
&\quad \le 2a_2\theta\Delta \sup_{t\in[0,t_k]} |y^{\Delta,v^\epsilon}(\lfloor t\rfloor_\Delta) - w^{\Delta,v}(\lfloor t\rfloor_\Delta)|^2
\end{aligned}
$$

$$
\begin{aligned}
&+K\int_0^{t_k}(|v^\epsilon(\lfloor t\rfloor_\Delta)|^2+1)\sup_{r\in[0,t]}\Big(|y^{\Delta,v^\epsilon}(\lfloor r\rfloor_\Delta)-w^{\Delta,v}(\lfloor r\rfloor_\Delta)|^2+|Z^{\Delta,v^\epsilon}(r)\\
&-z^{\Delta,v}(r)|^2\Big)\mathrm{d}t+\int_0^{t_k}\epsilon|\sigma(y^{\Delta,v^\epsilon}_{\lfloor t\rfloor_\Delta})|^2\mathrm{d}t\\
&+2\sqrt{\epsilon}\int_0^{t_k}\Big\langle b(y^{\Delta,v^\epsilon}_{\lfloor t\rfloor_\Delta})-b(w^{\Delta,v}_{\lfloor t\rfloor_\Delta}),\sigma(y^{\Delta,v^\epsilon}_{\lfloor t\rfloor_\Delta})(W(t)-W(\lfloor t\rfloor_\Delta))\Big\rangle\mathrm{d}t\\
&+2\sqrt{\epsilon}\int_0^{t_k}\langle Z^{\Delta,v^\epsilon}(t)-z^{\Delta,v}(t),\sigma(y^{\Delta,v^\epsilon}_{\lfloor t\rfloor_\Delta})\mathrm{d}W(t)\rangle\\
&+K\sup_{r\in[0,t_k]}|E^\epsilon(r)|^2+\frac{1}{4}|y^{\Delta,v^\epsilon}(t_k)-w^{\Delta,v}(t_k)|^2\\
&+\frac{1}{2}\sup_{r\in[0,t_k]}|Z^{\Delta,v^\epsilon}(r)-z^{\Delta,v}(r)|^2\\
&+\sup_{r\in[0,t_k]}2\sqrt{\epsilon}\int_0^r\langle -E^\epsilon(t),\sigma(y^{\Delta,v^\epsilon}_{\lfloor t\rfloor_\Delta})\mathrm{d}W(t)\rangle.
\end{aligned}
$$

Under the condition $\Delta\in\big(0,\frac{1}{4a_2\theta}\big]$, and noting that the above inequality remains valid when the left-hand side is evaluated at any t_l, $l=0,1,\ldots,k$, we have

$$
\begin{aligned}
&\frac{1}{2}\sup_{t\in[0,t_k]}|y^{\Delta,v^\epsilon}(\lfloor t\rfloor_\Delta)-w^{\Delta,v}(\lfloor t\rfloor_\Delta)|^2\\
&\quad\le K\int_0^{t_k}(|v^\epsilon(\lfloor t\rfloor_\Delta)|^2+1)\sup_{r\in[0,t]}\Big(|y^{\Delta,v^\epsilon}(\lfloor r\rfloor_\Delta)-w^{\Delta,v}(\lfloor r\rfloor_\Delta)|^2+|Z^{\Delta,v^\epsilon}(r)\\
&\qquad-z^{\Delta,v}(r)|^2\Big)\mathrm{d}t+\int_0^{t_k}\epsilon|\sigma(y^{\Delta,v^\epsilon}_{\lfloor t\rfloor_\Delta})|^2\mathrm{d}t\\
&\qquad+2\sqrt{\epsilon}\int_0^{t_k}\Big\langle b(y^{\Delta,v^\epsilon}_{\lfloor t\rfloor_\Delta})-b(w^{\Delta,v}_{\lfloor t\rfloor_\Delta}),\sigma(y^{\Delta,v^\epsilon}_{\lfloor t\rfloor_\Delta})(W(t)-W(\lfloor t\rfloor_\Delta))\Big\rangle\mathrm{d}t\\
&\qquad+2\sqrt{\epsilon}\int_0^{t_k}\langle Z^{\Delta,v^\epsilon}(t)-z^{\Delta,v}(t),\sigma(y^{\Delta,v^\epsilon}_{\lfloor t\rfloor_\Delta})\mathrm{d}W(t)\rangle\\
&\qquad+K\sup_{r\in[0,t_k]}|E^\epsilon(r)|^2+\frac{1}{4}|y^{\Delta,v^\epsilon}(t_k)-w^{\Delta,v}(t_k)|^2\\
&\qquad+\frac{1}{2}\sup_{r\in[0,t_k]}|Z^{\Delta,v^\epsilon}(r)-z^{\Delta,v}(r)|^2\\
&\qquad-2\sqrt{\epsilon}\int_0^{t_k}\langle E^\epsilon(t),\sigma(y^{\Delta,v^\epsilon}_{\lfloor t\rfloor_\Delta})\mathrm{d}W(t)\rangle.
\end{aligned}
$$

Combining the estimate

$$\begin{aligned}
&\frac{1}{2}\sup_{t\in[0,t_k]}|Z^{\Delta,v^\epsilon}(t)-z^{\Delta,v}(t)|^2\\
&\le\int_0^{t_k}K(|v^\epsilon(\lfloor t\rfloor_\Delta)|^2+1)\sup_{r\in[0,t]}\Big(|y^{\Delta,v^\epsilon}(\lfloor r\rfloor_\Delta)-w^{\Delta,v}(\lfloor r\rfloor_\Delta)|^2+|Z^{\Delta,v^\epsilon}(r)\\
&\quad-z^{\Delta,v}(r)|^2\Big)\mathrm{d}t+\int_0^{t_k}\epsilon|\sigma(y^{\Delta,v^\epsilon}_{\lfloor t\rfloor_\Delta})|^2\mathrm{d}t\\
&\quad+2\sqrt{\epsilon}\sup_{r\in[0,t_k]}\Big(\int_0^r\big\langle b(y^{\Delta,v^\epsilon}_{\lfloor t\rfloor_\Delta})-b(w^{\Delta,v}_{\lfloor t\rfloor_\Delta}),\sigma(y^{\Delta,v^\epsilon}_{\lfloor t\rfloor_\Delta})(W(t)-W(\lfloor t\rfloor_\Delta))\big\rangle\mathrm{d}t\\
&\quad+\int_0^r\langle Z^{\Delta,v^\epsilon}(t)-z^{\Delta,v}(t),\sigma(y^{\Delta,v^\epsilon}_{\lfloor t\rfloor_\Delta})\mathrm{d}W(t)\rangle\Big)\\
&\quad+K\sup_{r\in[0,t_k]}|E^\epsilon(r)|^2+\frac{1}{4}\sup_{s\in[0,t_k]}|y^{\Delta,v^\epsilon}(\lfloor s\rfloor_\Delta)-w^{\Delta,v}(\lfloor s\rfloor_\Delta)|^2\\
&\quad+2\sqrt{\epsilon}\sup_{r\in[0,t_k]}\int_0^r\langle -E^\epsilon(t),\sigma(y^{\Delta,v^\epsilon}_{\lfloor t\rfloor_\Delta})\mathrm{d}W(t)\rangle,
\end{aligned}$$

we therefore arrive at

$$\begin{aligned}
&\frac{1}{2}\sup_{t\in[0,t_k]}|Z^{\Delta,v^\epsilon}(t)-z^{\Delta,v}(t)|^2+\frac{1}{4}\sup_{t\in\{0,\ldots,t_k\}}|y^{\Delta,v^\epsilon}(t)-w^{\Delta,v}(t)|^2\\
&\le K\int_0^{t_k}(|v^\epsilon(\lfloor t\rfloor_\Delta)|^2+1)\Big(\sup_{r\in[0,t]}\big(|y^{\Delta,v^\epsilon}(\lfloor r\rfloor_\Delta)-w^{\Delta,v}(\lfloor r\rfloor_\Delta)|^2+|Z^{\Delta,v^\epsilon}(r)\\
&\quad-z^{\Delta,v}(r)|^2\big)\Big)\mathrm{d}t+2\epsilon\int_0^{t_k}|\sigma(y^{\Delta,v^\epsilon}_{\lfloor t\rfloor_\Delta})|^2\mathrm{d}t\\
&\quad+4\sqrt{\epsilon}\sup_{r\in[0,t_k]}\Big(\int_0^r\langle b(y^{\Delta,v^\epsilon}_{\lfloor t\rfloor_\Delta})-b(w^{\Delta,v}_{\lfloor t\rfloor_\Delta}),\sigma(y^{\Delta,v^\epsilon}_{\lfloor t\rfloor_\Delta})(W(t)-W(\lfloor t\rfloor_\Delta))\rangle\mathrm{d}t\\
&\quad+\int_0^r\langle Z^{\Delta,v^\epsilon}(t)-z^{\Delta,v}(t),\sigma(y^{\Delta,v^\epsilon}_{\lfloor t\rfloor_\Delta})\mathrm{d}W(t)\rangle\Big)\\
&\quad+K\sup_{t\in[0,t_k]}|E^\epsilon(t)|^2+4\sqrt{\epsilon}\sup_{r\in[0,t_k]}\int_0^r\langle -E^\epsilon(t),\sigma(y^{\Delta,v^\epsilon}_{\lfloor t\rfloor_\Delta})\mathrm{d}W(t)\rangle.
\end{aligned}$$

Applying the Gronwall inequality and $v^\epsilon(\cdot)\in\mathcal{A}_M$, we arrive at

$$\sup_{t\in[0,t_k]}|Z^{\Delta,v^\epsilon}(t)-z^{\Delta,v}(t)|^2+\sup_{t\in[0,t_k]}|y^{\Delta,v^\epsilon}(\lfloor t\rfloor_\Delta)-w^{\Delta,v}(\lfloor t\rfloor_\Delta)|^2$$

$$
\begin{aligned}
&\leq K e^{K(M)t_k}\Big[\epsilon \int_0^{t_k} |\sigma(y^{\Delta,v^\epsilon}_{\lfloor t\rfloor_\Delta})|^2 \mathrm{d}t \\
&+ \sqrt{\epsilon} \sup_{r\in[0,t_k]} \Big(\int_0^r \big\langle b(y^{\Delta,v^\epsilon}_{\lfloor t\rfloor_\Delta}) - b(w^{\Delta,v}_{\lfloor t\rfloor_\Delta}), \sigma(y^{\Delta,v^\epsilon}_{\lfloor t\rfloor_\Delta})(W(t) - W(\lfloor t\rfloor_\Delta))\big\rangle \mathrm{d}t \\
&+ \int_0^r \big\langle Z^{\Delta,v^\epsilon}(t) - z^{\Delta,v}(t), \sigma(y^{\Delta,v^\epsilon}_{\lfloor t\rfloor_\Delta})\mathrm{d}W(t)\big\rangle\Big) \\
&+ \sup_{t\in[0,t_k]} |E^\epsilon(t)|^2 + \sqrt{\epsilon} \sup_{r\in[0,t_k]} \int_0^r \langle -E^\epsilon(t), \sigma(y^{\Delta,v^\epsilon}_{\lfloor t\rfloor_\Delta})\mathrm{d}W(t)\rangle\Big].
\end{aligned}
$$

Utilizing Lemmas 6.3 and 6.5, the Burkholder–Davis–Gundy inequality, and the Young inequality, we obtain

$$
\begin{aligned}
&\mathbb{E}\Big[(e^{\mathfrak{a}t_k}\Delta^{-1}) \wedge \sup_{t\in[0,t_k]} |y^{\Delta,v^\epsilon}(\lfloor t\rfloor_\Delta) - w^{\Delta,v}(\lfloor t\rfloor_\Delta)|^2\Big] \\
&\leq K e^{K(M)t_k}\Big\{\epsilon \int_0^{t_k} \mathbb{E}[1 + \|y^{\Delta,v^\epsilon}_{\lfloor t\rfloor_\Delta}\|^2]\mathrm{d}t + \sqrt{\epsilon}\int_0^{t_k} \mathbb{E}\Big[|b(y^{\Delta,v^\epsilon}_{\lfloor t\rfloor_\Delta}) - b(w^{\Delta,v}_{\lfloor t\rfloor_\Delta})| \\
&\times |\sigma(y^{\Delta,v^\epsilon}_{\lfloor t\rfloor_\Delta})||W(t) - W(\lfloor t\rfloor_\Delta)|\Big]\mathrm{d}t \\
&+ \sqrt{\epsilon}\mathbb{E}\Big[\Big(\int_0^{t_k} |Z^{\Delta,v^\epsilon}(t) - z^{\Delta,v}(t)|^2 |\sigma(y^{\Delta,v^\epsilon}_{\lfloor t\rfloor_\Delta})|^2 \mathrm{d}t\Big)^{\frac{1}{2}}\Big] \\
&+ \sup_{t\in[0,t_k]} \mathbb{E}\big[|E^\epsilon(t)|^2 \wedge (e^{\mathfrak{a}t_k}\Delta^{-1})\big] + \sqrt{\epsilon}\mathbb{E}\Big[\Big(\int_0^{t_k} |E^\epsilon(t)|^2 |\sigma(y^{\Delta,v^\epsilon}_{\lfloor t\rfloor_\Delta})|^2 \mathrm{d}t\Big)^{\frac{1}{2}}\Big]\Big\} \\
&\leq K(M)e^{K(M)t_k}\Big[(1 + \|\xi\|^{2(\beta+1)})(\epsilon + \sqrt{\epsilon}) + \mathbb{E}\big[\sup_{t\in[0,t_k]} |E^\epsilon(t)|^2 \wedge (e^{\mathfrak{a}t_k}\Delta^{-1})\big]\Big].
\end{aligned}
$$

It follows from $v^\epsilon \xrightarrow[\epsilon\to 0]{d} v$ that

$$
\lim_{\epsilon\to 0} \mathbb{E}\Big[\sup_{t\in[0,t_k]} |E^\epsilon(t)|^2 \wedge (e^{\mathfrak{a}t_k}\Delta^{-1})\Big] = 0.
$$

Consequently, letting $\epsilon \to 0$ finishes the proof. □

6.2 Logarithmic Estimate for Numerical Density Function

This section is devoted to the logarithmic estimate on the finite time horizon for the density function of the θ-EM method (6.2), based on the smoothness of the numerical density function and the technique of the Malliavin calculus. The constant $T > 0$ is fixed in this section. For the numerical solution $y^{\Delta,\epsilon}(\cdot)$, under Assumptions 2.1, 5.1–5.3, it follows from Theorem 5.2 that for any $\Delta \in (0, \frac{1}{4\theta a_5})$ and $k \in \mathbb{N}_+$, $y^{\Delta,\epsilon}(t_k)$ admits a density function denoted by $\mathfrak{p}^{\Delta,\epsilon}(t_k, \cdot)$. We introduce the following assumption to ensure the smoothness of $\mathfrak{p}^{\Delta,\epsilon}(t_k, \cdot)$.

Assumption 6.1 *Assume that coefficients b and σ are smooth with bounded derivatives of arbitrary orders.*

The proof of the smoothness of the density $\mathfrak{p}^{\Delta,\epsilon}(t_k, \cdot)$ depends on the estimate of the Malliavin covariance matrix γ_k^ϵ of the numerical solution $y^{\Delta,\epsilon}(t_k)$ and the result that $y^{\Delta,\epsilon}(t_k) \in \mathbb{D}^\infty(\mathbb{R}^d)$.

Lemma 6.7 *Let Assumptions 2.1, 5.1, 5.3, and 6.1 hold. Then for any $p \geq 1$ and $\Delta \in (0, \frac{1}{4\theta(a_5 \vee (2\|\mathcal{D}b\|_{\mathcal{L}(\mathbb{C}^d;\mathbb{R}^d)}))})$ with a_5 given in Assumption 5.1,*

$$\| \det(\gamma_{N\Delta}^\epsilon)^{-1} \|_{L^p(\Omega)} \leq K\epsilon^{-d}\Delta^{-d}.$$

Proof Since $\det(\gamma_{N\Delta}^\epsilon)^{-1} = \prod_{i=1}^d \lambda_i^{-1} \leq (\lambda_{\min}^{-1})^d$, where $\lambda_i, i = 1, \ldots, d$ are the eigenvalues of $\gamma_{N\Delta}^\epsilon$ and $\lambda_{\min} := \min\{\lambda_i, i = 1, \ldots, d\}$, it suffices to estimate the smallest eigenvalue of $\gamma_{N\Delta}^\epsilon$. For the Malliavin covariance matrix γ_k^ϵ of the numerical solution $y^{\Delta,\epsilon}(t_k)$, similar to the recursive relation (5.17), for any $\Delta \in (0, \frac{1}{4\theta a_5})$, using the equality

$$\begin{aligned} A_{1,N\Delta-1}^\epsilon &:= \left(\mathrm{Id}_{d\times d} - \theta \mathcal{D}b(y_{t_{N\Delta}}^{\Delta,\epsilon})(I^{[0]}\mathrm{Id}_{d\times d})\Delta\right)^{-1} \\ &= \mathrm{Id}_{d\times d} + A_{1,N\Delta-1}^\epsilon \theta \mathcal{D}b(y_{t_{N\Delta}}^{\Delta,\epsilon})(I^{[0]}\mathrm{Id}_{d\times d})\Delta, \end{aligned} \tag{6.20}$$

we derive

$$\begin{aligned} \lambda_{\min}(\gamma_{N\Delta}^\epsilon) &= \min_{u\in\mathbb{R}^d, |u|=1} u^\top \gamma_{N\Delta}^\epsilon u \\ &\geq \epsilon\Delta \min_{u\in\mathbb{R}^d, |u|=1} u^\top A_{1,N\Delta-1}^\epsilon \sigma(y_{t_{N\Delta-1}}^{\Delta,\epsilon})\sigma(y_{t_{N\Delta-1}}^{\Delta,\epsilon})^\top (A_{1,N\Delta-1}^\epsilon)^\top u \\ &= \epsilon\Delta \min_{u\in\mathbb{R}^d, |u|=1} u^\top \Big[\sigma(y_{t_{N\Delta-1}}^{\Delta,\epsilon})\sigma(y_{t_{N\Delta-1}}^{\Delta,\epsilon})^\top \\ &\quad + A_{1,N\Delta-1}^\epsilon \theta \mathcal{D}b(y_{t_{N\Delta}}^{\Delta,\epsilon})(I^{[0]}\mathrm{Id}_{d\times d})\sigma(y_{t_{N\Delta-1}}^{\Delta,\epsilon}) \\ &\quad \times \sigma(y_{t_{N\Delta-1}}^{\Delta,\epsilon})^\top \Delta + \sigma(y_{t_{N\Delta-1}}^{\Delta,\epsilon})\sigma(y_{t_{N\Delta-1}}^{\Delta,\epsilon})^\top \\ &\quad \times (A_{1,N\Delta-1}^\epsilon \theta \mathcal{D}b(y_{t_{N\Delta}}^{\Delta,\epsilon})(I^{[0]}\mathrm{Id}_{d\times d})\Delta)^\top \\ &\quad + A_{1,N\Delta-1}^\epsilon \theta \mathcal{D}b(y_{t_{N\Delta}}^{\Delta,\epsilon})(I^{[0]}\mathrm{Id}_{d\times d})\Delta\sigma(y_{t_{N\Delta-1}}^{\Delta,\epsilon})\sigma(y_{t_{N\Delta-1}}^{\Delta,\epsilon})^\top \\ &\quad \times (A_{1,N\Delta-1}^\epsilon \theta \mathcal{D}b(y_{t_{N\Delta}}^{\Delta,\epsilon})(I^{[0]}\mathrm{Id}_{d\times d})\Delta)^\top \Big] u. \end{aligned} \tag{6.21}$$

According to Assumption 5.3, the fact $\|A^{\epsilon}_{1,N\Delta-1}\|_{\mathcal{L}(\mathbb{R}^d;\mathbb{R}^d)} \leq 2$ showed in the proof of Theorem 5.2, and the inequality

$$2\left|A^{\epsilon}_{1,N\Delta-1}\theta\mathcal{D}b(y^{\Delta,\epsilon}_{t_{N\Delta}})(I^{[0]}\mathrm{Id}_{d\times d})\right|\Delta \leq \frac{1}{2}$$

for $\Delta \in (0, \frac{1}{4\theta(a_5\vee\|\mathcal{D}b\|_{\mathcal{L}(C^d;\mathbb{R}^d)})})$, we obtain

$$\begin{aligned}
&\lambda_{\min}(\gamma^{\epsilon}_{N\Delta})\\
&\geq \frac{1}{2}\epsilon\Delta \min_{u\in\mathbb{R}^d,|u|=1} u^{\top}\sigma(y^{\Delta,\epsilon}_{t_{N\Delta-1}})\sigma(y^{\Delta,\epsilon}_{t_{N\Delta-1}})^{\top}u\\
&\quad+\epsilon\Delta \min_{u\in\mathbb{R}^d,|u|=1} u^{\top}\Big[\Big(\frac{1}{4}\mathrm{Id}_{d\times d}+A^{\epsilon}_{1,N\Delta-1}\theta\mathcal{D}b(y^{\Delta,\epsilon}_{t_{N\Delta}})(I^{[0]}\mathrm{Id}_{d\times d})\Delta\Big)\\
&\quad\times\sigma(y^{\Delta,\epsilon}_{t_{N\Delta-1}})\sigma(y^{\Delta,\epsilon}_{t_{N\Delta-1}})^{\top}\\
&\quad+\sigma(y^{\Delta,\epsilon}_{t_{N\Delta-1}})\sigma(y^{\Delta,\epsilon}_{t_{N\Delta-1}})^{\top}\Big(\frac{1}{4}\mathrm{Id}_{d\times d}+A^{\epsilon}_{1,N\Delta-1}\theta\mathcal{D}b(y^{\Delta,\epsilon}_{t_{N\Delta}})(I^{[0]}\mathrm{Id}_{d\times d})\Delta\Big)^{\top}\Big]u\\
&\geq \frac{1}{2}\epsilon\Delta \min_{u\in\mathbb{R}^d,|u|=1} u^{\top}\sigma(y^{\Delta,\epsilon}_{t_{N\Delta-1}})\sigma(y^{\Delta,\epsilon}_{t_{N\Delta-1}})^{\top}u+\epsilon\Delta\sigma_0\min_{u\in\mathbb{R}^d,|u|=1}u^{\top}\Big(\frac{1}{2}\mathrm{Id}_{d\times d}\\
&\quad+A^{\epsilon}_{1,N\Delta-1}\theta\mathcal{D}b(y^{\Delta,\epsilon}_{t_{N\Delta}})(I^{[0]}\mathrm{Id}_{d\times d})\Delta\\
&\quad+\Big(A^{\epsilon}_{1,N\Delta-1}\theta\mathcal{D}b(y^{\Delta,\epsilon}_{t_{N\Delta}})(I^{[0]}\mathrm{Id}_{d\times d})\Big)^{\top}\Delta\Big)u\\
&\geq \frac{1}{2}\epsilon\Delta\inf_{\phi\in C^d}\min_{u\in\mathbb{R}^d,|u|=1}u^{\top}\sigma(\phi)\sigma(\phi)^{\top}u\\
&\quad+\epsilon\Delta\sigma_0\min_{u\in\mathbb{R}^d,|u|=1}\Big(\frac{1}{2}|u|^2-2|u|^2\left|A^{\epsilon}_{1,N\Delta-1}\theta\mathcal{D}b(y^{\Delta,\epsilon}_{t_{N\Delta}})(I^{[0]}\mathrm{Id}_{d\times d})\right|\Delta\Big)\\
&\geq \frac{1}{2}\epsilon\Delta\sigma_0,
\end{aligned}\tag{6.22}$$

where in the first step we dropped the last term in (6.21) and σ_0 is given in Assumption 5.3. This gives that $\det(\gamma^{\epsilon}_{N\Delta})^{-1} \leq (\frac{1}{2}\sigma_0\epsilon\Delta)^{-d}$. □

Similar to proofs of Lemmas 5.1 and 6.3 with $\iota = 0$, we obtain the following regularity estimates of $y^{\Delta,\epsilon}(\cdot)$ and $w^{\Delta,v}(\cdot)$.

Proposition 6.8 *Under the conditions in Lemma 6.7, for any $T > 0$, $p \in \mathbb{N}_+$, and $\Delta \in (0, \frac{1}{4\theta a_5})$,*

$$\sup_{0\leq t_k\leq T}|w^{\Delta,\epsilon}(t_k)|^{2p} \leq K_T, \qquad \mathbb{E}\Big[\sup_{0\leq t_k\leq T}|y^{\Delta,\epsilon}(t_k)|^{2p}\Big] \leq K_T.$$

Moreover, for $n \in \mathbb{N}_+$*,*

$$\sup_{r_1,\dots,r_n\in[0,T]} \mathbb{E}\Big[\sup_{r_1\vee\cdots\vee r_n\le t_k\le T} |D_{r_1,\dots,r_n} y^{\Delta,\epsilon}(t_k)|^p\Big] \le K_T.$$

Therefore, $y^{\Delta,\epsilon}(t_k) \in \mathbb{D}^\infty(\mathbb{R}^d)$ *for* $k = 1, 2, \dots, N^\Delta$.

Proposition 6.9 *Under the conditions in Lemma 6.7, for any* $k \in \mathbb{N}_+$ *and* $\Delta \in (0, \frac{1}{4\theta(a_5\vee(2\|\mathcal{D}b\|_{\mathcal{L}(\mathcal{C}^d;\mathbb{R}^d)}))})$, $y^{\Delta,\epsilon}(t_k)$ *admits a smooth density* $\mathfrak{p}^{\Delta,\epsilon}(t_k,\cdot)$.

Proof The proof follows from Lemma 6.7, Proposition 6.8, and Theorem C.9. □

The limit of the logarithmic estimate on the finite time horizon of the numerical density function is closely related to the rate function of the LDP of $\{y^{\Delta,\epsilon}(t)\}_{\epsilon\in(0,1)}$ for $t \in [0, T]$. To proceed, we list the LDPs of $\{y^{\Delta,\epsilon}(\cdot)\}_{\epsilon\in(0,1)}$ and $\{y^{\Delta,\epsilon}(t)\}_{\epsilon\in(0,1)}$ in the following proposition, whose proofs are similar to those of Theorem 6.1 and Corollary 6.2 and thus are omitted. Denote by $C_{\xi^{Int}}([-\tau, T]; \mathbb{R}^d)$ the space of all continuous functions $u : [-\tau, T] \to \mathbb{R}^d$ with $u(r) = \xi^{Int}(r)$ for $r \in [-\tau, 0]$, endowed with the norm $\|u\|_{C_{\xi^{Int}}([-\tau,T];\mathbb{R}^d)} := \sup_{t\in[-\tau,T]} |u(t)|$. Let

$$L^2_\Delta([0, T]; \mathbb{R}^m) := \Big\{v : [0, T] \to \mathbb{R}^m \text{ is measurable} \;\Big|\; \int_0^T |v(\lfloor s\rfloor_\Delta)|^2 \mathrm{d}s < \infty\Big\}.$$

Proposition 6.10 *Under the conditions in Lemma 6.7, it holds that:*

(i) The family of random variables $\{y^{\Delta,\epsilon}(\cdot)\}_{\epsilon\in(0,1)}$ *in* $C_{\xi^{Int}}([-\tau, T]; \mathbb{R}^d)$ *satisfies the LDP with the good rate function given by*

$$I_T(f) = \inf_{\{v\in L^2_\Delta([0,T];\mathbb{R}^m):\, f=\mathcal{G}^\Delta(\int_0^\cdot v(\lfloor s\rfloor_\Delta)\mathrm{d}s)\}} \frac{1}{2}\int_0^T |v(\lfloor s\rfloor_\Delta)|^2 \mathrm{d}s.$$

(ii) For any $t \in [0, T]$*, the family of random variables* $\{y^{\Delta,\epsilon}(t)\}_{\epsilon\in(0,1)}$ *satisfies the LDP in* $\mathbb{R}^d$ *with the good rate function given by*

$$\tilde{I}_t(z) = \inf\big\{I_T(f) : f \in C_{\xi^{Int}}([-\tau, T]; \mathbb{R}^d),\, z = f(t)\big\}. \tag{6.23}$$

Now we state our main result as follows.

Theorem 6.11 *Under the conditions in Lemma 6.7, for any* $k \in \mathbb{N}_+$ *and sufficiently small* $\Delta > 0$,

$$\lim_{\epsilon\to 0} \epsilon \log \mathfrak{p}^{\Delta,\epsilon}(t_k, y) = -\tilde{I}_{t_k}(y), \quad y \in \mathbb{R}^d.$$

We prove Theorem 6.11 by introducing the following equation

$$\begin{cases} y^{\Delta,\epsilon,v}(t_{k+1}) = y^{\Delta,\epsilon,v}(t_k) + (1-\theta)b(y_{t_k}^{\Delta,\epsilon,v})\Delta + \theta b(y_{t_{k+1}}^{\Delta,\epsilon,v})\Delta + \sqrt{\epsilon}\sigma(y_{t_k}^{\Delta,\epsilon,v})\delta W_k \\ \qquad + \sigma(y_{t_k}^{\Delta,\epsilon,v})v(t_k)\Delta, \quad k \in \mathbb{N}, \\ y^{\Delta,\epsilon,v}(t_k) = \xi(t_k), \quad k = -N, \ldots, 0, \end{cases}$$

where $y_{t_k}^{\Delta,\epsilon,v}$ is the linear interpolation with respect to $y^{\Delta,\epsilon}(t_{k-N}), \ldots, y^{\Delta,\epsilon}(t_k)$, and $v \in L^2_\Delta([0,T];\mathbb{R}^m)$. Define $\mathbb{Z}^{\Delta,v}$ as follows:

$$\begin{cases} \mathbb{Z}^{\Delta,v}(t_{k+1}) = \mathbb{Z}^{\Delta,v}(t_k) + (1-\theta)\mathcal{D}b(w_{t_k}^{\Delta,v})\mathbb{Z}_{t_k}^{\Delta,v}\Delta + \theta\mathcal{D}b(w_{t_{k+1}}^{\Delta,v})\mathbb{Z}_{t_{k+1}}^{\Delta,v}\Delta \\ \qquad + \mathcal{D}\sigma(w_{t_k}^{\Delta,v})\mathbb{Z}_{t_k}^{\Delta,v}v(t_k)\Delta + \sigma(w_{t_k}^{\Delta,v})\delta W_k, \quad k \in \mathbb{N}, \\ \mathbb{Z}^{\Delta,v}(t_k) = 0, \quad k = -N, \ldots, 0, \end{cases}$$

where $w^{\Delta,v}$ is the solution of (6.4), and $\mathbb{Z}_{t_k}^{\Delta,v}$ is the linear interpolation with respect to $\mathbb{Z}^{\Delta,v}(t_{k-N}), \ldots, \mathbb{Z}^{\Delta,v}(t_k)$.

Introduce the auxiliary processes associated to $y^{\Delta,\epsilon,v}$ and $\mathbb{Z}^{\Delta,v}$, respectively as

$$\begin{cases} Z^{\Delta,\epsilon,v}(t_k) = \xi(t_k), \quad k = -N, \ldots, -1, \\ Z^{\Delta,\epsilon,v}(t_k) = y^{\Delta,\epsilon,v}(t_k) - \theta b(y_{t_k}^{\Delta,\epsilon,v})\Delta, \quad k \geq 0, \end{cases}$$

and

$$\begin{cases} \widetilde{\mathbb{Z}}^{\Delta,v}(t_k) = 0, \quad k = -N, \ldots, -1, \\ \widetilde{\mathbb{Z}}^{\Delta,v}(t_k) = \mathbb{Z}^{\Delta,v}(t_k) - \theta\mathcal{D}b(w_{t_k}^{\Delta,v})\mathbb{Z}_{t_k}^{\Delta,v}\Delta, \quad k \geq 0. \end{cases}$$

For $k \in \mathbb{N}$, they satisfy that

$$Z^{\Delta,\epsilon,v}(t_{k+1}) = Z^{\Delta,\epsilon,v}(t_k) + b(y_{t_k}^{\Delta,\epsilon,v})\Delta + \sqrt{\epsilon}\sigma(y_{t_k}^{\Delta,\epsilon,v})\delta W_k + \sigma(y_{t_k}^{\Delta,\epsilon,v})v(t_k)\Delta,$$

and

$$\begin{aligned} \widetilde{\mathbb{Z}}^{\Delta,v}(t_{k+1}) &= \widetilde{\mathbb{Z}}^{\Delta,v}(t_k) + \mathcal{D}b(w_{t_k}^{\Delta,v})\mathbb{Z}_{t_k}^{\Delta,v}\Delta \\ &\quad + \mathcal{D}\sigma(w_{t_k}^{\Delta,v})\mathbb{Z}_{t_k}^{\Delta,v}v(t_k)\Delta + \sigma(w_{t_k}^{\Delta,v})\delta W_k, \end{aligned}$$

respectively. Moreover, define the continuous versions $Z^{\Delta,\epsilon,v}(\cdot)$ and $\widetilde{\mathbb{Z}}^{\Delta,v}(\cdot)$ by

$$\begin{aligned} Z^{\Delta,\epsilon,v}(t) := Z^{\Delta,\epsilon,v}(t_k) + b(y_{t_k}^{\Delta,\epsilon,v})(t-t_k) + \sqrt{\epsilon}\sigma(y_{t_k}^{\Delta,\epsilon,v})(W(t) - W(t_k)) \\ + \sigma(y_{t_k}^{\Delta,\epsilon,v})v(t_k)(t-t_k), \end{aligned}$$

$$\widetilde{\mathbb{Z}}^{\Delta,v}(t) := \widetilde{\mathbb{Z}}^{\Delta,v}(t_k) + \mathcal{D}b(w_{t_k}^{\Delta,v})\mathbb{Z}_{t_k}^{\Delta,v}(t-t_k) + \mathcal{D}\sigma(w_{t_k}^{\Delta,v})\mathbb{Z}_{t_k}^{\Delta,v}v(t_k)(t-t_k)$$
$$+ \sigma(w_{t_k}^{\Delta,v})(W(t) - W(t_k))$$

for $t \in [t_k, t_{k+1})$ with $k \in \mathbb{N}$, and as usual, the initial data are defined by the linear interpolation.

Similar to the proof of Lemma 5.1, we obtain the following regularity estimates of $y^{\Delta,\epsilon,v}(\cdot)$.

Proposition 6.12 *Let Assumptions 2.1, 5.1, 5.3, and 6.1 hold. Then for any* $T > 0$, $p \in \mathbb{N}_+$, *and* $\Delta \in (0, \frac{1}{4\theta a_5})$,

$$\mathbb{E}\Big[\sup_{0\le t_k\le T} |y^{\Delta,\epsilon,v}(t_k)|^{2p}\Big] \le K_T.$$

Proof of Theorem 6.11 We first prove

$$\limsup_{\epsilon\to 0} \epsilon \log \mathfrak{p}^{\Delta,\epsilon}(t_k, y) \le -\tilde{I}_{t_k}(y). \tag{6.24}$$

By Lemma 6.7 and the boundedness of the pth moment of $y^{\Delta,\epsilon}(\cdot)$, we obtain

$$\|(\gamma_k^{\epsilon})^{-1}\|_{L^p(\Omega)} \le K\|\det(\gamma_k^{\epsilon})^{-1}\|_{L^{2p}(\Omega)} \le K\epsilon^{-d}\Delta^{-d}.$$

This, combining (6.23), Proposition 6.8, and [13, Proposition 4.1] finishes the proof of (6.24).

Next, we show that $\liminf_{\epsilon\to 0}\epsilon\log \mathfrak{p}^{\Delta,\epsilon}(t_k, y) \ge -\tilde{I}_{t_k}(y)$, whose proof is divided into the following five steps.

Step 1. Prove that for any $p \ge 1$ and $\Delta \in (0, \frac{1}{8\theta a_5}] \wedge \Big(0, \frac{1}{4\theta\|\mathcal{D}b\|_{\mathcal{L}(\mathbb{C}^d;\mathbb{R}^d)}}\Big]$,

$$\lim_{\epsilon\to 0}\mathbb{E}\Big[\sup_{0\le k\le N^{\Delta}} |y^{\Delta,\epsilon,v}(t_k) - w^{\Delta,v}(t_k)|^p\Big] = 0.$$

By the definitions of $Z^{\Delta,\epsilon,v}$ and $z^{\Delta,v}$, we have that for $p \ge 2$,

$$|Z^{\Delta,\epsilon,v}(t) - z^{\Delta,v}(t)|^p \le K\Big|\int_0^t b(y_{\lfloor s\rfloor_\Delta}^{\Delta,\epsilon,v}) - b(w_{\lfloor s\rfloor_\Delta}^{\Delta,v})\mathrm{d}s\Big|^p$$
$$+ K\Big|\int_0^t \sigma(y_{\lfloor s\rfloor_\Delta}^{\Delta,\epsilon,v})v(\lfloor s\rfloor_\Delta)$$
$$- \sigma(w_{\lfloor s\rfloor_\Delta}^{\Delta,v})v(\lfloor s\rfloor_\Delta)\mathrm{d}s\Big|^p + K\Big|\sqrt{\epsilon}\int_0^t \sigma(y_{\lfloor s\rfloor_\Delta}^{\Delta,\epsilon,v})\mathrm{d}W(s)\Big|^p.$$

It follows from the boundedness of derivatives of b, σ, and the Burkholder–Davis–Gundy inequality (see Proposition A.5) that

$$\begin{aligned}
&\mathbb{E}\Big[\sup_{t\in[0,T]}|Z^{\Delta,\epsilon,v}(t)-z^{\Delta,v}(t)|^p\Big]\\
&\le K_T\int_0^T\mathbb{E}[\|y^{\Delta,\epsilon,v}_{\lfloor s\rfloor_\Delta}-w^{\Delta,v}_{\lfloor s\rfloor_\Delta}\|^p]\mathrm{d}s+K_T\int_0^T\mathbb{E}[\|y^{\Delta,\epsilon,v}_{\lfloor s\rfloor_\Delta}\\
&\quad-w^{\Delta,v}_{\lfloor s\rfloor_\Delta}\|^p]\mathrm{d}s\Big(\int_0^T|v(\lfloor s\rfloor_\Delta)|^2\mathrm{d}s\Big)^{\frac{p}{2}}+K_T\epsilon^{\frac{p}{2}}\int_0^T(1+\mathbb{E}[\|y^{\Delta,\epsilon,v}_{\lfloor s\rfloor_\Delta}\|^p])\mathrm{d}s\\
&\le K_T\int_0^T\mathbb{E}[\sup_{r\in[0,s]}|y^{\Delta,\epsilon,v}(\lfloor r\rfloor_\Delta)-w^{\Delta,v}(\lfloor r\rfloor_\Delta)|^p]\mathrm{d}s\\
&\quad+K_T\epsilon^{\frac{p}{2}}\int_0^T(1+\mathbb{E}[\|y^{\Delta,\epsilon,v}_{\lfloor s\rfloor_\Delta}\|^p])\mathrm{d}s.
\end{aligned}$$

Moreover, we have

$$\begin{aligned}
&|y^{\Delta,\epsilon,v}(t_{k+1})-w^{\Delta,v}(t_{k+1})|^2\\
&\le|Z^{\Delta,\epsilon,v}(t_{k+1})-z^{\Delta,v}(t_{k+1})|^2\\
&\quad+2\theta\Delta\langle y^{\Delta,\epsilon,v}(t_{k+1})-w^{\Delta,v}(t_{k+1}),b(y^{\Delta,\epsilon,v}_{t_{k+1}})-b(w^{\Delta,v}_{t_{k+1}})\rangle\\
&\le|Z^{\Delta,\epsilon,v}(t_{k+1})-z^{\Delta,v}(t_{k+1})|^2+4\theta a_5\Delta\sup_{0\le k\le N^\Delta}|y^{\Delta,\epsilon,v}(t_k)-w^{\Delta,v}(t_k)|^2.
\end{aligned}$$

By $\Delta\le\frac{1}{8\theta a_5}$, we obtain

$$\sup_{0\le k\le N^\Delta}|y^{\Delta,\epsilon,v}(t_k)-w^{\Delta,v}(t_k)|^2\le 2\sup_{0\le k\le N^\Delta}|Z^{\Delta,\epsilon,v}(t_k)-z^{\Delta,v}(t_k)|^2, \tag{6.25}$$

which implies that

$$\begin{aligned}
&\mathbb{E}\Big[\sup_{0\le k\le N^\Delta}|y^{\Delta,\epsilon,v}(t_k)-w^{\Delta,v}(t_k)|^p\Big]\\
&\le K_T\int_0^T\mathbb{E}[\sup_{r\in[0,s]}|y^{\Delta,\epsilon,v}(\lfloor r\rfloor_\Delta)-w^{\Delta,v}(\lfloor r\rfloor_\Delta)|^p]\mathrm{d}s\\
&\quad+K_T\epsilon^{\frac{p}{2}}\int_0^T(1+\sup_{s\in[0,T]}\mathbb{E}[\|y^{\Delta,\epsilon,v}_{\lfloor s\rfloor_\Delta}\|^p])\mathrm{d}s.
\end{aligned}$$

According to Proposition 6.12 and the Gronwall inequality, one has

$$\mathbb{E}\Big[\sup_{0\le k\le N^\Delta}|y^{\Delta,\epsilon,v}(t_k)-w^{\Delta,v}(t_k)|^p\Big]\le K_T\epsilon^{\frac{p}{2}},$$

which finishes the proof of Step 1 by taking $\epsilon \to 0$.

Step 2. Prove that for any $p \geq 1$,

$$\lim_{\epsilon\to 0}\mathbb{E}\Big[\sup_{0\leq k\leq N^{\Delta}}|Y^{\Delta,\epsilon,v}(t_k)-\mathbb{Z}^{\Delta,v}(t_k)|^p\Big]=0,$$

where $Y^{\Delta,\epsilon,v}(t_k) := \frac{1}{\sqrt{\epsilon}}(y^{\Delta,\epsilon,v}(t_k) - w^{\Delta,v}(t_k))$.

By the definitions of $Z^{\Delta,\epsilon,v}(t)$, $z^{\Delta,v}(t)$, and $\widetilde{\mathbb{Z}}^{\Delta,v}(t)$, we arrive at

$$\frac{1}{\sqrt{\epsilon}}(Z^{\Delta,\epsilon,v}(t)-z^{\Delta,v}(t))-\widetilde{\mathbb{Z}}^{\Delta,v}(t)=\mathscr{A}_1(t)+\mathscr{A}_2(t)+\mathscr{A}_3(t),$$

where

$$\mathscr{A}_1(t):=\int_0^t\Big[\frac{1}{\sqrt{\epsilon}}\big(b(y^{\Delta,\epsilon,v}_{\lfloor s\rfloor_\Delta})-b(w^{\Delta,v}_{\lfloor s\rfloor_\Delta})\big)-\mathcal{D}b(w^{\Delta,v}_{\lfloor s\rfloor_\Delta})\mathbb{Z}^{\Delta,v}_{\lfloor s\rfloor_\Delta}\Big]\mathrm{d}s,$$

$$\mathscr{A}_2(t):=\int_0^t\Big[\frac{1}{\sqrt{\epsilon}}\big(\sigma(y^{\Delta\epsilon,v}_{\lfloor s\rfloor_\Delta})v(\lfloor s\rfloor_\Delta)-\sigma(w^{\Delta,v}_{\lfloor s\rfloor_\Delta})v(\lfloor s\rfloor_\Delta)\big)$$
$$-\mathcal{D}\sigma(w^{\Delta,v}_{\lfloor s\rfloor_\Delta})\mathbb{Z}^{\Delta,v}_{\lfloor s\rfloor_\Delta}v(\lfloor s\rfloor_\Delta)\Big]\mathrm{d}s,$$

$$\mathscr{A}_3(t):=\int_0^t\Big[\sigma(y^{\Delta,\epsilon,v}_{\lfloor s\rfloor_\Delta})-\sigma(w^{\Delta,v}_{\lfloor s\rfloor_\Delta})\Big]\mathrm{d}W(s).$$

For the term $\mathscr{A}_3$, utilizing the Burkholder–Davis–Gundy inequality, we deduce that for $p \geq 2$,

$$\begin{aligned}\mathbb{E}\Big[\sup_{t\leq T}|\mathscr{A}_3(t)|^p\Big]&\leq K_T\mathbb{E}\Big[\int_0^T\|y^{\Delta,\epsilon,v}_{\lfloor s\rfloor_\Delta}-w^{\Delta,v}_{\lfloor s\rfloor_\Delta}\|^p\mathrm{d}s\Big]\\&\leq K_T\int_0^T\mathbb{E}\Big[\sup_{0\leq k\leq N^{\Delta}}|y^{\Delta,\epsilon,v}(t_k)-w^{\Delta,v}(t_k)|^p\Big]\mathrm{d}s\leq K_T\epsilon^{\frac{p}{2}},\end{aligned}$$

where in the last step we used Step 1.

For the term $\mathscr{A}_1$, by the Taylor formula, we obtain

$$\begin{aligned}b(y^{\Delta,\epsilon,v}_{\lfloor s\rfloor_\Delta})-b(w^{\Delta,v}_{\lfloor s\rfloor_\Delta})&=\mathcal{D}b(w^{\Delta,v}_{\lfloor s\rfloor_\Delta})(y^{\Delta,\epsilon,v}_{\lfloor s\rfloor_\Delta}-w^{\Delta,v}_{\lfloor s\rfloor_\Delta})\\&\quad+\int_0^1(1-\varsigma)\mathcal{D}^2b\big((1-\varsigma)w^{\Delta,v}_{\lfloor s\rfloor_\Delta}\\&\quad+\varsigma y^{\Delta,\epsilon,v}_{\lfloor s\rfloor_\Delta}\big)\mathrm{d}\varsigma(y^{\Delta,\epsilon,v}_{\lfloor s\rfloor_\Delta}-w^{\Delta,v}_{\lfloor s\rfloor_\Delta},y^{\Delta,\epsilon,v}_{\lfloor s\rfloor_\Delta}-w^{\Delta,v}_{\lfloor s\rfloor_\Delta}),\end{aligned}$$

which implies

$$\begin{aligned}
&\frac{1}{\sqrt{\epsilon}}\big(b(y^{\Delta,\epsilon,v}_{\lfloor s\rfloor_\Delta})-b(w^{\Delta,v}_{\lfloor s\rfloor_\Delta})\big)-\mathcal{D}b(w^{\Delta,v}_{\lfloor s\rfloor_\Delta})\mathbb{Z}^{\Delta,v}_{\lfloor s\rfloor_\Delta}\\
&\quad=\mathcal{D}b(w^{\Delta,v}_{\lfloor s\rfloor_\Delta})(Y^{\Delta,\epsilon,v}_{\lfloor s\rfloor_\Delta}-\mathbb{Z}^{\Delta,v}_{\lfloor s\rfloor_\Delta})+\int_0^1(1-\varsigma)\mathcal{D}^2b\big((1-\varsigma)w^{\Delta,v}_{\lfloor s\rfloor_\Delta}\\
&\qquad+\varsigma y^{\Delta,\epsilon,v}_{\lfloor s\rfloor_\Delta}\big)\mathrm{d}\varsigma\big(Y^{\Delta,\epsilon,v}_{\lfloor s\rfloor_\Delta},y^{\Delta,\epsilon,v}_{\lfloor s\rfloor_\Delta}-w^{\Delta,v}_{\lfloor s\rfloor_\Delta}\big),
\end{aligned}\tag{6.26}$$

where $Y^{\Delta,\epsilon,v}_{\lfloor s\rfloor_\Delta}=\frac{1}{\sqrt{\epsilon}}(y^{\Delta,\epsilon,v}_{\lfloor s\rfloor_\Delta}-w^{\Delta,v}_{\lfloor s\rfloor_\Delta})$. It follows from Assumption 6.1 and Step 1 that for $p\geq 2$,

$$\begin{aligned}
&\mathbb{E}\Big[\sup_{t\leq T}|\mathscr{A}_1(t)|^p\Big]\\
&\leq K_T\mathbb{E}\Big[\int_0^T(\|Y^{\Delta,\epsilon,v}_{\lfloor s\rfloor_\Delta}-\mathbb{Z}^{\Delta,v}_{\lfloor s\rfloor_\Delta}\|^p+\|Y^{\Delta,\epsilon,v}_{\lfloor s\rfloor_\Delta}\|^p\|y^{\Delta,\epsilon,v}_{\lfloor s\rfloor_\Delta}-w^{\Delta,v}_{\lfloor s\rfloor_\Delta}\|^p)\mathrm{d}s\Big]\\
&\leq K_T\int_0^T\mathbb{E}\Big[\sup_{r\leq s}|Y^{\Delta,\epsilon,v}(\lfloor r\rfloor_\Delta)-\mathbb{Z}^{\Delta,v}(\lfloor r\rfloor_\Delta)|^p\Big]\mathrm{d}s\\
&\quad+K_T\epsilon^{-\frac{p}{2}}\int_0^T\mathbb{E}\Big[\sup_{0\leq k\leq N^\Delta}|y^{\Delta,\epsilon,v}(t_k)-w^{\Delta,v}(t_k)|^{2p}\Big]\mathrm{d}s\\
&\leq K_T\int_0^T\mathbb{E}\Big[\sup_{r\leq s}|Y^{\Delta,\epsilon,v}(\lfloor r\rfloor_\Delta)-\mathbb{Z}^{\Delta,v}(\lfloor r\rfloor_\Delta)|^p\Big]\mathrm{d}s+K_T\epsilon^{\frac{p}{2}}.
\end{aligned}$$

Similarly, for the term $\mathscr{A}_2$, by virtue of the Taylor formula, we have that for $p\geq 2$,

$$\begin{aligned}
&\mathbb{E}\Big[\sup_{t\leq T}|\mathscr{A}_2(t)|^p\Big]\\
&\leq\mathbb{E}\Big[\Big(\int_0^T\Big|\mathcal{D}\sigma(w^{\Delta,v}_{\lfloor s\rfloor_\Delta})(Y^{\Delta,\epsilon,v}_{\lfloor s\rfloor_\Delta}-\mathbb{Z}^{\Delta,v}_{\lfloor s\rfloor_\Delta})+\int_0^1(1-\varsigma)\\
&\quad\times\mathcal{D}^2\sigma\big((1-\varsigma)w^{\Delta,v}_{\lfloor s\rfloor_\Delta}+\varsigma y^{\Delta,\epsilon,v}_{\lfloor s\rfloor_\Delta}\big)\mathrm{d}\varsigma(Y^{\Delta,\epsilon,v}_{\lfloor s\rfloor_\Delta},y^{\Delta,\epsilon,v}_{\lfloor s\rfloor_\Delta}-w^{\Delta,v}_{\lfloor s\rfloor_\Delta})\Big||v(\lfloor s\rfloor_\Delta)|\mathrm{d}s\Big)^p\Big]\\
&\leq K_T\int_0^T\mathbb{E}\Big[\sup_{r\leq s}|Y^{\Delta,\epsilon,v}(\lfloor r\rfloor_\Delta)-\mathbb{Z}^{\Delta,v}(\lfloor r\rfloor_\Delta)|^p\Big]\mathrm{d}s\Big(\int_0^T|v(\lfloor s\rfloor_\Delta)|^2\mathrm{d}s\Big)^{\frac{p}{2}}\\
&\quad+K\epsilon^{-\frac{p}{2}}\int_0^T\mathbb{E}\Big[\sup_{0\leq k\leq N^\Delta}|y^{\Delta,\epsilon,v}(t_k)-w^{\Delta,v}(t_k)|^{2p}\Big]\mathrm{d}s\Big(\int_0^T|v(\lfloor s\rfloor_\Delta)|^2\mathrm{d}s\Big)^{\frac{p}{2}}\\
&\leq K_T\int_0^T\mathbb{E}\Big[\sup_{r\leq s}|Y^{\Delta,\epsilon,v}(\lfloor r\rfloor_\Delta)-\mathbb{Z}^{\Delta,v}(\lfloor r\rfloor_\Delta)|^p\Big]\mathrm{d}s+K_T\epsilon^{\frac{p}{2}}.
\end{aligned}$$

Combining estimates of terms $\mathscr{A}_1$, $\mathscr{A}_2$, and $\mathscr{A}_3$, we arrive at that for $p \geq 2$,

$$\mathbb{E}\Big[\sup_{t\leq T}\Big|\frac{1}{\sqrt{\epsilon}}(Z^{\Delta,\epsilon,v}(t)-z^{\Delta,v}(t))-\widetilde{\mathbb{Z}}^{\Delta,v}(t)\Big|^p\Big]$$
$$\leq K_T\int_0^T\mathbb{E}\Big[\sup_{r\leq s}|Y^{\Delta,\epsilon,v}(\lfloor r\rfloor_\Delta)-\mathbb{Z}^{\Delta,v}(\lfloor r\rfloor_\Delta)|^p\Big]\mathrm{d}s+K_T\epsilon^{\frac{p}{2}}. \tag{6.27}$$

Notice that

$$\Big|\frac{1}{\sqrt{\epsilon}}(Z^{\Delta,\epsilon,v}(t_k)-z^{\Delta,v}(t_k))-\widetilde{\mathbb{Z}}^{\Delta,v}(t_k)\Big|^2$$
$$\geq\Big|\frac{1}{\sqrt{\epsilon}}(y^{\Delta,\epsilon,v}(t_k)-w^{\Delta,v}(t_k))-\mathbb{Z}^{\Delta,v}(t_k)\Big|^2-2\theta\Delta\Big\langle\frac{1}{\sqrt{\epsilon}}(y^{\Delta,\epsilon,v}(t_k)-w^{\Delta,v}(t_k))$$
$$-\mathbb{Z}^{\Delta,v}(t_k),\frac{1}{\sqrt{\epsilon}}(b(y^{\Delta,\epsilon,v}_{t_k})-b(w^{\Delta,v}_{t_k}))-\mathcal{D}b(w^{\Delta,v}_{t_k})\mathbb{Z}^{\Delta,v}_{t_k}\Big\rangle,$$

and from (6.26) and Assumption 5.1 that

$$2\theta\Delta\Big\langle\frac{1}{\sqrt{\epsilon}}(y^{\Delta,\epsilon,v}(t_k)-w^{\Delta,v}(t_k))-\mathbb{Z}^{\Delta,v}(t_k),$$
$$\frac{1}{\sqrt{\epsilon}}(b(y^{\Delta,\epsilon,v}_{t_k})-b(w^{\Delta,v}_{t_k}))-\mathcal{D}b(w^{\Delta,v}_{t_k})\mathbb{Z}^{\Delta,v}_{t_k}\Big\rangle$$
$$=2\theta\Delta\Big\langle Y^{\Delta,\epsilon,v}(t_k)-\mathbb{Z}^{\Delta,v}(t_k),\mathcal{D}b(w^{\Delta,v}_{\lfloor s\rfloor_\Delta})(Y^{\Delta,\epsilon,v}_{\lfloor s\rfloor_\Delta}-\mathbb{Z}^{\Delta,v}_{\lfloor s\rfloor_\Delta})\Big\rangle$$
$$+2\theta\Delta\Big\langle Y^{\Delta,\epsilon,v}(t_k)-\mathbb{Z}^{\Delta,v}(t_k),\int_0^1(1-\varsigma)\mathcal{D}^2b\big((1-\varsigma)w^{\Delta,v}_{\lfloor s\rfloor_\Delta}$$
$$+\varsigma y^{\Delta,\epsilon,v}_{\lfloor s\rfloor_\Delta}\big)\mathrm{d}\varsigma\big(Y^{\Delta,\epsilon,v}_{\lfloor s\rfloor_\Delta},y^{\Delta,\epsilon,v}_{\lfloor s\rfloor_\Delta}-w^{\Delta,v}_{\lfloor s\rfloor_\Delta}\big)\Big\rangle$$
$$\leq 4a_5\theta\Delta\sup_{0\leq k\leq N^\Delta}|Y^{\Delta,\epsilon,v}(t_k)-\mathbb{Z}^{\Delta,v}(t_k)|^2$$
$$+2\theta\Delta\Big\langle Y^{\Delta,\epsilon,v}(t_k)-\mathbb{Z}^{\Delta,v}(t_k),\int_0^1(1-\varsigma)\mathcal{D}^2b\big((1-\varsigma)w^{\Delta,v}_{\lfloor s\rfloor_\Delta}$$
$$+\varsigma y^{\Delta,\epsilon,v}_{\lfloor s\rfloor_\Delta}\big)\mathrm{d}\varsigma\big(Y^{\Delta,\epsilon,v}_{\lfloor s\rfloor_\Delta},y^{\Delta,\epsilon,v}_{\lfloor s\rfloor_\Delta}-w^{\Delta,v}_{\lfloor s\rfloor_\Delta}\big)\Big\rangle$$
$$\leq 5a_5\theta\Delta\sup_{0\leq k\leq N^\Delta}|Y^{\Delta,\epsilon,v}(t_k)-\mathbb{Z}^{\Delta,v}(t_k)|^2$$
$$+K\Delta\epsilon^{-1}\sup_{0\leq k\leq N^\Delta}|y^{\Delta,\epsilon,v}(t_k)-w^{\Delta,v}(t_k)|^4,$$

where in the last step we used the Young inequality. Hence we derive that

$$\begin{aligned}&\Big|\frac{1}{\sqrt{\epsilon}}(y^{\Delta,\epsilon,v}(t_k)-w^{\Delta,v}(t_k))-\mathbb{Z}^{\Delta,v}(t_k)\Big|^2\\&\le\Big|\frac{1}{\sqrt{\epsilon}}(Z^{\Delta,\epsilon,v}(t_k)-z^{\Delta,v}(t_k))-\widetilde{\mathbb{Z}}^{\Delta,v}(t_k)\Big|^2+5\theta a_5\Delta\sup_{0\le k\le N^\Delta}|Y^{\Delta,\epsilon,v}(t_k)\\&\quad-\mathbb{Z}^{\Delta,v}(t_k)|^2+K\Delta\epsilon^{-1}\sup_{0\le k\le N^\Delta}|y^{\Delta,\epsilon,v}(t_k)-w^{\Delta,v}(t_k)|^4.\end{aligned}$$

This indicates that for $\Delta\le\frac{1}{8\theta a_5}$,

$$\begin{aligned}&\sup_{0\le k\le N^\Delta}\Big|Y^{\Delta,\epsilon,v}(t_k)-\mathbb{Z}^{\Delta,v}(t_k)\Big|^2\\&\le K\sup_{0\le k\le N^\Delta}\Big|\frac{1}{\sqrt{\epsilon}}(Z^{\Delta,\epsilon,v}(t_k)-z^{\Delta,v}(t_k))-\widetilde{\mathbb{Z}}^{\Delta,v}(t_k)\Big|^2\\&\quad+K\epsilon^{-1}\sup_{0\le k\le N^\Delta}|y^{\Delta,\epsilon,v}(t_k)-w^{\Delta,v}(t_k)|^4,\end{aligned}$$

which together with (6.27) yields that for $p\ge 2$,

$$\begin{aligned}&\mathbb{E}\Big[\sup_{0\le k\le N^\Delta}\Big|Y^{\Delta,\epsilon,v}(t_k)-\mathbb{Z}^{\Delta,v}(t_k)\Big|^p\Big]\\&\le K_T\int_0^T\mathbb{E}\Big[\sup_{r\le s}|Y^{\Delta,\epsilon,v}(\lfloor r\rfloor_\Delta)-\mathbb{Z}^{\Delta,v}(\lfloor r\rfloor_\Delta)|^p\Big]\mathrm{d}s+K_T\epsilon^{\frac{p}{2}}\\&\quad+K\Big(\mathbb{E}\Big[\sup_{0\le k\le N^\Delta}|Y^{\Delta,\epsilon,v}(t_k)|^{2p}\Big]\mathbb{E}\Big[\sup_{0\le k\le N^\Delta}|y^{\Delta,\epsilon,v}(t_k)-w^{\Delta,v}(t_k)|^{2p}\Big]\Big)^{\frac12}\\&\le K_T\int_0^T\mathbb{E}\Big[\sup_{r\le s}|Y^{\Delta,\epsilon,v}(\lfloor r\rfloor_\Delta)-\mathbb{Z}^{\Delta,v}(\lfloor r\rfloor_\Delta)|^p\Big]\mathrm{d}s+K_T\epsilon^{\frac{p}{2}}.\end{aligned}$$

Applying the Gronwall inequality completes the proof of Step 2.

Step 3. Prove that for any $p\ge 1$,

$$\lim_{\epsilon\to0}\mathbb{E}\Big[\sup_{\alpha\le t_k\le T}|D_\alpha y^{\Delta,\epsilon,v}(t_k)|^p\Big]=0.$$

It follows from the definition of $Z^{\Delta,\epsilon,v}$ that $D_\alpha Z^{\Delta,\epsilon,v}$ for $\alpha \le t$ satisfies

$$D_\alpha Z^{\Delta,\epsilon,v}(t) = \int_0^t \mathcal{D}b(y^{\Delta,\epsilon,v}_{\lfloor s\rfloor_\Delta}) D_\alpha y^{\Delta,\epsilon,v}_{\lfloor s\rfloor_\Delta} \mathrm{d}s + \int_0^t \sqrt{\epsilon}\mathcal{D}\sigma(y^{\Delta,\epsilon,v}_{\lfloor s\rfloor_\Delta}) D_\alpha y^{\Delta,\epsilon,v}_{\lfloor s\rfloor_\Delta} \mathrm{d}W(s)$$
$$+ \int_0^t \mathcal{D}\sigma(y^{\Delta,\epsilon,v}_{\lfloor s\rfloor_\Delta}) D_\alpha y^{\Delta,\epsilon,v}_{\lfloor s\rfloor_\Delta} v(\lfloor s\rfloor_\Delta)\mathrm{d}s + \sqrt{\epsilon}\sigma(y^{\Delta,\epsilon,v}_{\lfloor \alpha\rfloor_\Delta})\mathbf{1}_{[0,t]}(\alpha). \tag{6.28}$$

We obtain from the Burkholder–Davis–Gundy inequality and the Hölder inequality that for $p \ge 2$,

$$\begin{aligned}
&\mathbb{E}\Big[\sup_{t\le T} |D_\alpha Z^{\Delta,\epsilon,v}(t)|^p\Big] \\
&\le K_T \int_0^T \mathbb{E}[\|D_\alpha y^{\Delta,\epsilon,v}_{\lfloor s\rfloor_\Delta}\|^p]\mathrm{d}s + K_T \epsilon^{\frac{p}{2}} \int_0^T \mathbb{E}[\|D_\alpha y^{\Delta,\epsilon,v}_{\lfloor s\rfloor_\Delta}\|^p]\mathrm{d}s \\
&\quad + K\epsilon^{\frac{p}{2}}\big(1 + \mathbb{E}[\|y^{\Delta,\epsilon,v}_{\lfloor \alpha\rfloor_\Delta}\|^p]\big) \\
&\le K_T \int_0^T \mathbb{E}\Big[\sup_{r\le s} |D_\alpha y^{\Delta,\epsilon,v}(\lfloor r\rfloor_\Delta)|^p\Big]\mathrm{d}s + K\epsilon^{\frac{p}{2}}\big(1 + \mathbb{E}[\|y^{\Delta,\epsilon,v}_{\lfloor \alpha\rfloor_\Delta}\|^p]\big).
\end{aligned}$$

Moreover, similar to the proof of (6.25), we have

$$\sup_{0\le k\le N^\Delta} |D_\alpha y^{\Delta,\epsilon,v}(t_k)|^2 \le 2 \sup_{0\le k\le N^\Delta} |D_\alpha Z^{\Delta,\epsilon,v}(t_k)|^2$$

for $\Delta \in (0, \frac{1}{8\theta a_5}]$, which together with the Gronwall inequality (see Proposition A.3) finishes the proof of Step 3.

Step 4. Prove that for any $p \ge 1$ and sufficiently small $\Delta > 0$,

$$\lim_{\epsilon\to 0} \mathbb{E}\Big[\sup_{\alpha\le t_k\le T} |D_\alpha Y^{\Delta,\epsilon,v}(t_k) - D_\alpha \mathbb{Z}^{\Delta,v}(t_k)|^p\Big] = 0.$$

Since $D_\alpha \widetilde{\mathbb{Z}}^{\Delta,v}(t)$ for $\alpha \le t$ satisfies

$$D_\alpha \widetilde{\mathbb{Z}}^{\Delta,v}(t) = \int_0^t \mathcal{D}b(w^{\Delta,v}_{\lfloor s\rfloor_\Delta}) D_\alpha \mathbb{Z}^{\Delta,v}_{\lfloor s\rfloor_\Delta}\mathrm{d}s + \int_0^t \mathcal{D}\sigma(w^{\Delta,v}_{\lfloor s\rfloor_\Delta}) D_\alpha \mathbb{Z}^{\Delta,v}_{\lfloor s\rfloor_\Delta} v(\lfloor s\rfloor_\Delta)\mathrm{d}s$$
$$+ \sigma(w^{\Delta,v}_{\lfloor \alpha\rfloor_\Delta})\mathbf{1}_{[0,t]}(\alpha),$$

we derive from (6.28) that

$$\frac{1}{\sqrt{\epsilon}} D_\alpha Z^{\Delta,\epsilon,v}(t) - D_\alpha \widetilde{\mathbb{Z}}^{\Delta,v}(t) = \tilde{\mathscr{A}}_1(t) + \tilde{\mathscr{A}}_2(t) + \tilde{\mathscr{A}}_3(t) + \tilde{\mathscr{A}}_4(t),$$

where

$$\tilde{\mathscr{A}_1}(t) := \int_0^t \Big(\frac{1}{\sqrt{\epsilon}}\mathcal{D}b(y^{\Delta,\epsilon,v}_{\lfloor s\rfloor_\Delta})D_\alpha y^{\Delta,\epsilon,v}_{\lfloor s\rfloor_\Delta} - \mathcal{D}b(w^{\Delta,v}_{\lfloor s\rfloor_\Delta})D_\alpha \mathbb{Z}^{\Delta,v}_{\lfloor s\rfloor_\Delta}\Big)\mathrm{d}s,$$

$$\tilde{\mathscr{A}_2}(t) := \int_0^t \Big(\frac{1}{\sqrt{\epsilon}}\mathcal{D}\sigma(y^{\Delta,\epsilon,v}_{\lfloor s\rfloor_\Delta})D_\alpha y^{\Delta,\epsilon,v}_{\lfloor s\rfloor_\Delta} - \mathcal{D}\sigma(w^{\Delta,v}_{\lfloor s\rfloor_\Delta})D_\alpha \mathbb{Z}^{\Delta,v}_{\lfloor s\rfloor_\Delta}\Big)v(\lfloor s\rfloor_\Delta)\mathrm{d}s,$$

$$\tilde{\mathscr{A}_3}(t) := (\sigma(y^{\Delta,\epsilon,v}_{\lfloor \alpha\rfloor_\Delta}) - \sigma(w^{\Delta,v}_{\lfloor \alpha\rfloor_\Delta}))\mathbf{1}_{[0,t]}(\alpha),$$

$$\tilde{\mathscr{A}_4}(t) := \int_0^t \mathcal{D}\sigma(y^{\Delta,\epsilon,v}_{\lfloor s\rfloor_\Delta})D_\alpha y^{\Delta,\epsilon,v}_{\lfloor s\rfloor_\Delta}\mathrm{d}W(s)$$

By the Taylor formula, we obtain

$$\begin{aligned}
&\big(\mathcal{D}b(y^{\Delta,\epsilon,v}_{\lfloor s\rfloor_\Delta}) - \mathcal{D}b(w^{\Delta,v}_{\lfloor s\rfloor_\Delta})\big)D_\alpha y^{\Delta,\epsilon,v}_{\lfloor s\rfloor_\Delta}\\
&\quad = \mathcal{D}^2 b(w^{\Delta,v}_{\lfloor s\rfloor_\Delta})(y^{\Delta,\epsilon,v}_{\lfloor s\rfloor_\Delta} - w^{\Delta,v}_{\lfloor s\rfloor_\Delta}, D_\alpha y^{\Delta,\epsilon,v}_{\lfloor s\rfloor_\Delta})\\
&\qquad + \int_0^1 \mathcal{D}^3 b((1-\varsigma)w^{\Delta,v}_{\lfloor s\rfloor_\Delta} + \varsigma y^{\Delta,\epsilon,v}_{\lfloor s\rfloor_\Delta})\mathrm{d}\varsigma\\
&\quad \big(y^{\Delta,\epsilon,v}_{\lfloor s\rfloor_\Delta} - w^{\Delta,v}_{\lfloor s\rfloor_\Delta}, y^{\Delta,\epsilon,v}_{\lfloor s\rfloor_\Delta} - w^{\Delta,v}_{\lfloor s\rfloor_\Delta}, D_\alpha y^{\Delta,\epsilon,v}_{\lfloor s\rfloor_\Delta}\big).
\end{aligned}$$

Hence, for $p \geq 2$,

$$\begin{aligned}
&\mathbb{E}\Big[\sup_{t\leq T}|\tilde{\mathscr{A}_1}(t)|^p\Big]\\
&\quad = \mathbb{E}\Big[\sup_{t\leq T}\Big|\int_0^t \frac{1}{\sqrt{\epsilon}}(\mathcal{D}b(y^{\Delta,\epsilon,v}_{\lfloor s\rfloor_\Delta}) - \mathcal{D}b(w^{\Delta,v}_{\lfloor s\rfloor_\Delta}))D_\alpha y^{\Delta,\epsilon,v}_{\lfloor s\rfloor_\Delta}\\
&\qquad + \mathcal{D}b(w^{\Delta,v}_{\lfloor s\rfloor_\Delta})\big(\frac{1}{\sqrt{\epsilon}}D_\alpha y^{\Delta,\epsilon,v}_{\lfloor s\rfloor_\Delta} - D_\alpha \mathbb{Z}^{\Delta,v}_{\lfloor s\rfloor_\Delta}\big)\mathrm{d}s\Big|^p\Big]\\
&\quad \leq K_T\int_0^T \mathbb{E}\Big[\|Y^{\Delta,\epsilon,v}_{\lfloor s\rfloor_\Delta}\|^p\|D_\alpha y^{\Delta,\epsilon,v}_{\lfloor s\rfloor_\Delta}\|^p + \epsilon^{-\frac{p}{2}}\|y^{\Delta,\epsilon,v}_{\lfloor s\rfloor_\Delta} - w^{\Delta,v}_{\lfloor s\rfloor_\Delta}\|^{2p}\|D_\alpha y^{\Delta,\epsilon,v}_{\lfloor s\rfloor_\Delta}\|^p\\
&\qquad + \|\frac{1}{\sqrt{\epsilon}}D_\alpha y^{\Delta,\epsilon,v}_{\lfloor s\rfloor_\Delta} - D_\alpha \mathbb{Z}^{\Delta,v}_{\lfloor s\rfloor_\Delta}\|^p\Big]\mathrm{d}s\\
&\quad \leq K_T\Big(\mathbb{E}\Big[\sup_{0\leq k\leq N^\Delta}|Y^{\Delta,\epsilon,v}(t_k)|^{2p}\Big]\mathbb{E}\Big[\sup_{0\leq k\leq N^\Delta}|D_\alpha y^{\Delta,\epsilon,v}(t_k)|^{2p}\Big]\Big)^{\frac{1}{2}}\\
&\qquad + K_T\Big(\mathbb{E}\Big[\sup_{0\leq k\leq N^\Delta}|Y^{\Delta,\epsilon,v}(t_k)|^{2p}\Big]\Big)^{\frac{1}{2}}\Big(\mathbb{E}\Big[\sup_{0\leq k\leq N^\Delta}|y^{\Delta,\epsilon,v}(t_k) - w^{\Delta,v}(t_k)|^{4p}\Big]\\
&\qquad \times \mathbb{E}\Big[\sup_{0\leq k\leq N^\Delta}|D_\alpha y^{\Delta,\epsilon,v}(t_k)|^{4p}\Big]\Big)^{\frac{1}{4}}
\end{aligned}$$

$$+ K_T \int_0^T \mathbb{E}\Big[\sup_{r\le s}\Big|\frac{1}{\sqrt{\epsilon}} D_\alpha y^{\Delta,\epsilon,v}(\lfloor r\rfloor_\Delta) - D_\alpha \mathbb{Z}^{\Delta,v}(\lfloor r\rfloor_\Delta)\Big|^p\Big]\mathrm{d}s.$$

Similarly,

$$\begin{aligned}
&\mathbb{E}\Big[\sup_{t\le T}|\tilde{\mathscr{A}}_2(t)|^p\Big]\\
&\quad\le K_T\Big(\mathbb{E}\Big[\sup_{0\le k\le N^\Delta}|Y^{\Delta,\epsilon,v}(t_k)|^{2p}\Big]\mathbb{E}\Big[\sup_{0\le k\le N^\Delta}|D_\alpha y^{\Delta,\epsilon,v}(t_k)|^{2p}\Big]\Big)^{\frac12}\\
&\qquad + K_T\Big(\mathbb{E}\Big[\sup_{0\le k\le N^\Delta}|Y^{\Delta,\epsilon,v}(t_k)|^{2p}\Big]\Big)^{\frac12}\Big(\mathbb{E}\Big[\sup_{0\le k\le N^\Delta}|y^{\Delta,\epsilon,v}(t_k) - w^{\Delta,v}(t_k)|^{4p}\Big]\\
&\qquad \times \mathbb{E}\Big[\sup_{0\le k\le N^\Delta}|D_\alpha y^{\Delta,\epsilon,v}(t_k)|^{4p}\Big]\Big)^{\frac14}\\
&\qquad + K_T\int_0^T \mathbb{E}\Big[\sup_{r\le s}|\frac{1}{\sqrt{\epsilon}}D_\alpha y^{\Delta,\epsilon,v}(\lfloor r\rfloor_\Delta) - D_\alpha\mathbb{Z}^{\Delta,v}(\lfloor r\rfloor_\Delta)|^p\Big]\mathrm{d}s.
\end{aligned}$$

Consequently, we arrive at

$$\begin{aligned}
&\mathbb{E}\Big[\sup_{t\le T}\Big|\frac{1}{\sqrt{\epsilon}}D_\alpha Z^{\Delta,\epsilon,v}(t) - D_\alpha\widetilde{\mathbb{Z}}^{\Delta,v}(t)\Big|^p\Big]\\
&\quad\le K_T\Big(\mathbb{E}\Big[\sup_{0\le k\le N^\Delta}|Y^{\Delta,\epsilon,v}(t_k)|^{2p}\Big]\mathbb{E}\Big[\sup_{0\le k\le N^\Delta}|D_\alpha y^{\Delta,\epsilon,v}(t_k)|^{2p}\Big]\Big)^{\frac12}\\
&\qquad + K_T\Big(\mathbb{E}\Big[\sup_{0\le k\le N^\Delta}|Y^{\Delta,\epsilon,v}(t_k)|^{2p}\Big]\Big)^{\frac12}\Big(\mathbb{E}\Big[\sup_{0\le k\le N^\Delta}|y^{\Delta,\epsilon,v}(t_k) - w^{\Delta,v}(t_k)|^{4p}\Big]\\
&\qquad \times \mathbb{E}\Big[\sup_{0\le k\le N^\Delta}|D_\alpha y^{\Delta,\epsilon,v}(t_k)|^{4p}\Big]\Big)^{\frac14} + K_T\int_0^T \mathbb{E}\Big[\sup_{r\le s}|\frac{1}{\sqrt{\epsilon}}D_\alpha y^{\Delta,\epsilon,v}(\lfloor r\rfloor_\Delta)\\
&\qquad - D_\alpha\mathbb{Z}^{\Delta,v}(\lfloor r\rfloor_\Delta)|^p\Big]\mathrm{d}s + \mathbb{E}\Big[\sup_{0\le k\le N^\Delta}|y^{\Delta,\epsilon,v}(t_k) - w^{\Delta,v}(t_k)|^p\Big]\\
&\qquad + K_T\int_0^T \mathbb{E}\Big[\sup_{0\le k\le N^\Delta}|D_\alpha y^{\Delta,\epsilon,v}(t_k)|^p\Big]\mathrm{d}s.
\end{aligned}\tag{6.29}$$

Moreover, we have

$$\begin{aligned}
&\Big|\frac{1}{\sqrt{\epsilon}}D_\alpha y^{\Delta,\epsilon,v}(t_k) - D_\alpha\mathbb{Z}^{\Delta,v}(t_k)\Big|^2\\
&\quad\le 2\Big|\frac{1}{\sqrt{\epsilon}}D_\alpha Z^{\Delta,\epsilon,v}(t_k) - D_\alpha\widetilde{\mathbb{Z}}^{\Delta,v}(t_k)\Big|^2
\end{aligned}$$

$$+2\theta^2\Delta^2\Big|\mathcal{D}b(y_{t_k}^{\Delta,\epsilon,v})\frac{1}{\sqrt{\epsilon}}D_\alpha y_{t_k}^{\Delta,\epsilon,v}-\mathcal{D}b(w_{t_k}^{\Delta,v})D_\alpha \mathbb{Z}_{t_k}^{\Delta,v}\Big|^2.$$

According to the Taylor expansion of $\mathcal{D}b$ (6.26), we have

$$\begin{aligned}
&\sup_{0\le k\le N^\Delta}\Big|\frac{1}{\sqrt{\epsilon}}D_\alpha y^{\Delta,\epsilon,v}(t_k)-D_\alpha \mathbb{Z}^{\Delta,v}(t_k)\Big|^2\\
&\le 2\sup_{0\le k\le N^\Delta}\Big|\frac{1}{\sqrt{\epsilon}}D_\alpha Z^{\Delta,\epsilon,v}(t_k)-D_\alpha \widetilde{\mathbb{Z}}^{\Delta,v}(t_k)\Big|^2\\
&\quad+K\Delta\sup_{0\le k\le N^\Delta}|Y^{\Delta,\epsilon,v}(t_k)|^2|D_\alpha y^{\Delta,\epsilon,v}(t_k)|^2\\
&\quad+K\Delta\sup_{0\le k\le N^\Delta}|Y^{\Delta,\epsilon,v}(t_k)|^2|y^{\Delta,\epsilon,v}(t_k)-w^{\Delta,v}(t_k)|^2|D_\alpha y^{\Delta,\epsilon,v}(t_k)|^2\\
&\quad+4\theta\Delta\|\mathcal{D}b\|^2_{\mathcal{L}(C^d;\mathbb{R}^d)}\sup_{0\le k\le N^\Delta}\Big|\frac{1}{\sqrt{\epsilon}}D_\alpha y^{\Delta,\epsilon,v}(t_k)-D_\alpha \mathbb{Z}^{\Delta,v}(t_k)\Big|^2.
\end{aligned}$$

Then, for sufficiently small Δ,

$$\begin{aligned}
&\sup_{0\le k\le N^\Delta}\Big|\frac{1}{\sqrt{\epsilon}}D_\alpha y^{\Delta,\epsilon,v}(t_k)-D_\alpha \mathbb{Z}^{\Delta,v}(t_k)\Big|^2\\
&\le K\sup_{0\le k\le N^\Delta}\Big|\frac{1}{\sqrt{\epsilon}}D_\alpha Z^{\Delta,\epsilon,v}(t_k)-D_\alpha \widetilde{\mathbb{Z}}^{\Delta,v}(t_k)\Big|^2\\
&\quad+K\Delta\sup_{0\le k\le N^\Delta}|Y^{\Delta,\epsilon,v}(t_k)|^2|D_\alpha y^{\Delta,\epsilon,v}(t_k)|^2\\
&\quad+K\Delta\sup_{0\le k\le N^\Delta}|Y^{\Delta,\epsilon,v}(t_k)|^2|y^{\Delta,\epsilon,v}(t_k)-w^{\Delta,v}(t_k)|^2|D_\alpha y^{\Delta,\epsilon,v}(t_k)|^2.
\end{aligned}$$

Making use of (6.29), the Gronwall inequality (see Proposition A.3) and Steps 1–3 finishes the proof of Step 4.

Step 5. Prove that for any $p\ge 1$, $l\in\mathbb{N}_+$, and sufficiently small $\Delta>0$,

$$\lim_{\epsilon\to 0}\mathbb{E}\Big[\sup_{\alpha_1\vee\cdots\vee\alpha_l\le t_k\le T}|D_{\alpha_1\cdots\alpha_l}Y^{\Delta,\epsilon,v}(t_k)-D_{\alpha_1\cdots\alpha_l}\mathbb{Z}^{\Delta,v}(t_k)|^p\Big]=0.$$

The case of $l=1$ is proved in Step 4, and the case of $l>1$ can be proved by the induction argument on l. We omit the proof.

Combining Steps 1–5, and according to [13, Proposition 5.1], we obtain

$$\liminf_{\epsilon\to 0}\epsilon\log\mathfrak{p}^{\Delta,\epsilon}(t_k,y)\ge-\tilde{d}_{t_k}(y)$$

with

$$\tilde{d}_t(y) := \inf\{I_T(f) : f \in C_{\xi^{Int}}([-\tau, T]; \mathbb{R}^d), y = f(t), \det(\bar{\gamma}_{w^{\Delta,v}(t)}) > 0\},$$

where

$$\bar{\gamma}_{\Phi(v)} = (\langle \mathscr{D}(\Phi(v))^i, \mathscr{D}(\Phi(v))^j \rangle_{L^2([0,T];\mathbb{R}^m)})_{1 \le i,j \le d}$$

with some Fréchet differentiable functional $\Phi : L^2([0,T]; \mathbb{R}^m) \to \mathbb{R}^d$, and $\mathscr{D}$ denotes the Fréchet derivative.

At last, we show that $\tilde{I}_t = \tilde{d}_t$, which is equivalent to proving that $\det(\bar{\gamma}_{w^{\Delta,v}(t)}) > 0$ for each $t \in (0, T]$. Similar to the argument in [13, Appendix, Page 39], the mapping $v \in L^2_\Delta([0,T]; \mathbb{R}^m) \to w^{\Delta,v}(t) \in \mathbb{R}^d$ is Fréchet differentiable, and the Fréchet derivative $\mathscr{D}w^{\Delta,v}(t) \in \mathcal{L}(L^2_\Delta([0,T]; \mathbb{R}^m); \mathbb{R}^d)$ admits the representation

$$\mathscr{D}w^{\Delta,v}(t)(h) = \int_0^T \mathscr{D}_\alpha w^{\Delta,v}(t) h(\lfloor \alpha \rfloor_\Delta) \mathrm{d}\alpha$$

for $h \in L^2_\Delta([0,T]; \mathbb{R}^m)$, where $\mathscr{D}_\alpha w^{\Delta,v}(t) \in \mathbb{R}^{d \times m}$. Noticing that

$$\begin{aligned}\mathscr{D}_\alpha w^{\Delta,v}(t_{k+1}) = {} & \mathscr{D}_\alpha w^{\Delta,v}(t_k) + (1-\theta)\mathscr{D}b(w^{\Delta,v}_{t_k})\mathscr{D}_\alpha w^{\Delta,v}_{t_k}\Delta \\ & + \theta \mathscr{D}b(w^{\Delta,v}_{t_{k+1}})\mathscr{D}_\alpha w^{\Delta,v}_{t_{k+1}}\Delta \\ & + \mathscr{D}\sigma(w^{\Delta,v}_{t_k})\mathscr{D}_\alpha w^{\Delta,v}_{t_k} v(t_k)\Delta + \sigma(w^{\Delta,v}_{t_k})\mathbf{1}_{[t_k,t_{k+1})}(\alpha)\Delta,\end{aligned}$$

the proof of $\det(\bar{\gamma}_{w^{\Delta,v}(t)}) > 0$ is similar to that of Theorem 5.2 and is omitted. Hence the proof of Theorem 6.11 is finished. □

6.3 Numerical Experiments

In this section, we present numerical algorithms and numerical experiments to verify the LDP for numerical solutions of SFDEs. Our approach is to estimate the probability that the solution hits a small neighborhood of a target value x_{target} at time T and investigate the exponential decay property of this hitting probability.

Algorithm 6 is designed to verify the large deviation behavior of numerical solutions for a family of small noise SFDEs. We aim to estimate the logarithm of the hitting probability of a closed set F, i.e., $\log \mathbb{P}(y^{\Delta,\epsilon}(T) \in F)$. For a sequence of decreasing noise intensities $\epsilon_1 > \cdots > \epsilon_J$, we simulate M_{trials} trajectories for each ϵ_j, using the θ-EM method, and the terminal values of the numerical solutions are collected. The empirical probability $\widehat{p}_j = \frac{1}{M_{\text{trials}}}\sum_{\tilde{m}=1}^{M_{\text{trials}}} \mathbf{1}_{\{y^{\Delta,\epsilon_j}_{\tilde{m}}(T) \in F\}}$ is used as an estimator of the hitting probability $\mathbb{P}(y^{\Delta,\epsilon_j}(T) \in F)$, where $y^{\Delta,\epsilon_j}_{\tilde{m}}(T)$ denotes the θ-EM solution at T for the $\tilde{m}$-th trajectory and noise intensity ϵ_j. Below we present two concrete numerical examples.

Algorithm 6 Verification of the numerical LDP for small noise SFDEs

Input: SFDE parameter (b, σ, ξ, τ), parameter θ, final time $T > 0$, stepsize Δ, small noise parameters $\{\epsilon_j\}_{j=1}^J$, number of Monte Carlo samples M_{trials}, closed set $F \subset \mathbb{R}$.

Output: Logarithm of hitting probabilities $\{\log \mathbb{P}(x^{\epsilon_j}(T) \in F)\}_{j=1}^J$ versus $\{1/\epsilon_j\}_{j=1}^J$.

Discretize time: $N = T/\Delta$, delay steps $k_\tau = \tau/\Delta$.
for $j = 1, \dots, J$ **do** ▷ Loop over noise levels
  Set $\epsilon = \epsilon_j$, and set count $= 0$.
  for $\tilde{m} = 1, \dots, M_{\text{trials}}$ **do** ▷ Monte Carlo sampling
    Initialize $y_0 = \xi^{Int}$ with ξ^{Int} denoting the linear interpolation of ξ.
    for $k = 0, \dots, N-1$ **do**
      Generate Gaussian increment $\delta W_k \sim \mathcal{N}(0, \Delta)$.
      Compute the numerical solution using the θ-EM method

$$y(t_{k+1}) = y(t_k) + (1-\theta)b(y_{t_k})\Delta + \theta b(y_{t_{k+1}})\Delta + \sqrt{\epsilon}\,\sigma(y_{t_k})\,\delta W_k.$$

    end for
    Let $x^{(\tilde{m})}(T) = y(t_N)$.
    if $x^{(\tilde{m})}(T) \in F$ **then**
      count = count + 1.
    end if
  end for
  Estimate the probability $\widehat{p}_j = \frac{\text{count}}{M_{\text{trials}}}$.
  Compute the LDP estimator $\log \widehat{p}_j$.
end for
return $\{\log \widehat{p}_j\}_{j=1}^J$.
Plot $\log \widehat{p}_j$ against $1/\epsilon_j$, and superimpose a linear regression fit.

Linear Stochastic Delay Differential Equation We first consider the linear stochastic delay differential equation with small noise:

$$dx^\epsilon(t) = (-\alpha x^\epsilon(t) + \beta x^\epsilon(t-\tau))dt + \sqrt{\epsilon}\sigma dW(t), \quad t > 0, \tag{6.30}$$

where in the numerical experiment, constants $\alpha = 2, \beta = 1, \sigma = 0.5, \tau = 0.5$, and initial datum $\xi(t) = 1$ for $t \in [-\tau, 0]$. Figure 6.1 illustrates the verification of the LDP of the EM method for this model. For each noise level $\epsilon_j = 0.5, 0.25, 0.125, 0.0625$, we compute the empirical probability that the terminal value falls inside the small closed interval $F = [x_{\text{target}} - \delta, x_{\text{target}} + \delta]$ with $x_{\text{target}} = 0.75$ and $\delta = 0.05$. The plot demonstrates that the values of $\log \mathbb{P}(y^{\Delta,\epsilon_j}(T) \in F)$ with $T = 1$ decrease approximately linearly as $\frac{1}{\epsilon_j}$ becomes larger, which serves as a verification of the numerical LDP.

Nonlinear Stochastic Functional Differential Equation We then verify the numerical LDP of the following nonlinear SFDE with small noise:

$$dx^\epsilon(t) = \Big(\frac{1}{\tau}\int_{-\tau}^0 x^\epsilon(t+s)ds - (x^\epsilon(t))^3 - 2x^\epsilon(t)\Big)dt$$

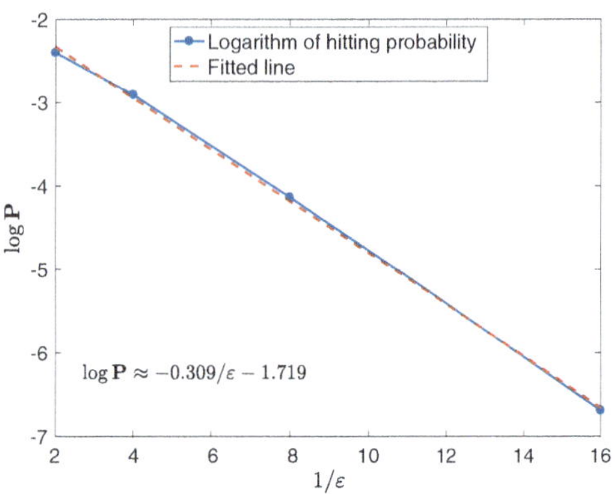

Fig. 6.1 Logarithm of hitting probability, i.e., $\log \mathbb{P}(|x^{\epsilon_j}(T) - x_{\text{target}}| \le \delta)$, $\delta = 0.05$, $x_{\text{target}} = 0.75$, for EM solutions of (6.30); $\alpha = 2$, $\beta = 1$, $\sigma = 0.5$, $\tau = 0.5$, $T = 1$, $M_{\text{trials}} = 50000$, $\Delta = 0.01$, $\xi \equiv 1$, $\epsilon_j = 0.5, 0.25, 0.125, 0.0625$

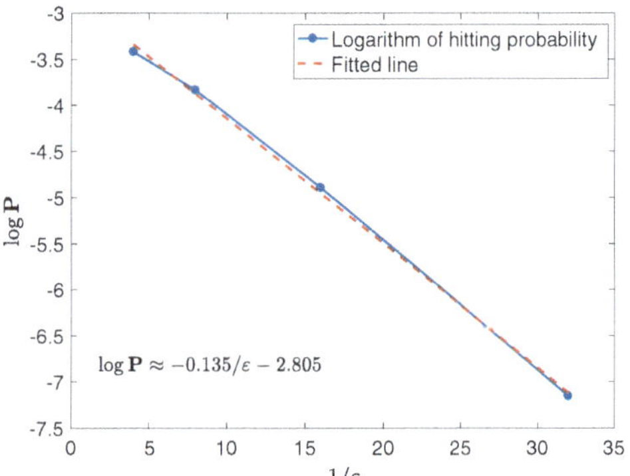

Fig. 6.2 Logarithm of hitting probability, i.e., $\log \mathbb{P}(|x^{\epsilon_j}(T) - x_{\text{target}}| \le \delta)$, $\delta = 0.05$, $x_{\text{target}} = 0.75$, for backward EM solutions of (6.31); $a = 1$, $\tau = 0.5$, $T = 1$, $M_{\text{trials}} = 50000$, $\Delta = 0.01$, $\xi \equiv 1$, $\epsilon_j = 0.25, 0.125, 0.0625, 0.03125$

$$+ \frac{a\sqrt{\epsilon}}{\tau} \int_{-\tau}^{0} x^{\epsilon}(t+s)\mathrm{d}s\mathrm{d}W(t), \tag{6.31}$$

where $\tau = 0.5$, $a = 1$ and the initial datum $\xi(t) = 1$ for $t \in [-\tau, 0]$. Figure 6.2 displays the verification of the LDP of the backward EM method for this nonlinear model. We take the noise level $\epsilon_j = 0.25, 0.125, 0.0625, 0.03125$, and the closed interval $F = [x_{\text{target}} - \delta, x_{\text{target}} + \delta]$ with $x_{\text{target}} = 0.75$ and $\delta = 0.05$. The numerical results again show a linearly decreasing trend of $\log \mathbb{P}(y^{\Delta,\epsilon_j} \in F)$ as $\frac{1}{\epsilon_j}$ becomes larger, demonstrating that the discretized system can also capture the expected large deviation behavior in the presence of nonlinear drift and multiplicative noise.

6.4 Summary and Outlook

This chapter has investigated Freidlin–Wentzell type LDPs for numerical methods applied to SFDEs driven by small noise. One classical approach to establishing such LDPs for stochastic systems is through the time discretization argument and the contraction principle, which requires certain subtle exponential estimates of small probabilities (see e.g. [14]). In addition to this approach, the monograph [12] develops a weak convergence approach to the large deviation theory utilizing the convergence of solutions to associated variational problems. In this chapter, we have adopted the latter approach to establish the Freidlin–Wentzell LDP of the θ-EM method for SFDEs with small noise over an infinite time horizon.

In the context of sample paths large deviations, and large deviations of the observables or invariant measures of stochastic systems, the LDPs of numerical methods have been studied in e.g., [4, 16, 17] for stochastic ordinary differential equations, in e.g., [2, 3, 5–7, 10] for stochastic partial differential equations, and in e.g. [8, 23] for stochastic delay differential equations. Nevertheless, there are still some unresolved issues in the study of numerical preservation of LDPs for SFDEs. For example, what is the relationship between the large deviation rate functions of the numerical solution and those of the exact solution? Can the rate function of the exact solution be exactly preserved by numerical methods? Moreover, if a numerical method successfully preserves the LDP of the SFDE, how can this property be exploited to reliably compute rare-event probabilities or extreme-event statistics in applications such as statistics, engineering, statistical mechanics, and applied probability?

The stochastic differential equations with small noise, such as (6.1) considered in this chapter, are single-scale models, for which the corresponding LDP characterizes the single-scale dynamics of the system. Many realistic models in physics and biology exhibit multiscale phenomena described by the interplay between slow and fast variables, commonly referred to as slow-fast systems or multiscale systems; see e.g. [18, 25, 26] for relevant studies of stochastic ordinary and partial differential equations. The averaging principle provides a fundamental approximation for the slow component, effectively serving as a functional LLN; see e.g. [1, 31]. For two-time-scale SFDEs, [28–30] establish the averaging principle based on the corresponding martingale problem formulation. It is of great interest to investigate whether numerical methods can preserve such an averaging principle for SFDEs. Furthermore, beyond the averaging limit, the LDP for multiscale stochastic ordinary and partial differential systems has garnered increasing attention; see e.g., [18, 19, 25, 30]. A natural and challenging problem is to establish the LDP for two-time-scale SFDEs, and further to analyze the large deviation properties of their numerical discretizations.

Another key result presented in this chapter is the logarithmic estimate for the density function of the θ-EM solution, assuming that drift and diffusion coefficients of the SFDE are smooth and have bounded derivatives of arbitrary order. We prove the smoothness of the numerical density function for the numerical solution and give the relation between the logarithmic limit and the rate function of the LDP for the θ-EM method of SFDE with small noise. The logarithmic asymptotics of the solution density for a stochastic system is also known as the Varadhan estimate; see e.g. [20, 21, 24] for the relevant study for exact solutions of stochastic systems. An important issue is whether, as the discretization parameter tends to zero, the logarithmic estimate of the numerical solution can recover the large deviation rate function of the exact solution. In addition, it would be interesting to extend the study of the logarithmic estimate for the numerical density function to SFDEs with superlinearly growing coefficients.

References

1. S. Cerrai, A Khasminskii type averaging principle for stochastic reaction-diffusion equations. Ann. Appl. Probab. **19**, 899–948 (2009)
2. C. Chen, A symplectic discontinuous Galerkin full discretization for stochastic Maxwell equations. SIAM J. Numer. Anal. **59**, 2197–2217 (2021)
3. C. Chen, Z. Chen, J. Hong, D. Jin, Large deviations principles of sample paths and invariant measures of numerical methods for parabolic SPDEs (2021). https://arxiv.org/abs/2106.11018
4. C. Chen, J. Hong, D. Jin, L. Sun, Asymptotically-preserving large deviations principles by stochastic symplectic methods for a linear stochastic oscillator. SIAM J. Numer. Anal. **59**, 32–59 (2021)
5. C. Chen, J. Hong, L. Ji, *Numerical Approximations of Stochastic Maxwell Equations—via Structure-Preserving Algorithms*, vol. 2341 of Lecture Notes in Mathematics (Springer, Singapore, 2023)
6. C. Chen, J. Hong, D. Jin, L. Sun, Large deviations principles for symplectic discretizations of stochastic linear Schrödinger equation. Potential Anal. **59**, 971–1011 (2023)
7. C. Chen, T. Dang, J. Hong, An adaptive time-stepping fully discrete scheme for stochastic NLS equation: strong convergence and numerical asymptotics. Stoch. Process. Appl. **173**, Paper No. 104373 (2024)
8. Z. Chen, D. Wang, L. Chen, Convergence order of one point large deviations rate functions for backward Euler method of stochastic delay differential equations with small noise. Appl. Numer. Math. **202**, 67–88 (2024)
9. J.B. Conway, *A Course in Functional Analysis*, vol. 96 of Graduate Texts in Mathematics (Springer, New York, 1985)
10. X. Dai, J. Hong, D. Sheng, Mittag–Leffler Euler integrator and large deviations for stochastic space-time fractional diffusion equations. Potential Anal. **60**, 1333–1367 (2024)
11. A. Dembo, O. Zeitouni, *Large Deviations Techniques and Applications*, vol. 38 of Applications of Mathematics, 2nd edn. (Springer, New York, 1998)
12. P. Dupuis, R.S. Ellis, *A Weak Convergence Approach to the Theory of Large Deviations*, Wiley Series in Probability and Statistics: Probability and Statistics (Wiley, New York, 1997)
13. M. Ferrante, C. Rovira, M. Sanz-Solé, Stochastic delay equations with hereditary drift: estimates of the density. J. Funct. Anal. **177**, 138–177 (2000)
14. M.I. Freidlin, A.D. Wentzell, *Random Perturbations of Dynamical Systems*, vol. 260 of Grundlehren der mathematischen Wissenschaften [Fundamental Principles of Mathematical Sciences] (Springer, New York, 1984)
15. P. Gao, Y. Liu, Y. Sun, Z. Zheng, Large deviations principle for stationary solutions of stochastic differential equations with multiplicative noise (2022). https://arxiv.org/abs/2206.02356
16. J. Hong, L. Sun, *Symplectic Integration of Stochastic Hamiltonian Systems*, vol. 2314 of Lecture Notes in Mathematics (Springer, Singapore, 2022)
17. J. Hong, D. Jin, D. Sheng, L. Sun, Numerically asymptotical preservation of the large deviations principles for invariant measures of Langevin equations (2020). https://arxiv.org/abs/2009.13336
18. W. Hong, S. Li, W. Liu, Freidlin–Wentzell type large deviation principle for multiscale locally monotone SPDEs. SIAM J. Math. Anal. **53**, 6517–6561 (2021)

19. W. Hong, M. Li, S. Li, W. Liu, Large deviations and averaging for stochastic tamed 3D Navier–Stokes equations with fast oscillations. Appl. Math. Optim. **86**, Paper No. 15 (2022)
20. A. Kohatsu-Higa, D. Márquez-Carreras, M. Sanz-Solé, Asymptotic behavior of the density in a parabolic SPDE. J. Theoret. Probab. **14**, 427–462 (2001)
21. A. Kohatsu-Higa, D. Márquez-Carreras, M. Sanz-Solé, Logarithmic estimates for the density of hypoelliptic two-parameter diffusions. J. Funct. Anal. **190**, 481–506 (2002)
22. A.N. Kolmogorov, S.V. Fomin, *Elements of the Theory of Functions and Functional Analysis. Vol. 1. Metric and Normed Spaces*, Russian edn. (Graylock Press, Rochester, 1957)
23. X. Ma, J. Liu, F. Xi, Large deviations for numerical approximation of stochastic differential delay equations. Commun. Nonlinear Sci. Numer. Simul. **152**, Paper No. 109389 (2026)
24. D. Márquez-Carreras, M. Mellouk, Density estimates on a parabolic SPDE. Publ. Mat. **46**, 77–96 (2002)
25. X. Sun, R. Wang, L. Xu, X. Yang, Large deviation for two-time-scale stochastic Burgers equation. Stoch. Dyn. **21**, Paper No. 2150023 (2021)
26. A.Y. Veretennikov, On large deviations for SDEs with small diffusion and averaging. Stoch. Process. Appl. **89**, 69–79 (2000)
27. C. Villani, *Optimal Transport: Old and New*, vol. 338 of Grundlehren der mathematischen Wissenschaften [Fundamental Principles of Mathematical Sciences] (Springer, Berlin, 2009)
28. F. Wu, G. Yin, An averaging principle for two-time-scale stochastic functional differential equations. J. Differ. Equ. **269**, 1037–1077 (2020)
29. F. Wu, G. Yin, Fast-slow-coupled stochastic functional differential equations. J. Differ. Equ. **323**, 1–37 (2022)
30. F. Wu, G. Yin, Two-time-scale stochastic functional differential equations: inclusion of infinite delay and coupled segment processes. J. Differ. Equ. **435**, Paper No. 113238 (2025)
31. X. Yin, Y. Tian, J.-L. Wu, Large deviation and averaging for multi-scale stochastic tidal dynamics equation. J. Math. Phys. **66**, Paper No. 072702 (2025)

Appendix A
Basic Inequalities and Some Tools from Martingale Theory

Below, we present the Young inequality, the Hölder inequality, several versions of the Gronwall inequality, Burkholder–Davis–Gundy inequality, and the martingale stopping theorem; see e.g. [18].

Proposition A.1 (Young Inequality) *Let ε be a positive real number, and $p, q > 1$ such that $\frac{1}{p} + \frac{1}{q} = 1$. Then*

$$|ab| \leq \varepsilon \frac{|a|^p}{p} + \varepsilon^{-\frac{q}{p}} \frac{|b|^q}{q}, \qquad a, b \in \mathbb{R}.$$

Proposition A.2 (Hölder Inequality) *Let $p, q > 1$ such that $\frac{1}{p} + \frac{1}{q} = 1$. Then*

$$\|fg\|_{L^1(\mathbb{R}^n)} \leq \|f\|_{L^p(\mathbb{R}^n)} \|g\|_{L^q(\mathbb{R}^n)}, \qquad f \in L^p(\mathbb{R}^n), \; g \in L^q(\mathbb{R}^n).$$

Proposition A.3 (Continuous Gronwall Inequality) *Let $T > 0, h$ be a non-negative function on $[0, T]$ with $\int_0^T h(s)\mathrm{d}s < \infty$, $f, g \in L^\infty(0, T)$, and g be a monotonically increasing and continuous function. If f satisfies*

$$f(t) \leq g(t) + \int_0^t f(s)h(s)\mathrm{d}s \quad a.e. \text{ in } t \in [0, T],$$

then

$$f(t) \leq g(t)e^{\int_0^t h(s)\mathrm{d}s} \quad a.e. \text{ in } t \in [0, T].$$

Proposition A.4 (Discrete Gronwall Inequality) *Let $\{a_n\}_{n\in\mathbb{N}}$ and $\{b_n\}_{n\in\mathbb{N}}$ be two sequences. Let $\delta \in [0, 1]$, $c_0 > 0$, $\iota > 0$, and $1 - \delta c_0 \iota > 0$. If $a_0 \leq b_0$ and*

C. Chen et al., *Numerical Analysis of Stochastic Functional Differential Equations*, Lecture Notes in Mathematics 2399, https://doi.org/10.1007/978-981-92-1592-8

$$a_{n+1} \le b_{n+1} + c_0 \iota \sum_{j=0}^{n} \big((1-\delta) a_j + \delta a_{j+1} \big) \quad \forall n \in \mathbb{N},$$

then for all $n \in \mathbb{N}$,

$$a_{n+1} \le b_{n+1} + \frac{c_0 \iota}{1 - \delta c_0 \iota} \sum_{j=0}^{n} \Big(\frac{1 + (1-\delta) c_0 \iota}{1 - \delta c_0 \iota} \Big)^{n-j} \big((1-\delta) b_j + \delta b_{j+1} \big).$$

Moreover, if $\{b_n\}_{n \in \mathbb{N}}$ *is monotonically increasing, it follows that for all* $n \in \mathbb{N}$,

$$a_n \le b_n \Big(\frac{1 + (1-\delta) c_0 \iota}{1 - \delta c_0 \iota} \Big)^{n}.$$

Proposition A.5 (Burkholder–Davis–Gundy Inequality) *Let* $g \in L^2(\mathbb{R}_+, \mathbb{R}^{d \times m})$. *Define, for* $t \ge 0$,

$$x(t) := \int_0^t g(s) \mathrm{d}W(s) \quad \text{and} \quad A(t) := \int_0^t |g(s)|^2 \mathrm{d}s.$$

Then for any $p > 0$, *there exist universal positive constants* c_p, C_p *such that*

$$c_p \mathbb{E}[|A(t)|^{\frac{p}{2}}] \le \mathbb{E}\Big[\sup_{0 \le s \le t} |x(s)|^p \Big] \le C_p \mathbb{E}[|A(t)|^{\frac{p}{2}}]$$

for all t. *In particular, one may take*

$$c_p = (\frac{p}{2})^p, \quad C_p = (\frac{32}{p})^{\frac{p}{2}} \quad \text{if } 0 < p < 2;$$

$$c_p = 1, \quad C_p = 4 \quad \text{if } p = 2;$$

$$c_p = (2p)^{-\frac{p}{2}}, \quad C_p = (\frac{p^{p+1}}{2(p-1)^{p-1}})^{\frac{p}{2}} \quad \text{if } p > 2.$$

Theorem A.6 (Martingale Stopping Theorem) *Let* $\{M_t\}_{t \ge 0}$ *be an* $\{\mathcal{F}_t\}_{t \ge 0}$-*adapted stochastic process, whose paths are right continuous and locally bounded. The following properties are equivalent:*

(i) $\{M_t\}_{t \ge 0}$ *is a martingale with respect to the filtration* $\{\mathcal{F}_t\}_{t \ge 0}$;
(ii) For any almost surely bounded stopping time $\mathcal{T}$ *such that* $\mathbb{E}[|M_{\mathcal{T}}|] < +\infty$, *we have* $\mathbb{E}[M_{\mathcal{T}}] = \mathbb{E}[M_0]$.

Appendix B
Markov Process, Markov Semigroup, Invariant Measure, Ergodicity

Denote by H a Banach space endowed with the norm $\|\cdot\|_H$. Denote by $C_b(H)$ (resp. $B_b(H)$) the Banach space of all continuous and bounded mappings (resp. Borel bounded mappings) $\varphi : H \to \mathbb{R}$ endowed with the norm $\|\varphi\|_0 := \sup_{u \in H} |\varphi(u)|$. Denote by $L(B_b(H))$ the space of all linear bounded operators from $B_b(H)$ into itself. Let $\mathcal{P}(H)$ be the collection of probability measures on H. The content of this appendix can be found in e.g. [3].

Definition B.1 (Markov Semigroup) A Markov semigroup P_t on $B_b(H)$ is a mapping

$$[0, +\infty) \to L(B_b(H)), \quad t \mapsto P_t,$$

such that

(i) $P_0 = 1,\ P_{t+s} = P_t P_s$ for all $t,\ s \geq 0$.
(ii) For any $t \geq 0$ and $u \in H$, there exists a probability measure $\pi_t(u, \cdot) \in \mathcal{P}(H)$ such that

$$P_t\varphi(u) = \int_H \varphi(v)\pi_t(u, \mathrm{d}v) \quad \text{for all } \varphi \in B_b(H).$$

(iii) For any $\varphi \in C_b(H)$ (resp. $B_b(H)$) and $u \in H$, the mapping $t \mapsto P_t\varphi(u)$ is continuous (resp. Borel measurable).

Definition B.2 (Feller Semigroup) Let P_t be a Markov semigroup. P_t is Feller if $P_t\varphi \in C_b(H)$ for any $\varphi \in C_b(H)$ and any $t \geq 0$.

Definition B.3 (Invariant Measure) Let P_t be a Markov semigroup. A probability measure $\pi \in \mathcal{P}(H)$ is said to be *invariant* for P_t if

$$\int_H P_t\varphi \mathrm{d}\pi = \int_H \varphi \mathrm{d}\pi \quad \text{for all } \varphi \in B_b(H) \text{ and } t \geq 0.$$

C. Chen et al., *Numerical Analysis of Stochastic Functional Differential Equations*, Lecture Notes in Mathematics 2399, https://doi.org/10.1007/978-981-92-1592-8

If P_t is Feller, the above definition is equivalent to

$$P_t^* \pi = \pi \quad \text{for all } t \geq 0,$$

where P_t^* is the dual operator of P_t.

If $\pi \in \mathcal{P}(H)$ is invariant for P_t, one has

$$\pi(A) = P_t^* \pi(A) = \int_H P_t \mathbf{1}_A(u) \pi(\mathrm{d}u), \quad A \in \mathcal{B}(H).$$

When the Markov semigroup P_t is Feller, a condition to guarantee that P_t exists at least one invariant measure is the so-called tightness property for the corresponding probability measures.

Definition B.4 (Tightness) A family of probability measures $\{\mathbb{P}_\alpha; \alpha \in \mathcal{I}\}$ is said to be tight if, for any $\varepsilon > 0$, there exists a compact set $\bar{H}_\varepsilon \subset H$ such that

$$\sup_{\alpha \in \mathcal{I}} \mathbb{P}_\alpha(H \setminus \bar{H}_\varepsilon) \leq \varepsilon.$$

Definition B.5 (Ergodicity) Let π be an invariant measure for the Markov semigroup P_t. π is said to be ergodic if

$$\lim_{T \to \infty} \frac{1}{T} \int_0^T P_t \varphi(u) \mathrm{d}t = \int_H \varphi(u) \pi(\mathrm{d}u) \quad \text{in } L^2(H, \pi)$$

for any $\varphi \in L^2(H, \pi)$, where $L^2(H, \pi)$ denotes the family of Borel measurable functions $\varphi : H \to \mathbb{R}$ such that $\int_H |\varphi(u)|^2 \pi(\mathrm{d}u) < \infty$.

Appendix C
Brief Introduction to Malliavin Calculus

Malliavin calculus is a powerful extension of classical calculus that is specifically designed for stochastic processes. It was developed in the late twentieth century (see [5, 6, 12, 15, 16, 25, 27]) and nowadays has evolved to offer profound insights into the properties and behaviors (such as the regularity and smoothness) of solutions of stochastic differential equations (see e.g. [4, 19, 21, 23, 26]). It has found applications in fields like mathematical finance and stochastic control; see [1, 13, 17, 20]. This appendix focuses on the basic elements of Malliavin calculus. We first give definitions of the Malliavin derivative and its adjoint, which can be viewed as infinite-dimensional analogs of the gradient and divergence operators in Euclidean space, respectively. We define the Malliavin derivative for a class of smooth functionals and give the closability of the Malliavin derivative operator. The divergence operator is defined as the dual of the Malliavin derivative. We also give the criteria for the existence and smoothness of density functions of random variables based on the Malliavin integration by parts formula.

C.1 Malliavin Derivative

Let $\mathbb{T} = [0, T]$ or $[0, +\infty)$ and H be the Hilbert space $L^2(\mathbb{T}; \mathbb{R}^m)$ endowed with the inner product $\langle g, h\rangle_H := \int_{\mathbb{T}} g(t)^\top h(t)\mathrm{d}t$ for $g, h \in H$. By identifying $W(t, \omega)$ with the value $\omega(t)$ at time t of an element $\omega \in C_0(\mathbb{T}; \mathbb{R}^m)$, we set $\Omega = C_0(\mathbb{T}; \mathbb{R}^m)$ as the Wiener space and $\mathbb{P}$ as the Wiener measure. For $g = (g^1, \ldots, g^m) \in H$, we set

$$W(g) := \sum_{k=1}^{m} \int_{\mathbb{T}} g^k(t)\mathrm{d}W^k(t).$$

C. Chen et al., *Numerical Analysis of Stochastic Functional Differential Equations*, Lecture Notes in Mathematics 2399, https://doi.org/10.1007/978-981-92-1592-8

Denote by $\mathcal{S}$ the class of smooth random variables such that $G \in \mathcal{S}$ has the form $G = f(W(g_1), \ldots, W(g_n))$, where $f \in C^\infty_{pol}(\mathbb{R}^n; \mathbb{R})$, $g_i \in H, i = 1, \ldots, n$. The Malliavin derivative of a smooth random variable G is an H-valued random variable given by

$$DG = \sum_{i=1}^{n} \frac{\partial f}{\partial x_i}(W(g_1), \ldots, W(g_n))g_i,$$

which is also an m-dimensional stochastic process $DG = \{D_r G, r \geq 0\}$ with

$$D_r G = \sum_{i=1}^{n} \frac{\partial f}{\partial x_i}(W(g_1), \ldots, W(g_n))g_i(r).$$

Unless specified otherwise, we assume throughout that the σ-algebra $\mathcal{F}$ is generated by the family $\{W(g), g \in H\}$ and all $\mathbb{P}$-null sets. The set of random variables $\{\exp\{W(g)\}, g \in H\}$ forms a total subset of $L^2(\Omega)$, and it is important to note that $\mathcal{S}$ is dense in $L^p(\Omega)$ for $p \geq 1$; see e.g. [11]. We introduce the following Malliavin integration by parts formula.

Lemma C.1 *For $G \in \mathcal{S}$ and $g \in H$, we have $\mathbb{E}[\langle DG, g\rangle]_H = \mathbb{E}[GW(g)]$.*

Applying this formula to a product FG, we obtain

$$\mathbb{E}[F\langle DG, g\rangle_H] = \mathbb{E}[FGW(g)] - \mathbb{E}[G\langle DF, g\rangle_H].$$

Then it can be shown that D is a closable operator and thus it has a closed extension.

Proposition C.2 *For any $p \geq 1$, the operator $D : \mathcal{S} \to L^p(\Omega; H)$ is closable.*

Definition C.1 For any $\{G_n\}_{n\geq 0} \subset \mathcal{S}$ satisfying $G_n \to G$ in $L^p(\Omega)$ as $n \to \infty$ and $\{DG_n\}_{n\geq 0}$ is a Cauchy sequence in $L^p(\Omega; H)$ for some $p \geq 1$, we define $DG := \lim_{n\to\infty} DG_n$ in $L^p(\Omega; H)$.

For any $p \geq 1$, we denote the domain of D in $L^p(\Omega)$ by $\mathbb{D}^{1,p}$, meaning that $\mathbb{D}^{1,p}$ is the closure of $\mathcal{S}$ with respect to the norm $\|G\|_{1,p} := (\mathbb{E}[|G|^p + \|DG\|^p_H])^{\frac{1}{p}}$.

The iterated derivative $D^k G$ is a random variable with values in $H^{\otimes k}$. Precisely, for $k \in \mathbb{N}_+$, $D^k G = \{D_{r_1,\ldots,r_k} G, \ r_i \geq 0, \ i = 1, \ldots, k\}$ is a measurable function on the product space $[0, +\infty)^k \times \Omega$. For any $p \geq 1$ and $k \in \mathbb{N}_+$, denote by $\mathbb{D}^{k,p}$ the completion of $\mathcal{S}$ with respect to the norm

$$\|G\|_{k,p} := \Big(\mathbb{E}\Big[|G|^p + \sum_{j=1}^{k} \|D^j G\|^p_{H^{\otimes j}}\Big]\Big)^{\frac{1}{p}}.$$

For $G \in \mathbb{D}^{k,p}$ and $g \in H$, define the directional derivative by $D^g G := \langle DG, g \rangle_H$ and by induction, $D^{g_1,\ldots,g_j} G := \langle DD^{g_1,\ldots,g_{j-1}} G, g_j \rangle_H$, $j = 2, \ldots, k$, where $g_i \in H$, $i = 1, \ldots, k$.

Define $L^{\infty-}(\Omega) := \bigcap_{p\geq 1} L^p(\Omega)$, $\mathbb{D}^{k,\infty} := \bigcap_{p\geq 1} \mathbb{D}^{k,p}$, $\mathbb{D}^\infty := \bigcap_{p\geq 1}\bigcap_{k\geq 1} \mathbb{D}^{k,p}$. Similarly, let V be a real separable Banach space and define the space $\mathbb{D}^{k,p}(V)$ as the completion of V-valued smooth random variables with respect to the norm

$$\|G\|_{k,p,V} = \left(\mathbb{E}\left[\|G\|_V^p + \sum_{j=1}^k \|D^j G\|_{H^{\otimes j}\otimes V}^p\right]\right)^{\frac{1}{p}}.$$

In this case, the corresponding spaces are denoted by $L^{\infty-}(\Omega; V)$, $\mathbb{D}^{k,\infty}(V)$ and $\mathbb{D}^\infty(V)$, respectively. We introduce the chain rule for the Malliavin derivative in the Banach space setting; see [22, Proposition 3.8].

Lemma C.3 *Let V be a separable Banach space and $p \geq 1$. Suppose $G : V \to \mathbb{R}^d$ is Fréchet differentiable and has a continuous and bounded derivative. If $X \in \mathbb{D}^{1,p}(V)$, then $G(X) \in \mathbb{D}^{1,p}(\mathbb{R}^d)$ with*

$$D(G(X)) = \mathscr{D}G(X)DX,$$

where $\mathscr{D}G$ is the Fréchet derivative of G.

As a corollary, we obtain the functional version of the chain rule for Malliavin derivatives; see also e.g. [2, Lemma 3.4] and [10].

Lemma C.4 *Let $G : C^d \to \mathbb{R}^d$ be Fréchet differentiable and have a continuous and bounded derivative. Suppose that for $t \geq 0$, the C^d-valued segment process x_t, driven by an m-dimensional Brownian motion, satisfies $x_t \in \mathbb{D}^{1,p}(C^d)$ for some $p \geq 1$. Then $G(x_t) \in \mathbb{D}^{1,p}(\mathbb{R}^d)$ with*

$$D(G(x_t)) = \mathscr{D}G(x_t)Dx_t,$$

where for each $s \geq 0$, $D_s x_t \in \mathbb{R}^m \otimes C^d$ denotes the mapping $r \mapsto D_s x_t(r) = D_s x(t+r)$.

C.2 Divergence Operator

A. V. Skorohod [24] introduces a specific type of stochastic integral, now widely referred to as the Skorohod integral. This integral facilitates the integration of non-adapted processes. As shown in [19], the Skorohod integral is equivalent to the adjoint of the Malliavin derivative operator, which is also called the divergence operator.

In this section, we restrict our presentation to the real-valued setting to introduce the divergence operator and the Skorohod integral. All the results presented here can be extended to the framework of separable Banach spaces; see [22, Section 4] for further details. Recall that the Malliavin derivative operator D is an unbounded linear operator from $\mathbb{D}^{1,2} \subset L^2(\Omega)$ to $L^2(\Omega; H)$, and it is densely defined.

Definition C.2 Let δ be the adjoint operator of D. That is, δ is an unbounded operator on $L^2(\Omega; H)$ with values in $L^2(\Omega)$ such that:

(i) The domain of δ, denoted by $\mathrm{Dom}(\delta)$, is the set of H-valued square integrable random variables $\phi \in L^2(\Omega; H)$ such that

$$|\mathbb{E}[\langle \phi, DX\rangle_H]| \leq K(\phi)\|X\|_{L^2(\Omega)}$$

for all $X \in \mathbb{D}^{1,2}$, where $K(\phi) > 0$ depends on ϕ.

(ii) If $\phi \in \mathrm{Dom}(\delta)$, then $\delta(\phi)$ is an element of $L^2(\Omega)$ characterized by

$$\mathbb{E}[\langle \phi, DX\rangle_H] = \mathbb{E}[X\delta(\phi)]. \tag{C.1}$$

The operator δ is called the *divergence operator* and is closed as the adjoint of the unbounded and densely defined operator D. The Malliavin integration by parts formula (C.1) establishes the duality between D and δ, which is analogous to the relationship between the gradient ∇ and the negative divergence $-\nabla\cdot$ in finite-dimensional spaces:

$$\int_{\mathbb{R}^d} \langle g, \nabla h\rangle \mathrm{d}x = -\int_{\mathbb{R}^d} h\nabla \cdot g\mathrm{d}x$$

for smooth functions g, h with compact support.

Recall that $\{\omega(t), t \geq 0\}$ is the coordinate process on the classical Wiener space. Denote by L^2_{ad} the closed subspace of $L^2(\mathbb{T} \times \Omega)$ formed by $\mathcal{F}_t$-adapted processes, where $\mathcal{F}_t$ is the σ-algebra generated by $\{\omega(s), s \leq t\}$. The following proposition reveals that the Skorohod integral is an extension of the Itô stochastic integral.

Proposition C.5 *It holds that* $L^2_{\mathrm{ad}} \subset \mathrm{Dom}(\delta)$, *and that the Skorohod integral restricted to* L^2_{ad} *coincides with the Itô stochastic integral, i.e.,*

$$\delta(\phi) = \sum_{k=1}^{m} \int_{\mathbb{T}} \phi_t^k \mathrm{d}W^k(t), \quad \phi \in L^2_{\mathrm{ad}}.$$

At the end of this section, we present the following property of the divergence operator, which allows us to factor out a scalar random variable in a divergence.

Proposition C.6 *Let $G \in \mathbb{D}^{1,2}$ and $\phi \in \mathrm{Dom}(\delta)$ such that $G\phi \in L^2(\Omega; H)$. Then $G\phi \in \mathrm{Dom}(\delta)$ and the following equality holds*

$$\delta(G\phi) = G\delta(\phi) - \langle DG, \phi \rangle_H,$$

provided the right-hand side of the equation is square integrable.

C.3 Existence and Smoothness of Density

In this section, we present the criteria for the existence and smoothness of the densities of random variables based on the Malliavin calculus technique. These criteria are essential for understanding the regularity of the corresponding probability distributions.

Let $C_b^\infty(\mathbb{R}^d)$ denote the class of functions $f : \mathbb{R}^d \to \mathbb{R}$ that are bounded and possess bounded derivatives of all orders. For $F := (F_1, \ldots, F_d) \in \mathbb{D}^{1,2}(\mathbb{R}^d)$, and $f \in C_b^\infty(\mathbb{R}^d)$, the chain rule yields

$$\langle Df(F), DF_l \rangle_H = \sum_{k=1}^{d} \partial_k f(F) \langle DF^k, DF^l \rangle_H, \quad l = 1, \ldots, d. \tag{C.2}$$

The Malliavin covariance matrix is the symmetric nonnegative random matrix defined by

$$\gamma_F := (\langle DF^i, DF^j \rangle_H)_{1 \le i,j \le d}.$$

When γ_F is invertible a.s., $\partial_k f(F)$ in (C.2) is solvable with

$$\partial_k f(F) = \sum_{l=1}^{d} \langle D(f(F)), (\gamma_F^{-1})_{kl} DF_l \rangle_H, \quad \text{a.s.}$$

As a result of the duality between D and δ,

$$|\mathbb{E}[\partial_k f(F)]| = \left| \sum_{l=1}^{d} \mathbb{E}\Big[f(F) \delta\Big((\gamma_F^{-1})_{kl} DF_l \Big) \Big] \right| \le K \sup_{z \in \mathbb{R}^d} |f(z)|,$$

provided

$$(\gamma_F^{-1})_{kl} DF_l \in \mathrm{Dom}(\delta) \text{ and } \delta\Big((\gamma_F^{-1})_{kl} DF_l \Big) \in L^1(\Omega) \ \forall k, l. \tag{C.3}$$

Based on the following lemma, one obtains that the probability distribution $\mathbb{P} \circ F^{-1}$ admits a density.

Lemma C.7 *Let μ be a finite measure on $\mathbb{R}^d$. Assume that for all $f \in C_b^\infty(\mathbb{R}^d)$, the following inequality holds:*

$$\left| \int_{\mathbb{R}^d} \partial_i f \mathrm{d}\mu \right| \leq K_i \sup_{z \in \mathbb{R}^d} |f(z)|, \quad 1 \leq i \leq d.$$

Then μ is absolutely continuous with respect to the Lebesgue measure.

In general, it is not easy to verify the conditions in (C.3). N. Bouleau and F. Hirsch obtained the following result on the existence of the densities of random variables.

Theorem C.8 *Let $F = (F_1, \ldots, F_d)$ be a random variable satisfying the following conditions:*

(i) $F \in \mathbb{D}^{1,p}(\mathbb{R}^d)$ for some $p > 1$;
(ii) The Malliavin covariance matrix γ_F is invertible a.s.

Then F admits a density, i.e., the law of F is absolutely continuous with respect to the Lebesgue measure on $\mathbb{R}^d$.

To study the smoothness of the densities of random variables, we introduce the following non-degeneracy condition.

Definition C.3 A random variable $F \in \mathbb{D}^\infty(\mathbb{R}^d)$ is called non-degenerate if its Malliavin covariance matrix γ_F is invertible a.s. and $\det(\gamma_F^{-1}) \in L^{\infty-}(\Omega)$.

The following proposition provides a criterion for the smoothness of the densities of random variables.

Theorem C.9 *Let $F \in \mathbb{D}^\infty(\mathbb{R}^d)$ be a non-degenerate random variable. Then F possesses a smooth density.*

Appendix D
Large Deviation Principle via Weak Convergence Approach

In this appendix, we present the fundamental aspects of the large deviation principle (LDP) within the weak convergence approach (see monograph [9]). This approach begins with the equivalence between the LDP and the Laplace principle (see Definition D.3 below), provided that the underlying space is Polish. This equivalence follows from Varadhan's Lemma and Bryc's converse to it, with proofs of the finite-dimensional case presented in [9] and the infinite-dimensional version given in [7]. One merit of the weak convergence method is that the LDPs can be derived under weaker assumptions, like superlinear growth conditions, on the coefficients of the considered stochastic systems. This section is devoted to giving some standard definitions and results from the LDP via the weak convergence approach.

Let $\mathcal{X}$ be a Polish space. Let $\epsilon > 0$ be an index parameter and $\{X^\epsilon\}_{\epsilon>0}$ a family of random variables from the probability space $(\Omega, \mathcal{F}, \mathbb{P})$ to $\mathcal{X}$. The following are definitions of the rate function and LDP.

Definition D.1 A function $I : \mathcal{X} \to [0, \infty]$ is called a rate function, if it is lower semi-continuous, i.e., for each $a \in [0, \infty)$, the level set $I^{-1}([0, a]) := \{x \in \mathcal{X} : I(x) \leq a\}$ is a closed subset of $\mathcal{X}$. If all level sets $I^{-1}([0, a])$ for $a \in [0, \infty)$ are compact, then I is called a good rate function.

Definition D.2 Let I be a rate function on $\mathcal{X}$. The family of random variables $\{X^\epsilon\}_{\epsilon>0}$ is said to satisfy the LDP on $\mathcal{X}$ with rate function I if the following two conditions hold:

(i) (**Large Deviation Upper Bound**) For each closed subset $F \subset \mathcal{X}$,

$$\limsup_{\epsilon\to 0} \epsilon \log \mathbb{P}\big(X^\epsilon \in F\big) \leq - \inf_{x\in F} I(x);$$

(ii) (**Large Deviation Lower Bound**) For each open subset $O \subset \mathcal{X}$,

$$\liminf_{\epsilon\to 0} \epsilon \log \mathbb{P}\big(X^\epsilon \in O\big) \geq - \inf_{x\in O} I(x).$$

C. Chen et al., *Numerical Analysis of Stochastic Functional Differential Equations*, Lecture Notes in Mathematics 2399, https://doi.org/10.1007/978-981-92-1592-8

Note that $\{\mathbb{P} \circ (X^\epsilon)^{-1}\}_{\epsilon>0}$ is a family of probability distributions. One can similarly define a family of probability measures $\{\mu^\epsilon\}_{\epsilon>0}$ that satisfy an LDP on $\mathcal{X}$ by considering the supremum limit and infimum limit of $\epsilon \log \mu^\epsilon(F)$ and $\epsilon \log \mu^\epsilon(O)$ instead.

A classical result is that the LDP is equivalent to the Laplace principle; see e.g. [9, Theorems 1.2.1 and 1.2.3].

Definition D.3 Let I be a rate function on $\mathcal{X}$. The family of random variables $\{X^\epsilon\}_{\epsilon>0}$ is said to satisfy the Laplace principle on $\mathcal{X}$ with rate function I if for all bounded and continuous functions $h : \mathcal{X} \to \mathbb{R}$,

$$\lim_{\epsilon\to 0} \epsilon \log \mathbb{E}\Big[\exp\Big\{-\frac{h(X^\epsilon)}{\epsilon}\Big\}\Big] = -\inf_{x\in\mathcal{X}}\{h(x) + I(x)\}.$$

Proposition D.1 *The family $\{X^\epsilon\}_{\epsilon>0}$ satisfies the LDP on $\mathcal{X}$ with rate function I if and only if $\{X^\epsilon\}_{\epsilon>0}$ satisfies the Laplace principle on $\mathcal{X}$ with the same rate function I.*

In view of this equivalent result, we will focus on the Laplace principle. To present a criterion for the Laplace principle, we set

$$L^2(\mathbb{T}; \mathbb{R}^m) := \Big\{v : \mathbb{T} \to \mathbb{R}^m \text{ is measurable} \,\Big|\, \int_{\mathbb{T}} |v(s)|^2 \mathrm{d}s < \infty\Big\}.$$

For any $M \in (0, +\infty)$, denote

$$\mathcal{S}_M := \Big\{v \in L^2(\mathbb{T}; \mathbb{R}^m) \Big| \int_{\mathbb{T}} |v(s)|^2 \mathrm{d}s \le M\Big\}$$

and

$$\mathcal{A}_M := \{v : \Omega \times \mathbb{T} \to \mathbb{R}^m | v \text{ is } \mathcal{F}_t\text{-measurable and } v \in \mathcal{S}_M \ \mathbb{P}\text{-a.s.}\}.$$

It follows from [14, Chapter III, Section 28, Theorem 1'] that $\mathcal{S}_M$ is a Polish space endowed with the weak topology. Set $\bar{\mathbb{T}} = [-\tau, T]$ or $[-\tau, \infty)$. Introduce the space $C_\xi(\bar{\mathbb{T}}; \mathbb{R}^d)$ consisting of continuous functions $u : \bar{\mathbb{T}} \to \mathbb{R}^d$ with $u(r) = \xi(r)$, $r \in [-\tau, 0]$. Let W be an m-dimensional Brownian motion.

The following assumption for the Laplace principle is proposed by A. Budhiraja and P. Dupuis in [7, Assumption 4.3].

Assumption D.1 *For each $\epsilon > 0$, let $\mathcal{G}^\epsilon : C(\mathbb{T}; \mathbb{R}^d) \to C_\xi(\bar{\mathbb{T}}; \mathbb{R}^d)$ be a measurable map. There exists a measurable map $\mathcal{G} : C(\mathbb{T}; \mathbb{R}^d) \to C_\xi(\bar{\mathbb{T}}; \mathbb{R}^d)$ such that the following two conditions hold:*

(i) For each $M > 0$, the set $\{\mathcal{G}(\int_0^\cdot v(s)\mathrm{d}s) | v \in \mathcal{S}_M\}$ is a compact subset of $C_\xi(\bar{\mathbb{T}}; \mathbb{R}^d)$.

(ii) *Let* $M > 0$ *and* $\{v^\epsilon\}_{\epsilon>0} \subset \mathcal{A}_M$ *such that* $v^\epsilon \xrightarrow[\epsilon\to 0]{d} v$ *as* S_M*-valued random variables. It holds that*

$$\mathcal{G}^\epsilon\Big(\sqrt{\epsilon}W + \int_0^\cdot v^\epsilon(s)\mathrm{d}s\Big) \xrightarrow[\epsilon\to 0]{d} \mathcal{G}\Big(\int_0^\cdot v(s)\mathrm{d}s\Big).$$

The following result is given in [7, Theorem 4.4], which provides sufficient conditions for the Laplace principle and thus the LDP.

Theorem D.2 *Let Assumption D.1 hold. Then* $X^\epsilon := \mathcal{G}^\epsilon(\sqrt{\epsilon}W)$ *satisfies the Laplace principle (hence LDP) on* $C_\xi(\bar{\mathbb{T}}; \mathbb{R}^d)$ *with good rate function* $I : C_\xi(\bar{\mathbb{T}}; \mathbb{R}^d) \to [0, \infty]$ *given by*

$$I(f) = \inf_{\left\{v\in L^2(\mathbb{T};\mathbb{R}^m):\, f=\mathcal{G}(\int_0^\cdot v(s)\mathrm{d}s)\right\}} \frac{1}{2}\int_{\mathbb{T}} |v(s)|^2 \mathrm{d}s.$$

At the end of this section, we introduce the contraction principle; see [8, Theorem 4.2.1].

Theorem D.3 *Let* $\mathcal{X}, \mathcal{Y}$ *be two Polish spaces,* $f : \mathcal{X} \to \mathcal{Y}$ *be a continuous function, and* $I : \mathcal{X} \to [0, \infty]$ *be a good rate function.*

(i) *For each* $y \in \mathcal{Y}$*, define* $\bar{I}(y) := \inf\{I(x) : x \in \mathcal{X}, y = f(x)\}$*. Then* $\bar{I}$ *is a good rate function on* $\mathcal{Y}$*.*
(ii) *If* I *controls the LDP associated with a family of probability measures* μ^ϵ *on* $\mathcal{X}$*, then* $\bar{I}$ *controls the LDP associated with the family of probability measures* $\mu^\epsilon \circ f^{-1}$ *on* $\mathcal{Y}$*.*

References

1. E. Alòs, D. García Lorite, *Malliavin Calculus in Finance—Theory and Practice*. Chapman & Hall/CRC Financial Mathematics Series (CRC Press, Boca Raton, 2021)
2. D.R. Baños, G. Di Nunno, H.H. Haferkorn, F. Proske, Stochastic functional differential equations and sensitivity to their initial path, in *Computation and Combinatorics in Dynamics, Stochastics and Control*, vol. 13. Abel Symp. (Springer, Cham, 2018), pp. 37–70
3. D. Bakry, I. Gentil, M. Ledoux, *Analysis and Geometry of Markov Diffusion Operators*, vol. 348. Grundlehren der mathematischen Wissenschaften [Fundamental Principles of Mathematical Sciences] (Springer, Cham, 2014)
4. D.R. Bell, *The Malliavin Calculus*, vol. 34. Pitman Monographs and Surveys in Pure and Applied Mathematics (Longman Scientific & Technical, Harlow; John Wiley & Sons, Inc., New York, 1987)
5. J.-M. Bismut, D. Michel, Diffusions conditionnelles. I. Hypoellipticité partielle. J. Funct. Anal. **44**, 174–211 (1981)
6. J.-M. Bismut, D. Michel, Diffusions conditionnelles. II. Générateur conditionnel. Application au filtrage. J. Funct. Anal. **45**, 274–292 (1982)

7. A. Budhiraja, P. Dupuis, A variational representation for positive functionals of infinite dimensional Brownian motion. Probab. Math. Stat. **20**, 39–61 (2000)
8. A. Dembo, O. Zeitouni, *Large Deviations Techniques and Applications*, vol. 38, 2nd edn. Applications of Mathematics (Springer, New York, 1998)
9. P. Dupuis, R.S. Ellis, *A Weak Convergence Approach to the Theory of Large Deviations*. Wiley Series in Probability and Statistics: Probability and Statistics (John Wiley & Sons, Inc., New York, 1997)
10. M. Ferrante, C. Rovira, M. Sanz-Solé, Stochastic delay equations with hereditary drift: estimates of the density. J. Funct. Anal. **177**, 138–177 (2000)
11. Z. Huang, J. Yan, *Introduction to Infinite Dimensional Stochastic Analysis*, vol. 502. Mathematics and Its Applications (Kluwer Academic Publishers, Dordrecht; Science Press Beijing, Beijing, Chinese ed., 2000)
12. N. Ikeda, S. Watanabe, An introduction to Malliavin's calculus, in *Stochastic Analysis (Katata/Kyoto, 1982)*, vol. 32. North-Holland Math. Library (North-Holland, Amsterdam, 1984), pp. 1–52
13. P. Imkeller, *Malliavin's Calculus and Applications in Stochastic Control and Finance*, vol. 1. IMPAN Lecture Notes (Polish Academy of Sciences, Institute of Mathematics, Warsaw, 2008)
14. A.N. Kolmogorov, S.V. Fomin, *Elements of the Theory of Functions and Functional Analysis. Vol. 1. Metric and Normed Spaces* (Graylock Press, Rochester, Russian ed., 1957)
15. S. Kusuoka, Dirichlet forms and diffusion processes on Banach spaces. J. Fac. Sci. Univ. Tokyo Sect. IA Math. **29**, 79–95 (1982)
16. P. Malliavin, Stochastic calculus of variation and hypoelliptic operators, in *Proceedings of the International Symposium on Stochastic Differential Equations* (Res. Inst. Math. Sci., Kyoto Univ., Kyoto, 1976), Wiley-Intersci. Publ. (John Wiley & Sons, New York-Chichester-Brisbane, 1978), pp. 195–263
17. P. Malliavin, A. Thalmaier, *Stochastic Calculus of Variations in Mathematical Finance*. Springer Finance (Springer, Berlin, 2006)
18. X. Mao, *Stochastic Differential Equations and Applications*, 2nd edn. (Horwood Publishing Limited, Chichester, 2008)
19. D. Nualart, *The Malliavin Calculus and Related Topics*, 2nd edn. Probability and Its Applications (Springer, Berlin, 2006)
20. D. Nualart, *Malliavin Calculus and Its Applications*, vol. 110. CBMS Regional Conference Series in Mathematics, Conference Board of the Mathematical Sciences, Washington (American Mathematical Society, Providence, 2009)
21. D. Nualart, E. Nualart, *Introduction to Malliavin Calculus*, vol. 9. Institute of Mathematical Statistics Textbooks (Cambridge University Press, Cambridge, 2018)
22. M. Pronk, M. Veraar, Tools for Malliavin calculus in UMD Banach spaces. Potential Anal. **40**, 307–344 (2014)
23. M. Sanz-Solé, Applications of Malliavin calculus to SPDE's, in *Stochastic Partial Differential Equations and Applications (Trento, 2002)*, vol. 227. Lecture Notes in Pure and Appl. Math. (Dekker, New York, 2002), pp. 429–442
24. A.V. Skorohod, On a generalization of the stochastic integral. Teor. Verojatnost. i Primenen. **20**, 223–238 (1975)
25. D.W. Stroock, The Malliavin calculus, a functional analytic approach. J. Funct. Anal. **44**, 212–257 (1981)
26. A.S. Üstünel, *An Introduction to Analysis on Wiener Space*, vol. 1610. Lecture Notes in Mathematics (Springer, Berlin, 1995)
27. S. Watanabe, *Lectures on Stochastic Differential Equations and Malliavin Calculus*, vol. 73. Tata Institute of Fundamental Research Lectures on Mathematics and Physics, Tata Institute of Fundamental Research, Bombay (Springer, Berlin, 1984)

Index

C. Chen et al., *Numerical Analysis of Stochastic Functional Differential Equations*, Lecture Notes in Mathematics 2399, https://doi.org/10.1007/978-981-92-1592-8

LECTURE NOTES IN MATHEMATICS Springer

Editorial Policy

1. Lecture Notes aim to report new developments in all areas of mathematics and their applications – quickly, informally and at a high level. Mathematical texts analysing new developments in modelling and numerical simulation are welcome.

   Manuscripts should be reasonably self-contained and rounded off. Thus they may, and often will, present not only results of the author but also related work by other people. They may be based on specialised lecture courses. Furthermore, the manuscripts should provide sufficient motivation, examples and applications. This clearly distinguishes Lecture Notes from journal articles or technical reports which normally are very concise. Articles intended for a journal but too long to be accepted by most journals, usually do not have this "lecture notes" character. For similar reasons it is unusual for doctoral theses to be accepted for the Lecture Notes series, though habilitation theses may be appropriate.

2. Besides monographs, multi-author manuscripts resulting from SUMMER SCHOOLS or similar INTENSIVE COURSES are welcome, provided their objective was held to present an active mathematical topic to an audience at the beginning or intermediate graduate level (a list of participants should be provided).

   The resulting manuscript should not be just a collection of course notes, but should require advance planning and coordination among the main lecturers. The subject matter should dictate the structure of the book. This structure should be motivated and explained in a scientific introduction, and the notation, references, index and formulation of results should be, if possible, unified by the editors. Each contribution should have an abstract and an introduction referring to the other contributions. In other words, more preparatory work must go into a multi-authored volume than simply assembling a disparate collection of papers, communicated at the event.

3. Manuscripts should be submitted either online at www.editorialmanager.com/lnm to Springer's mathematics editorial in Heidelberg, or electronically to one of the series editors. Authors should be aware that incomplete or insufficiently close-to-final manuscripts almost always result in longer refereeing times and nevertheless unclear referees' recommendations, making further refereeing of a final draft necessary. The strict minimum amount of material that will be considered should include a detailed outline describing the planned contents of each chapter, a bibliography and several sample chapters. Parallel submission of a manuscript to another publisher while under consideration for LNM is not acceptable and can lead to rejection.

4. In general, **monographs** will be sent out to at least 2 external referees for evaluation.

   A final decision to publish can be made only on the basis of the complete manuscript, however a refereeing process leading to a preliminary decision can be based on a pre-final or incomplete manuscript.

   Volume Editors of **multi-author works** are expected to arrange for the refereeing, to the usual scientific standards, of the individual contributions. If the resulting reports can be forwarded to the LNM Editorial Board, this is very helpful. If no reports are forwarded or if other questions remain unclear in respect of homogeneity etc, the series editors may wish to consult external referees for an overall evaluation of the volume.

5. Manuscripts should in general be submitted in English. Final manuscripts should contain at least 100 pages of mathematical text and should always include

   - a table of contents;
   - an informative introduction, with adequate motivation and perhaps some historical remarks: it should be accessible to a reader not intimately familiar with the topic treated;
   - a subject index: as a rule this is genuinely helpful for the reader.
   - For evaluation purposes, manuscripts should be submitted as pdf files.

6. Careful preparation of the manuscripts will help keep production time short besides ensuring satisfactory appearance of the finished book in print and online. After acceptance of the manuscript authors will be asked to prepare the final LaTeX source files (see LaTeX templates online: https://www.springer.com/gb/authors-editors/book-authors-editors/manuscriptpreparation/5636) plus the corresponding pdf- or zipped ps-file. The LaTeX source files are essential for producing the full-text online version of the book, see http://link.springer.com/bookseries/304 for the existing online volumes of LNM). The technical production of a Lecture Notes volume takes approximately 12 weeks. Additional instructions, if necessary, are available on request from lnm@springer.com.

7. Authors receive a total of 30 free copies of their volume and free access to their book on SpringerLink, but no royalties. They are entitled to a discount of 33.3 % on the price of Springer books purchased for their personal use, if ordering directly from Springer.

8. Commitment to publish is made by a *Publishing Agreement*; contributing authors of multiauthor books are requested to sign a *Consent to Publish form*. Springer-Verlag registers the copyright for each volume. Authors are free to reuse material contained in their LNM volumes in later publications: a brief written (or e-mail) request for formal permission is sufficient.

Addresses:
Professor Jean-Michel Morel, CMLA, École Normale Supérieure de Cachan, France
E-mail: moreljeanmichel@gmail.com

Professor Bernard Teissier, Equipe Géométrie et Dynamique,
Institut de Mathématiques de Jussieu – Paris Rive Gauche, Paris, France
E-mail: bernard.teissier@imj-prg.fr

Springer: Ute McCrory, Mathematics, Heidelberg, Germany,
E-mail: lnm@springer.com

GPSR Compliance

The European Union's (EU) General Product Safety Regulation (GPSR) is a set of rules that requires consumer products to be safe and our obligations to ensure this.

If you have any concerns about our products, you can contact us on ProductSafety@springernature.com

In case Publisher is established outside the EU, the EU authorized representative is:

Springer Nature Customer Service Center GmbH
Europaplatz 3
69115 Heidelberg, Germany

Batch number: 10391219

Printed by Printforce, the Netherlands